Lecture N[illegible]
Computer [illegible]

Edited by G. Goos and J. Hartmanis

357

T. Mora (Ed.)

Applied Algebra, Algebraic Algorithms and Error-Correcting Codes

6th International Conference, AAECC-6
Rome, Italy, July 4–8, 1988
Proceedings

Springer-Verlag
Berlin Heidelberg New York London Paris Tokyo Hong Kong

Editor

Teo Mora
Dipartimento di Matematica, Universitá di Genova
Via L.B. Alberti 4, I-16132 Genova, Italy

CR Subject Classification (1987): E.4, E.3, I.1, F.2, G.2

ISBN 3-540-51083-4 Springer-Verlag Berlin Heidelberg New York
ISBN 0-387-51083-4 Springer-Verlag New York Berlin Heidelberg

Printing and binding: Druckhaus Beltz, Hemsbach/Bergstr.
2145/3140-543210 – Printed on acid-free paper

FOREWORD

The two annual conferences

International Symposium on Symbolic and Algebraic Computation (ISSAC)

International Conference on Applied Algebra, Algebraic Algorithms and Error Correcting Codes (AAECC)

have taken place for the first time as a Joint Conference (FIJC) in Rome, July 4-8, 1988.

The Applied Algebra, Algebraic Algorithms and Error Correcting Codes conferences began in 1983, AAECC-6 being the sixth one.
The first International Colloquium on "Algebra and Error Correcting Codes: Theory and Applications" (AAECC-1) took place at the University Paul Sabatier in Toulouse (France), June 1983. The subjects of principal interest were: multivariate polynomials, applications of algebraic geometry, codes (group codes, self-dual, SAB, convolutional) and their cosets, code decomposition and complexity, discrete transforms, algebraic algorithms and software simulations.
The AAECC-2 Conference was again held in Toulouse (France), October 1984. Main subjects covered were: decomposition of algebras, multivariate codes, covering radius, construction-automorphisms of codes, applied algebra, cryptography, computer algebra.
The AAECC-3 Conference was organized by the Laboratoire d'Informatique Fondamentale et d'Intelligence Artificielle (LIFIA) in Grenoble (France), July 1985. The main motivation for this conference was to gather researchers in error-correcting codes, applied algebra and algebraic algorithms. The last topic was extended to computer algebra in general.
After three conferences, it appeared that they had filled a communication gap. It was thus natural that the AAECC conferences were going to be held in different countries annually. For this reason a permanent organizing committee was set up, consisting of: Thomas Beth, Jacques Calmet, Anthony C. Hearn, Joos Heintz, Hideki Imai, Heinz Lüneburg, H.F. Mattson Jr., Alain Poli.
The successive AAECC-4 was held in Karlsruhe (FRG), September 1985. The main topics covered were: applied and applicable algebra, algorithms and combinatorics, error correcting codes and cryptography, computer algebra, complexity.
The following AAECC-5 was held in Mahon (Menorca Island, Spain), June 1986. The main topics covered were: applied algebra, error correcting codes, complexity theory.

In 1988, for the first time, the two international conferences ISSAC'88 and AAECC-6 were organized jointly. The conference topics are in fact widely related with each other and the FIJC represented a good occasion for the two research communities to meet and share scientific experiences and results.

The FIJC program included fourteen invited papers on subjects of common interest for the two conferences, some of which are included in the proceedings and divided between the two volumes.

The FIJC was organized by:

Centro Interdipartimentale di Calcolo Scientifico (CICS), University of Roma, "La Sapienza"

Dipartimento di Informatica e Sistemistica, University of Roma, "La Sapienza"

Istituto di Analisi dei Sistemi ed Informatica, CNR, Roma

sponsored by:

University of Roma, "La Sapienza"

National Research Council (CNR): Mathematics, Engineering and Technological Committee

and received contributions by:

Fondazione Ugo Bordoni

IBM Italia

Olivetti

Raggruppamento Selenia Elsag

Scuola Superiore G. Reiss Romoli.

I would like to give particular recognition to the National Research Council (CNR) which supported the preparation of the present volume. I also like to acknowledge the fruitful cooperation with Maria Giulia Santechi during the entire period of the conference organization.

A.Miola
(General Chairman)

March 1989

PREFACE

The extended abstracts submitted to AAECC-6 were evaluated by two international referees each. Out of 61 contributions, 46 were accepted for oral presentation at the conference. The respective authors had then the option either to publish an extended abstract of their contribution or to submit a full paper version, again to be refereed by two international referees. After the second screening, this volume contains 5 invited contributions, 32 full papers and 5 extended abstracts.

The topics of the AAECC-6 conference were:

Applied Algebra
Theory and Applications of Error-Correcting Codes
Cryptography
Complexity
Algebra Based Methods and Applications in Symbolic Computing and Computer Algebra
Algebraic Methods and Applications for Advanced Information Processing.

I would like to express my thanks to the referees, whose help was essential to have this volume published in a short time. My thanks also go to the editor of the joint volume, Patrizia Gianni, who handled the contacts with the publisher.

Teo Mora
(Proceedings Editor)

March 1989

CONFERENCE OFFICERS

General Chairman
Alfonso Miola
Local Organizers
Gianna Cioni (IASI, Roma)
Mirella Schaerf (University of Roma, "La Sapienza")
AAECC Conference Co-Chairman
Alain Poli
AAECC Program Committee
A. Miola (Chairman), A. Poli (Co-Chairman), T. Mora (Proceedings Editor), T. Beth, J. Calmet, J. Diaz, A. C. Hearn, J. Heintz, L. Huguet, H. Imai, R. Loos, H. Lüneburg, H.F. Mattson, E. Olcayto.

REFEREES

S. Abhyankar, E. F. Assmus, G. Ausiello, E. R. Berlekamp, T. Beth, M. Bossert, E. Brickell, P. Bundschuh, P. Camion, J. Calmet, M. Carral, M. Clausen, G. Cohen, B. Courteau, R. Desq, G. Dettori, J. Díaz, P. Duhammel, R. Dvornicich, A. Ferro, W. Geiselmann, R. Goodman, A. C. Hearn, J. Heintz, L. Huguet, H. Imai, E. Kaltofen, H. Lüneburg, J. L. Massey, H. F. Mattson, J. F. Michon, A. Miola, T. Mora, G. Norton, A. Odlyzko, E. Olcayto, F. Parisi-Presicce, P. Piret, M. Pohst, A. Poli, M. Protasi,T. Recio, L. Robbiano, S. Sakata, A. Salwicki, R. Segal, M. Singer, P. Solé, E. Strickland, C. Traverso, J. H. Van Lint, H. N. Ward, V. Weispfenning, J. K. Wolf, J. Wolfmann, W. Wolfowicz, B. Wyman.

CONTENTS

INVITED CONTRIBUTIONS

FULL PAPERS

EXTENDED ABSTRACTS

The Coding Theory of Finite Geometries and Designs

BY

E.F. ASSMUS, JR.

LEHIGH UNIVERSITY

BETHLEHEM, PA 18015

USA

I would like to summarize a few of the more important ideas and theorems contained in a series of four papers, the result of joint work with J.D. Key over the past year. The ideas are developed in "Affine and projective planes", but the genesis of these ideas is in "Arcs and ovals in the hermitian and Ree unitals" and the work centers around a simple, as yet unanswered, question of J.D. Key: what is the nature of the minimal-weight vectors in the code generated by the incidence matrix of a finite affine plane?

The ideas will, hopefully, lead to a wide ranging theory for affine and projective planes and, indeed, for designs in general; the current results of the theory depend heavily on deep and important work in algebraic coding theory done by Philippe Delsarte and others. For details and references the reader should consult [**2**].

For planes of prime order the theory merely recasts the classification problem into the language of algebraic coding theory and we therefore begin the discussion with the first interesting example, the affine plane of order four. This plane is built in exactly the same way as the familiar euclidean plane but using $GF(4)$, the field with four elements, in place of the field of real numbers. Thus the plane has sixteen points and twenty lines. It can be completely described by its incidence matrix,

$$A = (a_{\ell,P}),$$

a 20×16 matrix of $0's$ and $1's$ with rows indexed by the lines of the plane and columns by the points with $a_{\ell,P} = 1$ precisely when the line ℓ goes through the point P.

Given any prime number p we may consider the row space, C_p, of the above matrix A over the field $GF(p)$. We have, then, an (n,k,d) code over $GF(p)$ with $n = 16$ and k and d to be determined. It is not difficult to see that $k = 16$ (and hence $d = 1$) unless we take $p = 2$; for the two-element field we obtain a binary $(16,9,4)$ code and, moreover, the minimal-weight vectors are the rows of A. The new idea we wish to inject here (the above result being well-known) is that the interesting object to study is not C_2 but $C_2 \cap C_2^{\perp}$, where $C_2^{\perp}$ denotes the code orthogonal to C_2 under the standard inner product. This code, which we call the hull of the plane and here denote by H_2, turns out to be the $(16,5)$ first-order Reed-Muller code. Its weight distribution is given by

$$x^0 + 30x^8 + x^{16}$$

and the 30 weight-8 vectors are simply the sums of any two parallel lines of the plane (i.e., any two rows of A with no 1's in common).

$H_2^\perp$ is, of course, the $(16,11)$ binary extended Hamming code. Since $H_2^\perp = C_2 + C_2^\perp$, amongst $H_2^\perp$'s minimal-weight vectors one finds the twenty lines of the plane (i.e. the rows of A) but besides these weight-4 vectors there are 120 others. Geometrically they correspond to Baer subplanes (i.e., Fano planes contained in the original plane of order four). Of course, one loses sight of the original plane once one has $H_2^\perp$ and, in fact, there are many ways (112 to be precise) of choosing 20 weight-4 vectors of $H_2^\perp$ to make up an incidence matrix, such as A, of an affine plane of order four.

These planes are all isomorphic, but they needn't have been and, even the very next case, that of the plane of order nine built from the field $GF(9)$, gives a hull whose orthogonal yields an affine (and hence projective) plane of order nine very different from the desarguesian plane one began with. Here the correct prime to choose is $p = 3$ and one gets a hull, H_3 say, which is a ternary $(81,27)$ code with minimum weight 18, the minimal-weight vectors being the scalar multiples of the difference of parallel lines (i.e., rows of the incidence matrix with no common $1's$). $H_2^\perp$ is a ternary $(81,54,9)$ code. It has $2 \cdot 1170$ minimal-weight vectors, $2 \cdot 90$ of them coming from the original desarguesian plane and the remaining $2 \cdot 1080$ corresponding, geometrically, to Baer subplanes (i.e., nine point affine planes of order three contained in the original desarguesian affine plane of order nine). Once again, one loses sight of the original plane in $H_3^\perp$ and, beside isomorphic copies of this plane to be found amongst the minimal-weight vectors of $H_3^\perp$, there is also another plane to be found: the so-called Hall plane, the non-desarguesian translation plane of order nine.

The theory applies quite generally to a class of combinatorial structures called "designs" and it groups designs into classes, two designs in the same class having isomorphic hulls; we call this "linear equivalence" and we next give an example (in fact the "genesis" example) of this phenomenon.

A collection of 4-subsets, called lines, of a 28 element set, the elements of which are called points, with the property that through any two points there passes a unique line is called a "unital of order eight". There are many unitals of order eight and, at present, no one has attempted to count or classify them; it is known, however, that there are exactly two (seemingly very different) unitals of order eight that have doubly-transitive automorphism groups–that is, subgroups of the symmetric group of the 28 point set that preserve the collection of lines and can move any ordered pair of points to any other ordered pair of points (and hence can carry any line to any other line). These two very special unitals are the hermitian unital of order eight and the Ree unital of order eight and their automorphism groups are, respectively, the unitary group $P\Gamma U_3(9)$ and the small Ree group, $P\Gamma L_2(8)$.

A unital of order eight has 63 lines so, just as for planes we get, from such a unital, a 63×28 matrix of $0's$ and $1's$ called the incidence matrix. Denoting the hermitian unital by $\mathbf{H}$ and the Ree unital by $\mathbf{R}$ we have $C_p(\mathbf{H})$ and $C_p(\mathbf{R})$, the row spaces of the respective incidence matrices. Once again these row spaces are $GF(p)^{28}$ whenever $p \neq 2$. But $C_2(\mathbf{H})$ is a $(28, 21, 4)$ binary code and $C_2(\mathbf{R})$ is a $(28, 19, 4)$ binary code. $C_2(\mathbf{H})$ has 315 weight-4 vectors but those of $C_2(\mathbf{H})$ are simply the 63 rows of its incidence matrix. These designs still look rather different! But the two hulls, $H_2(\mathbf{H}) = C_2(\mathbf{H}) \cap C_2(\mathbf{H})^\perp$ and $H_2(\mathbf{R}) = C_2(\mathbf{R}) \cap C_2(\mathbf{R})^\perp$ are isomorphic $(28, 7)$ binary codes with weight distribution

$$x^0 + 63(x^{12} + x^{16}) + x^{28}$$

and we say that $\mathbf{H}$ and $\mathbf{R}$ are "linearly equivalent at the prime 2". In fact, by a proper choice of the incidence matrices of $\mathbf{H}$ and $\mathbf{R}$ one can arrange matters so that one has the following Hasse diagram:

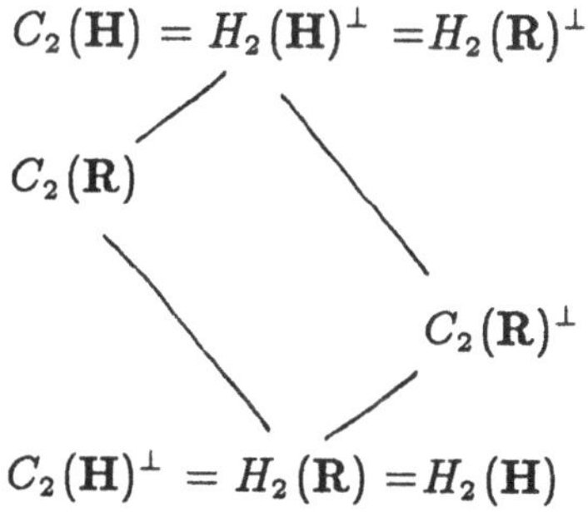

Observe that, in this case, $C_2(\mathbf{H})^\perp \subset C_2(\mathbf{H})$; the automorphism group of $C_2(\mathbf{H})$ is much larger than that of $\mathbf{H}$; it is the symplectic group $Sp_6(2)$. Put another way, both $\mathbf{H}$ and $\mathbf{R}$ can be extracted from a larger design given by the minimal- weight vectors of $C_2(\mathbf{H})$, a design whose automorphism group is $Sp_6(2)$ in its doubly-transitive action on 28 points.

We now have introduced the notion of the "hull at p" of a design and what it means to say that two designs are "linearly equivalent at p". We return now to the case of affine and projective planes and the development of these notions with a view toward the classification of finite projective planes.

A projective plane of order n can be given in terms of its incidence matrix, an $N \times N$ matrix (where $N = n^2 + n + 1$) of $0's$ and $1's$ with every row having exactly $n + 1$ $1's$ and every two rows having exactly one common 1. It follows that these two properties hold for the columns also and this "duality" has usually made projective planes, rather than affine planes, the object of study. As we shall see, the projective case obscures the importance of the notions of hull and linear equivalence.

If p is a prime dividing n and Π is a projective plane of order n, then $C_p(\Pi)$, the row space over $GF(p)$ of $\Pi's$ incidence matrix, is an $(N, k, n+1)$ code over $GF(p)$ with $k \leq \frac{1}{2}(N+1)$. $Hull_p(\Pi) = C_p(\Pi) \cap C_p(\Pi)^\perp$ is simply $\{c \in C_p(\Pi) / \sum_P c_P = 0\}$. That is, it is the subcode of codimension 1 in $C_p(\Pi)$ consisting of those vectors the sum of whose coordinates is 0. $Hull_p(\Pi)^\perp$ is simply $C_p(\Pi)^\perp \oplus \mathbf{F}_p J$ where J is the vector all of whose coordinates are 1. But, most importantly, $Hull_p(\Pi)^\perp$ has minimum weight $n+1$ AND its minimal-weight vectors are precisely the scalar multiples of the rows of $\Pi's$ incidence matrix. It follows that linear equivalence at p is the same as isomorphism of the projective planes; thus the hull, which is generated by the collection of differences of the rows of $\Pi's$ incidence matrix, does not play a significant rôle and will not allow one to relate non-isomorphic projective planes. There is, however, one important property that the hulls of desarguesian projective planes possess: if the desarguesian projective plane Π is built using $GF(p^s)$, then the minimum weight of $Hull_p(\Pi)$ is $2p^s$ (i.e., twice the order) AND the minimal-weight vectors are the scalar multiples of the differences of the rows of $\Pi's$ incidence matrix.

An arbitrary projective plane of order n whose hull at p, p dividing n, possesses these two properties we call "tame at p".

Now, given a projective plane Π via its incidence matrix, $\bar{A}$ say, and given a line L of Π we automatically get an affine plane, which we denote by Π^L, by declaring L the "line at infinity"; in terms of incidence matrices, that of Π^L, A say, is obtained from $\bar{A}$ by suppressing those columns where the row corresponding to L has $1's$. Thus, when Π is of order n, $n+1$ columns will be suppressed and, suppressing also the row corresponding to L which now consists entirely of $0's$, we have an $(n^2+n) \times n^2$ matrix A which is the incidence matrix of Π^L. The order of this affine plane is n and, for p dividing n, we have a very close connection between $C_p(\Pi)$ and $C_p(\Pi^L)$, $Hull_p(\Pi)$ and $Hull_p(\Pi^L)$, and $Hull_p(\Pi)^\perp$ and $Hull_p(\Pi^L)^\perp$. In particular $k = \dim C_p(\Pi^L) = \dim C_p(\Pi) - 1$, $\dim Hull_p(\Pi^L) = k - n$, and, most importantly, $Hull_p(\Pi^L)^\perp = C_p(\Pi^L) + C_p(\Pi^L)^\perp$ has, usually, more minimal-weight vectors than merely those that are scalar multiples of the rows of A, $\Pi^{L'}s$ incidence matrix. We saw that phenomenon in our opening examples and we saw that it was possible to create new affine planes even when one began with the desarguesian affine plane.

The Hasse diagram for any order n affine plane π is as follows:

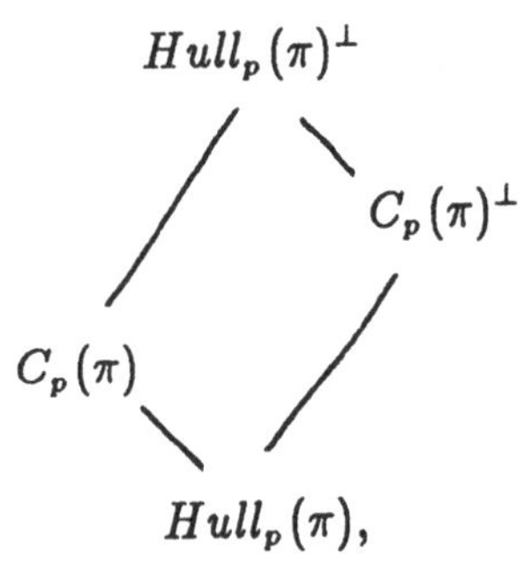

and, for any code B with

$$C_p(\pi) \subseteq B \subseteq Hull_p(\pi)^{\perp},$$

we have that B has minimum weight n. Of course $B's$ minimal-weight vectors include the scalar multiples of the rows of $\pi's$ incidence matrix. But B can, and frequently does, have more minimal-weight vectors; moreover, affine planes other than π may be hidden amongst them.

Fortunately for finite geometers, for each $q = p^s$ there is a class of such $B's$ that has been thoroughly studied by algebraic coding theorists, in particular by Philippe Delsarte. If fact there is a Galois correspondence between the subfields of $GF(p^s)$ and these $B's$ all of which are (q^2, k, q) codes over $GF(p)$. The smallest such B, corresponding to $GF(p^s)$, is the code generated by the desarguesian affine plane built using $GF(p^s)$; it has $(p-1)(q^2+q)$ weight-q vectors, namely the scalar multiples of the rows of this plane's incidence matrix. The largest, corresponding to the field $GF(p)$, accommodates, of course, the most affine planes; it has been intensively studied and its dimension and minimal-weight vectors are known. We denote this largest B by B_p. Many affine planes (e.g., the so-called "translation planes") have the property that

$$C_p(\pi) \subseteq B_p \subseteq Hull_p(\pi)^{\perp}.$$

This gives an immediate upper bound on $\dim C_p(\pi)$ since the dimension of B_p is known. We have the following

THEOREM. *For an affine plane π of order $q = p^s$ with $C_p(\pi) \subseteq B_p \subseteq Hull_p(\pi)^{\perp}$, $\dim C_p(\pi) \le q^2 + q - \dim B_p$. Moreover, any two such affine planes that meet the bound are linearly equivalent.*

REMARK. *For $s = 2$ there is a pretty formula for $\dim B_p$ and the bound becomes $\dim C_p(\pi) \le \frac{1}{2}p^4 - \frac{1}{3}p^3 + p^2 - \frac{1}{6}p$. The only previously known upper bound was $\frac{1}{2}(p^4 + p^2)$.*

In the other direction, that is lower bounds for $\dim C_p(\pi)$, there is the Hamada-Sachar Conjecture: $\dim C_p(\pi) \ge \binom{p+1}{2}^s$ with equality if and only if π is isomorphic to the desarguesian plane built using the field $GF(p^s)$. We cannot prove this conjecture for an arbitrary affine plane of order p^s but we can prove it for a certain class of those planes. It is here that the notion of tameness is brought in. That because of our next

THEOREM. *Let Π be a tame projective plane of order p^s with p odd. If an affine plane π of order p^s is linearly equivalent to Π^L for some line L of Π, then the projective plane associated with π is isomorphic to Π.*

We end this brief summary with a result along the lines of the Hamada-Sachar conjecture.

THEOREM. *Let δ be the desarguesian affine plane of order p^s with p odd. Then if π is any affine plane of order p^s with*

$$Hull_p(\delta) \subseteq C_p(\pi) \subseteq B_p,$$

we have that $\dim C_p(\pi) \geq \binom{p+1}{2}^s$ *with equality if and only if* π *isomorphic to* δ.

REMARKS. *(1) The fact that* $\dim C_p(\delta) = \binom{p+1}{2}^s$ *has been known for many years. (2) Our restriction to odd p may seem strange since $p = 2$ is the prime of choice in algebraic coding theory; in this context, however, the prime 2, as in algebraic number theory and algebraic geometry, plays an irksome rôle.*

REFERENCES

1. E. F. Assmus, Jr. and J. D. Key, *Arcs and ovals in the hermitian and Ree unitals.*
2. ————————, *Affine and projective planes*, Discrete Mathematics, Special Coding Issue (to appear).
3. ————————, *Baer subplanes, ovals, and unitals*, IMA Proceedings on Coding Theory and Design Theory, Springer-Verlag, 1988 (to appear).
4. ————————, *Translation planes and derivation sets*, (in preparation).

RECENT RESULTS ON COVERING PROBLEMS

J.H. van Lint*

1. Introduction

The aim of this paper is to do three things:

(i) To list the papers on covering radius problems that have come to my attention since the appearance of the survey paper by Cohen et al. in 1985 (cf. [10]) and to give some idea of the type of problems treated in these papers. We point out that many of these papers have not appeared yet;

(ii) To discuss a few of the most interesting new ideas occurring in the papers mentioned above;

(iii) To survey the recent (≥ 1983) developments on the so-called football pool problem.

We shall assume that the reader is familiar with coding theory and the covering radius problem (see the survey [10]). We shall use the following terminology. We consider an alphabet Z (in this paper the alphabet will nearly always be one of the fields $\mathbb{F}_2$ or $\mathbb{F}_3$).

A *code* C of length n is a subset of Z^n. We use the following notation. A k-dimensional linear subspace of $\mathbb{F}_q^n$ is called an $[n, k]$ code. If the minimum distance of the code C is d, then we shall call C an $[n, k, d]$ code. If C is not linear and $|C| = K$, then we use (n, K) code resp. (n, K, d) code. If $\mathbf{x} \in Z$ then we define the *sphere* of radius R around $\mathbf{x}$ by

(1.1) $$B_R(x) := \{y \in Z^n : d(\mathbf{x}, \mathbf{y}) \le R\}.$$

If $R = 1$ such a sphere is often called a *rook-domain* (because the case $k = 8$, $n = 2$, $R = 1$ corresponds to the positions a rook on a chessboard can reach in at most one move).
Let $|Z| = q$. Then we have

(1.2) $$V_q(n, R) := |B_R(\mathbf{x})| = \sum_{i=0}^{R} \binom{n}{i} (q-1)^i.$$

We shall say that C is an *R-covering* of Z^n if every word in Z^n has distance at most R to some codeword, i.e.

Department of Mathematics and Computing Science, Eindhoven University of Technology, Eindhoven, Netherlands.
This research was supported in part by the Institute of Mathematics and its Applications with funds provided by the National Science Foundation.

(1.3) $$Z^N = \bigcup_{c \in C} B_R(\mathbf{c}).$$

The minimal value of R for which this is true is called the *covering radius* of C. For a linear code the covering radius is equal to the weight of a coset leader of maximal weight.

Another very useful definition is given by using the parity check matrix of the code. If H is the $n-k$ by n parity check matrix of an $[n, k]$ code C, then the covering radius R of C is the smallest integer t such that every syndrome is a linear combination of at most t columns of H. We point out that some authors use $t(C)$ for the covering radius of C, some others use $CR(C)$, and furthermore that the notation $(n, K)\,R$ code is used for an (n, K) code with covering radius R.

2. Functions related to the covering radius.

Several functions have been introduced to express information about the covering radius of codes. The best known are

(2.1) $t\,[n, k] :=$ minimal covering radius for $[n, k]$ codes.

(2.2) $t\,(n, K) :=$ minimal covering radius for (n, K) codes.

(2.3) $K_q\,(n, R) := \min\,\{\,|\,C\,|\, : C$ is an R-covering of $Z^n\}$.

(We usually omit the index q if it is 2).

(2.4) $k(n, R) :=$ minimal dimension of a linear code of length n and covering radius R.

The following obvious inequality is known as the *sphere covering bound*:

(2.5) $$K_q\,(n, R) \geq \frac{q^n}{V_q(n, R)}.$$

In a recent paper ([4]) Brualdi, Pless and Wilson introduced a new function, which they called the *length function*, as follows:

(2.6) $l(m, r) :=$ smallest length of a binary code of codimension m and covering radius r.

At first, this does not seem too useful because one can simply translate most of the known results for $t[n, k]$ to obtain the equivalent statements for $l(m, r)$. However, in tables the function $l(m, r)$ saves a lot of space. E.g. the information $t[k+8, 8] = 2$ for $k = 18, 19, \cdots, 246$ takes many entries 2 in a table, where $l(8, 1) = 255$, $l(8, 2) \leq 26$ tells us the same thing. The translation of (2.5) is

(2.7) $$l\,(m, r) \geq \min\left\{ n : \sum_{i=0}^{r} \binom{n}{i} \geq 2^m \right\}$$

and the expression on the right hand side is called the *first feasible length* for a code of codimension m and covering radius r.

3. Normal codes and the ADS construction.

Two recent long papers on covering radius with many new results and methods are [14] by Graham and Sloane and [11] by Cohen, Lobstein and Sloane. From these we quote the following.

DEFINITION 3.1. Let C be a binary $[n, k]\,R$ code. For $1 \le i \le n$ denote by $C_0^{(i)}$ resp. $C_1^{(i)}$ the subcode of C consisting of the codewords with i-th coordinate 0 resp. 1. We assume that both are nonempty, i.e. they each contain half of the codewords. Define

(3.1) $\qquad N^{(i)} := \max\,\{d(\mathbf{x}, C_0^{(i)}) + d(\mathbf{x}, C_1^{(i)}) : \mathbf{x} \in \mathbb{F}_2^n\}$

and call this the norm of C with respect to position i. Then define

(3.2) $\qquad N := \min\,\{N^{(i)} : 1 \le i \le n\}.$

We call N the norm of the code and coordinate positions i for which $N = N^{(i)}$ are called *acceptable*. The code C is called a *normal* code if

(3.3) $\qquad N \le 2R + 1.$

DEFINITION 3.2. For a code C over a q-ary alphabet we define "normal" in a similar way. In (3.1) we now have q subcodes $C_j^{(i)}$ depending on the value of the i-th coordinate and (3.3) must be replaced by

(3.4) $\qquad N \le qR + (q-1).$

So, a code is normal if for some i and every word $\mathbf{x}$ the average distance from $\mathbf{x}$ to the subcodes $C_j^{(i)}$, $0 \le j \le q-1$, is less than $R + 1$.

Whereas most binary codes seem to be normal (cf. [14]) it is quite difficult to find nonbinary normal codes!

EXAMPLE 3.1. Let C be a perfect q-ary Hamming code ($R = 1$). If we take $\mathbf{x} = 0$ in the q-ary analog of (3.1), then we find

$$N = 0 + (q-1) \times 3 \;=\; 3q - 3,$$

whereas $qR + (q-1) = 2q - 1$, which is less unless $q = 2$. (In fact $N^{(i)} = 3q - 3$ for all $\mathbf{x}$.)

DEFINITION 3.3. Let C_i be an $[n_i, k_i]\,R_i$ code ($i = 1, 2$). We assume that the last coordinate of C_1 is acceptable and that the first coordinate of C_2 is acceptable. The *amalgamated direct sum* $C_1 \dot{+} C_2$ of C_1 and C_2 is the $[n_1 + n_2 - 1, k_1 + k_2 - 1]$ code consisting of the words $(c_1, c_2, \cdots, c_{n_1+n_2-1})$ for which the word $(c_1, c_2, \cdots, c_{n_1})$ is in the code C_1 and $(c_{n_1}, c_{n_1+1}, \cdots, c_{n_1+n_2-1})$ is in C_2.

This construction is called the "ADS-construction".

THEOREM 3.1. *If* C_1 *and* C_2 *are normal then the covering radius of* $C_1 \dot{+} C_2$ *is* $R_1 + R_2$.

In [14] this is called "saving a coordinate" (compared to the usual direct sum construction). Methods are given that save more than one coordinate.
The paper [11] is devoted to bounds for $K(n, R)$ for binary codes. Some of the results are based on the (obvious) generalization of Definition 3.1 to nonlinear codes. The lower bounds are improvements of (2.5) obtained by estimating how many codewords are "covered" more than once by the code, i.e. have distance $\le R$ to more than one codeword. Among the interesting results are the following

$$K(5, 1) = 7, \quad K(6, 1) = 12, \quad K(11, 1) \le 192,$$

and

$$K(2R+3, R) = 7, \quad K(2R+4, R) \le 12.$$

One of the useful ideas in the constructions is the concept of a *piecewise constant code*. This is a code of length $n_1 + n_2 + \cdots + n_t$ for which the coordinates are partitioned into blocks of size n_i, $(1 \le i \le t)$, and any permutation of any of the blocks is an automorphism of the code.

EXAMPLE 3.2. $K(5, 1) \le 7$ is demonstrated by the piecewise constant code

```
0 0  0 0 0
0 0  1 1 1
1 0  0 0 0
0 1  0 0 0
1 1  0 1 1
1 1  1 0 1
1 1  1 1 0
```

The following construction is given in [11] and attributed to Katsman and Litsyn, and to Mollard. Actually this idea already occurred in a slightly different form in [22]. (See Example 7.1.)

THEOREM 3.2. *Let* C *be an* (n, K) *code with covering radius 1. We define* C^* *to be the code with codewords* $(c_0, c_1, \cdots, c_n, c'_1, \cdots, c'_n)$, *for which* $c_0 + c_1 + \cdots + c_n = 0$ *and* $(c'_1 - c_1, \cdots, c'_n - c_n) \in C$. *Then* C^* *is a* $(2n+1, 2^n K)$ *code with covering radius 1.*

The idea of the proof is the same as in Example 7.1.

4. New bounds.

It is not surprising that many of the recent papers in this area are on bounds for the covering radius functions. Quite often they concern ad hoc results that we shall only briefly mention. However, some nice new ideas have also been introduced in these papers. In Section 3 we mentioned some improvements of the bound (2.5). The most significant recent improvements to the sphere covering bound are due to van Wee [36]. We do not mention all the results (for these see the tables for $K(n, R)$ in the paper) but illustrate the main idea by one simple example.

THEOREM 4.1. *If n is even then* $K(n, 1) \geq 2^n/n$.

Proof. Let n be even and let C be a code with covering radius 1 in $\mathbb{F}_2^n$, $C' := \mathbb{F}_2^n \setminus C$, and let A be the set of words in $\mathbb{F}_2^n$ that have distance ≤ 1 to at least two words of C. Note that if $d(\mathbf{x}, \mathbf{c}) = 1$ or 2, then $|B_1(\mathbf{x}) \cap B_1(\mathbf{c})| = 2$ and since $|B_1(\mathbf{x})| = n+1$ is odd, it follows that if $\mathbf{x} \in C'$, then $|B_1(\mathbf{x}) \cap A| \geq 1$. Now count in two ways the pairs $(\mathbf{x}, \mathbf{y})$ with $\mathbf{x} \in C'$, $\mathbf{y} \in A$, $d(\mathbf{x}, \mathbf{y}) \leq 1$. As we saw above, this number is at least $|C'| = 2^n - |C|$. On the other hand, this number is trivially at most $|A| \cdot (n-1)$. Therefore $|A| \geq (2^n - |C|)/(n-1)$. Since $|C|(n+1) \geq 2^n + |A|$ the result follows. ▯

Most of the improved bounds in [36] are based on the generalization of Theorem 4.1 to the case of $R > 1$.

The next result also concerns a significant improvement of earlier bounds. In fact, the following construction due to Wilson [4] gives an exponential improvement on the upper bounds found by methods using generalizations of the ADS-construction. Again we only illustrate the idea by giving one example. We are interested in covering radius $R = 2$. Using the terminology of the length function $l(m, r)$, we wish to find a subset S of $\mathbb{F}_2^m$ of minimal size such that any element of $\mathbb{F}_2^m$ is the sum of at most two elements of S. Considering everything projectively, this can be reformulated as follows. Find a set S of minimal cardinality in the space $\mathbf{P} = PG(m-1, 2)$ such that every point of $\mathbf{P}$ is on a secant of S. In general, a set S in $PG(n, q)$ such that every point of the space is on a secant of S is called a "secant covering set".

LEMMA 4.1. *Let* $q = 2^a$. *Then there is a secant covering set in* $PG(3, q)$ *with* $2q+1$ *points.*

Proof. Let π be a plane in $PG(3, q)$ and let O be an oval in π. (So $|O| = q+1$.) We consider a line l through the nucleus N of O but not in π. It is trivial to see that $O \cup l \setminus N$ is a secant covering set with $2q+1$ points: for points in π use the definition of oval and the fact that N is on l; for a point not on π use the fact that the plane through such a point and l intersects π in a line that must meet O. ▯

DEFINITION 4.1. Let S be the 4 by $2q+1$ matrix that has as columns the points of the secant covering set of Lemma 4.1. Let α be a primitive element of $\mathbb{F}_q$ and let M be the matrix

$(1\ \alpha\ \cdots\ \alpha_{q-2})$. We define a *binary* matrix S^* as follows: in the Kronecker product $S \otimes M$ replace each entry (an element of $\mathbb{F}_q$) by its representation as a column vector in $\mathbb{F}_2^a$.

THEOREM 4.2. *The columns of S^* are a secant covering set in $PG(4a-1, 2)$.*

Proof. An arbitrary column vector in $\mathbb{F}_2^{4a}$ can be interpreted as a column vector in $\mathbb{F}_q^4$, which by the definition of S is a linear combination of two columns of S and therefore the sum of two columns of S^*, since every nonzero multiple of a column of S is a column of S^*. □

The following theorem (from [4]) is an immediate consequence.

THEOREM 4.3. *For $a \geq 1$*

(4.1) $\qquad l(4a, 2) \leq (2^{a+1}+1)(2^a-1)$.

Note that an ADS-type construction saving e columns and using two Hamming codes would give the bound $l(4a, 2) \leq 2^{2a+1} - 2 - e$ and the two bounds differ by $2^a - e - 1$.
In [4] it is shown that some of the results of Graham and Sloane [14] can be translated as follows for the length function:

(4.2) $\qquad l(2s+1, s) = 2s+5 \quad \text{for} \quad s \geq 1,$

(4.3) $\qquad l(2s, s-1) = 2s+6 \quad \text{for} \quad s \geq 4.$

Since $l(m, 1)$ follows from the Hamming codes, the only difficult values of $l(m, r)$ are in the interval $2 \leq r < \frac{1}{2}m - 1$. The most recent update on these values can be found in a preprint by Brualdi and Pless [6]. The lower bounds

$$l(9, 2) \geq 33, \quad l(10, 2) \geq 46, \quad l(12, 2) \geq 91$$

are from [4], the result

$$l(9, 3) \geq 16$$

due to Simonis [31] was reported at this meeting, for $4 \leq r \leq 8$ there are several bounds due to van Wee as mentioned above, and finally

$$l(7, 2) \geq 18, \quad l(8, 2) \geq 24$$

occur in a recent preprint of Calderbank and Sloane [9]; the second one also occurs in [31]. Besides the upper bounds already mentioned in this section there are a few unpublished results found by computer. These are: $l(8, 2) \leq 26$, $l(9, 2) \leq 41$, $l(9, 3) \leq 18$, and $l(12, 3) \leq 38$. In [6] these are quoted as private communications by D. Ashlock and R. Kibler.

A lower bound on $K(n, R)$ from [11] is improved by van Lint, jr. in [27] as follows. Let C have length $n+2$, covering radius $R+1$, and assume that $|C| = M < K(n, P)$. Then for all

$1 \le i < j \le n+2$, the pair (c_i, c_j), where **c** runs through C, takes on all four possible values. Let A_j be the subset of C consisting of codewords with $c_j = 1$. Then by the assertion above, these sets have the property that if $k \ne l$ then all four sets $A_k \cap A_l$, $A_k \cap \bar{A}_l$, $\bar{A}_k \cap A_l$, $\bar{A}_k \cap \bar{A}_l$ are nonempty. A well known result from extremal set theory then states that

$$n+2 \le \binom{M-1}{\lfloor \frac{M}{2} \rfloor - 1} .$$

A consequence of this is the bound

$$K(2R+5) \ge 8 \quad \text{for} \quad R \ge 6.$$

As we saw above, the lower bounds on $K(n, R)$ of [14] and [36] were found by estimating the number of words that are covered more than once. More results of this type are given by Honkala in [17], e.g. $K(8, 2) \ge 10$, $K(11, 3) \ge 11$, $K(14, 4) \ge 13$.

In [18] Honkala and Hämäläinen construct new covering codes (and improve some upper bounds for $K(n, R)$) by first constructing the code of length n consisting of all words with weight w, where $w \in W$, W being a set satisfying a number of special conditions, and then doubling the length by the rules $0 \to 00$, $1 \to 11$ or $0 \to 10$, $1 \to 01$.

The well known *Griessmer bound* for binary codes states that if C is an $[n, k, d]$ code, then

$$n \ge \sum_{i=0}^{k-1} \lceil \frac{d}{2^i} \rceil .$$

The covering radius of codes meeting this bound is considered in [7] by Busschbach, Gerretzen and van Tilborg.

In [19] Janwa proves the following similar inequality for the covering radius. Let C be as above and assume that C has covering radius R. Then

$$R \le n - \sum_{i=1}^{k} \lceil \frac{d}{2^i} \rceil .$$

Let G be a generator matrix for an $[n, k]$ binary code C. Let G have a distinct columns, none of which is zero, and let these columns have multiplicities $m_1, m_2, \cdots, m_a$. In [33] Sloane defines the *normalized covering radius* ρ of C to be

$$\rho = R - \sum_{i=1}^{a} \lfloor \frac{m_i}{2} \rfloor .$$

In this paper and in Kilby and Sloane [23] the covering radius of codes of low dimension is studied and upper bounds are obtained for the normalized covering radius.

5. Subcodes.

We mention four papers on the covering radius of subcodes, three of which are related. In [2] Adams shows that if C_0 is a subcode of codimension 1 of a binary linear code C with covering radius R, then the covering radius R_0 of C_0 satisfies $R_0 \le 2R + 1$. In [5] this is generalized to $R_0 \le (i+1)R + i$ in case the codimension is i. Calderbank [8] shows that $R_0 \le 2R + 2^i - 1$.

In [20] Janwa and Mattson consider even-weight subcodes and their covering radius.

6. Special topics.

For the sake of making this survey as complete as the present author can do at this time, we list a number of papers that we know about but have not seen or have not studied sufficiently.
The references [12], [20], [28] and [34] are about special codes, such as cyclic codes, BCH codes, RM codes, etc.
The references [16] and [27] concern codes with *mixed alphabets*, i.e. some symbols are from $\mathbb{F}_2$, others from $\mathbb{F}_3$.

The first sections of this paper have clearly demonstrated that very much has been going on recently concerning the covering radius of binary codes and also that the present author cannot keep up with the reading of all this interesting material! In the following sections we turn to the problem of covering radius 1 for ternary codes.

7. The football pool problem.

In one entry in a football pool one forecasts the outcome of n football matches (win, lose or draw). To win first prize all forecasts have to be correct. If one wishes to guarantee winning the first prize, no matter what the outcome of the matches is, then one obviously has to submit 3^n entries. To win second prize one of the forecasts may be incorrect. In order to guarantee second prize, we consider each entry as a word in $\mathbf{F}_3^n$. If the set of entries is a 1-covering of the space, then winning at least the second prize again does not depend on the outcome of the matches. The number of entries needed is usually denoted by $\sigma(n, 3)$. So, in the terminology of (2.3) we have

$$\sigma(n, 3) = K_3(n, 1). \tag{7.1}$$

Since the ternary Hamming codes $\mathbb{H}_4$, $\mathbb{H}_{13}$ of length 4 resp. 13 (both perfect codes) yield $\sigma(4, 3) = 9$, $\sigma(13, 3) = 3^{10}$, we have

$$\sigma(n, 3) \le 3^{n-2} \quad \text{for} \quad 5 \le n \le 12. \tag{7.2}$$

The research in this area (in the last 25 years!) has been concerned with the values in (7.2). In [21] it was shown that equality holds in (7.2) for $n = 5$; the proof is very long. Of course trivially

(7.3) $\sigma(n+1, 3) \le 3\sigma(n, 3)$,

so an improvement of the upper bound for one value of n also could lead to an improvement for larger values.

We start by explaining a simple construction (cf. [22]) that is related to the one mentioned in Theorem 3.2. It still holds the world record for $n = 9$.

EXAMPLE 7.1. $\sigma(9, 3) \le 2 \cdot 3^6$

The proof of this assertion is by construction. The code C is

$$C := \{(x_0, x_1, x_2, x_3, x_4, y_1, y_2, y_3, y_4\},$$

where

(i) $\mathbf{x} \in \mathbb{F}_3^4$,

(ii) $\sum_{i=0}^{4} x_i \ne 0$,

(iii) $\mathbf{x} - \mathbf{y} \in \mathbb{H}_4$.

Actually the description of C in [22] was as follows. Each point of $\mathbb{F}_3^5$ was given one of 9 colors or the color "blank" in such a way that each blank point had distance 1 to a point of each of the 9 colors. The points were then replaced by an empty copy of $\mathbb{F}_3^4$ in the case of blank and a copy containing one of the nine cosets of $\mathbb{H}_4$ (one corresponding to each color) otherwise. This description of C is generalized in [3].

It is easy to see (from either description) that C has covering radius 1.

Using combinatorial constructions several new upper bounds for $\sigma(n, 3)$ for $6 \le n \le 8$ were found in the last five years. In 1983 Weber [35] proved $\sigma(6, 3) \le 79$ and Fernandes and Rechtschaffen [13] showed that $\sigma(7, 3) \le 225$ and $\sigma(8, 3) \le 567$. These results were obtained using 2-step coverings (like our second description of C in Example 7.1).

In 1984 Blokhuis and Lam [3] generalized the idea of Example 7.1 as follows.

DEFINITION 7.1. Let $A = (I\ M) = (\mathbf{a}_1, \mathbf{a}_2, \cdots, \mathbf{a}_n)$ be a $r \times n$ matrix, where I is the $r \times r$ identity matrix and M has entries from $\mathbb{F}_3$. A subset S of $\mathbb{F}_3^r$ is said to *cover* $\mathbb{F}_3^r$ *using* A if

$$\mathbb{F}_3^r = \{\mathbf{s} + \alpha\,\mathbf{a}_i : \mathbf{s} \in S,\ \alpha \in \mathbb{F}_3,\ 1 \le i \le n\}.$$

THEOREM 7.1. *If S is a covering of $\mathbb{F}_3^r$ using A, then $W = \{\mathbf{w} \in \mathbb{F}_3^n : A\mathbf{w} \in S\}$ is a 1-covering of $\mathbb{F}_3^n$ with $|W| = |S|\,3^{n-r}$.*

Proof. Let $\mathbf{x} \in \mathbb{F}_3^n$. Then $A\,\mathbf{x} = \mathbf{s} + \alpha\,\mathbf{a}_i$. Therefore, if $\mathbf{e}_i$ is the i-th basis vector in $\mathbb{F}_3^n$, $A(\mathbf{x} - \alpha\,\mathbf{e}_i) \in S$, so $\mathbf{x} - \alpha\,\mathbf{e}_i \in W$. □

Using Theorem 7.1 Blokhuis and Lam proved that $\sigma(7\,,\,3) \le 216$ and by a similar construction $\sigma(10\,,\,3) \le 5 \cdot 3^6$.
The idea of Definition 7.1 was generalized by van Lint, jr. in [27].

The most recent result is a combinatorial construction showing that $\sigma(8\,,\,3) \le 6 \cdot 3^4$. Again Theorem 7.1 is used, with $r = 4$, and

$$M = \begin{bmatrix} 0 & 1 & 1 & 1 \\ 1 & 0 & 1 & 1 \\ 1 & 1 & 0 & 1 \\ 0 & 0 & 0 & 0 \end{bmatrix} = \begin{bmatrix} M_1 \\ 0000 \end{bmatrix}.$$

Two points $\mathbf{x}$ and $\mathbf{y}$ are said to form a "*pair*" if $x_i = y_i$ for one value of i and $x_i = y_i - 1$ for the other two values of i. It is easy to find three such pairs such that each of them covers $\mathbb{F}_3^3$ using $(I\ M_1)$ with the exception of the other two pairs. This then immediately yields a set of six points in $\mathbb{F}_3^4$ covering that space using $(I\ M)$. This result is given in [26] with several other new bounds for $\sigma(n\,,\,3)$, all found using simulated annealing (to be discussed in the next section). In the present case a close analysis of the computer output led to the idea described above.

8. Simulated annealing.

In [26] van Laarhoven, Aarts, van Lint and Wille present three new upper bounds: the bound for $\sigma(8\,,\,3)$ mentioned in the previous section and furthermore $\sigma(6\,,\,3) \le 73$ and $\sigma(7\,,\,3) \le 186$. These bounds were obtained using a fairly new technique known as *simulated annealing* or *statistical cooling*. For an extensive treatment of this technique we refer to the book [25] by van Laarhoven and Aarts or the survey paper [1] by the same authors. A good introduction to applications in coding theory is given by van der Ham [15]. The essential difference between simulated annealing and the standard iterative improvement algorithms used to minimize certain functions is easily described in one sentence. In the standard algorithm a transition from one state (or configuration) to some neighbouring state is accepted only if the value of the function to be minimized decreases due to this transition; in simulated annealing transitions that increase this function are sometimes accepted and the decision is governed by a probabilistic algorithm. To give a somewhat more detailed idea of the background we quote from the introduction of [15].

Annealing is a technique from statistical mechanics; solids are annealed by raising the temperature to a maximal value at which the particles randomly arrange in the liquid phase, followed by cooling to force the particles into a low energy state of a regular lattice. At each temperature T, the solid is allowed to reach *thermal equilibrium*, characterized by a probability distribution of being in a state with energy E given by the *Boltzmann distribution*:

$$Pr\{\mathbf{E}=E\} = \frac{1}{Z(T)} \cdot \exp\left(-\frac{E}{k_B T}\right)$$

where $Z(T)$ is a normalization factor and k_B the Boltzmann constant. As the temperature decreases, the Boltzmann distribution concentrates on states with lowest energy and finally, when the temperature approaches zero, only minimum energy states have a non-zero probability of occurrence. However, if the cooling is too rapid, i.e. if the solid is not allowed to reach thermal equilibrium for each temperature value, defects can be "frozen" into the solid and a minimum energy state cannot be reached.

To simulate the evolution to thermal equilibrium of a solid for a fixed value of the temperature T, Metropolis et al. [29] proposed a *Monte Carlo method*, which generates sequences of states of the solid in the following way. Given the current state of the solid, characterized by the positions of its particles, a small, randomly generated perturbation is applied by a small displacement of a randomly chosen particle. If the difference in energy, ΔE, between the slightly perturbed state and the current one is negative, i.e. if the perturbation results in a lower energy for the solid, then the process is continued with the new state (the new state is accepted). If $\Delta E > 0$, then the probability to accept the perturbed state is given by

$$\exp(-\Delta E / k_B T).$$

This acceptance rule for new states is referred to as the *Metropolis criterion*. Following this criterion, the system eventually evolves into thermal equilibrium, i.e. after a large number of perturbations the probability distribution of the states approaches the Boltzmann distribution. This method is known as the *Metropolis Algorithm.*

Kirkpatrick et al. [24] used this Metropolis Algorithm to solve combinatorial optimization problems by using a *cost function* in place of the energy and configurations in place of the states of the solid; the temperature then assumes the role of a control parameter that is no longer fixed on one value.

We illustrate how the method can be used to show that $\sigma(6,3) \leq 73$. The following algorithm was executed a number of times using a code C with $|C| = \sigma$, starting with $\sigma = 80$ and then decreasing the value of σ until the algorithm failed to find a covering code. This happened for $\sigma = 72$.

The algorithm starts by taking a random code C of the prescribed size. A control parameter β that plays the role of the temperature is initialized. At certain points in the algorithm the temperature is lowered and for this a "cooling rule" has to be prescribed, e.g. one could take $\beta' = \alpha\,\beta$, say with $\alpha = 0,9$. (Of course, more sophisticated cooling rules have been proposed.) At a *fixed* temperature, several trials to decrease the chosen cost function are made. The number L of trials is prescribed (usually in such a way that it is reasonably likely that all neighbours of a configuration are tried). In our case the cost function was chosen as the number of points in $\mathbb{F}_3^6$ with distance

> 1 to the code. As a stopping rule one takes (a) the cost function has become 0 (a covering has been found) or (b) for h successive temperatures no trial was accepted (the system is "frozen"). It remains to explain how one trial works. First one picks a random codeword and then a random word at distance 1, not already in the code. The code C' is obtained by replacing the codeword by this neighbour. One calculates Δc, the difference in the cost function of C' and C. Then a random generator picks a number y in the interval $[0, 1]$. (Now the analog with the Metropolis criterion becomes apparent.) The "trial" C' is accepted as new code if and only if $\exp(-\Delta c / \beta) > y$. Note that if the cost function has decreased, i.e. $\Delta c < 0$, then the transition is always accepted, but if the cost function has increased, then the trial is accepted with probability $\exp(-\Delta c / \beta)$ and this probability becomes smaller as the "temperature" decreases.

As the numbers show, this technique has been remarkably successful for the football pool problem (and several others). The present author was even more pleased by the fact that the combinatorial construction that holds the present record for $\sigma(8, 3)$ was found by staring at the output of one of the simulated annealing algorithms.

REFERENCES

[1] E.H.L. AARTS AND P.J.M VAN LAARHOVEN, *Statistical Cooling: A General Approach to Combinatorial Optimization Problems*, Philips J. of Research, 40 (1985), pp. 193-226.

[2] M.J. ADAMS, *Subcodes and covering radius*, IEEE Trans. Information Theory, IT 32 (1986), pp. 700-701.

[3] A. BLOKHUIS AND C.W.H. LAM, *Coverings by Rook Domains*, J. Combinatorial Theory, A35 (1984), pp. 240-244.

[4] R.A. BRUALDI, V.S. PLESS AND R.M. WILSON, *Short Codes with a Given Covering Radius*, IEEE Trans. Information Theory (to appear).

[5] R.A. BRUALDI AND V.S. PLESS, *On the covering radius of a code and its subcodes*, preprint.

[6] ______________, *On the length of codes with a given covering radius*, preprint.

[7] P.B. BUSSCHBACH, M.G.L. GERRETZEN AND H.C.A. VAN TILBORG, *On the Covering Radius of Binary, Linear Codes Meeting the Griesmer Bound*, IEEE Trans. Information Theory, IT 31 (1985), pp. 465-468.

[8] A.R. CALDERBANK, *Covering Radius and the Chromatic Number of Kneser Graphs, J. Combinatorial Theory (to appear).*

[9] A.R. CALDERBANK AND N.J. SLOANE, *Inequalities for covering codes*, preprint.

[10] G.D. COHEN, M.G. KARPOVSKY, H.F. MATTSON, JR. AND J.R. SCHATZ, *Covering Radius-Survey and Recent Results*, IEEE Trans. Information Theory, IT 31 (1985), pp. 328-343.

[11] G.D. COHEN, A.C. LOBSTEIN AND N.J.A. SLOANE, *Further Results on the Covering Radius of Codes*, IEEE Trans. Information Theory, IT 32 (1986), pp. 680-694.

[12] D.E. DOWNEY AND N.J.A. SLOANE, *The covering radius of cyclic codes of length up to 31*, IEEE Trans. Information Theory, IT 31 (1985), pp. 446-447.

[13] H. FERNANDES AND E. RECHTSCHAFFEN, *The Football Pool Problem for 7 and 8 Matches*, J. Combinatorial Theory, A35 (1985), pp. 109-114.

[14] R.L. GRAHAM AND N.J.A. SLOANE, *On the Covering Radius of Codes*, IEEE Trans. Information Theory, IT 31 (1985), pp. 385-401.

[15] M.W. VAN DER HAM, *Simulated Annealing applied in Coding Theory*, Master's Thesis, Eindhoven University of Technology, (1988).

[16] H.O. HÄMÄLÄINEN, *Upper bounds for football pool problems and mixed covering codes*, preprint.

[17] I.S. HONKALA, *Lower Bounds for Binary Covering Codes*, IEEE Trans. Information Theory, IT 34 (1988), pp. 326-329.

[18] I.S. HONKALA AND H.O. HÄMÄLÄINEN, *A New Family of Covering Codes*, IEEE Trans. Information Theory (to appear).

[19] H. JANWA, *Some New Upper Bounds on the Covering Radius of Codes*, IEEE Trans. Information Theory (to appear).

[20] H. JANWA AND H.F. MATTSON, JR., *Covering Radii of Even Subcodes of t-dense Codes*, Proc. AAECC 3, Grenoble, Lect. Notes Comput. Sci. 229 (1986), pp. 120-130.

[21] H.J.L. KAMPS AND J.H. VAN LINT, *The Football Pool Problem for 5 Matches*, J. Combinatorial Theory, A3 (1967), pp. 315-325.

[22] ______________, *A covering problem*, Colloquia Mathematica Societatis János Bolyai, 4 (1970), pp. 679-685.

[23] K.E. KILBY AND N.J.A. SLOANE, *On the Covering Radius Problem for Codes I, II*, SIAM J. Algebraic and Discrete Methods, 8 (1987), pp. 604-627.

[24] S. KIRKPATRICK, C.D. GELATT, JR. AND M.P. VECHI, *Optimization by Simulated Annealing*, Science 220 (1983), pp. 671-680.

[25] P.J.M. VAN LAARHOVEN AND E.H.L. AARTS, *Simulated Annealing: Theory and Applications*, D. Reidel Publishing Company, Kluwer Academic Publishers, Dordrecht, The Netherlands, (1987).

[26] P.J.M. VAN LAARHOVEN, E.H.L. AARTS, J.H. VAN LINT AND L.T. WILLE, *New Upper Bounds for the Football Pool Problem for 6, 7 and 8 Matches*, J. Combinatorial Theory (to appear).

[27] J.H. VAN LINT, JR, *Covering Radius Problems*, Master's Thesis, Eindhoven University of Technology, (1988).

[28] H.F. MATTSON, JR., *An Improved Upper Bound on Covering Radius*, Lecture Notes in Computer Science, 228 (1986), pp. 90-106.

[29] N. METROPOLIS, A. ROSENBLUTH, M. ROSENBLUTH, A. TELLER AND E. TELLER, *Equation of State Calculations by Fast Computer Machines*, J. of Chem. Physics, 21 (19853), pp. 1087-1092.

[30] J. PACH AND J. SPENCER, *Explicit codes with low covering radius*, IEEE Trans. Information Theory (to appear).

[31] J. SIMONIS, *The minimal covering radius $t[15, 6]$ of a 6-dimensional binary linear code of length 15 is equal to 4*, IEEE Trans. Information Theory (to appear).

[32] A.N. SKOROBOGATOV, *On the covering radius of BCH codes*, Proc. Third Int. Workshop on Information Theory, Sochi (1987), pp. 308-309.

[33] N.J.A. SLOANE, *A new approach to the covering radius of codes*, J. Combinatorial Theory, A24 (1986), pp. 61-86.

[34] E. VELIKOVA, *Bounds on covering radius of linear codes*, Comptes Rendus de l'Academie bulgare des Sciences, **41** (1988), pp. 13-16.

[35] E.W. WEBER, *On the Football Pool Problem for 6 Matches: A New Upper Bound*, J. Combinatorial Theory, A35 (1983), pp. 109-114.

[36] G.J.M. VAN WEE, *Improved Sphere Bounds on the Covering Radius of Codes*, IEEE Trans. Information Theory, IT 34 (1988), pp. 237-245.

[37] L.T. WILLE, *The Football Pool Problem for 6 Matches: A New Upper Bound Obtained by Simulated Annealing*, J. Combinatorial Theory, A45 (1987), pp. 171-177.

CODES AND CURVES

by

Jean Francis MICHON
Département de Mathématiques
Université PARIS 7

This is a "horizontal" survey of the activity developped around the algebro-geometric codes introduced by Goppa in 1975. The aim of my talk is to give a quick introduction for non specialist mathematicians of the different and recent directions of this research.

I give in the appendix the most complete bibliography I have on the subject.

1. KEY FACTS AND CONCEPTS FROM ALGEBRAIC GEOMETRY

The base field of any computation is $\mathbf{F}$,the algebraic closure of $\mathbf{F}_p$, where p is a prime.
A *curve* is an absolutely irreducible algebraic set of dimension 1.

Plane curve : C is a plane curve if

$$C=\{(x,y,z) \in \mathbf{P}^2(\mathbf{F}) | P(x,y,z)=0\}$$

where P is a homogeneous irreducible polynomial (for example Fermat curve $X^n+Y^n-Z^n=0$) . A curve (resp. a point of a curve) is said to be defined over the finite field $\mathbf{F}_q$, or *rational over* $\mathbf{F}_q$ if it is stable under every $\mathbf{F}_q$-automorphism of $\mathbf{F}$. For a plane curve it is the case when the polynomial $P \in \mathbf{F}_q[X,Y,Z]$.

Space curve : defined for example by n equations in n+2 variables.

Abstract curves : for example modular curves.

The study of curves is proved to be equivalent to the study of the *function fields* (algebraic extensions of finite degree of the rational fractions field $\mathbf{F}_q(X)$). This point of view was developped by the "german school" of geometry during the period from nineteenth century to 1940. It seems that many aspects of their works are not known by a large part of the mathematical community : who knows that F. K. Schmidt proved the Riemann-Roch formula in finite caracteristic case?. The lack of a good treatment of our subject (curves and their jacobians in finite caracteristics) at a graduate level is evident. I refer to "Algebraic curves" by W. Fulton for beginners, this reading may be followed by the 1951 treatment of Chevalley : "Introduction to the theory of algebraic functions of one variable".

Riemann introduced a fundamental "invariant" of a curve : the genus g. This integer "caracterizes" the geometric complexity of the curve.

HASSE-WEIL FORMULA

The rational points of a curve play a fundamental role in Goppa construction. Recall the Hasse-Weil estimate for the number of rational points (over $\mathbf{F}_q$) of a curve : let $C(\mathbf{F}_q)$ the set of points of the curve C rational over $\mathbf{F}_q$ whose genus is g, then

$$q+1-2g\sqrt{q} \leq \mathrm{Card}(C(\mathbf{F}_q)) \leq q+1+2g\sqrt{q}.$$

We know that this estimation is bad in plenty of cases . For details I refer to the study of this problem by Serre. The connection with Weierstrass points is studied in Stöhr and Voloch.

THE THEOREM OF RIEMANN AND ROCH

Let C a curve of genus g. A *divisor* on this curve is a formal linear combination of points of the curve with coefficients in **Z**. The divisor of a rational function f on the curve is only the combination of its zeroes and poles with multiplicity, this divisor is denoted (f). The notion of differential form on a curve can be introduced purely algebraically and in the same manner the divisor of a differential form on C can be defined. The *degree* of a divisor G is the algebraic sum of its coefficients and is noted deg(G).

Riemann introduced the following vectorial spaces associated with a divisor G on C:

$$L(G)=\{f\in \mathbf{F}_q(C) \mid (f)\geq -G\}\cup\{0\}$$
$$\Omega(G)=\{\omega\in\Omega^1(C) \mid (\omega)\geq G\}$$

One can prove that

L(G) and $\Omega(G)$ are finite dimensional vectorial spaces,
if f is a function $\deg((f))=0$,
if ω is a differential form $\deg((\omega))=2g-2$,
$\deg(G)<0 \Rightarrow L(G)=\{0\}$
$\deg(G)>2g-2 \Rightarrow \dim(L(G))=\deg G+1-g$
$\dim(\Omega(0))=g$
$\dim L(G)-\dim L((\omega)-G)=\deg G+1-g$ (Riemann-Roch formula).

2. GOPPA CONSTRUCTION OF CODES (1975)

Take n points $P_1,\ldots,P_n$ rational over $\mathbf{F}_q$ on a curve C. Let D the divisor $P_1+\ldots+P_n$ and G another divisor on C, disjoint from D, rational over $\mathbf{F}_q$. Suppose that the genus of C is g then the images of the maps

$$L(G)\rightarrow(\mathbf{F}_q)^n$$
$$f\rightarrow(f(P_1),\ldots,f(P_n))$$

$$\Omega(G-D)\rightarrow(\mathbf{F}_q)^n$$
$$\omega\rightarrow(\mathrm{Res}_{P_1}(\omega),\ldots,\mathrm{Res}_{P_n}(\omega))$$

define two linear codes of length n : C_L and C_Ω dual to each other for the standard scalar product over $(\mathbf{F}_q)^n$. These two families of codes form what I call *geometric* codes (of course they have nothing to do with codes obtained from finite geometries). Several authors use the terminology D,G-code to be more precise.

Their parameters verify the following inequalities :

$\dim(C_L)\geq\deg G+1-g$ $\qquad$ $\dim(C_\Omega)\geq n-\deg G-1+g$

$\mathrm{dist}(C_L)\geq n-\deg G$ $\qquad$ $\mathrm{dist}(C_\Omega)\geq\deg G+2-2g$

I call these inequalities the **Goppa bound** of the D,G-code, they are simple consequences of Riemann-Roch formula.
The older construction of Goppa code is a special case of this construction. One must just replace the curve C by a projective line.

There is a more general construction of Goppa codes from invertible bundles. This was introduced by Manin and is well explained in the paper by Vladut.

3. EXTENDED CLASSICAL GOPPA CODES

Certain extended classical (projective line or genus 0) Goppa codes are cyclic (Berlekamp, Moreno, Tzeng, Zimmerman, Thiong Ly, Yu).
The fact that there are no more such codes was proved by Stichtenoth.

Other cases of genus 0 : an exhaustive study of all classical codes (Reed-Müller,cyclic, quadratic residue, etc...) would be interesting ! The difficulty is there is more than one way (curve or divisors) to construct a code. One must choose the curve (or the variety) and the divisors that brings or reflects the deepest properties of the code.

4. BOUNDS

I define the *Goppa line* associated to the curve and the divisor D to be the line whose equation is :

$$x + y = 1 - (g-1)/n$$

Classically a [n,k,d] code is associated with the point (d/n,k/n) of the unit-square. The Goppa bound says that D,G-code is always above or on its Goppa-line.

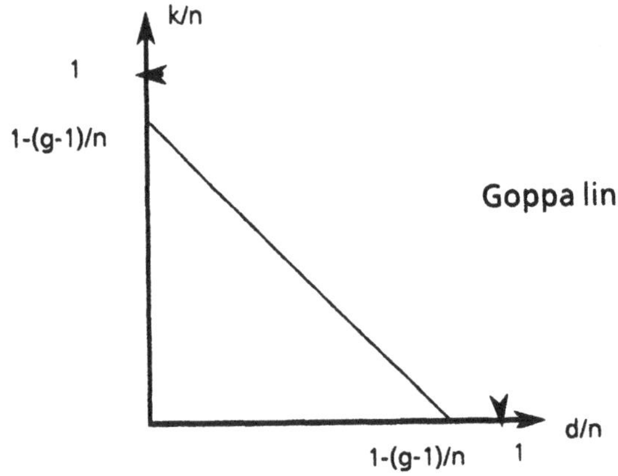

Varshamov - Gilbert bound :

If we try to construct an infinite family of Goppa codes which is asymptotically better than Varshamov -Gilbert bound we must look at the asymptotic behaviour of the quotient g/n. We must find an infinite family of curves such that the quotient g/n is very small. The problem is how grows the number of rational points on curves when genus grows ? The first light on this question was given by Stepanov and Ihara.
In 1981, Tsfasman, Vladut and Zink applied this to geometric codes and constructed a family of codes asymptotically better than the Varshamov-Gilbert bound under hypothesis for the finite field F_q. The trick is to use Shimura curves and Drinfeld's modular curves. I cannot tell in this short survey what is a Shimura curve, it is a whole subject in itself. Nevertheless I can tell a little about the underlying philosophy explaining the choice of such curves. They (in general the

Shimura varieties) are *modular* objects : that is to say they parametrize a family of *abelian varieties*. For example the projective line is a modular curve because it parametrizes the family of all elliptic curves (by the classical j function). All these notions exists over the complex numbers as well as over finite fields and you may go beetween finite and zero charateristics with the reduction mod p ("good" or "bad"). The genus of such curve can be computed by use of coverings, uniformisation, reduction and the number of rational points on the curve reflects the number of rational abelian varieties. Sometimes it's easier to count the number of rational abelian varieties than the number of rational points on the modular curve. For some families of Shimura curves over $\mathbf{F}_q$ there exists a formula wich gives a lower bound for the number of rational points over $\mathbf{F}_{q^2}$. Surprisingly the quotient g/n apppears to be quite small ! If $q^2 \geq 49$ this method gives a family of codes better than VG-bound thus improving this bound. I call this result the TVZ-bound. For the explicit construction of these codes TVZ proved that the construction is polynomial. Unfortunately the degree of the polynomial is high so explicit such codes are not known as far as I know.

Varshamov - Gilbert for binary codes :

The direct computation in the case of base field $\mathbf{F}_2$ shows easily that Goppa bound and Weil estimates give nothing interesting (the Goppa line goes to negative coordinates). Tsfasman discovered a good way to succeed : construct Goppa codes over $\mathbf{F}_{2^m}$,take the intersection with $\mathbf{F}_{2^n}$ and use a "Sugiyama - like" estimates. If one chooses Shimura curves then we get by this method of intersection a family of codes which is asymptotically convergent towards the V-G bound when d/n is small. This is a partial but important progress.

The Sugiyama-like estimate was first proved by Wirth (and independently rediscovered by Tsfasman I suppose) :

Theorem : A D,G-code C_Ω defined over $\mathbf{F}_{q^m}$ when intersected over $(\mathbf{F}_q)^n$ defines a $\mathbf{F}_q$ code whose dimension verifies

$$\dim(C_\Omega \cap (\mathbf{F}_q)^n) \geq n - m(\deg G - \deg G')$$

where $G' = \Sigma[m_Q/q]Q$ if $G = \Sigma m_Q Q$

The open question in this area is : are there curves over $\mathbf{F}_2$ that generate codes better than VG-bound without intersecting operation ?

Tighter than Goppa bound :

One can improve the Goppa's bound in special cases. This can be done by studying the role of Weierstrass points of the curve. Choosing G on Weierstrass points and using Frobenius automorphism one can get closer bound of the dimension of the code.

Open question : Can we choose a family of curves (Shimura curves for example) with known "good" Weierstrass points and weights such that asymptotically (when the genus grows) we get an amelioration of the TVZ bound ?

5. ELLIPTIC CODES

This is a very interesting family of codes I investigated with Driencourt. Elliptic codes come from the first non "classical" case of Goppa construction: take an elliptic curve (it is a synonim for genus one curve). The first known elliptic code is the example given by Goppa himself (the Fermat curve of degree 3).

I was interested in the impact of the well known group law on this curve on the properties of the code. For example does parameters depend from group properties of the curve? The answer is affirmative and translates problems of coding in combinatorial problèms in abelian finite groups.

Elliptic codes are also the ideal field to conceive and test tools for the study of codes coming from curves of higher genus. In this çase the group law must be replaced by the one on the jacobian variety of the curve.

The codes arising from elliptic codes are "MDS-1" that is to say:

$$\text{dimension} + \text{minimal distance} \geq \text{length}$$

The question of the existence of full MDS elliptic codes is not completely solved. But Tsfasman gave a (quite heavy) proof for the non existence of such codes if one takes as divisor D the full set of rational points of the curve over the base field. He described also completely the structure of the group of rational points $C(F_q)$ for any elliptic curve.

6. AUTO DUALITY

This field is the most mysterious. Up to now nobody gave a convincing algebro-geometric understanding of this phenomenon.

The only thing I can say is :

Theorem (Driencourt-Stichtenoth): Given a D,G-code on a curve C if there exists a differential form W on C such that $(W) \geq 2G\text{-}D$ and $\mathrm{Res}_{P_i}(W)$ is a constant independant of i then $C_L \subset (C_L)^{\perp}$. If $(W) = 2G\text{-}D$ then CL is self dual ($C_L = (C_L)^{\perp}$).

Open question : reflect the Mc Williams identity in geometric terms. I suppose that this is something like Riemann-Roch identity, or Zeta functional equation (which is a Poisson formula for number theoricist). The mysterious correspondance lattice-code could receive some light from this study.

7. DECODING

Up to 1987 nobody gave good reason to trust in the existence of a general (valid for all genus) decoding algorithm for geometric codes. A very important progress was made by Justesen and his collaborators establishing a "Petersson - like" algorithm for decoding geometric codes arising from plane curves. This decoding does not recover d/2 errors in all cases but this is going to be ameliorated by Skorobogatov and Krachkovsky. A corollary of this work is that elliptic codes can be decoded completely (ie with maximal d/2 errors recovering).

Open question : Is there an Euclidean algorithm of decoding ? (Hassner trusts in such a method)

7. SPECIAL CURVES

Of course many geometric codes were constructed up to now. They use Hermitian curves (a special case of Fermat), Artin-Schreier curves, hyperelliptic curves.

The study of hyperelliptic case seems (to me) very interesting because of the tools we have on it (one can compute in the jacobian following Cantor work).

BIBLIOGRAPHY

BARG A.M., KATSMAN G.L, TSFASMAN M.A.	Algebraic geometric codes got from curves of small genus	Preprint		
CARTIER P.	Sur la rationalité des diviseurs en Géométrie Algébrique	Bull. Soc. Math. France,	177-251	1958
CANTOR D.G.	Computing in the Jacobian of a Hyperelliptic Curve	Math. of Comp, 48,n°177	95-101	1987
CHAO MING LIU, KUMAR P.V.	On the maximum length of Goppa codes on elliptic curves	Preprint		1987
CHEVALLEY C.	Introduction to the theory of algebraic functions of one variable	Mathematical Surveys n°6, American Mathematical Society		1951
DRIENCOURT Y.	Some properties of elliptic codes over a field of characteristic 2	Proc 3rd international conf., AAECC-3, Lect. Notes in Comp Sc. Springer,229		1985
DRIENCOURT Y.	Codes elliptiques auto-duaux sur un corps de caractéristique 2	Preprint		
DRIENCOURT Y.MICHON J.F.	Elliptic codes over fields of characteristic 2	Journal of pure and applied algebra		1987
DRIENCOURT Y., MICHON J.F.	Remarques sur les codes géométriques	C.R. Acad. Sc.Paris, 301	15-17	1985
DRIENCOURT Y., STICHTENOTH J.F.	A criterium for self duality of geometric codes	Communications in algebra (to appear)		
DRINFELD V.G., VLADUT S.G.	Number of points of an algebraic curve,	Funct. Anal.,17	53-54	1983
FULTON W.	Algebraic curves	(Benjamin) New-York		1969
GOPPA V.D.	Code on algebraic curves	Sov. Math Dokl, 24	170-172	1981
GOPPA V.D.	Algebraico-geometric codes	Math Ussr Izvestiya, vol 21	75-91	1983
GOPPA V.D.	Codes and information	Russian Math. Surveys, vol 39-1	87-141	1984
GOPPA V.D.	Codes associated with divisors	Prob. Pered. Inform.Vol 13, N°1,	33-39	1977
HANSEN J.P.	Codes on the Klein quartic,ideals and decoding	Preprint), Aarhus university		1987
HARTSHORNE R.	Algebraic Geometry	Graduate Texts in Math. , n°52, Springer		1977
HASSNER M., BURGE W., WATT S.M.	Symbolic computation of Error Control Codes on the Elliptic Riemann Surface	Talk in this AAECC-conference		1988
IHARA Y.	Some remarks on the number of rational points of algebraic curvesover finite fields	J. Fac. Sci. Tokyo,IA 28	721-724	1981

JUSTESEN J., LARSEN K.J., ELBRØND JENSEN H., HAVEMOSE A. and HØHOLDT T.	Construction and decoding of a class of algebraic geometry codes	The technical University of Denmark, Mat-report n°1988-10		1988
KATSMAN G.L., TSFASMAN M.A.	Spectra of algebraic geometric codes	Prob. Pered. Informa. VOL 23, n°4, 19-34		1987
KATSMAN G.L., TSFASMAN M.A.,	A remark on algebraic geometric codes	preprint (to appear in Contemp. Math.)		1987
KATSMAN G.L., TSFASMAN M.A., VLADUT S.G.	Modular curves and codes with polynomial complexity of construction	Probl. Info. Trans.,20	35-42	1984
KATSMAN G.L., TSFASMAN M.A., VLADUT S.G.	Modular curves and codes with polynomial complexity of construction,	IEEE Trans., Inf. Theory, 30, N°2,	353-355	1984
KRACHKOVSKY V. Yu.,	A decoding method for algebraic-geometric codes,	to appear in the proceedings of the IX-th All Union conference on coding theory and information transmission, Moscow-Odessa		1988
KUMAR P.V., C.M. LIU	On the maximum length of MDS Goppa Codes on elliptic curves	preprint		1987
LACHAUD G.	Les codes géométriques de Goppa	Séminaire Bourbaki, n° 641		1985
LACHAUD G.	Sommes d'Eisenstein et nombre de points de certaines courbes algébriques sur les corps finis	Cr. Acad. sci. Paris,305	729-732	1987
LACHAUD G.	Codes de Reed et Müller	Preprint		
LANG S.	Elliptic functions	Addison-Wesley Reading Ma.		1973
LEBRIGAND D., RISLER J.J.	Algorithme de Brill-Noether et construction de codes de Goppa	Bull. Soc. Math. France (à paraître)		
LITSYN S.N., TSFASMAN M.A.	Constructive high-dimensional sphere packings	Duke Math. Journal, Vol. 54, n° 1,	147-161	1987
LITSYN S.N., TSFASMAN M.A.	A note on lower bounds	Preprint		
LITSYN S.N., TSFASMAN M.A.	Algebraic geometric and number theoric packings of spheres in Rn	Uspekhi Math. Nauk.,40	185-186	1985
MAC WILLIAMS F.,SLOANE M.J.A	The theory of error correcting codes	North Holland, Amsterdam		1977

LITSYN S.N., TSFASMAN M.A.	Construction of dense high dimensional spheres packings	Duke Math. J., vol 54, n°1	147-161	1987
MAC RAE R.E.	On unique factorization in certain rings of algebraic functions	J. of algebra,17	243-261	1971
MANIN Y.I.	What is the maximum number of points on a curve over F2	J. Fac. Sci. Tokyo, IA 28	715-720	1981
MANIN Y.I., VLADUT S.G	Linear codes and modular curves	Soverem. probl. Math. Viniti.,25	209-257	1984
MATZAT H.	Ein Vortag über Weierstrasspunkte	Thesis Universität Karlsruhe		1975
MICHON J.F.	Amélioration des paramètres des codes de Goppa	Preprint		1985
MICHON J.F.	Les codes BCH comme codes géométriques	Preprint		1985
MICHON J.F.	Codes de Goppa	Sém. Th. Nombres Bordeaux, Exp N°7		1983
QUEBBEMANN H.G.	Cyclotomic Goppa codes	Preprint		1987
QUEBBEMANN H.G.	On even codes	Preprint		1987
SCHARLAU W.	Selbstduale Goppa-Codes	Preprint		1987
SCHMIDT F.K.	Uber die Erhaltung der Kettensätze der Idealtheorie bei beliebigen endlichen Körpererweiterungen	Math Z.	443-450	
SCHMIDT F.K.	Die Wronskische Determinante in beliebigen differenzierbaren Funktionenkörpern	Math. Z. 45,	62-74	1939
SCHMIDT F.K.	Zur arithmetischen Theorie der algebraischen Funktionen. II	Math. Z. 45,	75-96	1939
SCHMIDT F.K.	Zur arithmetischen Theorie der algebraischen Funktionen. I	Math. Z. 41	415-438	
SCHMIDT F.K.	Beweis des Riemann-Rochschen Satzes für algebraische Funftionen mit beliebigen Konstantenkörper	Math. Zeit., 41	415-438	1936

SERRE J.P.	Sur le nombre des points rationnels d'une courbe algébrique sur un corps fini	Cr. Acad. Sci. Paris,296	397-402	1983
SERRE J.P.	Nombre des points des courbes algébruqes sur Fq	Sem. Th Nombres Bordeaux Exp. 22		1982-1983
VAN LINT J.H. SPRINGER T.A.	Generalized Reed-Solomon codes from Algebraic Geometry	IEEE Transactions in Information Theory, Vol IT-33, N°3,	305-309	1987
VLADUT S.G.	An exhaustion bound for algebraic - geometric "modular" codes,	Probl. Pered. Infor., Vol 23, N°1,	28-41	1987
WATERHOUSE W.C.	Abelian varieties over finite fields,	Ann. Sci. Ec. Norm. Sup.,4ème série, t.2,	521-560	1969
WIRTZ M.	On the parameters of Goppa codes	Preprint à paraître dans IEEE Information Theory		1987
WOLFMANN J.	Nombre de points rationnels de courbes algébriques sur des corps finis associées à des codes cycliques	Cr. Acad. Sci. Paris, 305	345-348	1987
ZINOVIEV V.A.	Generalized Cascade codes	Probl. Info. Trans., 12	2-9	1976

COMPUTER AND COMMUTATIVE ALGEBRA

by

LORENZO ROBBIANO
Univ. of Genova (Italy)

My purpose is to explain how Computer Algebra and Commutative Algebra are mutually influenced. I am a mathematician working in a group of research, which has a strong background in Commutative Algebra, so it would be more precise to say that I am going explain how our research in Commutative Algebra was influenced by the new techniques and ideas coming from Computer Algebra.
In May 86 we had in Genova the first meeting COCOA I (COmputer and COmmutative Algebra), which will be followed by COCOA II in May 89; its success induced Bruno Buchberger, the editor of JSC (Journal of Symbolic Computation) to propose the edition of a special volume of JSC on "Computational Aspects of Commutative Algebra". Being the editor of this volume, I have been in a proper position to observe what is going on in this field of research.
This conference is addressed to a wide and mostly non specialized audience, so it is my intention to draw attention more to ideas and themes; but to give a full flavor I will show also some technical aspects.

Commutative Algebra is a relatively young section of Mathematics, strongly developing in our century, with main roots at the end of the last century; it deals primarily with the commutative polynomial ring $A := k[x_1,\dots,x_n]$ over a field k.
To start with, let us consider the following opening theme:
How to describe the solutions of a system of polynomial equations

$$(*) \quad \{f_i(x_1,\dots,x_n) = 0 \ / \ i \in I\}$$

The first remark is that the same problem can be considered as the basic one for Algebraic Geometry; but of course Commutative Algebra focuses on algebraic properties of $(*)$, more than geometric or topological ones.
Another remark is that if the f_i's are linear, then $(*)$ is equivalent to a finite subsystem $(**)$ and then the existence and the size of the set of solutions depends only on two numerical invariants, namely the rank of the two matrices associated to $(**)$, so it is in some sense independent of k. Of course if $f_1, f_2, \dots, f_r$ are not all linear, then k matters; indeed it is sufficient to consider the different behavior of

$$x^2 + y^2 + 1 = 0 \text{ when } k = \mathbb{R} \text{ and when } k = \mathbb{C}.$$

Let us start our trip in the land of polynomials by considering the following theorem

THEOREM 1. (Basissatz, Hilbert 1890) *Every system of algebraic equations over a field k is equivalent to a subsystem of a finite number of equations.*

In modern language the theorem can be rephrased by saying that $k[x_1,\dots,x_n]$ is noetherian. The consequence is that our opening theme can be modified to become: how to describe the solutions of a finite system of polynomial equations

$$(**) \quad f_1(x_1,\dots,x_n) = f_2(x_1,\dots,x_n) = \cdots = f_r(x_1,\dots,x_n) = 0$$

After recalling that a term is a finite commutative word in n symbols, I can state the next fundamental

THEOREM 2. (Dickson 1913) *Given a sequence of terms* $\{t_r\}$, $r \in \mathbb{N}$, *there exists* $N \in \mathbb{N}$ *such that for every* $r > N$, t_n *is multiple of some term in the set* $\{t_1, \ldots, t_N\}$.

These two apparently different results are related; namely one can give a simultaneous proof or one can derive one from the other (see for instance Robbiano 1988).
The idea is that if T^n denotes the set of (commutative) terms in n symbols, then T^n has the algebraic structure of monoid. More precisely

T^n is isomorphic to $\mathbb{N}^n := \{ (a_1, \ldots, a_n) \ / \ a_i \in \mathbb{N}\}$
and the isomorphism is called <u>log</u> and given by

$$\log(x_1^{a_1} \cdots x_n^{a_n}) = (a_1, \ldots, a_n)$$

One can see that $A = k[x_1, \ldots, x_n]$ and $\mathbb{N}^n$ share the property that "ideals " are finitely generated, where of course the word ideal has different meaning for the ring A and the monoid $\mathbb{N}^n$.

Now we can push the connection between A and $\mathbb{N}^n$ a little further. Namely we recall that if R is a ring then the ring $R[x]$ is defined to be the set of finitely supported sequences of elements of R, with the usual sum and the Cauchy product.
Then the polynomial ring $k[x_1, \ldots, x_n]$ is defined recursively as

$$k[x_1, \ldots, x_n] := k[x_1, \ldots, x_{n-1}][x_n]$$

but it can be given a different interpretation as a monoid ring. Namely it is easy to see that

$k[x_1, \ldots, x_n] \simeq k\langle \mathbb{N}^n \rangle$ where $k\langle \mathbb{N}^n \rangle$ is the monoid k-algebra generated by $\mathbb{N}^n$

Without going into the details, we can say that while the first interpretation gives rise to representations of polynomials of type

$$y^3 - (1 + x^3)y^2 + xy$$

the second one gives representations of type

$$x^3y^2 + y^3 - \qquad y^2 + xy$$

An obvious remark is that the type or representation of polynomials may become essential for actual implementations of algorithms.

So far we have seen that the main object to deal with in Commutative Algebra and Algebraic Geometry is $k[x_1, \ldots, x_n]$ and k matters; but of course to make actual computation one needs computable structures, hence k has to be computable. This leads apparently to a splitting, since Algebraic Geometry tends to consider $\mathbb{C}$ as the fundamental field, while computability requires that the cardinality is denumberable. However in principle every specific problem over $\mathbb{C}[x_1,\ldots,x_n]$ usually involves only a finite number of coefficients, hence it can be handled over a finite extension of $\mathbb{Q}$, e.g. $\mathbb{Q}(x)(\sqrt{2})$.

Henceforth I am going to place myself in a context where k is a computable field, for instance $\mathbb{Q}$, $\mathbb{Z}_p$, $\mathbb{Q}(x_1,\ldots,x_n)$, $\mathbb{Q}(\alpha)$ where α is algebraic over $\mathbb{Q}$.
We have seen that sets of equations in $k[x_1,\ldots,x_n]$ give rise to ideals which are finitely generated by Theorem 1, so we are ready to introduce the main theme, namely:

What can we compute and what kind of problems can we try to solve in $k[x_1,\ldots,x_n]$?

Already G. Hermann (1926), following ideas of Kronecker, started the study of constructive aspects of Commutative Algebra, dealing for instance with the following fundamental question

QUESTION 1 (Ideal Membership)
Given $A := k[x_1,\ldots,x_n]$, $f_1,\ldots,f_r$, $f \in A$, how to decide whether or not $f \in (f_1,\ldots,f_r)$ i.e. there exist $a_1,\ldots,a_r \in A$ such that $f = \sum_i a_i f_i$?

Related to Question 1 is the following

QUESTION 2 (Syzygies)
Given $A := k[x_1,\ldots,x_n]$, $f_1,\ldots,f_r \in A$, how to compute a fundamental set of n-tuples $(a_1,\ldots,a_r) \in A^r$ such that $\sum_i a_i f_i = 0$ i.e. a set of generators for the first module of syzygies ?
Moreover one would like to estimate the size of the solutions to these questions.

The idea of looking at syzygies of elements of A extends in the following way; if we can prove that also the syzygies are finitely generated, then we can start the procedure again by looking for a set of fundamental relations among the first syzygies. These are called second syzygies and so on.
For instance let $f_1 := x$, $f_2 := y \in k[x,y]$ and let $I := (x,y)$, the ideal generated by $\{x,y\}$. It has two generators, x, y and the syzygies are generated by $(-y, x)$, as it is easily seen. In modern language we say that I has the finite free resolution

$$0 \longrightarrow A \xrightarrow{\alpha} A^2 \xrightarrow{\beta} I \longrightarrow 0$$

where $\alpha(f) = (-yf, xf)$ $\qquad$ $\beta(f,g) = (xf + yg)$

THEOREM 3 (Hilbert 1890) *For every ideal I in $k[x_1,\ldots,x_n]$, there are only a finite number of non trivial modules of syzygies and each of them is finitely generated i.e. I has a finite free resolution.*

QUESTION 3 (Radical membership)
Given $A := k[x_1,\ldots,x_n]$, $f_1,\ldots,f_r$, $f \in A$, how to decide whether or not the solutions of $f_1 = \cdots = f_r = 0$ are also solutions of $f = 0$?

We usually denote by $Z(f_1,\ldots,f_r)$ the set of common zeroes of the f_i 's, so the question can be rephrased by: how to decide whether or not $Z(f_1,\ldots,f_r) \subseteq Z(f)$?

This question has a subtlety in it because if $f_1 := x^2 + 1$, then $Z(f_1)$ is empty, if we look for solutions in $\mathbb{Q}$. In this case of course $Z(f_1)$ is contained in $Z(f)$ for every f; however if we enlarge $\mathbb{Q}$ to $\mathbb{Q}(i)$ then $Z(f_1) = \{i, -i\}$, and now the question becomes relevant.
To face this situation we need another fundamental theorem by Hilbert

THEOREM 4 (Nullstellensatz, Hilbert 1893) *Let* k *be algebraically closed and let* $A := k[x_1, \ldots, x_n]$, $f_1, \ldots, f_r$, $f \in A$.
Then $Z(f) \supseteq Z(f_1, \ldots, f_r)$ iff $f \in \mathrm{Rad}(f_1, \ldots, f_r)$ i.e. *there exists an integer* N *such that* $f^N \in (f_1, \ldots, f_r)$

After this Theorem, Question 3 reduces in some sense to Question 1, but of course we cannot test ideal membership for all the powers of f. However we can do the following reduction, which uses some small tools of Commutative Algebra
$f \in \mathrm{Rad}(I)$ iff $IA_f = A_f$ iff $I(A[z]/(1-zf)) = (1)$ iff $1 \in (I^e, 1-zf)$
where I^e means the ideal I extended to the ring $A[z]$.
In conclusion the radical membership of f with respect to $(f_1, \ldots, f_r)$ is equivalent to the ideal membership of 1 with respect to $(f_1, \ldots, f_r, 1-zf)$ in $k[x_1, \ldots, x_n, z]$, so it becomes indeed a subproblem of Problem 1.
Before going on, let me just mention that this subproblem was solved by G. Hermann; she used elimination theory and methods of linear algebra to find the solutions, which were showed to have degrees bounded by a doubly exponential function of the number of indeterminates. Very recently this question has been reconsider by Brownawell 1987, Caniglia-Galligo-Heintz 1987 and Kollar 1988, where essentially a simply exponential bound in the number of indeterminates is found. This connection between classical themes of Commutative Algebra and new research on computational aspects of them is in my opinion one of the most relevant aspects of the impact that Computer Algebra has on traditional sections of Mathematics.

Let us now consider the following

QUESTION 4 Given $f_1, \ldots, f_r$ homogeneous polynomials and $I := (f_1, \ldots, f_r)$ we have that $(A/I)_d$ is a finitely generated k- vectorspace for every $d \in \mathbb{N}$.
Let $H_{A/I}$ denote the function defined by

$$H_{A/I}(d) := \dim_k (A/I)_d$$

the so called Hilbert function of I.
Is it possible to compute it ?

QUESTION 5 Is it possible to compute a basis of A/I as a k-vectorspace ?

First of all we may associate to the function $H_{A/I}$ a series $\mathbb{P}_{A/I}(z) \in \mathbb{Q}[[z]]$, the so called Hilbert-Poincarè-Serre series, defined by

$$\mathbb{P}_{A/I}(z) := \textstyle\sum_i H_{A/I}(i) \cdot z^i$$

Then it is possible to show that $\mathbb{P}_{A/I}(z)$ is rational of type $Q(z)/(1-z)^d$ and this implies that $H_{A/I}(i) = P(i)$ for $i >> 0$ where $P(x) \in \mathbb{Q}[x]$.

Therefore the information of the Hilbert function is given by a finite number of invariants, namely the coefficients of the polynomial $Q(z)$ and the integer d, which turns out to have a fundamental meaning.
It gives the "size" of the set of solutions of our original system, i.e its dimension.

Now we are really in the middle of the game and we are ready to come to the main point. We have already seen a structural relation between the polynomial ring and $\mathbb{N}^n$, and now we are going to push this investigation much further.
We have seen that when $k[X_1, \ldots, X_n]$ is interpreted as $k\langle\mathbb{N}^n\rangle$ then the polynomials are represented as finite sums of monomials. A monomial is something of the following type

$c \cdot T$ where $c \in k$ and $T = x_1^{a_1} \cdots x_n^{a_n}$ is a term.

So our polynomial is given by a finite set of terms and coefficients and the question is only how to arrange them in a suitable ordering.
A priori there is no obvious rule for choosing the representation $3x^2 + yz$ instead of $yz + 3x^2$. Of course we would like to consider representations such that once a polynomial is written as $c_1 T_1 + \cdots + c_s T_s$, if we multiply it by a term T, then the representation of the product is $c_1 T \cdot T_1 + \cdots + c_s T \cdot T_s$.
These considerations lead to the notion of term-ordering on the monoid of terms T^n
DEFINITION A term-ordering on T^n is a total ordering which is compatible with the monoid structure of T^n and such that for every $T \in T^n$, $T \geqslant 1$. We denote by O the set of total orderings on T^n which are compatible with the monoid structure, and by TO the subset of term-orderings.

A classification of $O(T^n)$ is easy and it can be found in Robbiano 1985. It turns out that total orderings on T^n can be described by finite arrays of real vectors. Namely, given $\sigma \in O$, it can be described by an array $(u_1, \ldots, u_s)$, where $u_i \in \mathbb{R}^n$ and the meaning of $\sigma = ord(u_1, \ldots, u_s)$ is that

$$t >_\sigma t' \text{ iff } (\log(t) \cdot u_1, \ldots, \log(t) \cdot u_s) >_{lex} (\log(t') \cdot u_1, \ldots, \log(t') \cdot u_s)$$

Here lex is the ordering defined on $\mathbb{R}^n$ by $(a_1, \ldots, a_n) >_{lex} 0$ iff the first coordinate from the left is positive.

Term-orderings are easily recognized among orderings by the following rule:
let $\sigma := ord(u_1, \ldots, u_s)$ and let M be the (n,s) matrix whose $(i,j)^{th}$ entry is the ith coordinate of u_j; then

FACT 1) $\sigma \in TO$ iff for every row of M the first (from the left) non zero element is positive.

Moreover we have the following fact, which shows the importance of term-orderings for computational purposes

FACT 2) Let $\sigma \in O$ then
$\sigma \in TO$ iff σ is a well-ordering on T^n.

Once we are given a term-ordering σ, we can do several things.
1) The first important achievement is that we have a fixed rule for representing polynomials as sum of monomials. As we said before this is a "must" for actual implementation of Computer Algebra systems.
2) The second thing is that we may associate to every non zero polynomial f its leading term $Lt_\sigma(f)$ and its leading monomial $Lm_\sigma(f)$.
3) We can use a non zero polynomial as an operator on $k[x_1,\ldots,x_n]$, by introducing the relation of rewriting, which essentially does the following: it divides the given polynomial by the coefficient of its leading monomial, it expresses the monic polynomial f as $f = Lt_\sigma(f) - R$ and rewrites $Lt_\sigma(f)$ as R.

Now, suppose that we are given a set of non zero polynomials $F := \{f_1,\ldots,f_r\}$ and let us denote by (F) or by I the ideal generated by them. Then we can associate to I its leading ideal $Lt_\sigma(I)$ i.e. the ideal generated by $\{Lt_\sigma(f) \,/\, f \in I\setminus\{0\}\}$, and we can define a rewriting relation on $k[x_1,\ldots,x_n]$ by taking the reflexive-transitive closure of the relation given by rewriting via F. If we order F and we rewrite, taking in account the order given, we have the so called DIVISION ALGORITHM with respect to the array $(f_1,\ldots,f_r)$;
it has the following feature: given $f \in A$, the algorithm produces a uniquely determined array $(a_1,\ldots,a_r, R)$ such that $f = \sum_i a_i f_i + R$.
The polynomial R is called the remainder of $f \bmod (f_1,\ldots,f_r)$.
This is a natural generalization of the usual division algorithm for univariate polynomials, but the great difference is that R depends on the order of the f_i's.

EXAMPLE Let $\sigma :=$ deglex, $f_1 := xy$; $f_2 := y^2 - z$, $f_3 := xz$
and let $I := (f_1, f_2)$. Then $xy^2 = y\cdot f_1 + 0\cdot f_2 + 0 = x\cdot f_2 + 0\cdot f_1 + xz$
Moreover $f_3 = y\cdot f_1 - x\cdot f_2 \in I$, but $Lt_\sigma(f_3) \notin (Lt_\sigma(f_1), Lt_\sigma(f_2))$.

In the above example we have checked the dependence of the remainder in the division algorithm on the ordering of the f_i's and we have seen how $Lt_\sigma(I)$ can be strictly bigger than $(Lt_\sigma(f_1), Lt_\sigma(f_2))$

We are ready to give the

FUNDAMENTAL THEOREM Let $\sigma \in TO(T^n)$, $F := \{f_1,\ldots,f_t\} \subset A\setminus\{0\}$, I the ideal generated by F in $A = k[x_1,\ldots,x_n]$; then the following conditions are equivalent

A_1) For every $f \in I\setminus\{0\}$, there exist $c_{ij} \in k^*$ and $T_{ij} \in T^n$

such that $f = \sum_{ij} c_{ij} T_{ij} f_i$ with $Lt_\sigma(f) \geq T_{ij}\cdot Lt_\sigma(f_i)$ for every i,j

A_2) For every $f \in I\setminus\{0\}$, there exist $g_i \in A\setminus\{0\}$

such that $f = \sum_i g_i f_i$ with $Lt_\sigma(f) \geq Lt_\sigma(g_i)\cdot Lt_\sigma(f_i)$ for every i

A_3) For every $f \in I\setminus\{0\}$, there exist $g_i \in A\setminus\{0\}$

such that $f = \sum_i g_i f_i$ with $Lt_\sigma(f) = \mathrm{Max}_i\{Lt_\sigma(g_i)\cdot Lt_\sigma(f_i)\}$

B_1) $\{Lt_\sigma(f_1),\ldots,Lt_\sigma(f_t)\}$ generates $Lt_\sigma\{I\}$ in T^n

B_2) $(Lt_\sigma(f_1),\ldots,Lt_\sigma(f_t)) = Lt_\sigma(I)$ in A

B_3) $(Lm_\sigma(f_1),\ldots,Lm_\sigma(f_t)) = Lm_\sigma(I)$ in A

C_1) $f \in I \Longleftrightarrow f \xrightarrow{F} 0$

C_2) If $r \in I$ is irreducible with respect to "$\xrightarrow{F}$" then $r=0$

C_3) For every $g \in A$ there exists a unique $h \in A$, irreducible with respect to "$\xrightarrow{F}$", such that $g \xrightarrow{F} h$

C_4) "$\xrightarrow{F}$" is confluent

D_1) Every homogeneous element of $Ker(\Lambda)$ extends to an element of $Ker(\lambda)$

D_2) There exists a homogeneous basis of $ker(\Lambda)$ which extends to elements of $Ker(\lambda)$.

The proof can be found for instance in Robbiano 1988.

FUNDAMENTAL DEFINITION: Let $\sigma \in TO(T^n)$, $F := \{f_1, ..., f_t\} \subset A \setminus \{0\}$, I the ideal generated by F in $A = k[x_1, ..., x_n]$. Then F is called a **Gröbner basis** of I if it satisfies the equivalent conditions of the FUNDAMENTAL THEOREM (some authors call them standard bases).

Needless to say, the notion of Gröbner basis and a specific algorithm for computing it is due to B. Buchberger (see Buchberger 1965 and Buchberger 1970).
I do not want to give any detail, but I only want to point out that the main point for producing the algorithm is D_2), since the syzygies of terms are trivial to compute. It should be mentioned that a lot of work has been done recently on the optimization of the algorithm itself and on the study of its complexity. Major improvements toward the optimization of the algorithm have been obtained by Bayer and Stillman in their well known system "Macaulay" and by M. Möller and R. Gebauer (1986), whose improvement has been already implemented in the "big" Computer Algebra systems such as Scratchpad II and Reduce as well as in "small" experimental systems, which are nowadays developed by several groups. Among them I like to mention the system ALPI developing under the direction of prof. Traverso at the University of Pisa, and the system COCOA, developing at the University of Genova with the cooperation of the Dept. of Maths (Robbiano, Mora, Niesi) and the Dept. of Computer Science (Giovini).

What these systems actually do, or are about doing, can be understood by showing the magic power of the notion of Gröbner basis. More specifically, I am going to explain how all the preceding QUESTIONS can be solved effectively.

QUESTION 1 (Ideal membership)
Given $A := k[x_1, ..., x_n]$, $f_1, ..., f_r$, $f \in A$, how to decide whether or not $f \in (f_1, ..., f_r)$ i.e. there exist $a_1, ..., a_r \in A$ such that $f = \sum_i a_i f_i$?
SOLUTION Let $I := (f_1, ..., f_r)$, let $\sigma \in TO$ and let $\{g_1, ..., g_t\}$ be a Gröbner basis of I with respect to σ. Condition C_3) implies that every polynomial f has a unique normal form $NF(f)$ with respect to σ and I; moreover $NF(f)$ is computable.
Now the solution is simply that $f \in I$ iff $NF(f) = 0$

QUESTION 2 (Syzygies)
Given $A := k[x_1, ..., x_n]$, $f_1, ..., f_r \in A$, how to compute a fundamental set of n-tuples $(a_1, ..., a_r) \in A^r$ such that $\sum_i a_i f_i = 0$ i.e. a set of generators for the first module of syzygies ?

SOLUTION Let $I := (f_1, \dots, f_r)$, let $\sigma \in TO$ and let $\{g_1, \dots, g_t\}$ be a Gröbner basis of I with respect to σ.
A result of Zacharias 1978 shows how to compute a generating set of syzygies of $f_1, \dots, f_r$ via a generating set of syzygies of $(g_1, \dots, g_t)$ and the latter can be easily achieved as a byproduct of Buchberger's algorithm, by using condition D_2).
It should be mentioned that the most complete system for Commutative Algebra, i.e. the system Macaulay by Bayer and Stillmann, among a lot of other things, allows the user to compute full resolutions of ideals, therefore bringing to practice the teaching of Theorem 3.

QUESTION 3 (Radical membership)
Given $A := k[x_1, \dots, x_n]$, $f_1, \dots, f_r$, $f \in A$, how to decide whether or not the solutions of $f_1 = \cdots = f_r = 0$ are also solutions of $f = 0$?
SOLUTION We have already seen that this problem reduces to a subquestion of question 1, namely to check whether or not 1 belongs to an ideal I. As a consequnce of what we have said about question 1, we get the following answer.
Let $I := (f_1, \dots, f_r)$, let $\sigma \in TO$ and let $\{g_1, \dots, g_t\}$ be a Gröbner basis of I with respect to σ; then $1 \in I$ iff $1 \in \{g_1, \dots, g_t\}$

QUESTION 4 Given $f_1, \dots, f_r$ homogeneous polynomials and $I := (f_1, \dots, f_r)$ is it possible to compute the Hilbert function of I ?
SOLUTION Already Macaulay (1927) showed that the Hilbert function of I is the same as that of $Lt_\sigma(I)$. Condition B_2) of the fundamental Theorem says that if $\sigma \in TO$ and $\{g_1, \dots, g_t\}$ is a Gröbner basis of I with respect to σ then $(Lt_\sigma(g_1), \dots, Lt_\sigma(g_t)) = Lt_\sigma(I)$.
At this point the computation of the Hilbert function of an ideal generated by terms boils down to a simple combinatorial problem.

QUESTION 5 Is it possible to compute a basis of A/I as a k- vectorspace ?
SOLUTION Let $\sigma \in TO$ and let $\{g_1, \dots, g_t\}$ be a Gröbner basis of I with respect to σ.
Combining C_3) and B_2) of the fundamental Theorem, it is easy to see that
$\{T^n \backslash$ Monoideal generated by $(Lt_\sigma(g_1), \dots, Lt_\sigma(g_t))\}$ is a basis of A/I.

What we have seen so far is how the computation of a G-basis of an ideal allows to solve some basic problems in Commutative Algebra. But of course the algebraist wants to do much more and indeed he can do much more, as we are going to see.
First of all let us see how the solution of the preceding problems allows to answer more advanced questions.

The possibility of computing syzygies and resolutions of ideals as well as Hilbert functions leads to the computation of fundamental invariants such as the dimension (see for instance tho papers Carrà 1986, Kredel Weisspfenning 1988 and Giusti 1988), the multiplicity, the degree and can test whether an ideal is Cohen-Macaulay, Gorenstein and so on.

A big shot is that elimination theory, an old and fundamental theme in Commutative Algebra and Algebraic Geometry, becomes not only effective but it can be performed efficiently.

Let us consider the following

EXAMPLE Let be given a curve C in the four dimensional affine space $\mathbb{A}^4$ by the following parametric equations
$x = t; \quad y = t^2; \quad z = t^3; \quad w = t^4.$

It is easy to see that the defining ideal of C i.e. the ideal of $k[x,y,z,w]$ of polynomials vanishing at C is

$I = (f,g,h)$, where $f := y - x^2$, $g = z - xy$, $h = w - xz$

In Algebraic Geometry it is important to consider compact varieties, hence one introduces the projective spaces $\mathbb{P}^n$ and one shows that the right compactification of C in $\mathbb{P}^4$ is $\overline{C} := Z({}^hI)$ where ${}^hI$ is the homogenization of I.

We recall that if $f(x_1,\dots,x_n) \in k[x_1,\dots,x_n]$ is a polynomial of degree d, then ${}^hf := x_0^d \cdot f(x_1/x_0,\dots,x_n/x_0)$ and if I is an ideal of $k[x_1,\dots,x_n]$, then ${}^hI$ is by definition the ideal of $k[x_0,x_1,\dots,x_n]$ generated by $\{{}^hf \;/\; f \in I\backslash\{0\}\}$
So the problem is merely algebraic and it is solved by the following

THEOREM *Let* I *be an ideal of* $A := k[x_1,\dots,x_n]$, *let* $\sigma \in TO$ *be such that its "first vector" is* $(1,1,\dots,1)$ *and let* $\{g_1,\dots,g_t\}$ *be a Gröbner basis of* I *with respect to* σ. *Then* ${}^hI = ({}^hg_1, {}^hg_2, \dots, {}^hg_t)$.

So let σ := degrevlex i.e. $\sigma = ord(u_1,u_2,u_3,u_4)$ where
$u_1 := (1,1,1,1)$, $u_2 := (0,0,0,-1)$, $u_3 := (0,0,-1,0)$, $u_4 := (0,-1,0,0)$.
A G-basis of I with respect to ß is given by the maximal minors of the following matrix

$$\begin{Vmatrix} 1 & x & y & z \\ x & y & z & w \end{Vmatrix}$$

It turns out that the equations of the closure of C are the maximal minors of the following matrix

$$\begin{Vmatrix} x_0 & x_1 & x_2 & x_3 \\ x_1 & x_2 & x_3 & x_4 \end{Vmatrix}$$

These are 6 equations for a 1-dimensional object in a space of dimension 4 and they are minimal in the sense that every set of generators of this ideal needs 6 elements.
But now it is clear that $I = Rad(I) = Rad(f, g^2, h^3)$ and g^2 is rewritten via f as $G := y^3 - 2xyz + z^2$ while h^3 is rewritten via f, g as $H := z^4 - 3yz^2w + 3xzw^2 - w^3$.
Therefore $Rad(I) = Rad(f,G,H)$.
We see that $Lt_\sigma(f) = x^2$, $Lt_\sigma(G) = y^3$, $Lt_\sigma(H) = z^4$ hence they are pairwise coprime, which implies that $\{f,G,H\}$ is a G-basis with respect to σ.
Therefore ${}^hI = {}^hRad(I) = {}^hRad(f,G,H) = Rad({}^h(f,G,H)) = Rad({}^hf, {}^hG, {}^hH)$
the last equality following from the theorem above.
So we have seen that $\overline{C}$ as a set can be defined by 3 equations i.e. we have proved that $\overline{C}$ is a set-theoretic complete intersection.
This point of view is exploited in Robbiano-Valla 1983.

Another way to attack the problem of determining $I(\overline{C})$ can be that of considering the ideal

$$(x_0 - s^4,\; x_1 - s^3t,\; x_2 - s^2t^2,\; x_3 - st^3,\; x_4 - t^4)$$

and then to eliminate s and t.
In this case this way is less convenient, because we are dealing with more indeterminates and the complexity of Buchberger's algorithm is known to be intrinsically doubly exponential in the number of indeterminates, at least in worst cases (see Mayr-Meyer 1982).
However this remark leads to another important application of Gröbner bases theory

THEOREM *Let* I *be an ideal of* $k[x_1,\dots,x_n][y_1,\dots,y_m]$, *let* σ *be a term-ordering for which every term in the y-indeterminates is bigger than every term in the x-indeterminates and let* $\{g_1,\dots,g_t\}$ *be a Gröbner basis of* I *with respect to* σ.
Then $\{g_1,\dots,g_t\}\cap k[x_1,\dots,x_n]$ *is a Grobner basis of* $I\cap k[x_1,\dots,x_n]$.

This is the easy and beautiful result that allows computational elimination theory.

Techniques of elimination, combined with more specific tools, yield algorithms for more advanced problems in Commutative Algebra, such as the determination of the primary decomposition of ideals (see for instance Lazard 1984, Gianni-Trager-Zacharias 1986), primality tests (Grieco-Zucchetti) .

We have seen that some problems require G-bases computations with respect to specific orderings, but in general every problem leaves a lot of freedom in choosing the appropriate ordering to solve it. Fundamental research on this question has been recently carried over by Bayer-Stillmann 1987a, where it is shown that in the generic case degrevlex is an optimal ordering. This means that the choice of degrevlex can be considered in principle a good one. But a more accurate study of the set O, shows that every specific ideal I yields a partition of O itself, or rather of another set, which is strictly connected to O, into a finite collection of polyhedral cones. This gives the theory of the so called Gröbner Fans developed by Mora-Robbiano 1987 and also from different points of view by Bayer-Morrison 1987, by Weispfenning 1987 and by Schwartz 1986.
Each polyhedral cone corresponds to an infinite set of orderings, which give the same reduced Gröbner basis of I, hence the same $Lt_\sigma(I)$. This research may lead to a more structural optimization, namely the optimization of the choice of the term-ordering for the solution of a given problem.

At this point it should be mentioned that some experience on Buchberger's algorithm shows how computations on $\mathbb{Q}[x_1,\dots,x_n]$ lead frequently to big sized coefficients. Sometimes big numbers only effect intermediate computations and not the results

EXAMPLE (De Micheli) Let us consider the field extension of $\mathbb{Q}$ given by
$\mathbb{Q}(x) = \mathbb{Q}[x]$ where x is a root of the irreducible polynomial $X^5 - X - 2$, and let
$y := \frac{1}{x}(1 - x - 2x^3) \in \mathbb{Q}(x)$.
The minimal polynomial of y can be obtained by computing a Gröbner basis of
$(X^5 - X \quad 2,\ XY + 2X^3 + X - 1)$ with respect to σ = lex.
The computation gives the Gröbner basis $\{f,\ g\}$ where

$$f := X - \frac{1438}{45887}Y^4 - \frac{2183}{45887}Y^3 + \frac{10599}{45887}Y^2 - \frac{8465}{45887}Y - \frac{101499}{45887}$$

$$g := Y^5 + \frac{11}{2}Y^4 + 4Y^3 - 5Y^2 + 95Y + 259.$$

The answer to our question is g, but we can see that the size of the coefficient of the "useless" polynomial f is much bigger.
So recently a good deal of work has been done on this subject for instance by Winkler 1987, Melenk-Möller-Neun and Galligo-Traverso. The idea is to cut the size of the coefficients by

computing mod p. In some sense we are at the same point when Berlekamp (1967) introduced his method for factorizing univariate polynomials over $\mathbb{Z}_p$.

Another important question to investigate is the following.
Let $f_1, \ldots, f_r \in k[x_1, \ldots, x_n]$ and let $B := k[f_1, \ldots, f_r] \subsetneq A = k[x_1, \ldots, x_n]$.
Given $f \in A$ is it possible to test the membership of f to B ?
This problem was investigated by Shannon-Sweedler 1987, where the method of "tag" indeterminates is used.

Namely $B = k[f_1, \ldots, f_r] \simeq k[f_1, \ldots, f_r][z_1, \ldots, z_r]/(z_1 - f_1, \ldots, z_r - f_r)$,
which sits inside $k[x_1, \ldots, x_n][z_1, \ldots, z_r]/(z_1 - f_1, \ldots, z_r - f_r) \simeq A$.
Let us take a term-ordering σ such that every term in the x-indeterminates is bigger than every term in the z-indeterminates and let $\{g_1, \ldots, g_t\}$ be a Gröbner basis of I with respect to σ.
Then $f \in B$ iff f reduces modulo $\{g_1, \ldots, g_t\}$ to a pure term in the z-indeterminates.

Connected with the preceding theme, we have other questions.
Let $\sigma \in TO$; then we know that A has a Gröbner filtration associated to it, such that its associated graded object is again A.
But in general this is no more true; without going into the details of this more sophisticated subject let me show an easy example.

EXAMPLE Let $f := x^2 + x$ and let $B := k[f] \subset A$. Then every polynomial in B is has certainly as leading term an even power of x.
We can express this by saying that $gr(B) = k[x^2]$.

This leads to the following question.
Given $B \subsetneq A$ and $\sigma \in TO$, how can we determine $gr(B)$ with respect to the induced filtration ? There is some work in progress on this problem by Robbiano-Sweedler

Another fundamental section of Commutative Algebra is Local algebra, which deals with local rings such as $k[[x_1, \ldots, x_n]]$.
Of course the question is how to compute in local rings.
Already Hironaka 1964 introduced the notion of standard basis for computing invariants of local rings and since then many progresses have been made by several authors like Galligo and Mora. In particular it should be mentioned that in Mora 1982 an algorithm is described to compute the tangent cone to a closed point of a scheme. A method by Lazard 1983 shows how to use G-bases for computing particular standard bases. In Robbiano 1986 the notions of graded structures and generalized standard bases were introduced . They are a tool, which has become useful for treating problems related to more general algebraic objects such as localizations of rings, power series, subrings and so on. Several papers of the above mentioned Special Volume of JSC on Computational Aspects of Commutative Algebra use these notions (see for instance Miola-Mora 1988 and Brundu-Rossi 1987) and more recently Mora has several results, that are collected in the preprint Mora 1988.

Before concluding my survey I would like to touch another point of great interest.
In Commutative Algebra and Algebraic Geometry it is very useful to consider not only single structures such as rings, but also families of structures.
A proof by Bayer 1982 together with an easy generalization by myself shows that given I and a term-ordering σ then I and $Lt_\sigma(I)$ are members of of the same "nice" family of ideals.

Here nice has the algebraic meaning of "flat" and the flat family is a suitable homogenization of I which can be proven to exist, for instance by using the theory of Gröbner Fans.
This is a special instance of flat families given by the so called Rees rings associated to filtrations. Several papers deal with the study of families (see for instance Bayer 1982 and Carrà 1986) but the following question in its wide generality seems to be still open

QUESTION Given R, S finitely generated k-algebras and an injective homomorphism $\varphi : R \longrightarrow S$, is it possible to test flatness of f effectively ?
Some work in progress is carried over by myself, Bayer-Galligo-Stillman and Logar.

EXAMPLE Let $R := k[z]$, $S := k[x,y,z]/(x^2z - y^2z + xz^2 - xy^2)$ and let φ be the canonical immersion of R into S. Then φ is flat since $Lt_\sigma(f) = xy^2$ where σ is a term-ordering such that every pure term in x,y is bigger that every term in z (e.g. deglex)

Of course several other problems are still unsolved, such as the problem of finding efficient algorithms for computing the normalization and for testing factoriality of finitely generated integral k-algebras .
Some progresses have been made on the first question by Traverso 1986 and Logar, following the lines of pioneering work by Seidenberg 1975 and Stolzenberg 1968 while, as far as I know, nothing has been done on the second problem.

It is now time for a conclusion. I hope that what I have written here can give an idea of the tremendous impact that Computer Algebra has on Commutative Algebra. It is very important to notice that this new area of research makes a connection between Computer Science and Mathematics and the history of Sciences says that every time a link between different disciplines is established, new and beautiful discoveries have become possible. There is fresh air now in our traditional field of research and there are new challenging problems for Computer Scientists. We do hope that all of us can take advantage of this exciting atmosphere.

REFERENCES

Bayer,D. (1982) The division algorithm and the Hilbert scheme.
Ph. D. Thesis Harvard

Bayer,D. - Morrison, J. (1986) Standard bases and geometric invariant theory I, Initial ideals and state polytopes.
Special Volume of JSC on "Computational Aspects of Commutative Algebra
To appear

Bayer,D. - Stillman, M. (1987, a) A criterion for detecting m-regularity.
Invent. Math. **87** 1-11

Bayer,D - Stillman, M. (1987, b) A theorem on refining division orders by the reverse lexicographic order.
Duke Math. J. **55** 321-328

Brownawell, W.D. (1987). Bounds for the degrees in Nullstellensatz.
Ann. of Math. **126** 577-592

Brundu, M. - Rossi, F. (1987) On the computation of generalized standard bases.
Special Volume of JSC on "Computational Aspects of Commutative Algebra
To appear

Buchberger, B. (1965) Ein Algorithmus zum Auffinden der Basiselemente des Restklassenringes nach einem nulldimensionalen Polynomideale.
Ph.D. Thesis Innsbruck, Math. Inst.

Buchberger, B. (1970). Ein algorithmisches Kriterium für die Lösbarkeit eines algebraischen Gleichungssystems.
Aeq. Math. **4** 374-383

Caniglia, L. - Galligo, A. - Heintz, J. Some effectivity bounds in computational geometry
Preprint

Carrà Ferro, G. (1986) Some upper bounds for the multiplicity of an autoreduced subset of $\mathbb{N}^m$ and their applications.
Proc. AAECC-3 Springer Lecture Notes in Computer Science **229** 306-315

Carrà Ferro, G. (1986) Gröbner bases and Hilbert schemes.
Special Volume of JSC on "Computational Aspects of Commutative Algebra"
To appear

Dickson, L.E. (1913). Finiteness of the odd perfect and primitive abundant numbers with n distinct prime factors.
Am. J. of Math. **35** 413-426

Gebauer, R. - Möller, M. (1987) On an Installation of Buchberger's Algorithm.
Special Volume of JSC on "Computational Aspects of Commutative Algebra"
To appear

Gianni, P. - Trager, B. - Zacharias, G. (1986) Gröbner bases and primary decomposition of polynomial ideals.
Special Volume of JSC on "Computational Aspects of Commutative Algebra"
To appear

Giusti, M. (1988) Combinatorial dimension theory of algebraic varieties.
Special Volume of JSC on "Computational Aspects of Commutative Algebra"
To appear

Hermann, G. (1926). Die Frage der endlischen vielen Schritte in der Theorie der Polynomideale.
Math. Ann. **95** 736-788

Hilbert, D. (1890). Über die Theorie der algebraischen Formen.
Math. Ann. **36** 473-534

Hilbert, D. (1893). Über die vollen Invariantensysteme.
Math. Ann. **42** 313-373

Hironaka, H. (1964) Resolution of singularities of an algebraic variety over a field of characteristic 0.
Ann. Math **79** 109-326

Kollar, J. (1988). Sharp effective Nullstellensatz.
Preprint.

Kredel, H. - Weisspfenning, V. (1988) Computing dimension and independent sets for polynomial ideals.
Special Volume of JSC on "Computational Aspects of Commutative Algebra"
To appear

Lazard, D. (1984) Ideal bases and primary decomposition: the case of two variables.
Preprint, Paris

Lazard, D. (1983) Gröbner bases, Gaussian elimination and resolution of systems of algebraic equations.
Proc. EUROCAL 83 Springer Lecture Notes in Computer Science **162** 146-156

Macaulay, F.S. (1927) Some properties of enumeration in the theory of modular systems.
Proc. London Math. Soc. **26** 531-555

Mayr, E.W. - Meyer A.R. (1982) The complexity of the word problems for commutative semigroups and polynomial ideals.
Adv. Math. **46** 305-329

Miola, A. - Mora, T. (1988) Constructive lifting in graded structures; a unified view of Buchberger and Hensel methods.
Special Volume of JSC on "Computational Aspects of Commutative Algebra"
To appear

Mora, T. (1982) An algorithm to compute the equations of tangent cones.
Proc. EUROCAM 82 Springer Lecture Notes in Computer Science **144** 24-31

Mora, T. (1988) Seven variations on standard bases.
Preprint
Mora, T. - Möller, M. (1986) New constructive methods in classical ideal theory.
J. Algebra **100** 138-178
Mora, T. - Robbiano, L. (1987) The Gröbner fan of an ideal.
Special Volume of JSC on "Computational Aspects of Commutative Algebra"
To appear
Möller, M. (1986) On the construction of Gröbner bases using syzygies.
Special Volume of JSC on "Computational Aspects of Commutative Algebra"
To appear
Robbiano, L. (1988) Introduction to the Theory of Gröbner Bases.
To appear on Queen's Papers, Kingston Canada
Robbiano, L. (1985) Term Orderings on the Polynomial Ring.
Proc. EUROCAL 85, II Springer Lecture Notes in Computer Science **204** 513-517
Robbiano, L. (1986) On the theory of graded structures.
J. Symb. Comp. **2** 139-170
Robbiano, L.- Valla, G. (1983) On set-theoretic complete intersections in the projective space.
Rend. Sem. Mat. Milano **Vol LIII** 333-346
Schwartz, N. (1986) Stability of Gröbner bases.
Preprint
Seidenberg, S. (1975) S. Construction of the integral closure of a finite integral domain.
Proc. Amer. Math. Soc. **74** 368-372
Shannon, D. - Sweedler, M. (1987) Using Gröbner bases to determine algebra membership, split surjective algebra homomorphisms and to determine birational equivalence.
Special Volume of JSC on "Computational Aspects of Commutative Algebra"
To appear
Stolzenberg, G. (1968) Constructive normalization of an algebraic variety.
Bull. Amer. Math. Soc. **74** 595-599
Traverso, C. (1986) A study on algebraic algorithms: the normalization.
Rend. Sem. Mat. Torino Proc. Int. Conf. Torino 1985 111-130
Weispfenning,V. (1987) Constructing universal Gröbner bases
Proc. AAECC 5
Winkler, F. (1987) A p-adic approach to the computation of Gröbner bases.
Special Volume of JSC on "Computational Aspects of Commutative Algebra
To appear
Zacharias, G. (1978) Generalized Gröbner Bases in commutative polynomial Rings.
Bachelor Thesis M.I.T.

The Multiplicative Complexity of Boolean Functions

C.P. SCHNORR
Universität Frankfurt
Fachbereich Mathematik/Informatik

Abstract Let the multiplicative complexity L(f) of a boolean function f be the minimal number of $\wedge$-gates (with two entries) that are sufficient to evaluate f by circuits over the basis $\wedge,\oplus,1$. We relate L(f) with the dimension of the dual domain D(f); D(f) is the minimal linear space of linear boolean forms such that f modulo linear functions can be written as a function which takes for input linear forms in D(f).

1. Introduction and Summary

Determining the circuit complexity of boolean functions with respect to unrestricted circuits is a major challenge to complexity theory. So far very little algebraic or geometric theory has been used for the complexity theory of boolean functions. In this paper we introduce a geometric notion the *dual domain* of a boolean function and we study the relationship between the dimension of the dual domain and the multiplicative complexity of boolean functions. The multiplicative complexity L(f) of a boolean function f is the minimal number of $\wedge$-gates (binary multiplications) that are sufficient to evaluate f by a circuit over the basis $\wedge,\oplus,1$. The constant 1 and $\oplus$, the addition modulo 2, are free of charge. The dual domain D(f) of a boolean function f is the smallest linear space of linear boolean forms such that f, modulo linear functions, can be written as a function in linear froms in D(f). The dual domain of a boolean function is an invariant that adjusts the notion of a boolean variable to isomorphism classes of boolean functions.

Mirwald and Schnorr (1987) have shown that the multiplicative complexity of a quadratic boolean form f equals half the dimension of its dual domain, i.e. $L(f) = \frac{1}{2}$ dim D(f). Here we consider boolean functions of higher degree. We show that $L(f) \geq \frac{1}{2}$ dim D(f) holds for all boolean functions and that the equality only holds for functions of degree at most 2. We also show that strict and weak computational independence of two quadratic boolean forms can be characterized in geometric terms using the notion of dual domain.

The paper is organized as follows. In section 2 we summarize basic concepts from complexity theory and from linear algebra that are used subsequently. In section 3 we introduce the dual domain of a boolean function and study relations to the multiplicative complexity. In sections 4 and 5 we summarize results of Mirwald, Schnorr (1987) which characterize the multiplicative complexity of quadratic, boolean forms and pairs of quadratic boolean forms in terms of their dual domain.

2. Preliminaries

Let B_n be the ring of n-ary boolean functions in the boolean variables $x_1,...,x_n$ with the operations $\oplus$ (addition modulo 2), $\wedge$ (multiplication, logical and). The ring B_n is isomorphic to the factor ring $\mathbb{Z}_2[x_1,...,x_n]/(x_1-x_1^2,...,x_n-x_n^2)$. Here $\mathbb{Z}_2[x_1,...,x_n]$ is the ring of formal polynomials in the variables $x_1,...,x_n$ with coefficients in the Galois field $\mathbb{Z}_2 = \mathbb{Z}/2\mathbb{Z}$, and $(x_1-x_1^2,...,x_n-x_n^2) \subset \mathbb{Z}_2[x_1,...,x_n]$ is the ideal generated by the polynomials $x_1-x_1^2,...,x_n-x_n^2$. We associate with the boolean functions $f_1,...,f_r$ the linear subspace $<f_1,...,f_r> \subset B_n$ generated by $f_1,...,f_r$, i.e. the *span* of $f_1,...,f_r$. Let $B_{n,d} \subset B_n$ be the subspace $<x_{i_1}\cdot...\cdot x_{i_d} \mid 1 \le i_1 < ... < i_d \le n>$ of n-ary *boolean forms* of degree d; $B_{n,2}$ ($B_{n,1}$, resp.) is the linear space of quadratic (linear, resp.) forms.

Every boolean function $f \in B_n$ can be uniquely written as

$$f = \oplus_{d=0}^{n} f_d \quad \text{with} \quad f_d \in B_{n,d} \, .$$

The maximal number d with $f_d \neq 0$ is called the *degree* of f, notation deg f.

All forms in this paper will be boolean forms. We write $f = g \pmod{B_{n,1}}$ if $f \oplus g \in B_{n,1}$. To abbreviate formulae we oppress all $\wedge$-symbols. We use the letters f,g,h,k for boolean functions, u,v,w for linear forms, x,y for boolean variables, a,b,c for boolean constants and z for boolean vectors.

The *multiplicative complexity* $L(f_1,...,f_r)$ of a set of boolean functions $f_1,...,f_r \in B_n$ is the minimal integer t for which there exist boolean functions g_i,h_i,k_i in B_n for $i=1,...,t$ such that

(1) $h_1,k_1 \in <x_1,...,x_n>$, $g_1 = h_1k_1$ and
$h_i,k_i \in <g_1,...,g_{i-1},x_1,...,x_n>$, $g_i = h_ik_i$ for $i=2,...,t$.

(2) $f_1,\dots,f_r \in \langle g_1,\dots,g_t,x_1,\dots,x_n,1\rangle$.

The recursion (1), (2) describes a *circuit* for $f_1,\dots,f_r$. The inputs of this circuit are linear forms. It is sufficient that the constant 1 can be added finally.

For quadratic forms $f_1,\dots,f_r \in B_{n,2}$ the *level-one multiplicative complexity* $L_1(f_1,\dots,f_r)$ is the minimal integer t for which there exist linear forms $u_i,v_i \in B_{n,1}$ for $i=1,\dots,t$ such that

(3) $f_1,\dots,f_r \in \langle u_1v_1,\dots,u_tv_t,x_1,\dots,x_n\rangle$.

Obviously $L(f_1,\dots,f_r) \leq L_1(f_1,\dots,f_r)$. $L(f_1,\dots,f_r)$ and $L_1(f_1,\dots,f_r)$ do not decrease by using variables that none of the functions $f_1,\dots,f_r$ depends on. In this sense the measures L and L_1 do not depend on n.

We can represent n-ary quadratic forms by boolean nxn-matrices. This is done by the vector space homomorphism

$$\psi:\ M(n,n,\mathbb{Z}_2) \to B_{n,2}$$
$$(a_{i,j})_{1\leq i,j\leq n} \mapsto \oplus_{i\neq j} a_{ij}x_ix_j$$

that transforms boolean nxn matrices into quadratic forms. We use the notation f_A for $\psi(A)$. The kernel of ψ is the subspace of symmetrical matrices including the diagonal matrices. ψ is surjective and maps rank-1 matrices T into forms f_T with $L(f_T) \leq 1$. For every pair of boolean nxn matrices A,B we have

$$f_A = f_B \Leftrightarrow A \oplus B \text{ is symmetrical} \Leftrightarrow A \oplus A^T = B \oplus B^T.$$

Thus the matrix $A \oplus A^T$ is uniquely defined by the form f_A. Here A^T is the transpose of matrix A.

We will use the following concepts from linear algebra. For a linear mapping $T : \mathbb{Z}_2^n \to \mathbb{Z}_2^m$ and a boolean function $f \in B_m$, let $fT \in B_n$ be the function defined by $(fT)(x) = f(Tx)$. If $f \in B_{m,2}$ is a quadratic form then fT must not be a form but there is a unique quadratic form $g \in B_{n,2}$ which differs from fT by some linear form, i.e. $fT = g \pmod{B_{n,1}}$. We say T *transforms* f into g, even though fT and g differ by a linear form.

For a linear transformation $T : \mathbb{Z}_2^n \to \mathbb{Z}_2^m$ the *dual transformation* $T^* : B_{m,1} \to B_{n,1}$ is defined by

$$T^*(u) = uT \quad \text{for all } u \in B_{m,1}\,.$$

Conversely T can be defined from T^* by the equations

$$x_i(T(z)) = (T^*(x_i))(z) \quad \text{for all } z \in \mathbb{Z}_2^n\,.$$

Sometimes it is convenient to define a transformation T by describing its dual T^*.

With a linear subspace $E \subset \mathbb{Z}_2^n$ we associate the *dual space* $E^* \subset B_{n,1}$ defined by

$$E^* = \{\oplus_{i=1}^n a_i x_i \in B_{n,1} \mid (a_1,\dots,a_n) \in E\} .$$

Conversely E is the space that is dual to E^*, i.e. $(E^*)^* = E$.

To represent linear mappings by matrices we consider $z \in \mathbb{Z}_2^n$ to be a n×1-matrix (i.e. a column vector). The linear mapping $T : \mathbb{Z}_2^n \to \mathbb{Z}_2^m$ can be represented by a boolean m×n matrix τ such that

$$T(z) = \tau z \quad \text{for all column vectors } z \in \mathbb{Z}_2^n .$$

Consider the m-ary quadratic form f_A that is represented by the m×m boolean matrix A. The mapping T transforms f_A into the quadratic form

$$f_B = f_A T \quad \text{with } B = \tau^T A \tau .$$

To describe the dual mapping $T^* : B_{m,1} \to B_{n,1}$ we consider a linear form $u = \oplus_{i=1}^m u_i x_i$ to be the column vector with coordinates u_i. We have

$$T^*(u) = \tau^T u \quad \text{for all } u \in B_{m,1} .$$

If $T : \mathbb{Z}_2^m \to \mathbb{Z}_2^n$ then every circuit for $f \in B_n$ can be transformed into a circuit for fT by replacing the variables x_i in $B_{n,1}$ by linear forms $T^*(x_i)$ in $B_{m,1}$. Thus we have $L(fT) \leq L(f)$. We call two boolean functions $f,g \in B_n$ *isomorphic*, if there is a linear isomorphism $T : \mathbb{Z}_2^n \to \mathbb{Z}_2^n$ such that $fT = g \bmod (B_{n,0} \oplus B_{n,1})$, i.e. f and g are isomorphic iff $fT \oplus g$ is linear. Clearly, if f and g are isomorphic then $L(f) = L(g)$.

3. The dual domain of a boolean function

A boolean function $f : \{0,1\}^n \to \{0,1\}$ can be written in a natural way as a function in the boolean variables $x_1,\dots,x_n$, i.e. we can write $f = f(x_1,\dots,x_n)$ as a composition of f with the variables $x_1,\dots,x_n$ which is defined by

$$f(x_1,\dots,x_n)(a) = f(x_1(a),\dots,x_n(a)) \qquad \text{for all } a \in \{0,1\}^n .$$

Boolean variables of a boolean function are not invariant with respect to isomorphisms. Isomorphic boolean functions may even depend on differently many boolean variables. E.g. the functions

$$x_1 x_2,\ (x_1 \oplus x_2)(x_3 \oplus x_4)$$

are isomorphic but depend on 2 (4, resp.) boolean variables.

We wish to introduce an invariant, *the dual domain*, of a boolean function that adjusts the concept of a boolean variable to an isomorphism class of boolean functions. We prove that for every boolean function $f \in B_n$ there is a smallest linear subspace $D(f) \subset B_{n,1}$ of linear forms such that f modulo linear functions can be written as a function of linear forms in D(f), i.e. there exists $u_1,...,u_m \in D(f)$ and $g \in B_m$ such that $f \oplus g(u_1,...,u_m)$ is linear. We call this smallest space of linear forms the *dual domain* D(f) of f.

We are going to prove the existence of D(f). A set of linear forms $u_1,...,u_m$ is called a *basis* for f if

(1) the forms $u_1,...,u_m$ are linearly independent and

(2) there exists a boolean function $g \in B_m$ such that $f \oplus g(u_1,...,u_m)$ is linear.

The following lemma implies that any two minimal rank bases for f generate the same linear space.

Lemma 3.1 Let the linear forms $u_1,...,u_r$ and $v_1,...,v_s$ be two bases for the boolean function f, and let $w_1,...,w_t$ be a basis for the linear space $\langle u_1,...,u_r\rangle \cap \langle v_1,...,v_s\rangle$. Then $w_1,...,w_t$ is a basis for f.

Proof. We can extend $w_1,...,w_t$ to a basis of the linear spaces $\langle u_1,...,u_r\rangle$ and $\langle v_1,...,v_s\rangle$. So let $w_1,...,w_t, \bar{u}_1,...,\bar{u}_{\bar{r}}$ be a basis of $\langle u_1,...,u_r\rangle$ and let $w_1,...,w_t, \bar{v}_1,...,\bar{v}_{\bar{s}}$ be a basis of $\langle v_1,...,v_s\rangle$. Then there exist boolean functions h,g such that the functions

$$f \oplus h(w_1,...,w_t, \bar{u}_1,...,\bar{u}_{\bar{r}})$$
$$f \oplus g(w_1,...,w_t, \bar{v}_1,...,\bar{v}_{\bar{s}})$$

are linear. It follows that the function $h(w_1,...,w_t, \bar{u}_1,...,\bar{u}_{\bar{r}}) \oplus g(w_1,...,w_t, \bar{v}_1,...,\bar{v}_{\bar{s}})$ is linear. We conclude that $h(w_1,...,w_t,\bar{u}_1,...,\bar{u}_{\bar{r}})$ linearly depends on $\bar{u}_1,...,\bar{u}_{\bar{r}}$ and $g(w_1,...,w_t,\bar{v}_1,...,\bar{v}_{\bar{s}})$ linearly depends on $\bar{v}_1,...,\bar{v}_{\bar{s}}$. This conclusion is obvious in case that $w_1,...,w_t,\bar{u}_1,...,\bar{u}_{\bar{r}},\bar{v}_1,...,\bar{v}_{\bar{s}}$ are pairwise distinct variables. Since the linear forms $w_1,...,w_t,\bar{u}_1,...,\bar{u}_{\bar{r}},\bar{v}_1,...,\bar{v}_{\bar{s}}$ are linearly independent we can transform them by an isomorphic transformation into pairwise distinct variables. This transformation reduces the general case to the special case in which $w_1,...,w_t,\bar{u}_1,...,\bar{u}_{\bar{r}},\bar{v}_1,...,\bar{v}_{\bar{s}}$ are pairwise distinct variables.

Since $h(w_1,...,w_t,\bar{u}_1,...,\bar{u}_{\bar{r}})$ is linear in $\bar{u}_1,...,\bar{u}_{\bar{r}}$ there exists a Boolean function $\bar{h}$ such that $h(w_1,...,w_t,\bar{u}_1,...,\bar{u}_r) \oplus \bar{h}(w_1,...,w_t)$ is linear. It follows that $f \oplus \bar{h}(w_1,...,w_t)$ is linear, and thus $w_1,...,w_t$ is a basis for f. □

As an immediate consequence of Lemma 3.1 every two minimal rank bases for f generate the same linear space of linear forms. We call this linear space the *dual domain* of f and we denote it D(f).

Corollary 3.2 For every boolean function f we have

$$D(f) = \bigcap \langle u_1,\dots,u_m\rangle$$

where the intersection ranges over all bases $u_1,\dots,u_m$ for f.

The dual domain of a boolean function f is an invariant with respect to isomorphic transformations. Let $f \in B_n$ be a boolean function and $T : \mathbb{Z}_2^n \to \mathbb{Z}_2^n$ an isomorphism. Then we have

$$D(fT) = T^* D(f).$$

This identity follows from the

Lemma 3.3 For every homomorphism $T : \mathbb{Z}_2^n \to \mathbb{Z}_2^n$ and every $f \in B_n$ we have

$$D(fT) \subset T^* D(f).$$

Proof. Let $u_1,\dots,u_m$ be a minimal rank basis for f. Then there exists a boolean function $h \in B_n$ such that $f \oplus h(u_1,\dots,u_m)$ is linear. It follows that $fT \oplus h(u_1T,\dots,u_nT)$ is linear and therefore

$$D(fT) \subset \langle u_1T,\dots,u_nT\rangle .$$

We have $T^*(u_i) = u_iT$ and thus $T^* D(f) = \langle u_1T,\dots,u_nT\rangle$. □

The inclusion of Lemma 3.3 may be proper, as is the case for $f = x_1x_2$ and $T(a_1,a_2) = (a_1,a_1)$. In this case $fT = x_1x_1$ is linear and thus D(fT) is the empty space whereas $T^* D(f) = \langle x_1\rangle$. However if T is an isomorphism we have $D(fT) \subset T^* D(f)$ and $D(f) = D(fTT^{-1}) \subset T^{-1*} D(fT) \subset D(f)$ which implies that all inclusions are improper, and thus $D(fT) = T^* D(f)$.

We clearly have dim D(f) = 0 iff f is linear, i.e. $f \in B_{n,1} \oplus B_{n,0}$. We can easily prove the

Lemma 3.4 *dim D(f) ≤ 2 L(f) for all boolean functions f.*

Proof. Consider an L-minimal computation of f. Let $u_1,\dots,u_r$ be the linear forms that are inputs to some $\wedge$-gate in this computation. This implies

$$D(f) \subset \langle u_1,\dots,u_r\rangle ,$$

$$\dim D(f) \le r$$

$$L(f) \geq r/2$$

Hence $L(f) \geq r/2 \geq \dim D(f)/2$. □

It has been shown by Mirwald and Schnorr (1987) that the inequality of Lemma 3.4 is exact for quadratic boolean forms.

Theorem 3.5 *(Mirwald, Schnorr 1987) For every quadratic, boolean form f we have $L(f) = \dim D(f)/2$.*

The equality $L(f) = \dim D(f)/2$ holds for all boolean functions f with degree ≤ 2. This follows from Theorem 3.5 and the observation that $\dim D(f) = 0$ iff f is linear. On the other hand the equality $L(f) = \dim D(f)/2$ cannot hold for boolean functions with degree greater than 2.

Lemma 3.6 *For every boolean function f with degree ≥ 3 we have $L(f) > \dim D(f)/2$.*

Proof. Suppose that $L(f) = \dim D(f)/2 = t$. Consider a circuit for f with t $\wedge$-gates. Since $\dim D(f) = 2t$ this circuit must take for inputs 2t linearly independent, linear boolean forms $u_1,...,u_{2t}$. All these forms must be inputs of $\wedge$-gates in the circuit. Since there are only t $\wedge$-gates in total we see that all inputs of $\wedge$-gates in the circuit are linear forms. This implies that the degree of f is at most 2. Consequently the equality $L(f) = \dim D(f)/2$ does not hold for boolean functions with degree ≥ 3. □

We now study the relationship between the degree of f, the dimension of D(f) and L(f). We will see that $\deg(f) - 1$ is a lower bound for the multiplicative complexity L(f). We will need the following Lemma.

Lemma 3.7 *Let $f = \bigwedge_{i=1}^{r} u_i$ be a product of r linear boolean forms that is nonzero. Then we have $L(f) = \dim D(f) - 1 = \deg f - 1$.*

Proof. W.l.o.g. we can assume that the linear forms $u_1,...,u_s$ are linearly independent and $u_{s+1},...,u_r \in \langle u_1,...,u_s \rangle$. We can also assume that $u_1,...,u_r$ are distinct boolean variables; by an isomorphic transformation we can transform $u_1,...,u_s$ into distinct boolean variables without changing deg f and dim D(f). We next prove by induction on i that

$$\bigwedge_{j=1}^{s} u_j = \bigwedge_{j=1}^{s+i} u_j .$$

For the induction step $i-1 \rightarrow i$ let $u_{s+i} = \oplus_{j\in I} u_j$ with $I \subset \{1,...,s\}$. We have

$$\left(\bigwedge_{j=1}^{s} u_j\right) u_{s+i} = \bigoplus_{j\in I} \bigwedge_{j=1}^{s} u_j = \begin{cases} 0 & \text{if } \#I \text{ even} \\ \bigwedge_{j=1}^{s} u_j & \text{if } \#I \text{ odd} \end{cases}.$$

We conclude from $f \neq 0$ that $(\bigwedge_{j=1}^{s} u_j)u_{s+i} = \bigwedge_{j=1}^{s} u_j$.

Now the claim of the lemma is obvious in the case that $f = \bigwedge_{j=1}^{s} u_j$ is product of s distinct boolean variables $u_1,\dots,u_s$. □

Proposition 3.8 *We have for all boolean function f $L(f) \geq \deg f - 1$.*

Proof. We prove the claim by induction on $d = \deg f$. The claim holds if $d \leq 1$. Now consider the induction step $d-1 \to d$. Consider a first $\wedge$-gate in an L-minimal circuit for f. W.l.o.g. we can assume that this $\wedge$-gate has variables x_1,x_2 for input. (In general the inputs of a first $\wedge$-gate are linear forms; these linear forms can be transformed into the variables x_1,x_2 by an isomorphic transformation.) Now consider the restriction of the circuit and the function f that arises by fixing x_1 to either 0 or 1. Each of these restrictions eliminates the $\wedge$-gate that computes x_1x_2. Below we show that

$$\max_{b=0,1} \deg f_{|x_1=b} \geq \deg f - 1 .$$

We apply the induction hypothesis to the b with $\deg f_{|x_1=b} \geq \deg f - 1$. This finishes the induction step since the induction hypothesis yields

$$L(f) \geq L(f_{|x_1=b}) + 1 \geq \deg f_{|x_1=b} \geq \deg f - 1.$$

It remains to prove that $\deg f_{|x_1=b} \geq \deg f - 1$ for either $b = 0$ or $b = 1$. We can write f as a sum

$$f = x_1 f_1 \oplus f_2$$

where f_2 does not depend on the boolean variable x_1. Then we have

$$\deg f_{x_1=0} \geq \deg f - 1 \quad \text{if} \quad \deg f_2 \geq \deg f - 1$$

$$\deg f_{x_1=1} \geq \deg f - 1 \quad \text{if} \quad \deg f_2 < \deg f - 1 .$$

This proves the claim. □

4. Complexity classes and isomorphism classes of quadratic forms

We summarize some of the main results in Mirwald, Schnorr (1987).We first analyse the L_1-complexity of quadratic forms and we lateron prove that the complexity measures L and L_1 coincide for quadratic forms.

Lemma 4.1 Let $u_1,\ldots,u_t, v_1,\ldots,v_t \in B_{n,1}$ and $f = \oplus_{i=1}^{t} u_i v_i$. If $L_1(f) = t$ then $u_1,\ldots,u_t,v_1,\ldots,v_t$ are linearly independent and f is isomorphic to $\oplus_{i=1}^{t} x_{2i-1}x_{2i}$.

Proof Suppose $L_1(\oplus_{i=1}^{t} u_i v_i) = t$ holds for the linearly dependent linear forms $u_1,\ldots,u_t,v_1,\ldots,v_t$. W.l.o.g. let $u_t = \oplus_{i<t} a_i u_i \oplus \oplus_{i=1}^{t} b_i v_i$ with $a_i, b_i \in \mathbb{Z}_2$. We have

$$\oplus_{i=1}^{t} u_i v_i = \oplus_{i=1}^{t-1}(u_i \oplus b_i v_t)(v_i \oplus a_i v_t) \pmod{v_t}$$

and thus $L_1(\oplus_{i=1}^{t} u_i v_i) < t$. This contradicts to the assumption.

If $u_1,v_1,\ldots,u_t,v_t$ are linearly independent there is a linear isomorphism T on $\mathbb{Z}_2^n$ such that $T^*(u_1) = x_1$, $T^*(v_1) = x_2$, ..., $T^*(u_t) = x_{2t-1}$, $T^*(v_t) = x_{2t}$. T transforms f into $\oplus_{i=1}^{t} x_{2i-1}x_{2i}$. □

Lemma 4.2 For every boolean nxn matrix $A = (a_{i,j})$ with $\oplus_{i=1}^{t} x_{2i-1}x_{2i} = f_A = \oplus_{i,j} a_{i,j} x_i x_j$ we have $\frac{1}{2}\operatorname{rank}(A \oplus A^T) = t$.

Proof We have $\frac{1}{2}\operatorname{rank}(A \oplus A^T) = t$ for the matrix $A = (a_{i,j})$ with $a_{2i-1,2i} = 1$ for $i=1,\ldots,t$ and $a_{i,j} = 0$ for all other entries. The claim follows from the equivalence $f_A = f_B$ iff $A \oplus A^T = B \oplus B^T$. □

Lemma 4.3 For every boolean nxn-matrix B we have $\frac{1}{2}\operatorname{rank}(B \oplus B^T) = L_1(f_B)$.

Proof Let $L_1(f_B) = t$. Then there exist linear forms $u_1,\ldots,u_t,v_1,\ldots,v_t$ such that $f_B = \oplus_{i=1}^{t} u_i v_i \bmod B_{n,1}$. By Lemma 4.1 the forms $u_1,\ldots,u_t,v_1,\ldots,v_t$ are linearly independent and f_B is isomorphic to $\oplus_{i=1}^{t} x_{2i-1}x_{2i}$. Let A be a boolean nxn matrix such that $f_A = \oplus_{i=1}^{t} x_{2i-1}x_{2i}$. Since f_A, f_B are isomorphic there exists a unimodular matrix τ such that $B = \tau^T A \tau$, and thus we have $B \oplus B^T = \tau^T(A \oplus A^T)\tau$. By Lemma 4.2 we have

$$t = \frac{1}{2}\operatorname{rank}(A \oplus A^T) = \frac{1}{2}\operatorname{rank}(B \oplus B^T). \ \square$$

We immediately obtain from Lemmata 4.1-4.3 the following Theorem 4.4.

Theorem 4.4 Let $f \in B_{n,2}$ and $u_1,\ldots,u_t,v_1,\ldots,v_t \in B_{n,1}$ such that $f = \oplus_{i=1}^{t} u_i v_i \bmod B_{n,1}$. Then the following conditions are equivalent:

(a) $L_1(f) = t$,

(b) $u_1,\ldots,u_t,v_1,\ldots,v_t$ are linearly independent,

(c) f is isomorphic to the *canonical form* $\oplus_{i=1}^{t} x_{2i-1}x_{2i}$,

(d) for every boolean n×n matrix A with $f_A = f$ we have $\frac{1}{2}\,\mathrm{rank}(A \oplus A^T) = t$.

The equality $L(f_A) = \frac{1}{2}\,\mathrm{rank}(A \oplus A^T)$ distinguishes n-ary boolean forms in $B_{n,2}$ from quadratic forms in $\mathbb{Z}_2[x_1,\ldots,x_n]$. By Theorem 4.4 (d) the rank of $A \oplus A^T$ is even for all boolean n×n matrices A. The matrices $A \oplus A^T$ range over all symmetrical, boolean matrices that have zero diagonal elements. These matrices $(b_{i,j}) = A \oplus A^T$ are skew symmetrical, i.e. $b_{i,j} = -b_{j,i}$. It is well known that skew symmetrical matrices have even rank.

Theorem 4.5 Every circuit that computes some quadratic form f with L(f) many ∧-gates has at most one level of ∧-gates.

By combining Theorems 4.4 and 4.5 we can in Theorem 4.4 replace the L_1-complexity by the L-complexity.

Theorem 4.6 The equality $L_1(f) = L(f)$ holds for all quadratic forms f, and thus the characterizations of $L_1(f)$ in Theorem 4.4 also apply to L(f).

In particular by Theorem 4.4 (d) the multiplicative complexity L(f) of f can be found in polynomial time from the coefficients a_{ij} of the quadratic form $f = \oplus_{i,j}\, a_{ij}\, x_i x_j$ since $L(f) = \frac{1}{2}$ rank $(A \oplus A^T)$ holds for the coefficient matrix $A = (a_{ij})_{1 \le i,j \le n}$. From the equation $L(f) = \frac{1}{2}$ rank $(A \oplus A^T)$ we can easily determine the maximal multiplicative complexity of n-ary quadratic forms:

$$\max\{L(f) \mid f \in B_{n,2}\} = \lfloor n/2 \rfloor .$$

For quadratic forms there is another characterization of the dual domain D(f) based on an L-minimal representation of f. If $f = \oplus_{i=1}^{t} u_i v_i$ (mod $B_{n,1}$) with linearly independent forms u_i, $v_i \in B_{n,1}$ for i=1,...,t, then $D(u_i v_i) = \langle u_i, v_i \rangle$ is the span of the forms u_i and v_i, and D(f) is the span of the linear spaces $D(u_i v_i)$,

$$D(f) = \langle u_1, v_1, \ldots, u_t, v_t \rangle .$$

Obviously D(f) is the direct sum of $D(u_i v_i)$ for i=1,...,t.

Corollary 4.7 Let f_1, f_2 be quadratic forms. Then $L(f_1 \oplus f_2) = L(f_1) + L(f_2)$ iff $D(f_1 \oplus f_2) = D(f_1) \times D(f_2)$ where × denotes the direct sum of linear spaces.

Proof. "⇒" It follows from $L(f_1 \oplus f_2) = L(f_1) + L(f_2)$ and $\dim D(f) = 2\,L(f)$ that

$$\dim D(f_1 \oplus f_2) = \dim D(f_1) + \dim D(f_2).$$

Since $D(f_1 \oplus f_2) \subset \mathrm{span}(D(f_1),D(f_2))$ we have $D(f_1 \oplus f_2) = D(f_1) \times D(f_2)$.

"⇐" $D(f_1 \oplus f_2) = D(f_1) \times D(f_2)$ implies $L(f_1 \oplus f_2) = (\dim D(f_1 \oplus f_2))/2 = (\dim D(f_1) + \dim D(f_2))/2 = L(f_1) + L(f_2)$. □

We call two quadratic forms f_1,f_2 *strictly computational independent* if $L(f_1 \oplus f_2) = L(f_1) + L(f_2)$. If the quadratic forms f_1,f_2 depend on disjoint sets of variables then Corollary 4.7 implies $L(f_1 \oplus f_2) = L(f_1) + L(f_2)$ and thus $L(f_1,f_2) = L(f_1) + L(f_2)$. The latter conclusion is an instance of Strassen's L-additivity conjecture. It is open whether $L(F \cup G) = L(F) + L(G)$ holds for sets G, G of quadratic boolean forms depending on disjoint sets of variables. Strassen conjectured that the latter equality holds for sets of formal quadratic forms $F,G \subset \mathbf{k}[x_1,\ldots,x_n]$ with coefficients in an infinite field $\mathbf{k}$.

5. The multiplicative complexity of pairs of quadratic forms

We give a polynomial time algorithm for determining the multiplicative complexity of pairs of quadratic forms. The complexity theory of pairs of quadratic forms is rather different from the theory of pairs of bilinear forms as developped in Ja' Ja' (1979, 1980). We first analyse the L_1-complexity and we lateron prove that the complexity measures L and L_1 coincide for pairs of quadratic forms. For every pair of quadratic forms (f_1,f_2) we establish in Theorem 5.1 the existence of a quadratic form g such that $L_1(f_1,f_2) = L(f_1 \oplus g) + L(f_2 \oplus g) + L(g)$ and $L(f_i) = L(f_i \oplus g) + L(g)$ for $i = 1,2$. These equations describe an L_1-minimal circuit for (f_1,f_2) consisting of two L-minimal circuits for f_1 and f_2 with a joint subcircuit for g. Proofs can be found in Mirwald, Schnorr (1987).

Theorem 5.1 For every pair of quadratic forms $f_1,f_2 \in B_{n,2}$ there exists $g \in B_{n,2}$ such that $L_1(f_1,f_2) = L(f_1 \oplus g) + L(f_2 \oplus g) + L(g)$ and $L(f_i) = L(f_i \oplus g) + L(g)$ for $i = 1,2$.

In order to find the function g of Theorem 5.1 it is important to have for fixed f a neat description of all pairs of linear forms u,v such that $L(f \oplus uv) = L(f)-1$. By Corollary 4.7 these forms u,v must satisfy $D(uv) \subset D(f)$, and thus $u,v \in D(f)$. Theorem 5.2 characterizes the set of these linear forms u,v in case that f is a canonical form.

We associate with the canonical form $f = \oplus_{i=1}^{t} x_{2i-1}x_{2i}$ the bilinear form

$$\Phi_f : B_{2t,1}^2 \rightarrow \mathbb{Z}_2 \;,\; \Phi_f(u,v) = \oplus_{i=1}^{t}(u_{2i-1}v_{2i} \oplus u_{2i}v_{2i-1})$$

where $u_i, v_i \in \mathbb{Z}_2$ are the coordinates of $u = \oplus_{i=1}^{2t} u_i x_i$ and $v = \oplus_{i=1}^{2t} v_i x_i$. $\Phi_f(u,v)$ is the parity of the number of monomials $x_{2i-1}x_{2i}$ with $1 \le i \le t$ of uv. By definition $x_i x_j$ is *monomial* of the quadratic form $\oplus_{i,j} c_{ij} x_i x_j$ iff $c_{ij} \oplus c_{ji} = 1$.

Theorem 5.2

(1) For the canonical form $f = \oplus_{i=1}^{t} x_{2i} x_{2i-1}$ and $u,v \in D(f)$ we have $L(f \oplus uv) = t-1$ iff uv has an odd number of the monomials $x_{2i-1}x_{2i}$ with $1 \le i \le t$, i.e. iff $\Phi_f(u,v) = 1$.

(2) For every quadratic form $f \in B_{n,2}$ and every subspace $U \subset \mathbb{Z}_2^n$ there exist $u,v \in U^*$ with $L(f \oplus uv) = L(f)-1$ iff f is non-linear on U.

We call two quadratic forms f_1,f_2 *weakly computational independent* if $L_1(f_1,f_2) = L(f_1) + L(f_2)$. Theorem 5.3 shows that f_1, f_2 are weakly computational independent iff either f_1 or f_2 is linear on $(D(f_1) \cap D(f_2))^*$. This means the following. Let $u_1,\ldots,u_n$ be any basis of $B_{n,1}$ that contains a basis $u_1,\ldots,u_r$ of $D(f_1) \cap D(f_2)$. Suppose we write f_1,f_2 as polynomials in $u_1,\ldots,u_n$. Then f_1,f_2 are weakly computational independent iff either f_1 or f_2 depends linearly on $u_1,\ldots,u_r$.

We give two examples. The functions

$$f_1 = x_1x_2 \oplus x_3x_4 \,,$$
$$f_2 = x_1x_3 \oplus x_2x_4$$

are both non-linear on $\mathbb{Z}_2^4 = (D(f_1) \cap D(f_2))^*$. By Theorem 5.3 (1) this implies $L_1(f_1,f_2) \le 3$. A corresponding circuit with 3 $\wedge$-gates is as follows.

$$f_1 = x_1(x_2 \oplus x_3) \oplus x_3(x_4 \oplus x_1)$$
$$f_2 = x_1(x_2 \oplus x_3) \oplus x_2(x_4 \oplus x_1) \,.$$

This example is quite similar to the well-known multiplication of two complex numbers within three real multiplications.

On the other hand by Theorem 5.3 (1) the functions

$$f_1 = x_1x_2 \oplus x_3x_4 \oplus \cdots\cdots \oplus x_{4t-1}x_{4t}$$
$$f_2 = x_2x_3 \oplus x_6x_7 \oplus \cdots\cdots \oplus x_{4t-2}x_{4t-1}$$

are weakly computational independent since f_1 depends linearly on the variables of f_2. This shows

$$L_1(f_1,f_2) = 3t \,.$$

The L_1-complexity of (f_1,f_2) is maximal for all pairs of forms $f_1,f_2 \in B_{4t,2}$, as follows from Theorem 5.5

Theorem 5.3 Let f_1, f_2 be quadratic forms in $B_{n,2}$, $E = D(f_1) \cap D(f_2)$. Then the following holds.

(1) $L_1(f_1,f_2) = L(f_1) + L(f_2)$ iff either f_1 or f_2 is linear on E.

(2) If $L_1(f_1,f_2) < L(f_1) + L(f_2)$ we can find in polynomial time linear forms $u, v \in B_{n,1}$ such that $L(f_i \oplus uv) = L(f_i)-1$ for $i=1,2$.

Determination of $L_1(f_1,f_2)$ for $f_1,f_2 \in B_{n,2}$

initiation $g := 0 \in B_{n,2}$, $\overline{f}_i := f_i$ for $i=1,2$, $E := D(f_1) \cap D(f_2)$.
while $\min_{i=1,2} L(\overline{f}_{i|E^*}) > 0$ do
begin according to Theorem 5.3 construct in polynomial time
some $\overline{g} \in B_{n,2}$ with $L(\overline{g}) = 1$ and
$L(\overline{f}_i \oplus \overline{g}) = L(\overline{f}_i) - 1$ for $i=1,2$.
$\overline{f}_i := \overline{f}_i \oplus \overline{g}$ for $i=1,2$.
$g := g \oplus \overline{g}$, $E := D(\overline{f}_1) \cap D(\overline{f}_2)$.
end
output $L_1(f_1,f_2) := L(f_1 \oplus g) + L(f_2 \oplus g) + L(g)$.

Correctness of the algorithm

On termination we have

$$L(f_i) = L(f_i \oplus g) + L(g) \quad \text{for } i=1,2 .$$

By Corollary 4.7 this implies

$$D(f_i) = D(f_i \oplus g) \times D(g) \quad \text{for } i=1,2 .$$

We conclude

$$L_1(f_1,f_2) = L_1(f_1 \oplus g, f_2 \oplus g) + L(g) .$$

On termination we also have

$$\min_{i=1,2} L(\overline{f}_{i|E^*}) = 0 \quad \text{for } E = D(\overline{f}_1) \cap D(\overline{f}_2) .$$

By Theorem 5.3 this implies

$$L_1(\overline{f}_1,\overline{f}_2) = L(\overline{f}_1) + L(\overline{f}_2) .$$

Alltogether we see that

$$L_1(f_1,f_2) = L_1(f_1 \oplus g, f_2 \oplus g) + L(g) = L(f_1 \oplus g) + L(f_2 \oplus g) + L(g).$$

We observe that the output formula

$$L_1(f_1,f_2) = L(f_1 \oplus g) + L(f_2 \oplus g) + L(g)$$

also yields an L_1-minimal circuit for f_1, f_2. This circuit separately computes $f_1 \oplus g$, $f_2 \oplus g$ and g. So far we have proved the following theorem.

Theorem 5.4 The function $L_1(f_1,f_2)$ is polynomial time computable from the coefficients of $f_1,f_2 \in B_{n,2}$. Given $f_1,f_2 \in B_{n,2}$ we can find in polynomial time some form $g \in B_{n,2}$ such that $L_1(f_1,f_2) = L(f_1 \oplus g) + L(f_2 \oplus g) + L(g)$ and $L(f_i) = L(f_i \oplus g) + L(g)$ for $i=1,2$.

Theorem 5.5 $\max\{L_1(f_1,f_2) \mid f_1,f_2 \in B_{n,2}\} = \lfloor 3n/4 \rfloor$.

Our final theorem shows that the complexities L, L_1 coincide for pairs of quadratic forms. Thus all the previous characterizations of $L_1(f_1,f_2)$ also hold for $L(f_1,f_2)$.

Theorem 5.6 $L(f_1,f_2) = L_1(f_1,f_2)$ for all $f_1,f_2 \in B_{n,2}$.

References

J. Ja' Ja' (1979): Optimal evaluation of pairs of bilinear forms. Siam J. Computing 8 (1979), 443-462.

J. Ja' Ja' (1980): On the complexity of bilinear forms. Siam J. Computing 9 (1980), 713-728.

R. Mirwald and C.P. Schnorr (1987): The multiplicative complexity of quadratic boolean forms. Symposium on 28th annual Symposium on Foundations of Computer Science, Los Angeles, pp. 141-149.

C.P. Schnorr (1986): A Gödel theorem on network complexity lower bounds. Zeitschrift für math. Logik und Grundlagen der Mathematik 32 (1986), 377-384.

C.P. Schnorr (1980): A 3n-lower bound on the network complexity of Boolean functions. Theor. Comp. Science 10 (1986), 83-92.

V. Strassen (1973): Vermeidung von Divisionen. Crelles Journal für die reine und angew. Mathematik 264 (1973), 184-202.

V. Strassen (1984): Algebraische Berechnungskomplexität. In Perspectives in Mathematics, Birkhäuser Verlag, Basel.

AN ALGORITHM ON QUASI-ORDINARY POLYNOMIALS

M.E. Alonso*, I. Luengo**
Dpto. de Algebra
Facultad de CC. Matemáticas
Universidad Complutense
28040 Madrid (Spain)

M. Raimondo***
Dpto. di Matematica
Via L. Battista Alberti, 4
Università di Genova
16132 Genova (Italy)

§0. Introduction

Let K be a field of characteristic 0, and let R denote K[X] or K[[X]]. It is well known that the roots of a polynomial F∈R[Z] are fractional powers series in $\bar{K}[[X^{1/d}]]$, where $\bar{K}$ is a finite extension of K and d∈ℕ, and they can be obtained by applying the Newton Puiseux algorithm (cf. [W]). Although this is not true for polynomials in more than one variable, there is an important class of polynomials F∈ R[Z] (R= $K[[X_1,...,X_n]]=K[[\underline{X}]]$, called quasi-ordinary (Q.O.) polynomials, for which the same property holds (i.e. their roots are fractional power series in $\bar{K}[[\underline{X}^{1/d}]]$. This result, known as Jung-Abhyankar theorem, plays an important role in the theory of resolution of singularities (cf. [Z]), and was proved by Jung for n=2 and K=ℂ and by Abhyankar (cf. [A]) in the general case. A new and more elementary proof was given by one of the authors (cf. [Lu]).

The goal of this paper is to give an algorithm to compute these fractional power series for K a computable field and n=2. The paper is divided in 5 sections: in §1 we give some notions and results about Newton Polyhedra, Q.O. and ν-Q.O polynomials. In §2 we state property of Newton polyhedra of Q.O. polynomials (cf. (2.3)). Since the situation turns out to be more complicated than for curves (cf. (2.4)) we need to develop a symbolic computation with algebraic power series (§3), which allows us to state the algorithm to determine a root (§4). Finally, in §5 we explain some consequences of the algorithm, from a geometric point of view.

---

*) Partially supported by CAYCIT-PB860062 (Spain) & CNR (Italy)
**) Partially supported by CAYCIT-PB870336 (Spain)
***) Partially supported by M.P.I. Funds (Italy)

§1. Preliminaries

(1.0) Let K be a field of characteristic 0, $X=(X_1,\dots,X_n)$ and $F\in K[[X,Z]]$. We call $\exp(F)=\{\alpha=(\alpha_1,\dots,\alpha_{n+1})\in \mathbb{N}^{n+1}$ such that there exists $F_\alpha\in K^*$ and $F_\alpha X_1^{\alpha_1}\dots X_n^{\alpha_n}Z^{\alpha_{n+1}}$ is a monomial of F}, and NP(F) will denote the convex hull of $\exp(F)+\mathbb{N}^{n+1}$. NP(F) is called the Newton polyhedron of F.

Since $\exp(F)+\mathbb{N}^{n+1}$ is additively stable, there are finitely many exponents of F, say $A_1,\dots,A_k$ such that: $\exp(F)+\mathbb{N}^{n+1}=\bigcup_{i=1}^{k}(A_i+\mathbb{N}^{n+1})$, and for every j, $A_j\notin\bigcup_{i\neq j} A_i+\mathbb{N}^{n+1}$. Moreover these exponents are uniquely determined by F.

(1.1) Assume now, $F\in K[[X,Z]]$ is regular of order d ($d\geq 1$), that is, $F(\underline{0},Z)=\lambda_0 Z^d+\lambda_1 Z^{d+1}+\dots$, $\lambda_0\neq 0$, $\lambda_i\in$ K and we write $\mathrm{ord}_Z F=d$. Thus $(0,0,\dots,d)\in\exp(F)$ and we denote it by A. Let $\Pi NP(F)$ the projection into $\mathbb{N}^n/d!$ with center at A of the set of points in NP(F) whose last coordinate is smaller than d; that is

$$\Pi NP(F)=\left\{\left(\frac{\alpha_1 d}{d-\gamma},\dots,\frac{\alpha_n d}{d-\gamma}\right)\text{ for }(\alpha_1,\dots,\alpha_n,\gamma)\in NP(F),\ \gamma\leq d-1\right\}$$

For every edge ℓ of the polyhedron drawn from A, we call characteristic equation of ℓ to the following:

$$F_\ell=\sum_{(i_1,\dots,i_{n+1})\in\ell} F_{i_1\dots i_n}X_1^{i_1}\dots X_n^{i_n}Z^{i_{n+1}}$$

Notice that all the exponents of F on such edges are some of the A_i's.

Finally we introduce:

Definition.- Let $F\in K[[X]][Z]$ be a regular polynomial of order d, and $D_Z F\in K[[X]]$ its discriminant w.r.t. Z. F is quasi-ordinary (Q.O.) w.r.t. Z if $D_Z F=X_1^{\alpha_1}\dots X_r^{\alpha_r}\xi$, $\xi\in K[[X]]$, $\xi(\underline{0})\neq 0$.

(1.2) Lemma Let $F\in K[[X,Z]]$ be regular with $\mathrm{ord}_Z F=d$ and let $u\in K[[X,Z]]$ be a unit; $u=\lambda+v$, $\lambda\in K^*$, $v(\underline{0})=0$. Then NP(uF)=NP(F), and for every edge ℓ drawn from A we have $F_\ell=\lambda(uF)_\ell$.

Proof.- Let $A_1,\dots,A_k$ be as in (1.0) and let $M_1,\dots,M_k$ be the corresponding monomials of F. We first observe that λM_i is a monomial of uF: in fact, since $uF=\lambda F+vF$ and since λM_i is a monomial of λF, it is enough to prove that no monomials with exponents A_i appear in vF. Otherwise , there would exist two monomials m and n of v and F respectively with exponents a and b such that $A_i=a+b$, and we would have $a\neq(0,\dots,0)$ since $v(\underline{0})=0$. Therefore $b<A_i$ in the partial ordering of $\mathbb{N}^{n+1}$, which is a contradiction with the choice of the A_j's.

Hence we have $\exp(F)\subset\exp(uF)$ and changing the role of F and uF we get $NP(F)=NP(uF)$. Moreover, the exponents lying on an edge ℓ are some of the A_j's; hence the above argument yields $F_\ell=\lambda(uF)_\ell$. ∎

The following notion, introduced by Hironaka, plays an important role in studying Newton Polyhedra of Q.O. polynomials.

(1.3) <u>Definition</u>.- Let $F\in K[[X]][Z]$ be a regular polynomial of order d; we say that F is ν-quasiordinary (ν-Q.O.) if:

i) $\Pi NP(F)$ has a unique vertex, say $R_1=\Pi(R)$, $R\in\exp(F)\setminus\{A\}$

ii) Setting S the edge of NP(F) containing A and R , F_S is a polynomial which is not a power of a linear form.

(1.4) <u>Remarks</u>.-

1) Let $F\in K[[X]][Z]$ be regular in Z; by lemma (1.2) if F is ν-Q.O. so is uF, for every unit $u\in K[[X]][Z]$.

2) There are ν-Q.O. polynomials which are not Q.O. (for instance: $Z^4-2X_1X_2^2Z^2+X_1^4X_2^4+X_1^2X_2^7$). However we have the following result (cf. [Lu])

(1.5.) <u>Theorem</u>.- Let $F\in K[[X]][Z]$ be a Q.O. Weierstrass polynomial of degree d having no monomials corresponding to Z^{d-1}. Then F is ν-Q.O. with respect to Z.

We shall use the following version of the above result.

(1.6) <u>Corollary</u>.- Let $F\in K[[X]][Z]$ be a Q.O. polynomial; then there exists a power series $h\in K[[X]]$ such that $F(X_0,\dots,X_n,Z-h)$ is ν-Q.O.

<u>Proof</u>.- By Weierstrass preparation theorem let $F=uF^*$ with F^* a Weierstrass polynomial and u a unit. Hence, u is a polynomial in Z and so, D_ZF^* divides D_ZF, hence F^* is Q.O. Let $F^*=Z^d+a_1Z^{d-1}+\dots+a_d$ and set $h=a_1/d$. Then by (1.5) $F^*(X_1,\dots,X_n,Z-h)$ is ν-Q.O. and by (1.4) so is $F(X_1,\dots,X_n,Z-h)$ ∎

§2. Newton polyhedra of Q.O. polynomials

(2.1) In this paragraph we find a property of Newton polyhedra of Q.O. polynomials, which will be essential in order to show that the main algorithm of §4 stops.

We restrict ourselves to the case of n=2, i.e. to the variables X,Y,Z, and $F\in K[[X,Y]][Z]$ regular of order d. We call α,β,γ the exponents w.r.t X,Y,Z. We keep the notations of the preceeding section, and we call $P_1,\dots,P_r$ the points of $\Pi(\exp(F))$ which are vertices (ie. extremal points) of $\Pi(NP(F))$. We write $P_i=(\alpha_i,\beta_i)\in \frac{1}{d!}\cdot\mathbb{N}^2$, and let $\ell_1,\dots,\ell_r$ denote the edges of NP(F) drawn from A determined by the lines joining A and $P_1,\dots,P_r$ respectively.

(2.1) <u>Lemma</u>.- F_{ℓ_i} is a d-power of a linear form (i.e. there exists $H\in K[X,Y]$ such that $F_{\ell_i}=(Z-H)^d$) if and only if $(\alpha_i/d,\ \beta_i/d)\in\mathbb{N}^2$, and $F_{\ell_i}(1,1,Z)=(Z-\lambda_i)^d$ with $\lambda_i\in K$. In such a case $F_{\ell_i}=(Z-\lambda_i X^{\alpha_i/d}Y^{\beta_i/d})^d$.

<u>Proof</u>. Write $F_{\ell_i}=Z^d+U_1Z^{d-1}X^{\alpha_i/d}Y^{\beta_i/d}+\dots+U_dX^{\alpha_i}Y^{\beta_i}$, where $U_j\in K$ and $U_j=0$ if $(\alpha_i/d,\beta_i/d)\notin j\mathbb{N}^2$. Then F_{ℓ_i} is an homogeneous polynomial in Z and $t=X^{\alpha_i/d}Y^{\beta_i/d}$ of degree d, so the lemma follows straightforward.∎

(2.2) <u>Lemma</u>.- Let $M=C_0X^{\alpha_0}Y^{\beta_0}$ be a monomial such that $(\alpha_0 d,\beta_0 d)\in\Pi NP(F)$, and let $Z=Z_1+M$ and $F_1=F(X,Y,Z_1+M)$; then:

a) F_1 is regular of order d and $\Pi NP(F_1)\subset\Pi NP(F)$.

b) For every P_i we have:

i) if $P_i=(\alpha_0 d,\beta_0 d)$, and $F_{\ell_i}(1,1,Z) = (Z-C_0)^d$, then P_i does not appear in $\Pi NP(F_1)$ and the inclusion in a) is strict,

ii) if F_{ℓ_i} is not a power of a linear form, P_i is again an extremal point of $NP(F_1)$ and $F_{1\ell_i}$ in not a power of a linear form.

<u>Proof</u>.

a) Let us show the inclusion. We write:

$$F=Z^d+\sum_{0\le\gamma\le d-1}F_{\alpha\beta\gamma}X^{\alpha}Y^{\beta}Z^{\gamma}+R$$

where R contains the monomials $X^{\alpha}Y^{\beta}Z^{\gamma}$ with $\gamma\ge d$ or $\gamma=d$ and $(\alpha,\beta)\ne(0,0)$. The monomial $F_{\alpha\beta\gamma}X^{\alpha}Y^{\beta}Z^{\gamma}$ provides, after the change $Z=Z_1+M$, monomials with exponents $(\alpha+k\alpha_0,\beta+k\beta_0,\gamma-k)$, $0\le k\le\gamma$. Their projections (for those with $\gamma-k\leq d-1$) are: $\omega=\left(\left(\frac{\alpha+k\alpha_o}{d-\gamma+k}\right)d, \left(\frac{\beta+k\beta_o}{d-\gamma+k}\right)d\right)$.

If $\gamma\geq d$, $\omega\in(\alpha_o d,\beta_o d)+\frac{1}{d!}\mathbb{N}^2\subset \Pi NP(F)$, and if $0\leq\gamma\leq d-1$, then $(\alpha_o d,\beta_o d)$ and $(\frac{\alpha d}{d-\gamma},\frac{\beta d}{d-\gamma})$ are in $\Pi NP(F)$ and ω is a convex combination of them.

b) From **a)** we deduce that, if for some i P_i still appears in $\Pi NP(F)$, then P_i is an extremal point.

We first prove the following claim:

(2.2.1) If a point ω as above coincides with P_i, then it comes from an exponent $(\alpha,\beta,\gamma)\in \ell_i$, i.e. from a point of NP(F) lying on P_i.

For this, we should show that $\omega=P_i$ cannot come from a monomial in R. Otherwise we would have $\left(\frac{\alpha+k\alpha_o}{d-\gamma+k}\right)d=\alpha_i$, $\left(\frac{\beta+k\beta_o}{d-\gamma+k}\right)d=\beta_i$, with $\gamma-k\leq d-1$ and $d\leq\gamma$, hence $k\neq0$ and $\alpha_i-\alpha_o d=\frac{d\alpha-\alpha_i(d-\gamma)}{k}$, $\beta_i-\beta_o d=\frac{d\beta-\beta_i(d-\gamma)}{k}$. If $d-\gamma<0$, since $(\alpha_i,\beta_i)\neq(0,0)$, we have $(\alpha_i,\beta_i)>(\alpha_o d,\beta_o d)$ in the partial order; while if $d=\gamma$, $(\alpha,\beta)\neq(0,0)$ since (α,β,γ) is an exponent in R, hence $(\alpha_i,\beta_i)>(\alpha_o d,\beta_o d)$. In both cases this is a contradiction with the fact P_i is an extremal point. So, we assume $\omega=\Pi((\alpha,\beta,\gamma))$ with $\gamma\leq d-1$. Now, for every i, $1\leq i\leq r$, let $H_i=A_iX+B_iY+C_i$ be the hyperplane such that $H_i(P_i)=0$ and $H_i|_{\Pi NP(F)-\{P_i\}}>0$. Then we have:

$$(2.2.2)\qquad H_i(\omega)=\frac{(A_i\alpha d+B_i\beta d+C_i(d-\gamma))+k(A_i\alpha_o d+B_i\beta_o d+C_i)}{d-\gamma+k}\geq 0.$$

Since $(\alpha_o d,\beta_o d)\in \Pi NP(F)$, both summands in the numerator are non negative, so, $H_i(\omega)=0$ if and only if both are so. As $\omega=P_i$, $H_i(\omega)=0$ and $H_i(\frac{\alpha d}{d-\gamma},\frac{\beta d}{d-\gamma})=0$ that is $(\alpha,\beta,\gamma)\in \ell_i$, which proves the claim.

Let us come to **b)**: **b i)** is easily deduced from (2.2.1) and (2.1), because in this case $F_{\ell_i}(X,Y,Z_1+M)=Z_1^d$.

For **b ii)**, assume F_{ℓ_i} is not a power of a linear form. We have two cases. If $P_i\neq(\alpha_o d,\beta_o d)$, by (2.2.2) we have that $H_i(\omega)=0$ (i.e. $\omega=P_i$) iff $(\alpha,\beta,\gamma)\in\ell_i$ and $k=0$. So, in this case, $P_i\in \Pi(\exp(F_1))\subset \Pi NP(F_1)$, and as we have shown before it is an extremal point of it. Moreover $F_{1\ell_i}=F_{\ell_i}$, and therefore it is not a power of a linear form. If $P_i=(\alpha_o d,\beta_o d)$, again by (2.2.2) we have that $H_i(\omega)=0$ for every ω such that $(\alpha,\beta,\gamma)\in\ell_i$. Thus $P_i\in \Pi NP(F_1)$ iff $F_{\ell_i}(X,Y,Z_1+M)\neq Z_1^d$, which holds true since F_{ℓ_i} is not a power of a linear form. Moreover, in this case, $F_{1\ell_i}=F_{\ell_i}(X,Y,Z_1+M)$, and therefore $F_{1\ell_i}$ is not a power of a linear form.∎

(2.3) Proposition. Let $F\in K[[X,Y]][Z]$ be a Q.O. regular polynomial of order d, then for all but at most one i, F_{ℓ_i} is a d-power of a linear form.

Proof. Let $F^*=uF$, where $u\in K[[X,Y,Z]]$ is a unit and F^* is a Weierstrass polynomial of order d in Z.

By (1.6), there exists $h=\sum_{k\geq 0} h_k = \sum_{k\geq 0} \lambda_{i_k j_k} X^{i_k} Y^{j_k}$, such that $F(X,Y,Z-h)$ is ν-Q.O. (so that $\Pi NP(F(X,Y,Z-h))$ has only one extremal point), and where the $h_k=\lambda_{i_k j_k} X^{i_k} Y^{j_k}$ are the non vanishing monomials numbered by any order compatible with the degree,

Let us set: $u_k=u(X,Y,Z-\sum_1^{k+1} h_i)$, $F_k^*=F^*(X,Y,Z-\sum_1^{k+1} h_i)$ and $F_k=F(X,Y,Z-\sum_1^{k+1} h_i)$ with $u_o=u$, $F_o^*=F^*$ and $F_o=F$. Then $F_k^*=F_k u_k$, u_k is a unit and $F_{k-1}^*=Z^d+(\sum_{m\geq 0} h_{k+m})Z^{d-1}+\ldots$, so: $(i_k d, j_k d)=\Pi(i_k, j_k, d-1)\in \Pi NP(F_{k-1}^*)=$ $=\pi NP(F_{k-1})$ by (1.2).

Let P be an extremal point of $\Pi NP(F)$ determining and edge ℓ from A, and assume that F_ℓ is not a power of a linear form. By (2.2) P is extremal in $\Pi NP(F_1)$ and $F_{1\ell}$ is not a power of a linear form. Again by lemma (2.2) P is extremal in $\Pi NP(F_2)$ and $F_{2\ell}$ is not a power of a linear form. Thus, continuing this way, we get that P is extremal of every $\Pi NP(F_j)$ and so P is an extremal point of $\Pi NP(F(X,Y,Z-h))$ (since $F_j\longrightarrow F$ in the (X,Y,Z)-adic topology). As $F(X,Y,Z-h)$ is ν-Q.O., the proposition is proved.∎

(2.4) Construction.- Given a Q.O. polynomial $F\in K[[X,Y]][Z]$ regular of order d in Z, following Abhyankar (cf. [A]) (see also [Lu]) and Zariski (cf [Z]) its roots w.r.t. Z at the origin have the form:

(2.4.1) $$\zeta_i=H+X^{\alpha}Y^{\beta}H_i(X^{1/e},Y^{1/e'}) \quad i=1,\ldots,d$$

where H, $H_i\in K[X,Y]]_{alg}$ (algebraic series over $K[X,Y]$), $H(0,0)=0$ and at least for some $i\neq j$ $H_i(0,0)\neq H_j(0,0)$ ((α,β) is called the first distinguished pair of ζ_i (cf. [Li])).

Now, if we try to calculate these roots using $NP(F)\subset \mathbb{N}^3$ as in Newton-Puiseux algorithm for plane curves, we easily see that this is possible if there exists an edge ℓ of NP(F) joining A=(0,0,d) and a point $B=(\alpha,\beta,\gamma)$ of the Newton polyhedron, with $\gamma<d$ and such that

$NP(F) \subset \ell + \mathbb{N}^3$. In fact: let c be a root of $F_\ell(1,1,Z)=0$ of multiplicity $d_1 \le d$, and let $\frac{\alpha}{d-\gamma} = \frac{\alpha_0}{e}$ and $\frac{\beta}{d-\gamma} = \frac{\beta_0}{e'}$ where $(\alpha_0,e)=(\beta_0,e')=1$. By making the change:

$$(2.4.2)\quad \begin{cases} X=X_1^{e} \\ Y=Y_1^{e'} \\ Z=X_1^{\alpha_0}Y_1^{\beta_0}(c+Z_1) \end{cases}$$

we get $F_1(X_1,Y_1,Z_1)=X^{-\alpha_0 d}Y^{-\beta_0 d}F(X_1^{e},Y_1^{e'},X_1^{\alpha_0}Y_1^{\beta_0}(c+Z_1))$ which is again Q.O and has order d_1 in Z. If $d_1<d$ (i.e. if F is ν-Q.O.) the order of F_1 has decreased.

When we try to go further, we need to get again the ν-Q.O. situation, and by (1.6) (or using (2.4.1)), we see this happens after a change of variables $Z'=Z+h(X,Y)$, with $h \in K[[X,Y]]_{alg.}$. Since, even if $F \in K[X,Y,Z]$ h is not in general a polynomial, this process cannot be carried on. For this reason we need to work with algebraic power series before stating the algorithm.

§3. Computation with algebraic power series

(3.1) Let $T=(U_1,\ldots,U_n)$ be a set of variables and K a computable field with charact.(K)=0. Given $F_1,\ldots,F_r$, $F_i \in K[T,T_1,\ldots,T_i]$ such that $F_i(\underline{0},\underline{0})=0$ and $F_i(0,0,\ldots,T_i)=f_iT_i+\ldots$, $f_i \in K^*$; by the formal implicit function theorem there exist <u>unique</u> formal power series $t_1(T),\ldots,t_r(T)$, with $t_j(\underline{0})=0$, $j=1,\ldots,r$ verifying the algebraic equations:

$$(3.1.1)(F)\quad \begin{cases} F_1(T,T_1)=0 \\ \ldots\ldots\ldots \\ F_r(T,T_1,\ldots,T_r)=0 \end{cases}$$

Such a system will be called a <u>local smooth system.</u>

(3.2) Let us consider the evaluation homomorphism $\sigma_F: K[T,T_1,\ldots,T_r] \longrightarrow K[[T]]$, where $\sigma_F(T_j)=t_j$, and let $R_F = \mathrm{im}\ \sigma_F = K[T,t_1,\ldots,t_r] \subset K[[T]]$.

(3.3) <u>Algorithm</u>. There is a finite procedure which determines, for a given polynomial P, whether $P \in \mathrm{Ker}\ \sigma_F$ i.e. $P(T,t_1,\ldots,t_r)=0$.

<u>Proof</u>.- Let be $I=(F_1,\ldots,F_r)K[T,T_1,\ldots,T_r]_{(T,T_1,\ldots,T_r)}$ and

$I'=I\,K[[T,T_1,\dots,T_r]]$. We first show that $P(T,t_1,\dots,t_r)=0$ iff $P\in I'$. For, by iterative use of Weierstrass division theorem we have $P=\Sigma Q_iF_i+R$ with $R\in K[[T]]$. Then $P\in I'$ iff $R=0$ iff $P(T,t_1,\dots,t_r)=0$.

Now, by faithfully flatness we deduce that $P\in I'$ iff $P\in I$. Finally, by the tangent cone algorithm (cf. [M]) we can decide whether $P\in I$. For, notice that $\{F_1,\dots,F_r\}$ is already a standard basis w.r.t. an ordering making $\dots T^2<T_1^2\dots<T_s^2<T<T_1<\dots<T_s$ ∎

(3.4) Algorithm.- Given $q\in R_F[Z]$, $q(\underline{0},\underline{0},Z)=Z^d+\dots+cZ^m$ $c\neq 0$, $m\geq d$ it is possible to decide whether q has a root $h\in K[[T]]$ of multiplicity d.

Proof.- Write $q(T,t_1,\dots,t_r,Z)=a_0+a_1Z+\dots+a_mZ^m$, $a_j\in R_F$. We check, using (3.3) whether $a_0=0$.

If $a_0=0$ the answer is yes iff all $a_j=0$, $j\leq d-1$ (again checked by (3.3)).

If $a_0\neq 0$, let $q^{(d-1)}=\partial^{d-1}q\,/\,\partial Z^{d-1}$; so $q^{(d-1)}(\underline{0},\underline{0},Z)=Z+\dots+cZ^{m-d+1}$. We define $F_{r+1}=\Sigma f_i(T,T_1,\dots,T_r)T_{r+1}^i\in K[T,T_1,\dots,T_r,T_{r+1}]$ such that $\mathrm{ord}_{T_{r+1}}F_{r+1}=1$ and $q^{(d-1)}=\Sigma\, f_i(T,t_1,\dots,t_r)Z^i$. Then we call $F'=(F)\cup(F_{r+1})$ which is also a local smooth system, and t_{r+1} the new algebraic function defined by (F'). Now, let $q'=q(T,t_1,\dots,t_r,Z+t_{r+1})\in$ $\in R_{F'}[Z]$, and as above $q'=a_0'+a_1'Z+\dots+a_n'Z^n$. Thus, q has a root of multiplicity d (which is t_{r+1}) iff $a_0'=\dots=a_{d-1}'=0$ in $R_{F'}$ ∎

(3.4.1) Remark.- In the above situation if the required root does exist, we can write $q=(\partial^{d-1}q\,/\,\partial Z^{d-1})^d\cdot u$, $u\in K[T,\underline{t},Z]_{(T,\underline{t},Z)}$ being a unit.

(3.5) From now on, we assume n=1, and we consider the following two local smooth systems, where T has been replaced by X or Y.

$$(H)=\{H_1(X,X_1)=\dots=H_r(X,X_1,\dots,X_r)=0\}$$

$$(G)=\{G_1(Y,Y_1)=\dots=G_r(Y,Y_1,\dots,Y_s)=0\}$$

Let us denote by $x_1,\dots,x_r$ and $y_1,\dots,y_r$ the unique solutions of (H) and (G) respectively. Also we denote R_H and R_G as in (3.1), and since $(H)\cup(G)$ is also a local smooth system, we denote $R_{H,G}=K[X,Y,x_1,\dots,x_r,y_1,\dots,y_r]\subset K[[X,Y]]$.

We observe that a system such as (H) defines a germ of algebraic curve $(X,x_1,\dots,x_r)$ through $X=0$ whose tangent direction has the form $(1:a_1:\dots:a_r)$, $a_i\in K$.

In the following algorithm, which is essential for the main algorithm in §4, we have to change several times the systems (H) and

(G), hence also the x_i's and y_j's.

(3.6) <u>Algorithm</u>.- Given $P\in R_{H,G}$, $P\neq 0$, we find after changing (H)⟶(H'), (G)⟶(G') (and hence the x_i's and y_j's), $P'\in R_{H',G'}$ and $(\alpha_o,\beta_o)\in \mathbb{N}^2$ such that

i) $P=X^{\alpha_o}Y^{\beta_o}P'$

ii) $P'(X,0,x,0)\neq 0$ in $R_{H'}$, $P'(0,Y,0,y)\neq 0$ in $R_{G'}$.

Notice that (3.6) allows us to compute NP(P): in fact, by i) we have that $NP(P)\subset(\alpha_o,\beta_o)+\mathbb{N}^2$, and (ii) says that $\alpha=\alpha_o$ and $\beta=\beta_o$ are the non-bounded faces of NP(P).

<u>Proof</u>.- We remark that the condition $P\neq 0$ can be checked by (3.3).

We consider $P_Y=P(0,Y,\underline{0},\underline{y})\in R_G$ and $P_X=P(X,0,\underline{x},\underline{0})\in R_H$.

If $P_X\neq 0$ and $P_Y\neq 0$ we take $(\alpha_o,\beta_o)=(0,0)$ and $P'=P$.

If $P_Y=0$, we can take out a factor X from P as follows: let $H'_i(X,X'_1,\dots,X'_r)=\frac{1}{X}\ H_i(X,X(X'_1+a_1),\dots,X(X'_r+a_r))$, where $(1:a_1:\dots:a_r)$ is the tangent direction at X=0 to the curve defined by (H) (c.f (3.5)), and set $P'=\frac{1}{X}\ P(X,Y,\underline{x}',\underline{y}')$. Then, return to the beginning with P' and (H').

If $P_X=0$, we apply the same argument as for P_Y above.

Finally notice that the algorithm halts since $P\neq 0$ ■

§4. The Main Algorithm

(4.1) The INPUT consists of

- a computable field K of characteristic 0
- two sets of algebraic functions $\underline{x}=(x_1,\dots,x_r)$ and $\underline{y}=(y_1,\dots,y_s)$ defined by systems (H) and (G) respectively as in (3.5).
- a polynomial $F\in R[Z]$ where $R=K[X,Y,\underline{x},\underline{y}]$ such that $D_ZF=X^aY^bu(X,Y,\underline{x},\underline{y})$ with $u(0,0,\underline{0},\underline{0})\neq 0$ and $F(0,0,\underline{0},\underline{0},Z)=Z^d+\dots+\varepsilon Z^m$, $\varepsilon\neq 0$.

We remark that these conditions can be checked by (3.3) and (3.6)

The algorithm RETURNS:

- a finite extension K' of K.
- a new pair of polynomial systems (H'), (G') as in (3.5).
- a series $z\in K'[[X,Y]]$ with $z(0,0)=0$ defined univocally as root of a polynomial $Q(X,Y,\underline{x}',\underline{y}',Z)\in R'[Z]$ with $ord_Z Q(0,0,\underline{0},\underline{0},Z)=1$.
- a polynomial $P(X,Y,\underline{x}',\underline{y}',Z)$ such that the input F verifies $F(X^e,Y^{e'},P(X,Y,\underline{x}',\underline{y}',z))=0$ $(e,e'\in\mathbb{N})$, that is, $(X^e,Y^{e'},P(X,Y,\underline{x}',\underline{y}',z))$ is a formal parametrization of F in K'.

(4.2) Let $F_X=F(X,0,\underline{x},\underline{0},Z)\in R_H[Z]$ ($F_Y=F(0,Y,\underline{0},\underline{y},Z)\in R_G[Z]$ respectively);

we shall use #3 to check in which of the following situations we are:

case $\underline{a}(X)$: F_X has a root of multiplicity d, $h\in K[[X]]$ $h\neq 0$, $h(0)=0$.

In that case we have $F_X=\left(\frac{\partial^{d-1}F_X}{\partial Z^{d-1}}\right)^d\cdot u$ (cf. (3.4) and (3.4.1))

case $\underline{b}(X)$: $F_X=Z^d\cdot u$ (cf. (3.4))

In this case we can calculate the equation of a plane $\Pi=\{b\gamma+d\beta=bd\}$ with $b\in\frac{1}{d!}\cdot\mathbb{N}$ such that $\exp(F)\cap\Pi\neq\emptyset$ and $NP(F)\subset\Pi+\mathbb{N}^3$. In fact, let $F=\Sigma P_k Z^k$ with $P_k\in R_{H,G}$. We can determine the non-bounded face $\beta=b_k$ of $NP(P_k)$ (for $P_k\neq 0$) parallel to the α-axis by using (3.6), and set $b=\min\{b_k\}$. We will distinguish two subcases which will be treated differently:

case $\underline{b1}(X)$: $\underline{b}(X)$ and $b<d$

case $\underline{b2}(X)$: $\underline{b}(X)$ and $b\geq d$

Finally

case $\underline{c}(X)$: otherwise.

We define in the same way $\underline{a}(Y)$, $\underline{b1}(Y)$, $\underline{b2}(Y)$ and $\underline{c}(Y)$ for F_Y.

(4.3) MAIN ALGORITHM.

Begin by checking conditions $\underline{a}(X)$, $\underline{b}(X)$ and $\underline{c}(X)$.

(4.3.1) If $\underline{a}(X)$, then let $M=\frac{\partial^{d-1}F_X}{\partial Z^{d-1}}$. We enlarge the system (H) adding $H_{r+1}:=M$ (cf (3.4)) and put $F:=F(X,Y,\underline{x},\underline{y},Z+x_{r+1})$.

(4.3.2) If $\underline{b1}(X)$, we consider F_Y and continue by examining the corresponding Y's conditions.

(4.3.3) If $\underline{b2}(X)$, we set $b'=b-kd$ with $0\leq b'<d$ and $k\in\mathbb{N}$. Then, using (3.6) as in (4.2), we can define

(4.3.3.1) $F_0:=Y^{-kd}F(X,Y,\underline{x},\underline{y},ZY^k)$.

Notice that the x_i's, y_j's and the system (H) and (G) have been changed. This case behaves differently according to $b'>0$ or $b'=0$.

(4.3.3.2) If $b'>0$ we do $F:=F_0$, and F_0 verifies $\underline{b1}(X)$ with $b=b'$, so again we continue by examining the Y's conditions.

(4.3.3.3) If $b'=0$, we take a root c of $F_0(0,0,\underline{0},\underline{0},Z)=0$ (which can be managed as in [D]) and put $F:=F_0(X,Y,\underline{x},\underline{y},Z+c)$. Let d_1 ($d_1\leq d$) be the order of F in Z.

If $d_1=1$ then stop.

If $d_1\neq 1$ then begin with the new F.

(4.3.4) If $\underline{c}(X)$ go to examine $\underline{a}(Y)$, $\underline{b}(Y)$ and $\underline{c}(Y)$.

(4.3.5) Comment.-

i) By (4.4) we cannot meet the cases $\underline{a}(X)$ or $\underline{b2}(X)$ after $\underline{b2}(X)$

indefinitely (resp. for Y instead of X). Moreover theorem (4.5) shows that we cannot simultaneously have c(X) and c(Y).

ii) When we pass from X's to Y's cases, and reciprocally, the order of F decreases unless when we go: from b1(X) to b1(Y) or c(Y) and from c(X) to b1(Y) (or viceversa)

(4.3.6) If we have passed from b1(X) to b1(Y) or c(Y) or from c(X) to b1(Y) (or viceversa), by theorem (4.5) below we are in the ν-Q.O. situation, so, we make the Newton-Puiseux change (cf. (2.4.2)) and the order of F_1 is smaller than d. Notice that, to carry on this change we use again (3.6), and the $\underline{x}$'s, $\underline{y}$'s, (H) and (G) have been changed.

Thus we continue the algorithm with $F:=F_1$. Hence, going on, we arrive to $\mathrm{ord}_Z F=1$.

With the previous notations we have:

(4.4) Proposition.- Situations a(X) after b2(X) and b2(X) after b2(X) (resp. for Y) cannot be met indefinitely.

Proof.- Assume we go from b2(X) to a(X) or b2(X). That is, we have $F_0=Y^{-kd}\ F(X,Y,\underline{x},\underline{y},ZY^k)$, $b'=0$, $F_1=F_0(X,Y,\underline{x},\underline{y},Z+c)$ for a root c of $F_0(0,0,\underline{0},\underline{0},Z)$, and we find that $(F_1)_X$ has a root of multiplicity d at X=0 (which happens exactly in cases a(X) and b(X)). Let us consider $T(Y,Z):=F$ as element of $R_H[Y][Z]$. Its Newton polygon is the projection of NP(F) onto the plane $\beta\gamma$. We call $\Pi_1=\{b\gamma+d\beta-bd=0\}$ and $\ell=\Pi_1\cap\beta\gamma$, (see fig. 1). Then it follows that $T_\ell(1,Z)=(F_0)_X$. Applying Newton-Puiseux algorithm to T, by a classical argument (cf. [W] page 105), we get, in a finite number of steps a T_ℓ without multiple roots. That means that, making the changes (4.3.3.1) and (4.3.3.2) we get $(F_1)_X$ without multiple roots. ∎

(4.5) Theorem.- Conditions c(X) and c(Y) cannot be both verified, while each of the following conditions implies that F is ν-Q.O.:

i) b1(X) and b1(Y).

ii) b1(X) and c(Y) (or b1(Y) and c(X)).

Proof.- Let us prove first the claim about i) and ii)

i) We know that there are no points of NP(F) under the planes $\Pi_1=\{b_Y\gamma+d\beta-b_Yd=0\}$ and $\Pi_2=\{b_X\gamma+d\alpha-b_Xd=0\}$ (i.e. $NP(F)\subset\Pi_1+\mathbb{N}^3$, $NP(F)\subset\Pi_2+\mathbb{N}^3$, see fig. 1), where b_Y and b_X have been determined in cases b(X) and b(Y) as in (4.2), and $\exp(F)\cap\Pi_i\neq\emptyset$, i=1,2. Let us show that there must be one edge of NP(F) lying on the line AP(fig. 1). Otherwise we would have two edges of NP(F) through A, say ℓ_1 and ℓ_2, $\ell_1\neq\ell_2$, $\ell_1\subset\Pi_1$, $\ell_2\subset\Pi_2$, such that F_{ℓ_1} and F_{ℓ_2} would not be a power of a linear form, since b_X and b_Y are smaller than d. This contradicts proposition

(2.3). Hence, calling ℓ the edge through A on the line AP, $\exp(F) \subset \ell + \mathbb{N}^3$ and F_ℓ is not a d-power.

ii) In this situation: there are no points of NP(F) under the plane $\Pi_1 = \{b_Y\gamma + d\beta - b_Y d = 0\}$, $\exp(F) \cap \Pi_1 \neq \emptyset$ and F_Y is not a d-power times a unit. So, there must be some points on the plane $\beta\gamma$ under the line $\gamma = d$, namely the exponents of F_Y. We consider the Newton polygon of F_Y and we call ℓ its edge through A (fig. 2) which is also an edge of NP(F). We want to show that $\ell \subset \overline{AQ}$, with $Q = (0, b_Y, 0)$. Assume the contrary, and let ℓ' be an edge of NP(F) on Π_1. Since $b_Y < d$, $F_{\ell'}$ is not a power of a linear form, so by (2.3) F_ℓ must be of the form $(Z-M)^d$. Making the change $Z = Z_1 + M$ and $F_1 = F(X, Y, \underline{x}, \underline{y}, Z_1 + M)$, by lemma (2.2) ℓ' is again an edge of $NP(F_1)$. Moreover, $F_{1Y} \neq Z_1^d u$, for u a unit since $F_Y \neq (Z-M)^d v$ for v a unit; thus $NP(F_1)$ has an edge say ℓ_1 on the plane $\beta\gamma$ under the line $\gamma = d$ (fig. 2). Since again by (2.2) $F_{1\ell'}$ is not a power of a linear form, $F_{1\ell_1}$ must be a d-power. Continuing this process we would find, by the same argument as in (4.4), that F_Y is a d-power times a unit. Contradiction. Hence we have $\ell \subset \overline{AQ}$, $NP(F) \subset \ell + \mathbb{N}^3$ and F_ℓ is not a power of a linear form.

Finally, first claim follows from a similar argument to that of case ii) ∎

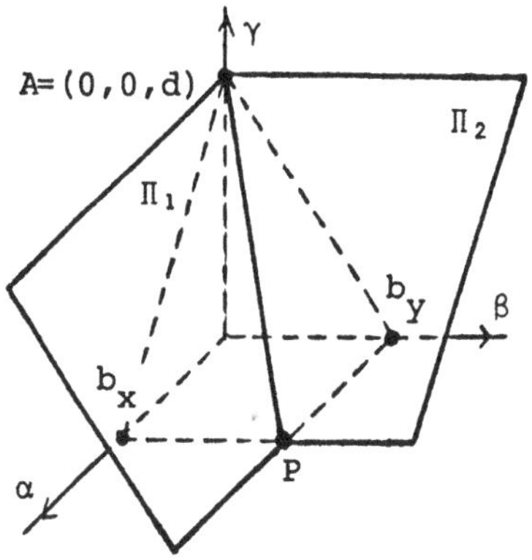

fig. 1

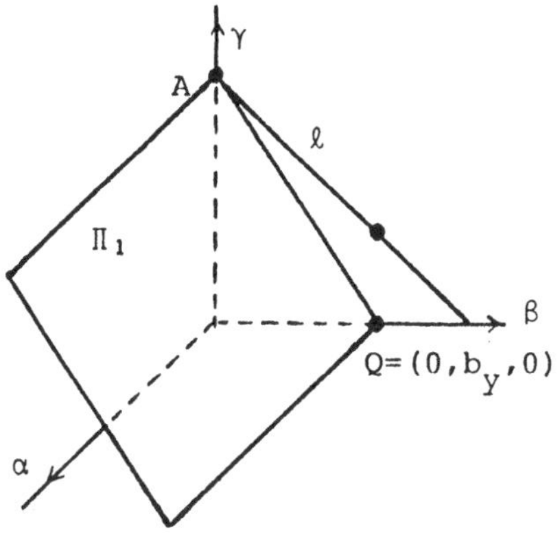

fig. 2

We finish this paragraph with an example:

<u>Ex</u>.- Let $F = X^{27}Y^{53} + X^{10}Y^{20}Z^3 + Z^5$, $D_Z F = X^{104}Y^{206}(3125 \cdot X^4 Y^6 + 108)$.

So, F is Q.O.

Looking at X's cases we are in $\underline{b}(X)$, so we calculate: $b = 25$(case $\underline{b2}(X)$), $k = 5$ and $b' = 0$. Then we make

$$Z = Z_1 Y^{10}, \quad F_1 = Y^{-50} F(X, Y, Z_1 Y^{10}) = X^{27}Y^3 + X^{10}Z_1^3 + Z_1^5.$$

Since the unique root of $F_1(0,0,Z)$ is 0 we begin with F_1 examining the

X's cases. We are in $\underline{c}(X)$, so we look at the Y's cases, and we find we are in $\underline{b}(Y)$. We compute: b=25, k=5, b'=0, and we make:

$$Z_1=Z_2X^5, \quad F_2=X^{-25}F_1(X,Y,Z_2X^5)=X^2Y^3+Z_2^5+Z_2^3.$$

Hence the order of F_2 in Z_2 has decreased. We choose a root e of $F_2(0,0,Z_2)=Z_2^5+Z_2^3=0$, for instance e=0, and begin with $F_2(X,Y,Z_2+e)$. It is again in case $\underline{b}(X)$ with b=3, k=1 and b'=0. We do

$$Z_2=Z_3Y, \quad F_3=Y^{-3}F_2(X,Y,Z_3Y)=X^2+Y^2Z_3^5+Z_3^3.$$

Now, starting with F_3 we are in $\underline{c}(X)$, so we look at the Y's. We find $\underline{b1}(Y)$; hence we have arrived to a ν-Q.O. situation with the edge $\ell=\overline{(0,0,3),(2,0,0)}$. We choose a root c of $Z^3+1=0$ and we make the Newton Puiseux change: $X=U^3$, $Z_3=(Z_4+c)U^2$ and

$$F_4=U^{-6}F_3(U^3,Y,(Z_4+c)U^2)=$$
$$10c^2U^4Y^2Z_4^3-c^2U^4Y^2+\underline{3c^2Z_4}+5cu^4Y^2Z_4^4-5cU^4Y^2Z_4+3cZ_4^2+U^4Y^2Z_4^5-10U^4Y^2Z_4^2+Z_4^3$$

The order of F_4 in Z_4 is 1, hence we have finished. Moreover the local analytic structure of F=0 at the origin is: two analytic complex conjugated branches and one real analytic branch (obtained by choosing $e=\pm i$).

§5. Consequences of the algorithm

Here we establish, without going into details, some consequences which follow from the algorithm of §4. We shall keep the notation of (4.1) assuming also $K\subset\mathbb{C}$.

First notice that F is not in general a Weierstrass polynomial in Z, so that our definition of Q.O. (cf. (1.1)) is stronger than the usual one (cf. [A], [Li], [Z]). In fact our F is Q.O. (with the usual meaning) at every point X=Y=0 $Z=\alpha_i$, α_i being a root of $q(Z)=F(0,0,\underline{0},\underline{0},Z)$, and at the infinite of the Z-axis.

(5.1) Local analytic branches. The algorithm (4.3) finds a formal algebraic (so analytic) parametrization of F=0 at X=Y=Z=0, that is, a local analytic branch. By considering all different roots in steps (4.3.3.3) and (4.3.6) we get other analytic branches. The rest of them are calculated with the following argument.

By (2.3) there is only one edge of NP(F), through A, say ℓ_1, which does not meet the plane $\gamma=0$. Let it be $\ell_1=AQ_1$ with $Q_1=(\alpha',\beta',d')\in$ exp(F) (d'<d). The we put: $X=X'^{d-d'}$, $Y=Y'^{d-d'}$, $Z=Z'X'^{\alpha'}Y'^{\beta'}$ and we consider the systems:

$$H_i' = H(X'^{d-d'}, X_1', \ldots, X_i'),\ i=1,\ldots,r$$

$$G_j' = G_j(Y'^{d-d'}, Y_1', \ldots, Y_j'),\ j=1,\ldots,s,$$

and their corresponding solutions $\underline{x}'$ and $\underline{y}'$. By using (3.6) we can take $X'^{\alpha' d}Y'^{\beta' d}$ as a factor of $F(X'^{d-d'}, Y'^{d-d'}, \underline{x}', \underline{y}', Z'X'^{\alpha'}Y'^{\beta'})$, so we can calculate

$$F_1 = X'^{-\alpha' d}Y'^{-\beta' d}F(X'^{d-d'}, Y^{d-d'}, \underline{x}', \underline{y}', Z'X'^{\alpha'}Y'^{\beta'})$$

(we may have changed (H'), (G') and hence the $\underline{x}'$ and $\underline{y}'$). Then applying the algorithm to F_1 (since it verifies (4.1)), we have a parametrization of F_1 at $X'=Y'=Z'=0$ which provides another one for F.

In particular if F is analytically irreducible, every edge ℓ through A meets the plane $\gamma=0$. Moreover, in this case, as in Newton Puiseux algorithm, it is easy to prove that for every edge ℓ in the ν-Q.O. situation (i.e. $NP(F) \subset \ell + \mathbb{N}^3$ and F_ℓ is not a power of a linear form), its characteristic equation $F_\ell=0$ has a root $cX^{\alpha_0}Y^{\beta_0}$ with $(\alpha_0, \beta_0) \in \mathbb{N}^2$ and $c \in K$, and the others differ of it in some root of the unity. Hence in that case the parametrization does not involve other algebraic numbers.

Finally we observe that, because of our definition of Q.O., we can compute also the local analytic structure of $F=0$ at the points $X=Y=0$, $Z=\alpha_i$, for any root α_i of $F(0,0,\underline{0},\underline{0},Z)$ by considering $F(X,Y,\underline{x},\underline{y},Z+\alpha_i)$ and, at the point $X=Y=0$, $Z=\infty$ by considering $Z^pF(X,Y,\underline{x},\underline{y},Z^{-1})$, for $p=\deg_Z F$.

(5.2) <u>Rational parametrizations</u>.- Instead of the Newton Puiseux change (cf. (2.4.2), we can make a similar change to the one made by Duval in the case of curves (cf. [D]) in order to obtain rational parametrizations.

(5.3) <u>Distinguished pairs, algebraic numbers and algebraic series</u>.- When we arrive to a ν-Q.O. situation, what happens in case (4.3.3.2) with $d_1<d$ and i) and ii) of (4.5), the order of F decreases and we eventually introduce (and only in these cases) algebraic numbers. Moreover in these cases the so called <u>distinguished pairs</u> of a Q.O. branch of the surface F=0 are calculated (cf. [Li]).

On the other hand, after the change (4.3.3) we have $\deg_Z F_0(X,0,\underline{x},\underline{0},Z) \le d$. Hence in case (4.3.3.3) with $d_1=d$, F is a Weierstrass polynomial in Z. Thus, if after that case we are in $\underline{a}(X)$, the new x_{r+1} determined in (4.3.1) is again in $R_H=K[X,x_1,\ldots,x_r]$. Therefore we conclude that at most two algebraic series have been introduced between two characteristic monomials (monomials corresponding to distinguished pairs): one w.r.t. X and another w.r.t. Y.

Finally we recall that distinguished pairs precisely determine a canonical resolution of the surface F=0 (cf. [Li]) and its topological type (cf. [Li], [G]). ∎

REFERENCES

[A] Abhyankar, S.S.; "On the ramification of algebraic Functions". Amer. J. Math. 77 (1955), 575-592.

[D] Duval, D.; "Diverses questions relatives au Calcul Formel avec des nombres algebriques". Thèse Univ. Grenoble 1987.

[G] Gau, Y.N.: "Topology of the quasi-ordinary surfaces singularities". Topology. Vol. 25 (1986), 495-519.

[Li] Lipman, J.; "Quasi-Ordinary singularities of surfaces in $\mathbb{C}^3$". Proceed. of Symp. in Pure Math. Vol. 40 (1983), 161-172.

[Lu] Luengo, I.; "A new proof of the Jung-Abhyankar Theorem". J. of Algebra vol. 85, (1983), 399-409.

[M] Mora, F.; "A constructive characterization of standard bases". Boll. U.M.I. Sez. D, 2 (1983), 41-50.

[W] Walker, R.J.; "Algebraic curves". Springer Verlag, New York-Heidelberg-Berlin, (1978).

[Z] Zariski, O.; "Exceptional singularities of an algebroid surface and their reduction". Acad. Naz. dei Lincei. Rendiconti. Classe di Scienze Fisiche, Matematiche e Naturale. Serie VIII, XLIII (1967), 135-146.

A COMPUTER-AIDED DESIGN FOR SAMPLING A NONLINEAR ANALYTIC SYSTEM

Jean-Pierre BARBOT
Laboratoire des Signaux et Systèmes*
Plateau du Moulon
91192 Gif-sur-Yvette, FRANCE

and

U.E.R. Sciences Fondamentales et Appliquées
Université d'Orléans
Rue de Chartres
ORLEANS Cedex 2 45046, FRANCE

ABSTRACT

In a recent paper by S. Monaco and D. Normand-Cyrot, a method for sampling a nonlinear continuous-time system was described. The object of this present work is to propose an algorithm for computing the sampled system. A symbolic language (REDUCE) is used for this.

* mailing address

INTRODUCTION

The Σ system is the object of a great deal of research on nonlinear control theory. This analytic system is described by the equation

$$\Sigma = \begin{cases} \dot{x} = f(x) + \sum_{i=1}^{m} g^i(x)\, u_i & \text{(A.1)} \\ y = h(x) & \text{(A.2)} \end{cases}$$

(where $x \in \mathbb{R}^n$, $y \in \mathbb{R}^p$, $U(= u_1, \ldots, u_m)^T \in \mathbb{R}^m$ and f, g_1, h are analytic functions of appropriate dimensions).

The assumption of linearity in the control U in Σ is made in general to simplify the analysis of feedback control strategies and is often verified in practice, for example, in robotics, spacecraft, etc. Much work has been concentrated on the design of state feedback control strategies modifying adequately the input-output behaviour of Σ (external linearization, input-output decoupling, etc.) [4].

The control law U(x) so defined is a continuous-time control but its digital implementation poses a difficult problem. The usual method for studying this problem is to apply the Euler discretization scheme to the Σ system:

$$x((k+1)\delta) = x(k\delta) + \delta\, [f(x(k\delta)) + \sum_{i=1}^{m} g^i\, (x(k\delta))\, u_i(k\delta)]$$

where δ is the sampling period, in general small.

In the last few years more sophisticated discrete control strategies acting on Σ have been introduced in [6, 7]. One of these solutions consists of discretizing the system Σ more accurately that with Euler's method, taking into account higher terms in δ.

Indeed, the sampling method proposed in [6] is based on the representation of the solution of the differential equation (A.1) in terms of a formal Lie exponential. A sampled system Σ_δ described by the following equations is obtained:

$$\Sigma_\delta = \begin{cases} x_\delta((k+1)\delta) = f_\delta(x_\delta(k\delta)\ U_\delta(k\delta)) & \text{(B.1)} \\ y_\delta(k\delta) = h_\delta(x_\delta(k\delta)) & \text{(B.2)} \end{cases}$$

(where x_δ, U_δ, y_δ, f_δ are defined as above).

Such state equations may be used to compute directly discrete feedback laws in order to achieve the desired control requirements (linearity, decoupling, etc.) [7]. The object of this work is to propose a formal algorithm for computing Σ_δ according to the definition given in [6].

This proposed algorithm can be used in order to compute the Volterra series [1] which characterize the sampled output map associated to Σ when it is driven by piecewise constant inputs. The state-dependent "coefficients" of the Volterra series are computed by means of iterative formulas. This is very useful in control theory in order to design discrete control strategies satisfying input-output objectives.

In nonlinear control theory, the computation of sophisticated nonlinear state feedback laws may introduce a non-negligible computing time which generates a delay (δm) in its digital implementation. To take this problem into account the present algorithm computes another choice of Σ_δ denoted Σ' which depends both on the control at the previous sampling time $U((k-1)\delta)$ and on the control $U((k)\delta)$. This system Σ' can be interpreted as a state predictor compensating the errors due to computing time delays.

From (B.1-2) it is clear that Σ_δ can be used to simulate the behaviour of the continuous-time system Σ submitted to a piecewise constant control. From this point of view Σ_δ can be thought of as an integration algorithm for the differential state equation A.1 which can be compared to usual numerical methods such as Runge-Kutta, Adams, etc. [2, 5]. This aspect will also be considered. From [6] the state and output (f_δ, $h(f_\delta)$) evaluation at times $k\delta$ are computed in terms of Lie series expansion, i.e. infinite expansion in powers of δ. For obvious practical reasons these series must be truncated.

It can be verified that the approximated discretization at the order k coincides with the Taylor expansion truncated at the same order of the function defining the Runge-Kutta solution.

The paper is organized as follows:

- The first section recalls the sampling method based on the fundamental formula of the Lie exponential expansion introduced in [6].

- The second section presents the four aspects of the discretization (state, output, Volterra series, state with delay) and lays stress on some mathematical recursive formulas.

- The third section presents the programme written in the symbolic language REDUCE [3]. This programme consists of some elementary procedures which are simply the computer translation of mathematical recursive formulas.

- The fourth section illustrates the usefulness of this programme with an example.

I. THE SAMPLING METHOD

Remarks

(i) For the sake of simplicity we will consider the single input-single output system (S.I.S.O.). The generalization to the multi-input - multi-output system (M.I.M.O.) is obvious.

(ii) We assume the single input U to be constant on $[k\delta.(k+1)\delta]$ for all $k \in N$.

We will denote Lf and Lg^1 the directional derivatives associated to f and g^1 respectively. They act on any analytic function $h : \mathbb{R}^N - \mathbb{R}$ as follows:

$$Lf.h\Big|_x = \sum_{I=1}^{N} \frac{\partial h}{\partial x_I} f_I(x) \qquad (I.1)$$

where $f_I(x)$ and x_I are respectively the Ith component of f(x) and x. We recall two properties verified by all vector fields.

Proposition I.A

Let f, g^1 be two C^1 vector fields with dimension N.

The composition of directional derivation gives

$$\text{(i) } Lf\,.\,Lg^1\,.h\Big|_{xo} = \sum_{i=1}^{N} f_i(x) \Big(\sum_{j=1}^{N} \frac{\partial g_j^1(x)}{\partial x_i} \frac{\partial h}{\partial x_j} + g_j^1 \frac{\partial h}{\partial x_i x_j}\Big|_{xo}\Big) \qquad (I.2)$$

(ii) If Id is the identity vector field

$$Lf.Id\Big|_{xo} = f(xo) \qquad (I.3)$$

Proofs

(i) $Lf\,.\,Lg\,.h = \sum_{i=1}^{N} f_i(x) \frac{\partial}{\partial x_i} (\sum_{j=1}^{N} g_j(x) \frac{\partial h}{\partial x_j}) \Big|_{xo}$

$$Lf\,.\,Lg\,.h = \sum_{i=1}^{N} f_i(x) (\sum_{j=1}^{N} (\frac{\partial g_j(x)}{\partial x_i} \frac{\partial h}{\partial x_j} + g_j \frac{\partial h}{\partial x_i x_j}) \Big|_{xo}) \qquad \square$$

(ii) is obvious (see for example $Lf_i\,.Id\Big|_{xo} = f_i(xo)$) $\square$

The exact discretization problem is to find a discrete-time system Σ_δ (B.1, 2) which will be equal to the continuous system Σ (A.1, 2) at each sampling time $k\delta$. Our problem is to compute $f_\delta(x_\delta(k\delta), U_\delta(k\delta))$ such that:

$$f_\delta(x_\delta(k\delta), U_\delta(k\delta)) = \int_0^\delta \dot{x}(t)\, dt + x_\delta(k\delta) \qquad (I.4)$$

where $\dot{x}$ verifies the differential equation (A.1).

The fundamental formula used throughout this paper is introduced in [6]. We recall this in the following proposition.

Proposition I.B

The solution of (I.4) at time $t = ((k+1)\delta)$ when it is initialized at $x_\delta(k\delta)$ is

$$x_\delta(k+1)\delta) = f_\delta(x_\delta(k\delta), U_\delta(k\delta)) = e^{(\delta\, L(f+g^1U))}\, Id\Big|_{x_\delta(k\delta)} \qquad (I.5)$$

where $e^{(\delta\, L(f+g^1U))}\, Id\Big|_{x_\delta(k\delta)}$ is the formal Lie exponential of $(\delta\, L(f+g^1U))$ applied to the identity function. It is computed as follows:

$$e^{(\delta L(f+g^1U))} \, Id\Big|_{x_\delta(k\delta)} = \sum_{I=0}^{\infty} \frac{\delta_I}{I!} L^I(f+g^1U) \; Id\Big|_{x_\delta(k\delta)} \qquad (I.6)$$

With the following recursive definition of the Ith iterate composition of Lie derivatives, we obtain

$$L^o(f+g^1U) \; Id\Big|_{x_\delta(k\delta)} = Id\Big|_{x_\delta(k\delta)}$$

and
$$L^I(f+g^1U) \, . \, Id\Big|_{x_\delta(k\delta)} = L(f+g^1U) \, . \, L^{(I-1)}(f+g^1U) \; Id\Big|_{x_\delta(k\delta)}$$

Proof

For the sake of simplicity we take k = 0

$$\frac{\partial f_\delta(x_\delta o, U(o))}{\partial \delta} = \frac{\partial}{\partial \delta} \Big(\sum_{I=0}^{\infty} \frac{\delta^I}{I!} L^I(f+g^1U) \; Id\Big)\Big|_{x_\delta o}$$

$$'' \qquad = L(f+g^1u)_o \, e^{(\delta L(f+g^1U))} \, Id\Big|_{x_\delta o}$$

$$'' \qquad = L(f+g^1u)_o \, f_\delta(x_\delta o, U(o)) \qquad (I.7)$$

The equation (A.1) is verified, and if $\delta = 0$ we obtain

$$e^{(\delta L(f+g^1U))} \, Id\Big|_{x_\delta o} = x_\delta o \qquad (I.8)$$

The equations (I.8) and (I.9) show that the solution of Proposition (I.A) verifies the ordinary differential equation (A.1).

Remarks

(i) For a linear system $\Sigma 1$:

$$\dot{x} = Ax + BU = f(x) + g(x) \, U$$

with $x \in \mathbb{R}^n$, $U \in \mathbb{R}$ and A.B a matrix of appropriate dimension. From Proposition I.A we have

with $x \in \mathbb{R}^n$, $U \in \mathbb{R}$ and A.B a matrix of appropriate dimension. From Proposition I.A we have

$$x(\delta) = e^{(\delta\, L(f(x)+g(x)U))}\, Id\Big|_{xo}$$

$$x(\delta) = \sum_{I=0}^{\infty} \frac{\delta^I}{I!} L^I(f+gU)\, .\, Id\Big|_{xo}$$

Performing some matrix algebra, we easily obtain

$$L(f+gU)\, .\, Id\Big|_{xo} = A\, xo + BU \text{ and } L^I(f+gU)\, .\, Id\Big|_{xo} = A^{I-1}(A\, xo + BU)$$

$$x(\delta) = xo + \sum_{I=0}^{\infty} \frac{\delta^I}{I!} A^{I-1}(A\, xo + BU)$$

$$x(\delta) = \sum_{I=0}^{\infty} \frac{\delta^I}{I!} A^I xo + \sum_{I=0}^{\infty} \frac{\delta^I}{I!} A^{I-1}\, BU$$

We thus recover the usual solution for the linear system.

$$x(\delta) = e^{\delta A}\, xo + \int_0^{\delta} e^{tA}\, BU(0)\, dt$$

(ii) The fundamental formula (I.5) gives a formal expression of the discrete state. A natural way for obtaining the output can be deduced from the following equation:

$$y_\delta(x_\delta(k\delta)) = h_\delta(x_\delta(k\delta)) = h(x_\delta(k\delta)) \tag{I.9}$$

This equation is verified when $f_\delta(x_\delta(k\delta), U(k\delta))$ is calculated exactly. This is not always possible, and we are obliged to restrict the development of the equation at a given order p in δ. It is obvious, however, that this restriction, when applied to equation (I.9), does not give the same approximation for the output. For this reason we introduce the following corollary.

Corollary I.C

$$y_\delta(k\delta) = e^{\delta\, L(f+gU)}\ h\Big|_{x_\delta((k-1)\delta)}$$

Proof

The proof is achieved as in Proposition I.B by substituting the output function h to the identity. For obvious technical reasons we now introduce the following definition:

Definition I.D

The state (respectively the output) behaviour of the continuous time sampled system Σ is approached at the order p in δ terms by the discrete time system Σ_δ if the computation of $e^{\delta\, L(f+gU)}\, Id$ (respectively $e^{\delta\, L(f+gU)}h$) is equal to the exact solution up to the order p in δ.

II. THE PRESENTATION PROGRAMME (Nonlinear Discretization)

The programme allows us to discretize a continuous nonlinear system at order p in δ for different objects.

(1) Input-state discretization (choice = 0)

(2) Input-output discretization (choice = 1)

(3) Input-output discretization in terms of the discrete Volterra series, output expansion in input terms (the first k nucleus) (choice = 2)

(4) Input-state discretization with delay (choice = 3)

Remarks

(i) The Volterra series are well-known to engineers and are only a different way of writing the corollary I.C (see [1], and for discrete time [6, 8]) where the words of the series are arranged in increasing order of the input u. We do not present the conversion here (see, for example, [6]). We simply recall the following formalism:

$$y_\delta((k+1)\delta) = \sum_{I=0}^{\infty} W_1(x_\delta(k\delta))\, U^I = \sum_{I=0}^{\infty} N_1(x_\delta(k\delta),U)$$

(where $N_1(x_\delta(k\delta),U) = W_1(x_\delta(k\delta))\, U^I$).

The kernels are given by exponential equations [6], and we present as an illustration the first three terms:

$$W_o(x(k\delta)) = e^{\delta\, Lf}\, h\Big|_{x(k\delta)} \tag{II.2}$$

$$W_1(x(k\delta)) = \int_o^{\delta} e^{(\delta-t_1)Lf}\,{}_oLg_o e^{t_1 Lf}\, h\Big|_{x(k\delta)}\, dt_1 \tag{II.3}$$

$$W_2(x(k\delta)) = \int_o^{\delta} e^{(\delta-t_1)Lf}\,{}_oLg_o e^{(t_1-t_2)Lf}\,{}_oLg_o e^{t_2 Lf}\, h\Big|_{x(k\delta)}\, dt_1 \tag{II.4}$$

(ii) In practice it is often necessary to take into account that the data acquisition instants $XS(k\delta)$ are not equal to then update control $u(k\delta+t_1)$ For this reason we have introduced the system Σ'_δ.

We write the system Σ'_δ as follows:

$$\Sigma'_\delta = \begin{cases} x'_\delta((k+1)\delta) = f'_\delta(x'_\delta(k\delta),\ U((k-1)\delta), U(k\delta)) \\ y'_\delta(k\delta) = h'_\delta(x'_\delta(k\delta)) \end{cases}$$

The system Σ'_δ which is equivalent, at the sampling time, to the continuous system Σ where $U(t) = U((k-1)\delta)$ for $t \in [\delta k, \delta k+t_1[$ and $U(t) = U(k\delta)$ for $t \in [\delta k+t_1, \delta(k+1)[$. Note that t_1 is the computing time.

The fundamental formula (I.5) gives in a straightforward way the following proposition.

Proposition II.A

$$f'_\delta(x'_\delta(k\delta), U((k-1)\delta).U(k\delta)) = e^{t_1 L(f+gU((k-1)\delta)}.e^{(\delta-t_1)L(f+gU(k\delta))}\, Id\Big|_{x'_\delta(k\delta)} \tag{II.5}$$

Proof

The proposition I.B gives

$$f'_\delta(x'_\delta(k\delta),U((k-1)\delta),U(k\delta)) = e^{(\delta-t_1)L(f+gU(k\delta))}\, Id\Big|_{x(k\delta+t_1)}$$

and, if in the corollary I.A we take $h = e^{(\delta-t_1)L(f+gU(k\delta))}\, Id$ we obtain the equation (II.5). Note that h can be seen as a function because the state space is given by $x(k\delta+t_1)$ and not by $x(k\delta)$. □

Remark

With the aid of the Cauchy product it is possible to write f'_δ as

$$f'_\delta(x'_\delta(k\delta),U((k-1)\delta),U(k\delta)) = \sum_{N=0}^{\infty}\sum_{I=0}^{N} L^{N-I}(f+gU((k-1)\delta)_0 L^I(f+gU(k\delta)_0 Id\ \delta^N m^{N-I}(1-m)^I$$

where $m = t_1/\delta$.

III. THE ORGANIZATION PROGRAMME AND SOME COMMENTS

The programme is composed of three parts:

1) A procedure subprogramme "Pronoli"
2) A data acquisition subprogramme "Donoli" (not presented in this paper)
3) The principal programme "Prinoli"

The system which we discretize is written by the user as follows. (Note that the system is now M.I.M.O.)

$$XPRIME := f(x) + \sum_{I=0}^{M} g^I(x)\ U_I$$

$$YSORTIE := h(x)$$

where XPRIME = $\dot{x}$ and YSORTIE = y. Note that x and y are the state derivatives and the output of the system defined in (A.1) and (A.2). We note

DIMX = the state space dimension
DIMU = the input space dimension
DIMY = the output space dimension

Moreover δ is written as TE and for the discretization with delay we denote U for $t \in [0.TE*m[$ and U_1 for $t \in [TE*m.TE[$.

We now focus on some specific procedures contained in "PRONOLI".

- The first procedure DERIDELIE calculates the Lie derivative L $\alpha.\beta$ with the following notations CHANVEC=α AND FONCADERI=β.

```
PROCEDURE DERIDELIE (FONCADERI.CHANVEC.DIMX);
        BEGIN
        RESU:=FOR I:=1 STEP 1 UNTIL DIMX
                         SUM (DF(FONCADERI.X(I))*CHANVEC(I);
        END;
```

Note

For the sake of simplicity we will forget the "declare" and "transfer" lines and abbreviate the STEP1 UNTIL to a colon (:).

The second procedure FLOT computes the formal exponential $e^{TS\ L\alpha}\beta$ at the given order p in TS with the following notation ORDRAP=P, FONCAL=α, CHANVEC=Lα and RESFLOT=$e^{TS\ L\alpha}\beta$.

```
PROCEDURE FLOT (FONCAL, CHANVEC, DIMX, ORDRAP, TS):
        BEGIN
        RESFLOT: = FONCAL;
        DERINT(0): = FONCAL;
        KFACT:=1;
        FOR K:=1: ORDRAP DO
              BEGIN
              DERINT(K):=DERIDELIE (DERINT(K-1), CHANVEC, DIMX) ;
```

```
        KFACT:=KFACT*K;
        RESFLOT:=RESFLOT+DERINT(K)*TS**K/KFACT;
        END;
    END;
```

The third procedure PROSERIE calculates the formal series product $e^{(TE*m)L\alpha} e^{TE(1-m)L\beta}h$ = RESPRO at the order ORDRAP in TE with $e^{TE(1-m)L\beta}h$=FONCFI, $L\alpha$=CHANF. The main advantage of PROSERIE is the deletion of the K order (K>ordrasup) word in TE. The terms of order K are wrong because the two series are truncated before this order.

```
PROCEDURE PROSERIE (FONCFI.CHANF.DIMX.ORDRAP):
    BEGIN
    RESPRO:=FLOT(FONCFI.CHANF.DIMX.ORDRAP,(TE*m);
    DERIVEE:=DF(RESPRO.TE.ORDRAP+1)
    FOR K:=1 : ORDRAP+1 DO
        BEGIN
        DERIVEE=INT (DERIVEE, TE);
        END:
    RESPRO=RESPRO-DERIVEE:
    END:
```

The fourth procedure CALNOY calculates the first k kernels of the input Volterra series. We do not give this procedure here, as it is just a rewriting of mathematical formulas.

With respect to the principal subprogramme, we give here only a logic diagram.

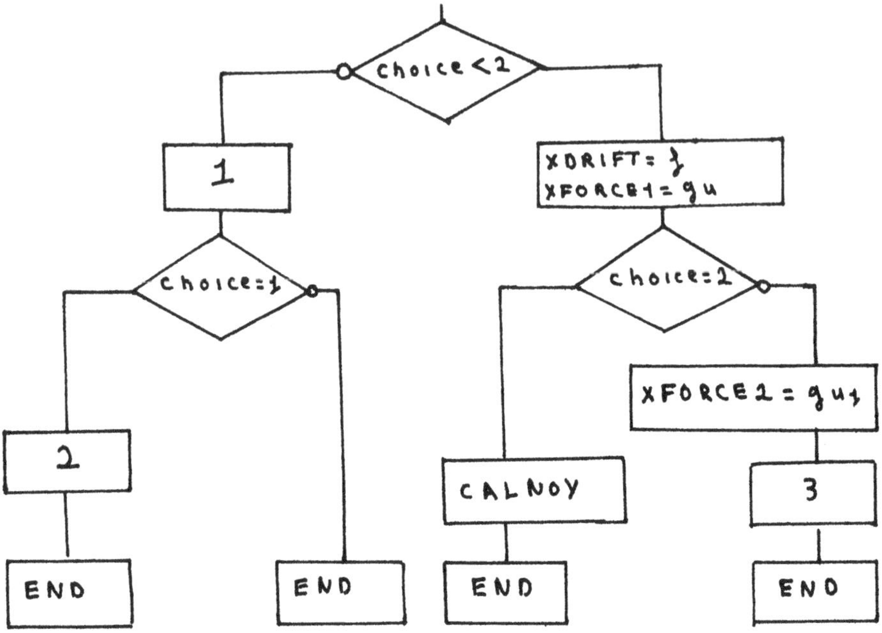

- Block 1 computes the equation (I.5) at one fixed order (k) in δ terms. This block solves the input-state discretization.

- Block 2 computes the equation (I.10) at one fixed order (k) in δ terms. This block solves the input-output discretization.

- Block 3 computes the equation (II.5) at one fixed order (k) in δ terms. This block is composed of two parts. The first is similar to block 1 with a sampling period equal to δm. The second part is similar to block 2 with the solution of the first part as input function (h) and a sampling period equal to $\delta(1\text{-}m)$. This block solves the input-state discretization.

IV. EXAMPLE AND COMMENTS

Let us now consider the system given in reference [6]. The continuous-time system Σ is the following:

$$\dot{x}_1(t) = u(t)$$
$$\dot{x}_2(t) = u(t) + x_1(t)\,(1+x_1(t))$$
$$y_1(t) = x_2(t) + x_1(t)^2$$

Looking for a third order approximated discretization of the state evolution, we obtain:

$$x_1((k+1)\delta) = \delta\, U(k\delta) + x_1(k\delta)$$
$$x_2((k+1)\delta) = (\delta^3 2\, U^2(k\delta) + \delta^2(3\, U(k\delta)\,(2\, x_1(k\delta) + 1)) + \delta\, 6(U(k\delta) + x_1(k\delta)\,(x_1(k\delta) + 1)) + 6\, x_2(k\delta))/6$$

It should be noted that the linearity in U is preserved until the second order in δ. This is convenient for feedback control purposes. The cpu time for this computation is 15 seconds on a μ-VAX. We note that this time takes into account the memorization, reading procedures time, etc.

CONCLUSION

The main aim of this short introduction to the DINOLI programme was to show that this way of computer discretization is very simple, and is only a rewriting of recursive mathematical equations (see paragraph II). The main advantages of the recursive equations and symbolic language REDUCE are to give a very short and legible programme. The programme allows us to compute discrete-time models. This is used to design a discrete control law. The "complicated" computation of the control law for nonlinear systems forces us to use a computer in the feedback. It is a discrete element and the link problem between continuous-time systems and discrete-time control can be solved by discretization of the system.

ACKNOWLEDGEMENTS

I wish to thank Dorothée Normand-Cyrot and Salvatore Monaco for many discussions and suggestions during the course of this study.

This work was supported by a European MRES grant from France.

REFERENCES

[1] M. FLIESS, M. LAMNABHI, F. LAMNABHI-LAGARRIGUE, "An algebraic approach to nonlinear functional expansions", IEEE Trans. Circuits Systems, v.29, p.554-570, 1983.

[2] S. GODOUNOV, V. RIABENKI, "Schémas aux différences", Editions MIR, 1977.

[3] A.C. HEARN, "REDUCE user's manual", The Rand Corporation, Santa Monica, 1984.

[4] A. ISIDORI, "Nonlinear control systems: an introduction", Lecture Notes in Control and Information Sciences, v.72, Springer-Verlag, 1985.

[5] G. MARCHOUK, V. SHAYDOUROV, "Raffinement des solutions des schémas aux différences", Editions MIR, 1983.

[6] S. MONACO, D. NORMAND-CYROT, "On the sampling of the linear control system", 24th IEEE CDC, Fort Lauderdale, p.1457-1482, 1984.

[7] S. MONACO, D. NORMAND-CYROT, S. STORNELLI, "On the linearizing feedback in nonlinear sampled data control schemes", 25th IEEE CDC, p.2056-2060, Athens, 1986.

[8] S. MONACO, D. NORMAND-CYROT, "Finite Volterra-series realisations and input-output approximations of nonlinear discrete-time systems", Int. J. Control, v.45, n°5, p.1771-1787, 1987.

Decoding of Generalized Concatenated Codes

Martin Bossert*

Abstract

A decoding algorithm for generalized concatenated codes is described which is a modification of the one proposed in [6]. It is shown that the decoding capability is superior and the complexity is less compared to the previous known decoding algorithms for generalized concatenated codes.

1 Introduction

Generalized concatenated (GC) codes have been introduced by Zinoviev [5] in 1976. The concept of GC allows the construction of codes to correct error bursts and independent errors simultaniously [6] and codes for unequal error protection, which can also be decoded in case of error bursts and independent errors [7]. Furthermore, in a recent paper (from 1987) Zinoviev, Zyablov and Portnoy studied in [8] generalized concatenation of codes over the euclidian space. A special case of this are *coded modulation* systems, see also [2].

The decoding of GC codes to half the minimum distance is due to Blokh and Zyablov and was improved by Zinoviev [6] (see also for earlier references). The decoding procedure for GC codes, which are based on a s–th order partition of the inner code, consists of s steps. Every step needs a decoder for an inner code and a decoder with error erasure correcting capability for the corresponding outer code. We first describe the concept of GC codes and then give the decoding procedure which consists of two different algorithms. One for the first step and one for all consecutive steps denoted by GCD–1 and GCD–I, respectively. The algorithm GCD–1 is a generalization of the algorithm used in [6] for all s steps. The result of the decoding of the previous steps gives some additional information to be used in the actual step. We show that introducing GCD–I which uses the information of the previous step will improve the decoding capability and decrease the decoding complexity compared to the previous algorithms.

2 Generalized Concatenated Codes

We will give only a short description of GC codes. For more details we refer e.g. to [6] and the literature cited there.

A code C will be denoted by $C(q; n, M, d)$, where n is the length, M the number of code words, d the minimum (Hamming–) distance and the alphabeth size is q. We say a code

[1]The author is with the AEG Olympia AG, Ulm, West Germany. The research was supported by the Deutsche Forschungsgemeinschaft (DFG).

$\mathcal{B}^{(1)}(q; n_b, M_b^{(1)}, d_b^{(1)})$ is partitioned into subcodes $\mathcal{B}_i^{(2)}(q; n_b, M_b^{(2)}, d_b^{(2)})$, $i = 1, 2, \ldots, q^{\mu_1}$ if the codes $\mathcal{B}_i^{(2)}$ are disjoint and their union is the code $\mathcal{B}^{(1)}$, i.e.:

$$\mathcal{B}^{(1)} = \bigcup_{i=1}^{q^{\mu_1}} \mathcal{B}_i^{(2)} \quad .$$

Every subcode $\mathcal{B}_i^{(2)}$ can be partitioned again into subcodes $\mathcal{B}_{i,j}^{(3)}$, $j = 1, 2, \ldots, q^{\mu_2}$, and so on. Thus, a s–th order partition of a code $\mathcal{B}^{(1)}(q; n_b, M_b^{(1)}, d_b^{(1)})$ can be described by an enumeration $i_1, i_2, \ldots, i_s$ and the mapping of the numbers onto the vectors $\mathbf{a}^{(i)} \in GF(q)^{\mu_i}$. The number i_1 enumerates the subcode of $\mathcal{B}^{(1)}$, i.e. $\mathcal{B}_{i_1}^{(2)}$ and i_2 enumerates the subcode of $\mathcal{B}_{i_1}^{(2)}$, i.e. $\mathcal{B}_{i_1,i_2}^{(3)}$ and so on. Finally, the number i_s determines a codeword of the subcode $\mathcal{B}_{i_1,i_2,\ldots,i_{s-1}}^{(s-1)}$. Therefore, every codeword is completely described by the set of numbers $i_1, i_2, \ldots, i_s$. The minimum distances $d_b^{(i)}$, $i = 1, \ldots, s$ of the partitions are known. Notice, that we have $d_b^{(i)} \geq d_b^{(i-1)}$, $i = 2, \ldots, s$. The $\mathcal{B}^{(j)}$ are called the inner codes.

In case of linear codes a partition can be described with cosets (confirm [4, p.15]). Furthermore, a method using direct sums can be derived, which has the advantage that partitions can be calculated (without decoding procedures) instead of beeing tabulated (see [2]).

The idea of GC is now to take a set of vectors $\mathbf{a}_1^{(i)}, \mathbf{a}_2^{(i)}, \ldots, \mathbf{a}_{k_i}^{(i)}$ and protect this set with a so called outer code $A^{(i)}(q^{\mu_i}; n_a, M_a^{(i)}, d_a^{(i)})$, where k_i depends on the size $M_a^{(i)}$ of the outer code. Given s outer code words arranged in a matrix with entries $\mathbf{a}_j^{(i)} \in GF(q)^{\mu_i}$, $j = 1, 2, \ldots, n_a$, $i = 1, 2, \ldots, s$ (s is the order of the partition) as follows:

$$\begin{pmatrix} \mathbf{a}_1^{(1)} & \mathbf{a}_1^{(2)} & \cdots & \mathbf{a}_1^{(s)} \\ \mathbf{a}_2^{(1)} & \mathbf{a}_2^{(2)} & \cdots & \mathbf{a}_2^{(s)} \\ \vdots & \vdots & \ddots & \vdots \\ \mathbf{a}_{n_a}^{(1)} & \mathbf{a}_{n_a}^{(2)} & \cdots & \mathbf{a}_{n_a}^{(s)} \end{pmatrix} \tag{1}$$

The matrix consists of n_a rows of the form

$$\left(\mathbf{a}_j^{(1)}, \mathbf{a}_j^{(2)}, \mathbf{a}_j^{(3)}, \ldots, \mathbf{a}_j^{(s)}\right) \quad .$$

Any row determines a code word of $\mathcal{B}^{(1)}$ uniquely according to the partition described above. After mapping any row onto a code word of the inner code $\mathcal{B}^{(1)}$ we have a code word of the GC code $\mathcal{C}$ which is a $(n_a \times n_b)$ matrix with entries from $GF(q)$.

$$\begin{pmatrix} c_{11} & c_{12} & \cdots & c_{1,n_b} \\ c_{21} & c_{22} & \cdots & c_{2,n_b} \\ \vdots & \vdots & \ddots & \vdots \\ c_{n_a,1} & c_{n_a,2} & \cdots & c_{n_a,n_b} \end{pmatrix} \tag{2}$$

In most cases the true minimum distance of GC codes is not known but it can be bounded by

$$d \geq \min_{i=1,\ldots,s} \left\{ d_b^{(i)} \cdot d_a^{(i)} \right\} \quad .$$

Assume we have two different code words $\mathbf{c} \neq \mathbf{c}'$ of a GC code. The matrices eq. (1) must differ in at least one column, say j. $\mathcal{A}^{(j)}$ has minimum distance $d_a^{(j)}$ and the matrices eq. (2) for $\mathbf{c}$ and $\mathbf{c}'$ differ in at least $d_a^{(j)}$ rows. These rows differ in at least $d_b^{(j)}$ symbols which gives the bound for the minimum distance of a GC code (confirm also [5]).

Definition 1 *A generalized concatenated code (GC code) $C(q;n,M,d)$ consists of s outer codes $\mathcal{A}^{(i)}(q^{\mu_i};n_a,M_a^{(i)},d_a^{(i)})$ and a s-th order partition of the inner code $\mathcal{B}^{(1)}(q;n_b,M_b^{(1)},d_b^{(1)})$. C is a code over the alphabet q with length $n = n_a \cdot n_b$, size $M = \prod_{i=1}^{s} M_a^{(i)}$ and minimum distance*

$$d \geq \min_{i=1,\ldots,s} \left\{ d_b^{(i)} \cdot d_a^{(i)} \right\} \quad .$$

Example 1 *As inner code we use a binary parity check code of length 4. Then we obtain:*

$$\begin{aligned}
\mathcal{B}_0^{(2)} &= \{(0000),(1111)\} \\
\mathcal{B}_1^{(2)} &= \{(0011),(1100)\} \\
\mathcal{B}_2^{(2)} &= \{(0101),(1010)\} \\
\mathcal{B}_3^{(2)} &= \{(0110),(1001)\}
\end{aligned}$$

$\mathcal{B}_i^{(2)}$, $i = 0,\ldots,3$ are repetition codes with minimum distance $d_b^{(2)} = 4$. Their union $\mathcal{B}^{(1)}$ is a parity check code with distance 2. Any pair of numbers i_1, i_2, $i_1 \in \{0,1,2,3\}$, $i_2 \in \{0,1\}$ determines a codeword of $\mathcal{B}^{(1)}$. If we encode a set of numbers i_1 by a repetition code $A^{(1)}(4;8,4,8)$ and i_2 by an extended Hamming code $A^{(2)}(2;8,2^4,4)$ we get a GC code

$$C(2;32,2^6,16)\,,$$

which has the same parameters as the first order Reed Muller code of length 32.

Before we describe the decoding algorithm we want to point out that a code with unequal error protection is obtained by choosing a sequence of the product $d_b^{(i)} \cdot d_a^{(i)}$ decreasing with i. The information symbols encoded with the outer code $\mathcal{A}^{(i)}$ are then protected against at least $(d_b^{(i)} \cdot d_a^{(i)})/2$ errors. The decoding procedure remains exactly the same. Furthermore, the inner code can be a code over the euclidean space and again the decoding procedure stays essentially the same (see [2]).

3 Decoding Algorithm GCD–1

Let $C(q,n,M,d)$ be a GC code constructed by a s–th order partition of the inner code. A code word of the GC code C will be denoted by $\mathbf{c} \in C$ and consists of n_a code words of the inner code $\mathcal{B}^{(1)}$.

Assuming a q–ary symmetric channel, every row $\mathbf{b}_j^{(1)}$ of the matrix $\mathbf{c}$ is distorted by an error $\mathbf{e}_j \in GF(q)^{n_b}$. The received rows are $\mathbf{r}_j = \mathbf{b}_j^{(1)} + \mathbf{e}_j$, $j = 1,2,\ldots,n_a$. In the first step $\mathbf{r}_j^{(1)} = \mathbf{b}_j^{(1)} + \mathbf{e}_j$, $j = 1,2,\ldots,n_a$ has to be decoded according to the inner code $\mathcal{B}^{(1)}$. The result of this decoding is an estimate of $\mathbf{b}_j^{(1)}$, i.e.

$$\hat{\mathbf{b}}_j^{(1)} = \begin{cases} \mathbf{b}_j^{(1)} + \mathbf{e}_j - \mathbf{f}_j \\ \otimes \end{cases} \tag{3}$$

where the symbol $\otimes$ denotes an erasure. Every $\hat{\mathbf{b}}_j^{(1)} \neq \otimes$ determines one symbol $\hat{\mathbf{a}}_j^{(1)}$ of the alphabet $GF(q)^{\mu_1}$ of the outer code $\mathcal{A}^{(1)}$. Since $\mathbf{a}_j^{(1)}$ are the transmitted symbols we have

$$\hat{\mathbf{a}}_j = \begin{cases} \mathbf{a}_j & \text{if } \mathbf{f}_j = \mathbf{e}_j \quad \text{(correct decoding)} \\ \neq \mathbf{a}_j & \text{if } \mathbf{f}_j \neq \mathbf{e}_j \wedge \mathbf{b}_j^{(1)} + \mathbf{e}_j - \mathbf{f}_j \in \mathcal{B}^{(1)} \quad \text{(decoding error)} \\ \otimes & \text{if } \hat{\mathbf{b}}_j = \otimes \quad \text{(decoding failure)} \end{cases} \tag{4}$$

We define sets of row numbers which have been decoded with the same number of errors (wt(...): Hamming weight):

$$\mathcal{J}_k := \{l \mid \text{wt}(\mathbf{f}_l) = k\,, \quad l = 1,2,\ldots,n_a\}\,, \quad k = 1,2,\ldots,\nu$$

and the erasure set

$$\mathcal{J}_\otimes := \{l \mid \hat{\mathbf{a}}_l = \otimes\,, \quad l = 1,2,\ldots,n_a\} \quad .$$

Note that ν ist not specified and depends on the output of the decoder. The set of wrong decoded rows (decoding errors) is

$$\mathcal{E} := \{l \mid \mathbf{f}_l \neq \mathbf{e}_l, \hat{\mathbf{a}}_l \neq \otimes\,, \quad l = 1,2,\ldots,n_a\} \tag{5}$$

and t_k and s_k are the numbers of decoding errors and correct decoded rows in the set $\mathcal{J}_k$, respectively

$$\begin{aligned} t_k &:= |\mathcal{J}_k \cap \mathcal{E}| \\ s_k &:= |\mathcal{J}_k| - t_k \end{aligned} \qquad k = 0,1,\ldots,\nu \quad . \tag{6}$$

A union of sets $\mathcal{J}_k$ is defined by:

$$\mathcal{X}_j := \bigcup_{l=j}^{\nu} \mathcal{J}_l, \;\; j = \nu,\; \nu-1,\ldots,0 \quad . \tag{7}$$

We will need also the number y describing the maximum number of erasures which can be declared according to the minimum distance $d_a^{(1)}$:

$$y: \; |\mathcal{X}_y| < d_a^{(1)}\,, \qquad |\mathcal{X}_{y-1}| \geq d_a^{(1)} \quad . \tag{8}$$

Note that:

$$\underset{j \in \mathcal{J}_k}{\forall} : \quad \begin{cases} \text{wt}(\mathbf{e}_j) \geq d_b^{(1)} - k\,, & j \in \mathcal{E} \\ \text{wt}(\mathbf{e}_j) = k\,, & j \notin \mathcal{E} \end{cases} \quad . \tag{9}$$

In order to prove that the algorithm GCD–1 (figure 1) can correct all channel errors up to half the minimum distance we will follow the main ideas of [6].

Theorem 1 *If* $\sum_{j=1}^{n_a} \text{wt}(\mathbf{e}_j) < \frac{d_a^{(1)} \cdot d_b^{(1)}}{2}$ *then for some* $j \in \{y, y+1, \ldots, \nu\}$ *:*

$$2 \cdot \left(|\mathcal{E}| - \sum_{i=j}^{\nu} t_i \right) + |\mathcal{X}_j| < d_a^{(1)}$$

where $\mathcal{X}_j$*,* $\mathcal{E}$ *and* t_i *are defined in equations (7), (5), and (6), respectively.*

Algorithm GCD–1

received: $r_j,\ j = 1, 2, \ldots, n_a$
obtain: $\hat{\mathbf{a}}_j,\ j = 1, 2, \ldots, n_a$
ν
$\mathcal{J}_k,\ k = 1, 2, \ldots, \nu,\ \mathcal{J}_\otimes$
$\mathcal{X}_j,\ j = \nu, \nu - 1, \ldots, y$
initialize: $j = \nu + 1,\ X = n_a \cdot d_b^{(1)},\ \mathcal{X}_{\nu+1} = \{\,\}$

Step 1: $\mathcal{R} = \mathcal{X}_j + \mathcal{J}_\otimes$
IF $j < y$ THEN Step 5

Step 2: $$\mathbf{b}_j = \begin{cases} \hat{\mathbf{a}}_j, & j \notin \mathcal{R} \\ \otimes, & j \in \mathcal{R} \end{cases} \qquad j = 1, 2, \ldots, n_a$$

Step 3: decode $(\mathbf{b}_1, \mathbf{b}_2, \ldots, \mathbf{b}_{n_a})$
according to the outer code $\mathcal{A}^{(1)}$ as
$(\hat{\mathbf{b}}_1, \hat{\mathbf{b}}_2, \ldots, \hat{\mathbf{b}}_{n_a})$

Step 4: $$\mathcal{T}_j := \left\{ l \,\middle|\, \hat{\mathbf{a}}_l \neq \hat{\mathbf{b}}_l, \hat{\mathbf{a}}_l \neq \otimes, \quad l = 1, 2, \ldots, n_a \right\}$$
$$\Gamma_j := \sum_{i=1}^{n_a} \gamma_i$$
where
$$\gamma_l = \begin{cases} k, & l \notin \mathcal{J}_k \cap \mathcal{T}_j \\ \max\left\{ d_b^{(1)} - k, \frac{d_b^{(1)}}{2} \right\}, & l \in \mathcal{J}_k \cap \mathcal{T}_j \end{cases} , k = 0, 1, \ldots, \nu$$
If $\Gamma_j < X$ THEN $X = \Gamma_j$, $\mathbf{a}^* = \hat{\mathbf{b}}$, $L = \mathcal{T}_j$
IF $X < \frac{d_b^{(1)}}{2} \cdot (d_a^{(1)} - |\mathcal{J}_\otimes|)$ GOTO Step 5
$j := j - 1$ GOTO Step 1

Step 5: Decoding decision is $\mathbf{a}^*, X, L$

Figure 1: The Decoding Algorithm GCD-1

Proof: Assume that for any step, two times the number of errors plus the number of erasures is larger than the minimum distance, i.e. for every j we have:

$$2 \cdot |\mathcal{E}| + \sum_{i=j}^{\nu}(s_i - t_i) \geq d_a^{(1)} - |\mathcal{J}_\otimes| \quad ,$$

or

$$2 \cdot \sum_{i=0}^{j-1} t_i + \sum_{i=j}^{\nu}(t_i + s_i) \geq d_a^{(1)} - |\mathcal{J}_\otimes| = d_a \quad .$$

The estimation of the channel errors gives (without those which have caused erasures)

$$\sum_{i=0}^{\nu} \sum_{j \in \mathcal{J}_i} \mathrm{wt}(\mathbf{e}_j) \geq \sum_{i=0}^{\nu} t_i \cdot (d_b^{(1)} - i) + \sum_{i=1}^{\nu} s_i \cdot i \quad .$$

We can write

$$t_i = \sum_{l=i}^{\nu} t_l - \sum_{l=i+1}^{\nu} t_l$$

and

$$s_i = \sum_{l=i}^{\nu} s_l - \sum_{l=i+1}^{\nu} s_l \quad .$$

With this the error estimation becomes

$$\sum_{i=0}^{\nu} \sum_{j \in \mathcal{J}_i} \mathrm{wt}(\mathbf{e}_j) \geq \sum_{i=0}^{\nu} \left(\sum_{l=i}^{\nu} t_l - \sum_{l=i+1}^{\nu} t_l \right) (d_b^{(1)} - i) + \sum_{i=1}^{\nu} \left(\sum_{l=i}^{\nu} s_l - \sum_{l=i+1}^{\nu} s_l \right) i \quad .$$

Reformulation yields

$$\sum_{i=0}^{\nu} \sum_{j \in \mathcal{J}_i} \mathrm{wt}(\mathbf{e}_j) \geq \sum_{i=0}^{\nu} \left(2 \cdot \sum_{i=0}^{j-1} t_i + \sum_{i=j}^{\nu}(t_i + s_i) \right) \quad .$$

Now we use the assumption and obtain

$$\sum_{i=0}^{\nu} \sum_{j \in \mathcal{J}_i} \mathrm{wt}(\mathbf{e}_j) \geq \frac{d_a \cdot d_b^{(1)}}{2}$$

or

$$\sum_{i=1}^{n_a} \mathrm{wt}(\mathbf{e}_i) \geq \sum_{i=0}^{\nu} \sum_{j \in \mathcal{J}_i} \mathrm{wt}(\mathbf{e}_j) + |\mathcal{J}_\otimes| \frac{d_b^{(1)}}{2} \geq \frac{d_a \cdot d_b^{(1)}}{2} + |\mathcal{J}_\otimes| \frac{d_b^{(1)}}{2} = \frac{d_a^{(1)} \cdot d_b^{(1)}}{2} \quad ,$$

which is a contradiction. ∎

There are many channel error patterns of weight larger than half the minimum distance which can also be corrected, especially in case of error bursts (see [6]).

In the algorithms described in [3] and [6] for ν holds: $\nu = d_b^{(1)}/2$. This means that if the inner decoder corrects one received row with an error of weight larger than half the minimum distance this row is viewed as a decoding failure immediately. It is obvious that this restriction may cause a decoding failure in case there exists an inner decoder that can decode also errors beyond half the minimum distance, or if a maximum likelihood decoder can be used. If a bounded minimum distance decoding scheme is used the outcome of the decoder will fullfill this restriction anyhow. However, the case of the algorithms in [3] and [6] are included in GCD–1,

$$\underbrace{\mathcal{J}_{\otimes}, \mathcal{J}_{\nu}, \ldots, \mathcal{J}_{j},}_{\text{erasure set of [3] and [6]}} \qquad \ldots, \mathcal{J}_0$$

since there is a step j from which both algorithms are identical.

4 Decoding Algorithm GCD–I

From the first (or previous) step we know the sets $J_{\otimes}, J_k, k = 0, 1, \ldots \nu$ and the set L. The set L consists of all row numbers, where a wrong decoding of the inner code was detected by the outer code. A decoded code word $\mathbf{a}^{(i)} = (a_1^{(i)}, a_2^{(i)}, \ldots, a_{n_a}^{(i)})$ determines the subcodes which are to be used to decode the rows $\mathbf{r}_j^{(i)}, j = 1, \ldots, n$ in the following step $i+1$; $(\mathbf{a}_j^{(i)} \rightarrow \mathcal{B}^{(i+1)})$.

Every row number which is element of L was a decoding error of the previous inner decoder. Thus we can state:

Lemma 1 *Let $d_b^{(i)} = d_b^{(i-1)} + \Delta$ be the minimum distance of the inner code $\mathcal{B}^{(i)}$. The decoding of all rows $\mathbf{r}_j^{(i)}$ might lead to a decoding error, whenever:*

$$j \in \mathcal{J}_k \cap L, k < d_b^{(i)} - \Delta - \frac{d_b^{(i)} - 1}{2} \qquad .$$

Proof: From the previous decoding we know:

$$\underset{j \in \mathcal{J}_k \cap L}{\forall} \quad : \quad \mathrm{wt}(\mathbf{e}_j) \geq d_b^{(i-1)} - k \qquad .$$

Thus, only if $d_b^{(i-1)} - k \leq \left\lfloor \frac{d_b^{(i)} - 1}{2} \right\rfloor$ correct decoding might be possible.

Therefore $k \geq d_b^{(i)} - \Delta - \left\lfloor \frac{d_b^{(i)} - 1}{2} \right\rfloor$. ∎

This fact has two consequences:

First: only such rows where decoding might be possible according to Lemma 1 are decoded with the inner code. Otherwise they are declared as an erasure.

Second: only such decoding outcomes which fullfill the condition predicted by Lemma 3 will be accepted. Otherwise they are declared as an erasure.

The decoding procedure in [6] decodes all rows $j = 1, 2, \ldots, n_a$ in every step. This is due to the fact that although the decoding of a row was wrong, the symbol of the previous outer code was correct, i.e. the decoding error of the inner code was not detected. We will reduce decoding complexity by the following statement:

Algorithm GCD–I

decode: $\mathbf{r}_j$, $\mathcal{M} = \{j \mid j \in \mathcal{J}'_{\otimes} \cup \{\mathcal{J}'_k \cap L'\},\ k \geq \lambda\}$

result is $\mathbf{f}_j$ or $\otimes$ and ν

obtain: $\mathcal{J}_k = \mathcal{J}'_k \setminus \{\mathcal{J}'_k \cap \mathcal{M}\}$

$\mathcal{J}_{\otimes} = \{j \in \{\mathcal{J}'_k \cap \mathrm{L}',\ k < \lambda\}\}$

from decoding: for $j \in \mathcal{M}$

$\mathcal{J}_k = \mathcal{J}_k \cup \left\{j \,\middle|\, \mathrm{wt}(\mathbf{f}_j) = k \geq d_b^{(i-1)} - v, v : j \in \mathcal{J}'_v\right\}$

$\mathcal{J}_{\otimes} = \mathcal{J}_{\otimes} \cup \left\{j \,\middle|\, \mathbf{r}_j \rightarrow \otimes,\ \mathrm{wt}(\mathbf{f}_j) = k < d_b^{(i-1)} - v, v : j \in \mathcal{J}'_v\right\}$

initialize: $j = \nu + 1$, $X = n_a \cdot d_b^{(i)}$, $\mathcal{X}_{\nu+1} = \{\ \}$

Step 1: $R = \mathcal{X}_j$

IF $j < y$ THEN Step 5

Step 2: $\mathbf{b}_j = \begin{cases} \hat{\mathbf{a}}_j, & j \notin \mathcal{R} \\ \otimes, & j \in \mathcal{R} \end{cases}, j = 1, 2, \ldots, n_a$

Step 3: decode $(\mathbf{b}_1, \mathbf{b}_2, \ldots, \mathbf{b}_{n_a})$

according to the outer code $\mathcal{A}^{(i)}$ as

$(\hat{\mathbf{b}}_1, \hat{\mathbf{b}}_2, \ldots, \hat{\mathbf{b}}_{n_a})$

Step 4: $\mathcal{T}_j := \left\{l \,\middle|\, \hat{\mathbf{a}}_l \neq \hat{\mathbf{b}}_l, \hat{\mathbf{a}}_l \neq \otimes, l = 1, 2, \ldots, n_a\right\}$

$\Gamma_j := \sum_{i=1}^{n_a} \gamma_i$ where

$\gamma_l = \begin{cases} k, & l \notin \mathcal{J}_k \cap \mathcal{T}_j \\ \max\left\{d_b^{(i)} - k, \frac{d_b^{(i)}}{2}\right\} & l \in \mathcal{J}_k \cap \mathcal{T}_j \end{cases}, k = 0, 1, \ldots, \nu$

IF $\Gamma_j < X$ THEN $X = \Gamma_j$, $\mathbf{a}^* = \hat{\mathbf{b}}$, $L = \mathcal{T}_j$

IF $X < \frac{d_b^{(i)}}{2}(d_{a1} - |\mathcal{J}_{\otimes}|)$ GOTO Step 5

ELSE $j := j - 1$ GOTO Step 1

Step 5: Decoding decision is $\mathbf{a}^*$, X, L

Figure 2: The Decoding Algorithm GCD-I

Proposition 1 *Let L be the set of row numbers where a detected decoding error occured. The decided errors of the previous decoding of the inner code, say $\mathcal{B}^{(i-1)}$ are $\mathbf{f}_j, j = 1,2,\ldots,n_a$. The decoding according to $\mathcal{B}^{(i)}$ will give $\mathbf{h}_j$, $j = 1,2,\ldots,n_a$ as decoding decisions. Then*

$$\begin{aligned} j \notin L &\implies \mathbf{h}_j = \mathbf{f}_j \\ j \in L &\implies \mathbf{h}_j \neq \mathbf{f}_j \end{aligned}$$

Proof: First, note that if $\mathbf{f}_j$ is an undetected decoding error, then $\mathbf{e}_j - \mathbf{f}_j \in \mathcal{B}^{(2)}$.

- If $j \in L$: $\mathbf{e}_j - \mathbf{f}_j \in \mathcal{B}^{(1)}$ but $\mathbf{e}_j - \mathbf{f}_j \notin \mathcal{B}^{(2)}$. Also $\mathbf{e}_j - \mathbf{h}_j \in \mathcal{B}^{(2)}$ thus $\mathbf{h}_j \neq \mathbf{f}_j$.
- If $j \notin L$: either $\mathbf{e}_j = \mathbf{f}_j = \mathbf{h}_j$ or $\mathbf{e}_j - \mathbf{f}_j \in \mathcal{B}^{(1)}$.

So in this case either the decoding was correct or the decoder of $\mathcal{B}^{(1)}$ was a decoder for $\mathcal{B}^{(2)}$, thus $\mathbf{h}_j = \mathbf{f}_j$. ∎

Therefore, in step i only those rows with numbers which are elements of L and those which are decoding failures have to be decoded. Clearly, only less than $d_a^{(i-1)}$ rows that have to be decoded in step i.

If the inner decoder is able to decode beyond half the minimum distance the result of Lemma 1 has to be modified. According to the decoders for $\mathcal{B}^{(i-1)}$ and the subcodes $\mathcal{B}^{(i)}$ we define:

$$\lambda = d_b^{(i-1)} - \left\lfloor \frac{d_b^{(i)}}{2} \right\rfloor - \text{const} \quad .$$

The value of the constant λ depends on the error correcting capability of the inner code. We will denote the outcome of the previous step by $\mathcal{J}'_{\otimes}$, $\mathcal{J}'_k$, $k = 0,1,\ldots,\nu'$ and L', respectively. We will also assume that the subcode, according to which we decode now the rows $\mathbf{r}_j$, was determined. The algorithm GCD–I is shown in figure 2.

In order to describe GCD we did not need to specify the decoding algorithms for the inner codes. Any algorithm which guarantees BMD could be used. The principle of GCD could even be used in case of convolutional inner codes. Thereby the reliability could be dependent on the number of corrected bits in a symbol of the outer code.

Furthermore, it should be noted that in case of a GC code with short inner codes table–look–up methods might be used. This guarantees maximum likelihood decoding as well as low complexity and easy implementation. Note that for other than GC codes with equivalent length and distance it is usually impossible to use table–look–up methods. In [5] various examples of such GC codes can be found. Almost trivial inner codes up to length 16 are used to construct GC codes up to length 200. Many of these codes can be found in the table of best known codes in [4, pp.675].

These codes are to be used in practice !

References

[1] **Blokh, E.L., Zyablov, V.V.**, *Coding of Generalized Cascade Codes.* Problemy Peredachi Informatsii, Vol. 10, 1974, pp. 45–50.

[2] **Bossert, M.**, *Concatenation of Block Codes.* DFG Report, 1988.

[3] **Ericson, T.**, *Concatenated Codes - Principles and Possibilities* AAECC–4, Karlsruhe 1986.

[4] **MacWilliams,F.J., Sloane,N.J.A.**, *The Theory of Error–Correcting Codes.* North Holland Publishing Comp., 1981.

[5] **Zinoviev, V.A.**, *Generalized Cascade Codes.* Problemy Peredachi Informatsii, Vol. 12, 1976, pp. 5–15.

[6] **Zinoviev, V.A.**, *Generalized Concatenated Codes for Channels with Error Bursts and Independent Errors.* Problemy Peredachi Informatsii, Vol. 17, 1981, pp. 53–56.

[7] **Zinoviev, V.A., Zyablov, V.V.**, *Codes with Unequal Error Protection of Symbols.* Problemy Peredachi Informatsii, Vol. 15, 1979, pp. 50–60.

[8] **Zinoviev, V.A., Zyablov, V.V., Portnoy, S.L.**, *Concatenated Methods for Construction and Decoding of Codes in Euclidean Space.* Preprint USSR Academy of Science, Institute for Problems of Information Transmission, 1987.

Backtrack Searching in the Presence of Symmetry

Cynthia A. Brown* and Larry Finkelstein*

College of Computer Science
Northeastern University
360 Huntington Ave.
Boston, Mass. 02115

Paul Walton Purdom, Jr.

Computer Science Dept.
Indiana University
101 Lindley Hall
Bloomingon, In. 47405

* Work supported in part by the National Science Foundation under grant number DCR-8603293.

Abstract. Methods from computational group theory are used to improve the speed of backtrack searching on problems with symmetry. The symmetry testing algorithm, which is similar to a color automorphism algorithm, takes the symmetry group as input and uses it to avoid searching equivalent portions of the search space. The algorithm permits dynamic search rearrangement in conjunction with symmetry testing. Experimental results confirm that the algorithm saves a considerable amount of time on some search problems.

1. Introduction

The development of efficient algorithms for computing basic algebraic primitives makes it possible to exploit algebraic structure in problems which were formerly solved by purely combinatorial methods. In this paper we present a general algorithm for backtrack search over domains where the candidate solutions are invariant under a group of symmetries. Given a description of the symmetry group, the algorithm generates the tests needed to avoid searching portions of the the search space that are equivalent under the symmetries. There are many applications of the method, such as searching for matrix multiplication algorithms over a finite field, and the embedding of combinatorial objects with automorphisms. More details, along with proofs of the algorithm's correctness, may be found in the longer version [3].

The matrix multiplication problem provides an example of a search problem with a rich symmetry group. One can search for matrix multiplication algorithms by searching for solutions to the equation

$$\sum_{1\le t\le L} x_{ijt}y_{klt}z_{mnt} = \delta_{jk}\delta_{lm}\delta_{ni}, \tag{1}$$

for $1 \le i, j, k, l, m, n \le N$, where N is the size of the matrices and L is the number of multiplies [18, 19]. (This equation actually yields an algorithm for computing the transpose of the matrix product.) When modulo 2 arithmetic is used, the $N = 2$, $L = 7$ problem has a symmetry group [12] with size $7! \cdot 6 \cdot 6^3 = 6,531,840$, while the $N = 3$, $L = 23$ problem has one with size $23! \cdot 6 \cdot 168^3 \approx 7 \times 10^{29}$. [15, 14, 21] Many of the approaches that are used in the matrix multiplication problem generalize to finding fast algorithms for other tensor products [2].

The idea of using symmetries to reduce searching is quite old; a paper on the eight queens problem from 1874 already takes advantage of it [10]. Up to now, the applications of symmetry in search programs have been specialized techniques developed for specific problems. These special techniques are often very effective [1,9]. The binary matrix multiplication search with $N = 2$, $L = 7$ [18, 19], without using symmetry, could not find any solutions in a reasonable time (a week). Use of a subgroup of size 7! of the complete symmetry group led to a solution in two hours of computer time; use of the complete symmetry group led to a solution in 30 seconds.

Backtrack searching may be speeded up using a number of enhancements, the most important of which is dynamic search rearrangement. Our algorithm incorporates symmetry testing in a way that is compatible with this and other techniques. Our method for handling search rearrangement uses a new fast algorithm for computing a representation of a permutation group relative to a new basis in time $O(n^3)$ [5], where n is the number of points being permuted.

The method for combining symmetry with searching described in this paper has characteristics similar to other intelligent search methods. In practice it may speed up searching greatly, but there is no guarantee of a speed-up on any particular problem, and the method does not improve the worst-case time for searching. Our experimental results show that this method can reduce the time for searching (as well as the size of the search space) by a significant amount on moderate size problems, and can make the difference in whether a large search can be run or not.

2. Backtracking

The symmetric backtracking algorithm uses backtracking to find all the inequivalent solutions to the problem specified by the predicate P and the symmetry group G. Let the problem variables be $v_1, v_2, \ldots, v_n$. For notational convenience we assume that each variable has m possible values, $1, \ldots, m$. At each node in the search space we maintain for each variable a set of possible values. At the root of the search tree each set contains the entire range 1 to m, but at the deeper nodes some of the sets are smaller. A variable is *set* when its set of possible values contains exactly one element. It is *unset* otherwise. A *partial assignment* is a list $X = X_1, X_2, \ldots, X_n$ of sets of possible values for the variables. Partial assignment Y is *more precise* than X if and only if $Y_i \subseteq X_i$ for all i. The backtracking algorithm uses an *intermediate predicate*, also called P, which agrees with the problem predicate on complete assignments and is false on a partial assignment only if the problem predicate is false on all complete extensions of the partial assignment.

The natural lexicographic order is seldom the best order for assigning values to variables. Bitner and Reingold [1] showed that search trees are usually much smaller when the variable with the fewest possible values is chosen as the next variable. In one version of dynamic search rearrangement, each value of each unset variable is tested to see whether the intermediate predicate would become false if the variable were set to that value. Such values are removed from future consideration. If some variable is left with no values, the algorithm backs up. Otherwise, a variable with a minimal number of remaining values is selected and set. Forcing of variables can be a multi-phase process. Once some variables are restricted, it may be possible to restrict more variables as a result.

The backtracking algorithm uses four functions. Function *Select*(X) chooses the next variable to be used in the backtrack search. Only variables with two or more remaining values (unset variables) may be selected. *Select* returns the value *none* if there are no such variables. If *Select* determines that some variable cannot take on a particular value, it indicates this by modifying X. Function $P(X)$ returns *false* when it is sure that no extension of the current partial assignment X can lead to a solution. For completely specified X, it must return *false* if X is not a solution to the problem. In all other cases it returns *true*. If P determines that some values for variables are impossible, it indicates this by modifying X. The algorithm uses a permutation π to assist in the interface between search rearrangement and symmetry checking. Function *Order*(π, X) returns a permutation π' that is similar to π, but that has all the set variables first, i.e., (1) $i^{\pi'} \leq i^{\pi}$ if X_i is set and (2) $i^{\pi'} < j^{\pi'}$ if X_i is set and X_j is not set. Function *Symtest*(X, G, π) returns *false* if the current partial assignment X is equivalent to a smaller one (using π) under the group G. If *Symtest* determines that some variable values lead to redundant partial assignments, it indicates this by modifying X. *Symtest* may also modify G as it traverses the search tree. *Symtest* is discussed in more detail in later sections. The algorithm is designed to work with a wide range of routines for checking the intermediate predicate and choosing the next variable. These supporting routines may be simple or very complex.

Backtracking Algorithm with Symmetry

1. [Initialize.]
   For $i \leftarrow 1$ to n, set $X_i \leftarrow \{1, \ldots, m\}$.
   Initialize the stack to empty.
   Do symmetry checking initialization.

2. [Test value.]
   Save X.
   If not $P(X)$, then go to Step 5. (The value did not pass the test.)
   If X has changed, then repeat Step 2.
   Set $\pi \leftarrow \mathit{Order}(\pi, X)$.
   If not $\mathit{Symtest}(X, G, \pi)$, then go to Step 5. (The value did not pass the test.)
   If X has changed, then repeat Step 2.
3. [Solution?]
   If each variable has exactly one possible value, then print the solution and go to Step 5.
4. [Search deeper.]
   Set $v \leftarrow \mathit{Select}(X)$ and $\pi \leftarrow \mathit{Order}(\pi, X)$.
   If $v \neq \mathit{none}$, then *Push* $[X, v, G]$ and go to Step 6.
5. [Next value.]
   If the stack is empty, then exit.
   Set X, v, and G to the top of the stack.
   Remove the first value from X_v, and make the same change in the top of the stack.
   If X_v is empty, then go to Step 7.
6. [Set value.]
   Set $w \leftarrow$ the first element of X_v and $X_v \leftarrow \{w\}$. (Set variable v to value w.)
   Go to Step 2.
7. [Back up.]
   Pop the stack and go to Step 5.

3. Problem Representation

We represent symmetry operations as permutations on the set of problem variables. Let S_n be the symmetric group on n elements. Let g be an element of S_n, and let $X = X_1, X_2, \ldots, X_n$ be an assignment. Then $Y = X^g$ is the assignment defined by setting $Y_j = X_i$, where $j = i^g$ represents the image of i under g. Equivalently, one may write Y as the assignment $X_{1^{g^{-1}}}, X_{2^{g^{-1}}}, \ldots, X_{n^{g^{-1}}}$. It is easy to check that according to this definition $X^{gh} = (X^g)^h$ for all sequences X and all elements $g, h \in S_n$. An element g of S_n that maps i to itself *fixes* i; two elements of S_n *agree* on i if they map i to the same value. It is often more convenient to consider $X^{g^{-1}}$ instead of X^g, because $X^{g^{-1}} = X_{1^g} X_{2^g} \ldots X_{n^g}$.

A *symmetry* is a permutation that leaves invariant the set of solution sequences to P; i.e., each symmetry maps solutions to solutions, and non-solutions to non-solutions. Since the product of two symmetries also preserves solutions, the set of all finite products of the given symmetries forms a group. The following examples illustrate methods for formulating a problem so that the symmetries are represented as permutations of the variables.

An important class of problems is determining whether one "combinatorial structure" can be embedded in another. In order to make the discussion specific, we consider the case of graph embedding. Let $\mathcal{G}_1$ and $\mathcal{G}_2$ be graphs. We say that $\mathcal{G}_1$ can be *embedded* in $\mathcal{G}_2$ if $\mathcal{G}_1$ can be identified with a subgraph of $\mathcal{G}_2$. An *embedding* is a 1-1 map ϕ of the vertices of $\mathcal{G}_1$ into a subset of the vertices of $\mathcal{G}_2$ such that if $\{i, j\}$ is an edge of $\mathcal{G}_1$, then $\{i^\phi, j^\phi\}$ is an edge of $\mathcal{G}_2$.

Suppose that g_1 and g_2 are automorphisms of $\mathcal{G}_1$ and $\mathcal{G}_2$ respectively. Then $\phi' = g_1 \phi g_2$ is clearly an embedding of $\mathcal{G}_1$ into $\mathcal{G}_2$ as well. The groups of automorphisms of $\mathcal{G}_1$ and $\mathcal{G}_2$ thus define symmetries of the embeddings of $\mathcal{G}_1$ in $\mathcal{G}_2$. In order to express the symmetry as a permutation of problem variables, we specify ϕ by the mn binary variables ϕ_{ij}, where $\mathcal{G}_1$ has m vertices, $\mathcal{G}_2$ has n vertices, and $\phi_{ij} = \mathit{true}$ when $\phi(i) = j$ and *false* otherwise. We use the convention $\mathit{true} < \mathit{false}$. Although we increase the size of the search space by using mn boolean variables, instead of m variables with each variable having n possible values, the practical effects are minimal since we incorporate in our predicate the fact that, for each i, exactly one entry $\phi_{ij} = true$ and the remaining values $\phi_{ij'} = false$ for all $j \neq j'$.

Let $G = G_1 \times G_2$ be the direct product of known subgroups G_i of $Aut(\mathcal{G}_i)$ for $i = 1, 2$. Let $g = g_1 g_2 \in G$ where $g_i \in G_i$ and let ϕ be a partial solution. We define the action of G on the set of partial solutions by setting $\phi^g{}_{ij} = \phi_{i^{g_1^{-1}} j^{g_2^{-1}}}$. It follows directly from this definition that $\phi^{(g_1 g_2)(g_1' g_2')} = \phi^{(g_1 g_1')(g_2 g_2')}$ for elements g_i, g_i' of G_i, $i = 1, 2$. Thus, this action is compatible with the group structure of G.

We now check that this action defines the desired transformation. Suppose that ϕ is an embedding and let $g_i \in G_i$ for $i = 1, 2$. Let $\phi' = \phi^{g_1 g_2^{-1}}$. Then $\phi'_{ij} = true$ precisely when $\phi_{i^{g_1} j^{g_2^{-1}}} = true$. If we interpret ϕ and ϕ' as maps from the vertices of $\mathcal{G}_1$ to the vertices of $\mathcal{G}_2$, then the previous equation shows that $i^\phi = j \iff i^{g_1 \phi} = j^{g_2^{-1}}$. Thus $\phi' = g_1 \phi g_2$.

For another example, consider the transpose of matrix multiplication, as described in the introduction. For the $N = 2$ problem, one set of symmetries comes from the fact that if x, y, and z form a solution (recall that in this case x, y, and z are lists of 2 by 2 binary matrices), then so do u, v, and z, where $u_{i,j,t} = x_{i,p(j),t}$ and $v_{k,l,t} = y_{p^{-1}(k),l,t}$, p is the permutation (12), and p^{-1} is the inverse of p (which is also (12)). If we think of the individual elements of x, y, and z as the problem variables (which is the most natural approach), then these transformations can be represented as a permutation of the variables.

Another set of symmetries for the same problem comes from the fact that if x, y, and z form a solution, so do g, h, and z, where $g_{i1t} = x_{i1t} + x_{i2t}$, $g_{i2t} = x_{i2t}$, $h_{1lt} = y_{1lt} - y_{2lt}$, and $h_{2lt} = y_{2lt}$. In effect, the matrix x is being postmultiplied by the matrix $\begin{pmatrix} 1 & 1 \\ 0 & 1 \end{pmatrix}$ and y is being premultiplied by its inverse. Such transformations cannot be represented by permutations of the original problem variables, but they can be represented as permutations if the original variables are expanded by adding variables for the nonsingular linear combinations of the original variables. In other words, solving eq. (1) modulo 2 is equivalent to solving

$$\bigoplus_{1 \leq t \leq L} u_{ijt} v_{klt} w_{mnt} = (j \otimes k) \wedge (l \otimes m) \wedge (n \otimes i) \tag{2}$$

$$u_{i_1,j,t} \oplus u_{i_2,j,t} = u_{i_1 \oplus i_2,j,t}$$

$$u_{i,j_1,t} \oplus u_{i,j_2,t} = u_{i,j_1 \oplus j_2,t}$$

$$v_{i_1,j,t} \oplus v_{i_2,j,t} = v_{i_1 \oplus i_2,j,t}$$

$$v_{i,j_1,t} \oplus v_{i,j_2,t} = v_{i,j_1 \oplus j_2,t}$$

$$w_{i_1,j,t} \oplus w_{i_2,j,t} = w_{i_1 \oplus i_2,j,t}$$

$$w_{i,j_1,t} \oplus w_{i,j_2,t} = w_{i,j_1 \oplus j_2,t},$$

for $1 \leq i, j, k, l, m, n \leq 2^N - 1$, where $a \oplus b$ is the sum modulo 2, and $a \otimes b$ is the *logical and* of the binary representation of a and b. Each variable in the new problem corresponds to a linear combination of variables in the old problem. The combination is given by the binary value of the first two indices. Thus, u_{11t} corresponds to x_{11t}, u_{12t} to x_{12t}, u_{13t} to $x_{11t} \oplus x_{12t}$, etc. With the new variables, the symmetries discussed above are permutation symmetries. Adding such variables does not expand the size of the search space significantly, since as soon as the original variables are set the new variables have their values forced.

4. Group Theory Methods

Many computational group theory algorithms are based on the idea of a set of *strong generators* for a permutation group G over a set of points $\Omega = \{1, \ldots, n\}$. Let G_i be the subgroup of G which fixes $1, \ldots, i$ pointwise, with $G_0 = G$. Then a set $S \subseteq G$ is a set of strong generators for G if

$G_i = \langle S \cap G_i \rangle$. A *base* Γ for G is a set of points with the property that the identity is the only element of G that fixes each element of Γ pointwise. The set of points i such that $G_i \neq G_{i+1}$ and $G_i \neq I$ forms a base. We use m for the size of this base; note that m depends on the ordering of the points of Ω. An alternate base and set of strong generators can be obtained by reordering the points of Ω. Any ordering, however, results in a base that is no larger than $\lg(n)$ times the size of the smallest base [4].

Let U_i be a set of coset representatives for G_{i-1} over G_i. The elements of U_i are in one to one correspondence with the points in the orbit $i^{G_{i-1}}$, with each element of U_i mapping i to a distinct point of the orbit. Each element g of G has a unique representation of the form $g = g_{n-1} \cdots g_2 g_1$, where $g_i \in U_i$. Moreover, given the U_i, this representation for g is easily found; each g_i is a coset representative from U_i that agrees with $g g_1^{-1} \ldots g_{i-1}^{-1}$ on i. The process of finding this representation (called *sifting*) takes time $O(mn)$.

It is straightforward to compute the sets of coset representatives U_i from a set of strong generators. There are up to n elements in each U_i, so the complete set of permutations occupies $O(n^3)$ space. Jerrum [13] introduced a compact data structure called a *labelled branching*, which uses only $O(n^2)$ space, and allows any coset representative to be computed in $O(n)$ time, which is the same asymptotic time needed to read a permutation that is stored explicitly. Since we will use labelled branchings in a search rearrangement context, we need to allow for a change of basis. Let G be a permutation group acting on $\Omega = \{1, 2, \ldots, n\}$, and let $\pi = \pi_1, \pi_2, \ldots, \pi_n$ be an ordering of the points of Ω. Define the point stabilizer chain of subgroups for G with respect to π by setting $G_i = G_{(\pi_1, \ldots, \pi_i)}$ for $i = 1, \ldots, n-2$. (That is, G_i is the subgroup of G that fixes $\pi_1, \ldots, \pi_i$.) A *branching* on Ω relative to π is a directed forest in which each edge has the form (π_i, π_j) for $i < j$. A branching $\mathcal{G}$ is said to be a *labelled branching for G* relative to π if each edge (π_i, π_j) is labelled by a permutation σ_{ij} so that (i) $\sigma_{ij} \in G_{i-1}$ and moves π_i to π_j, and (ii) the set of edge labels of $\mathcal{G}$ generates G. A labelled branching $\mathcal{G}$ is *complete* if, in addition, (iii) if π_k is in the G_{i-1} orbit of π_i, then there is a directed path in $\mathcal{G}$ from π_i to π_k. Criterion (iii) ensures that the edge labels of $\mathcal{G}$ form a strong generating set for G relative to the ordering π.

Although we have described the labelled branching in terms of the edge labels σ, the actual permutations stored in the branching data structure are products of the edge labels. Let π_r be the root of the connected component of $\mathcal{G}$ containing π_j. Associate with each node a *node label* $\tau(j)$, where $\tau(j)$ is the product of the edge labels from the root π_r to π_j if $\pi_r \neq \pi_j$, and is the identity if $\pi_r = \pi_j$. Since $\tau(j)^{-1}$ moves π_j to its root π_r, we can recover the label for edge (π_i, π_j) as $\sigma_{ij} = \tau(i)^{-1}\tau(j)$. Furthermore, if there is a path from π_k to π_j in $\mathcal{G}$, then $\tau(k)^{-1}\tau(j)$ is the product of the edge labels along the path from π_k to π_j. This means that coset representatives can also be recovered from the data structure at the cost of one multiply, as opposed to $O(n)$ multiplies if edge labels were stored. Jerrum proved that a complete labelled branching for a permutation group can be computed in $O(n^5)$ time using $O(n^2)$ storage.

5. Comparison Order

The basic idea behind symmetry checking is to divide the solution space into symmetry classes, and to explore only one element of each class. To identify the unique element that is to be explored, we use a partial order $\prec$ on the set of all partial assignments. Let X_i^+ be the largest of the possible values for variable v_i (under assignment X), and let X_i^- for the smallest. If X_i contains only one element, we call it x_i. The ordering is defined by $X \prec Y$ if there exists a k $(1 \leq k \leq v)$ such that

1. for $1 \leq i < k$, $X_i^+ = Y_i^-$,
2. $X_k^+ < Y_k^-$, and
3. for $1 \leq i \leq k$, Y_i contains exactly one value.

When X and Y are complete assignments, $\prec$ is the usual lexicographic ordering. When X and Y are not complete, each possible way of increasing the precision of X leads to a result that is lexicographically less than any result from increasing the precision of Y. As the ordering is used in *Symtest*, Y is a partial solution and X is the same partial solution transformed by applying a group element. *Symtest* uses comparisons under the ordering to test for a non-minimal partial solution and to possibly eliminate variable values that would lead to non-minimal partial solutions. Parts 1 and 2 of the definition suffice in an environment where a fixed search order is used; part 3 is needed in a search rearrangement setting.

The function *Select* chooses the next variable for the backtracking algorithm to investigate. In general, *Select* may choose any unset variable. In addition, variables may be forced out of their normal order by having their values restricted by the predicate, by *Select*, or by *Symtest*. *Symtest* is most powerful when the comparison order is close to the search order, so if the search order is dynamic, the comparison order should also be dynamic. In the generalization of $\prec$, the comparison order π determines the order in which the components of partial assignments are compared: in the above definition of $\prec$, every subscript i or k is replaced by i^π or k^π. When using a comparison order other than the identity with $\prec$, we indicate its use by a subscript, as in $X \prec_\pi Y$.

A dynamic search order is determined as the search tree is explored. At each node of the search tree, the order of the set variables is determined by the history of the search, while the order of the unset variables is undetermined. Let Π be the set of search orders that agree with π on the set variables. *Symtest* will work correctly if $\prec$ is stable under variations of π that belong to Π, i.e., for $\pi_1, \pi_2 \in \Pi$, if $X \prec_{\pi_1} Y$ then also $X \prec_{\pi_2} Y$. *Symtest* only compares $X \prec Y$ when $X = Y^g$ for some symmetry transformation g. Since all the comparison orders in Π agree on the set variables, the comparisons between Y and Y^g are stable if each Y_{i^π} contains one value. This justifies part 3 of the definition.

The function *Order* maintains the comparison order consistent with the search order. Initially, no variables are set and π is the identity. In general, suppose that k variables have been set, and X_{i^π} is set for $1 \le i \le k$. If p additional variables are set before the next call to *Order*, then a new representative comparison ordering must be computed so that $X_{i^{\pi'}}$ is set for $1 \le i \le k+p$ and such that $i^\pi = i^{\pi'}$ for $1 \le i \le k$. The transformation proceeds in at most p steps. In each step the current value of π is transformed so that the index of the next newly set variable (proceeding in order from left to right) is placed at the end of the current prefix of indices of set variables. Specifically, suppose that r is the smallest index such that X_{r^π} is unset and s is the smallest index such that $r < s$ and X_{s^π} is set. We construct π' from π by means of a right cyclic shift of the values of π with indices in the range r to s, so that π' has the form $\pi' = \pi_1, \ldots, \pi_{r-1}, \pi_s, \pi_r, \ldots, \pi_{s-1}, \pi_{s+1} \ldots \pi_n$.

Symtest is called with a labelled branching $\mathcal{G}$ for G relative to the comparison order π as it was defined at the time of the previous call. *Symtest* must transform $\mathcal{G}$ so that it is relative to the current π. Each time *Symtest* is called a series of cyclic shifts are computed that transform the previous π into the current one. Simultaneously, the new base change algorithm [5] is used to transform $\mathcal{G}$ relative to π into $\mathcal{G}$ relative to π'. One transformation takes time $O(n^2)$. If p variables have been set since the last call to *Symtest*, at most p transformations are needed. Either the transformations are proceeding rapidly ($O(n^2)$) or the backtracking program is making rapid progress (setting lots of variables without doing any branching). At any time, then, and with relatively little expense, we can obtain a labelled branching relative to the comparison order defined above. The advantage of having such a labelled branching is that it makes symmetry checking under the current search order straightforward.

We now discuss how to use the relative labelled branching to do symmetry checking. Let X be the current partial solution, and let π be a comparison order. Assume $\mathcal{G}$ is a labelled branching for the symmetry group G relative to π. *Symtest* expects an ordinary labelled branching: one that is relative to the identity. Rather than change the code for *Symtest* to handle labelled branchings relative to an arbitrary permutation, it is more efficient to transform X and $\mathcal{G}$ before calling *Symtest*. This is easily done. We first copy X to $Y = X^{\pi^{-1}}$, taking X_i to $Y_{i^{\pi^{-1}}}$. Considering the variables of Y in consecutive order, we now have a version of the partial solution with the variables renamed so that the set ones are first.

The next step is to make a copy of $\mathcal{G}$, altering it so that it represents $G^{\pi^{-1}}$ instead of G. Conjugating by π^{-1} has the effect of renaming the points of Ω in accordance with the names used in Y. We call the new branching $\mathcal{G}^{\pi^{-1}}$; it is obtained from $\mathcal{G}$ by keeping the same graph and mapping each vertex label i to i^{π} and each edge label σ to $\pi\sigma\pi^{-1}$. The branching $\mathcal{G}^{\pi^{-1}}$ is a labelled branching for $G^{\pi^{-1}}$ relative to the identity.

Now *Symtest* can be called on inputs $\mathcal{G}^{\pi^{-1}}$ and Y with no modifications to the code. To see that the correct tests are performed, observe that elements of $G^{\pi^{-1}}$ are of the form $\pi g \pi^{-1}$, where $g \in G$. Now, $Y^{\pi} = X$. Applying π^{-1} to X renames the variables and returns us to Y. Thus, $Y^{\pi g \pi^{-1}} = X^{g\pi^{-1}}$ and $Y = X^{\pi^{-1}}$, so comparisons between $Y^{\pi g \pi^{-1}}$ and Y give the same results as comparisons between $X^{g\pi^{-1}}$ and $X^{\pi^{-1}}$.

6. Symmetry Checking

Function *Symtest* is the main symmetry checking routine. *Symtest* returns *false* if there is a $g' \in G$ such that $X^{g'^{-1}} \prec X$, and *true* otherwise. If *Symtest* returns *false*, the main backtracking algorithm backs up.

Symtest is itself a backtracking algorithm, which attempts to construct a transformation $g \in G$ incrementally, with the property that $X^{g^{-1}} \prec X$. At all times, *Symtest* has a permutation g in G which is the current candidate for g'. Given the partial solution $X = X_1 \ldots X_n$, the image of X under $g^{-1} \in G$ is $Y = X_{1^g} X_{2^g} \ldots X_{n^g}$. If the search for g' gets to level r, then we have a prefix $\{1, \ldots, r-1\}$ of $\Omega = \{1, \ldots, n\}$ such that $X_1, \ldots, X_{r-1}$ are set, $Y_1, \ldots, Y_{r-1}$ are set, and $x_i = y_i$ for $1 \le i \le r-1$. Let $T = (1^g, \ldots, (r-1)^g)$. T represents a partial permutation which we are attempting to extend. On level r, *Symtest* performs the following tests, using indexes t_r such that there exists an element of G with prefix (T, t_r).

1. If $x_r > X_{t_r}^+$, then return *false*. The current partial solution is symmetric to a smaller one.

2. If $x_r < X_{t_r}^+$, then remove from X_{t_r} those values smaller than x_r. The removed values would only lead to *Symtest* returning false later. Do not consider this way of extending T any further. It transforms the current partial solution into a larger one.

3. Otherwise, $x_r = X_{t_r}^+$. Remove from X_{t_r} those values smaller than x_r. Variable v_{t_r} is now set. Test this extension further, by computing an appropriate g and trying the next value of r.

Let k be the index of the first unset variable. Using the tests given above, the partial image T can be extended, at most, to the point where $r = k$. Much of the useful work done by *Symtest* occurs during the attempt to extend T. Each time a variable is set or its domain of definition is restricted, a large portion of the search space is removed.

Symtest uses symmetries to prune the search space of permutations. We first consider the computation of the candidate points for extending a partial image T. Suppose the points $\{1, \ldots, m\}$ form a base for G, and consider a partial permutation $T = (t_1, \ldots, t_i)$, $i \le m$, of distinct points of Ω such that there exists an element $g \in G$ with $t_j = j^g$ for $1 \le j \le i$. If $i = m$, T specifies a unique $g \in G$. Define $G(T) = \{g \in G : (1^g, \ldots, i^g) = T\}$. Given a set of strong generators for G, it is straightforward to calculate the elements of $G(T)$ once one element is known: $G(T) = (G_i)g$. [6] Now define $\Omega_G(T) = \{y \in \Omega : G(T \cup y) \neq \emptyset\}$. One may think of $\Omega_G(T)$ as the possible ways of extending a partial permutation by one more point. The set $\Omega_G(T)$ may be calculated using $\Omega_G(T) = (i^{G_{i-1}})^g$.

[6] Thus, knowledge of the sets G_i and an element of $G(T)$ is sufficient to calculate all the elements of $\Omega_G(T)$. A point that *Symtest* uses to extend a partial permutation T must be in $\Omega_G(T)$.

Symtest is basically a color automorphism algorithm that has a side effect of restricting the values of variables. Every element of G is an automorphism of Ω. For purposes of *Symtest*, there are two useful colorings of Ω. If variable i is set, the color of i is the value of that variable. For unset variables, the first coloring, C_1, is obtained by using X_i^+ as the color of i. The second coloring uses a special color u, which is not a possible value, and which is incomparable with the possible values. When using C_2, we assume that the variables are indexed so that the set variables occur before the unset ones, as would normally be the case when search rearrangement is used. The colorings are determined before *Symtest* starts to work, and do not change during the course of a run of *Symtest*, even if a variable is set by having all but one of its values eliminated. Also, the order of the variables does not change during a run of *Symtest*.

For efficiency reasons, C_1 is used only in the case where all the unset variables have the same maximum value. *Symtest* checks the unset variables, setting variable *samecolor* to *true* if all points with index greater than k have the same upper value. If *samecolor* is *true*, then *Symtest* can proceed in a way appropriate for C_1. With this proviso, elements $k+1,\ldots,n$ of Ω all have the same color under either coloring, and so we have $G_k = H_k$, where H is the color automorphism subgroup of G. With coloring C_2, once a permutation maps a variable into an index larger than k, it can no longer be color preserving. Using this coloring, the group H consists of permutations that stabilize setwise and preserve colors on the first k points.

At this time, it is not known if there is a polynomial time algorithm to solve the color automorphism problem, but it must be easier than NP-complete unless the polynomial hierarchy collapses to Σ_2^p [11]. Also, there do exist special cases (which depend on the structure of G) for which polynomial time algorithms are known [17]. Our method for testing for symmetry is based on a general method for computing with permutation groups developed by Sims [20] and later refined by Bulter[6] and Leon[16]. Sims' method has a wide range of applications and forms the basis of many of the group theory algorithms implemented in the CAYLEY system[8]. In the case where G is a p-group, i.e. the order of G is a power of a prime p, it is possible to adapt Luks' method [17] to symmetry testing. The algorithm presented in [7] can be modified to perform *Symtest*, so that, for very large classes of symmetry groups, *Symtest* can be made to work in polynomial time. In conjunction with our efficient base change algorithm [5], this approach could lead to a *Symtest* algorithm that was efficient for a large class of groups and also compatible with search rearrangement. We plan to explore this type of algorithm in future research.

At any time during a run of *Symtest*, the group K generated by the elements of G which have passed all the tests in *Symtest* is a subgroup of the color automorphism group H of Ω. H is constructed by systematically building a list $\mathcal{H}$ of generators for the chain of subgroups

$$G_k = H_k \subseteq H_{k-1} \subseteq \ldots \subseteq H_0 = H,$$

where $k+1$ is the index of the first unset variable. At any given time, we know a set $\mathcal{K}$ of generators for a K such that $H_s \subseteq K \subseteq H_{s-1}$. (Initially, $s = k$ and $K = G_k$.) Once we have examined all possible elements of G which fix each element of the set $\{1,\ldots,s-1\}$, we decrement s, and continue in this way until $s = 0$.

Let $T = (1,\ldots,s-1,t_s,\ldots,t_r)$ be the partial permutation which we are attempting to extend. Thus t_i is the same color as i for each i, $1 \le i \le r$. Since *Symtest* accumulates an element $g \in G(T)$ and updates this element each time T is extended, it is unnecessary to store T explicitly; T is simply the first r elements of g. If we are able to extend T so that $r = k$, then g is color preserving on Ω, and we may add g to K. Furthermore, we may reset r to s and choose a new value for t_s! The justification for this is that any two elements of H_{s-1} which agree on s belong to the same coset of

H_s, and we need only one representative for each such coset. This pruning can save a very significant amount of searching.

There is a problem with pruning, however. A partial solution might be redundant for finding new elements of H, but still lead to restrictions of variable values, or even to *Symtest* returning *false*. In order to compensate for pruning, we retain the information necessary to discover all the potential side effects of examining each partial solution, including those that are never explicitly looked at. We use a list of ordered pairs of the form $(\mathcal{O}_i, m_i)$, where the sets $\mathcal{O}_i$ are the orbits of the current value of K acting on Ω. A lower bound that is valid for some variable is also valid for all the variables in its K orbit; m_i is the largest value which has been used to restrict the elements of an orbit. Initially, $m_i = 1$ (the smallest possible value) for each orbit. As K is enlarged, orbits may be combined. This gives a further opportunity for restrictions.

The opportunity for applying these ideas comes when we are attempting to extend a partial permutation $T = (t_1, \ldots, t_{r-1})$, for some r, $1 \leq r \leq k$, within a call to *Symtest*. We first *reduce* the set $\Omega_G(T)$. The reducing process removes unsuitable points from $\Omega_G(T)$ and at the same time checks for opportunities to restrict variables or return *false*. This is accomplished by examining the implications of each $t \in \Omega_G(T)$ for our backtracking environment. If $t \leq k$, then t is added to the set Y_r of candidate extensions if and only if t is the same color as r. Otherwise, $t > k$ and t belongs to some K-orbit $\mathcal{O}_i$. If $x_r > m_i$ then we set $m_i \leftarrow x_r$ and check to see if there exists an element $\bar{t} \in \mathcal{O}_i$ such that $X_{\bar{t}}^+ < m_i$. If such a $\bar{t}$ exists, then X is equivalent under symmetry to a smaller partial solution and *Symtest* will return false.

If we do not find a reason to return *false*, we choose $t_r = \min(Y_r)$ and update both Y_r and T by setting $Y_r \leftarrow Y_r - \{t_r\}$ and $T \leftarrow (t_1, \ldots, t_r)$. (We actually set T by modifying g appropriately.) If $r < k$, then we increment r and continue trying to extend T. Otherwise, $r = k$, and we adjoin g to $\mathcal{K}$, the set of generators for K. The orbit information has to be updated as well. Each new orbit $\mathcal{O}_{i'}$ will be a union of certain of the old orbits $\mathcal{O}_i$, which may be identified using the action of g on Ω. In addition, we set $m_{i'}$ to be the maximum of the values m_i of the constituent suborbits $\mathcal{O}_i$. It may now be possible to find a $t \in \mathcal{O}_{i'}$ such that $X_t^+ < m_{i'}$. If this occurs, then again *Symtest* returns *false*.

Pruning by resetting r to s when an element of H is found can be very effective. *Symtest* also uses a special form of symmetry testing to cut down on the amount of searching it does. This symmetry testing uses a lexicographic ordering $\ll$ on the set of permutations of Ω. *Symtest* examines partial permutations in an order consistent with $\ll$, since we start with g equal to the identity and always choose the smallest element from Y_r to extend T. In the course of extending K to H_{s-1}, we reject any partial permutation which is not the smallest element in its left coset of K. (Technical reasons prevent the use of right cosets.) To identify such elements, we apply the following result [6]: g is first in $gK \iff t = j^g$ is first in $t^{K_{t_1,\ldots,t_{j-1}}}$, for all $j, 1 \leq j \leq m$, where $t_i = i^g$ and $K_{t_1,\ldots,t_{j-1}}$ is the pointwise stabilizer in K of the set $t_1, \ldots, t_{j-1}$. This pruning is done in Step 2.

Symtest

Input Parameters: The current partial solution X and a list $\mathcal{G}$ of generators for G as described above. Output Parameters: X and the list $\hat{\mathcal{G}}$ of generators that can be used for the next call to *Symtest*. Returned Value: A Boolean value which is *false* if there is an element $b \in G$ such that $X^{b^{-1}} \prec X$, and *true* otherwise. Local Variables: Y_r is the set of possible candidates for t_r; s is the largest integer for which $t_i = i$, for all i, $1 \leq i \leq s-1$; K is the subgroup of H that is currently known, with $H_s \subseteq K \subseteq H_{s-1}$; $\mathcal{K}$ is the current set of generators for K (each time a generator is added to $\mathcal{K}$, the group K changes to conform with $\mathcal{K}$); g is an element of G such that $(1, \ldots, r)^g = T$; *reduced* is an array of Boolean values where *reduced*$_r$ is *true* if $\Omega_G(T)$ has been reduced as described above, and *false* otherwise; $\mathcal{O}$ is a list of ordered pairs of the form $(\mathcal{O}_i, m_i)$ as described above. The Boolean variable *samecolor* is true if all unset variables have the same color.

1. [Initialize]
   Transform the representation of G and X with π.
   Set k to the largest value such that the first k variables are set.
   Set *samecolor* $\leftarrow$ *true* if $X_j^+ = X_{k+1}^+$, for $k < j \le n$.
   Set $\mathcal{K} \leftarrow$ {generators for $H_k = G_k$} and $\mathcal{O} \leftarrow \emptyset$.
   For each orbit $\mathcal{O}_i$ of K acting on $\{1, \ldots, n\}$, set $m_i \leftarrow 1$ and $\mathcal{O} \leftarrow \mathcal{O} \cup (\mathcal{O}_i, m_i)$.
   For $i \leftarrow 1$ to k set $Y_i \leftarrow \Omega_G(1, \ldots, i-1)$ and $reduced_i \leftarrow false$.
   Set $s \leftarrow k$, $r \leftarrow k$, and $g \leftarrow e$ (the identity permutation).
2. [Backtrack]
   If $r < s$, then go to Step 7.
   Set $Y_r \leftarrow Y_r - \{t_r{}^{K_{t_1,\ldots,t_{r-1}}}\}$
   If $reduced_r$, then go to Step 4.
3. [Test position r]
   For each $t \in Y_r$ do the following:
   - if $X_t^+ < x_r$, then return(*false*);
   - let t belong to the K-orbit $\mathcal{O}_i$;
   - set $m_i \leftarrow \max(m_i, x_r)$;
   - if $X_t^+ > x_r$, then
     - $Y_r \leftarrow Y_r - \{t\}$;
     - if not(*samecolor*), then
       - if there exists a $t' \in \mathcal{O}_i$ such that $X_{t'}^+ < m_i$, then return(*false*);
   - if $X_t^+ = x_r$, then
     - if not(*samecolor*) and $t > k$, then
       - $Y_r \leftarrow Y_r - \{t\}$;
       - if there exists a $t' \in \mathcal{O}_i$ such that $X_{t'}^+ < m_i$, then return(*false*).

   Finally, set $reduced_r \leftarrow true$.
4. [Next t_r]
   If $Y_r = \emptyset$, then set $r \leftarrow r - 1$ and go to Step 2.
   Set $t_r \leftarrow Min(Y_r)$, h to some element in G_{r-1} such that $t_r = r^{hg}$, and $g \leftarrow hg$.
5. [Search deeper]
   Set $r \leftarrow r + 1$.
   If $r \le k$, then set $Y_r \leftarrow \Omega_G(T)$, $reduced_r \leftarrow false$, and go to Step 3.
6. [Image is complete]
   Set $\mathcal{K} \leftarrow \mathcal{K} \cup \{g\}$. (At this point, we know that $g \in K$)
   For each K-orbit $\mathcal{O}_{i'}$, (Update $\mathcal{O}$)
   - set $m_{i'} \leftarrow 0$;
   - for each $(\mathcal{O}_i, m_i) \in \mathcal{O}$ with $\mathcal{O}_i \subseteq \mathcal{O}_{i'}$,
     - set $\mathcal{O} \leftarrow \mathcal{O} - (\mathcal{O}_i, m_i)$;
     - if $m_i > m_{i'}$, then set $m_{i'} \leftarrow m_i$;
   - set $\mathcal{O} \leftarrow \mathcal{O} \cup (\mathcal{O}_{i'}, m_{i'})$.

   If not(*samecolor*), then if there exists a $t' \in \mathcal{O}_{i'}$ such that $X_{t'}^+ < m_{i'}$, then return(*false*).
   Finally, set $r \leftarrow s$ and go to Step 2.
   (It is only necessary to consider one representative for each coset of H_s in H_{s-1}.)
7. [Now $K = H_s$]
   Set $s \leftarrow s - 1$. (Now $r = s$.)
   If $r > 0$, then go to Step 2.
8. [Restrict remaining values and return]
   For each $(\mathcal{O}_i, m_i) \in \mathcal{O}$ for each $t \in \mathcal{O}_i$, set $X_t \leftarrow X_t - \{w \in X_t : w < m_i\}$.
   Compute $\hat{\mathcal{G}}$ and set $\mathcal{G} = \hat{\mathcal{G}}$
   Return(*true*).

	Time		Nodes		Solutions		Calls
	with	without	with	without	with	without	with
3	0	0	3	5	0	0	2
4	33	16	7	11	1	2	6
5	116	66	11	27	2	10	11
6	266	233	19	75	1	4	16
7	733	1033	47	209	8	40	35
8	2716	4833	145	783	12	92	81
9	10966	24333	549	3005	46	352	242
10	46150	113466	2173	11809	92	724	818
11	284116	629066	9603	52389	341	2680	3342
12	1445966	3777283	45877	250551	1787	14200	14884
13	8763433	—	229523	—	9233	—	70233
14	51813166	—	1236927	—	45752	—	345780

Table 1. Timings, nodes, and solutions, for n-queens with and without symmetry checking; also calls to *Symtest* (with case only). Time units are milliseconds. A dash indicates the measurement was not done because excessive running time was anticipated.

We have not discussed the method of accumulating the generators for H. They can be inserted into a labelled branching as they arise. Since they are already strong generators, it is cheap to run Jerrum's algorithm and produce a complete labelled branching for H. Maintaining a labelled branching for H as it is accumulated also allows the pruning in Step 2 to be done conveniently, using the fast base change algorithm [5]. In addition, the group H can sometimes replace G in future calls of *Symtest* on the same branch of the search space, for example, if there are no generators that preserve colors on some prefix of set variables and map a set variable to an unset one. This condition is easy to test for. H can be considerably smaller than G. If we let n_k be the number of orbits of G_k acting on Ω, and let m_k be the number of generators of G_k that we know upon entry to *Symtest*, then $|\mathcal{H}| \leq n_k - t + m_k$, where t is the number of orbits of H on Ω. In particular, if G_k is the identity subgroup, then $|\mathcal{H}| \leq n - t$.

7. Experimental Results

We tested our algorithm on the n-queens problem, for n ranging from 3 to 14. The n-queens problem is a rather unfavorable problem for this algorithm, because the size of the symmetry group is only 8. Nevertheless, the data in Table 1 show that an algorithm that combined symmetry checking with search rearrangement saved considerable time over an algorithm that used dynamic search rearrangement but no symmetry checking.

8. Conclusions

The algorithm presented gives a general way to use symmetry groups to reduce the size of the search space for problems with symmetries. It appears to be practical for use with large problems that have known symmetries and where it is not worth designing a special algorithm for the problem. (It can be extremely difficult to design a special algorithm that applies symmetry in a way that does not interfere with search rearrangement.) Although the algorithm is similar to algorithms that have been used for color automorphism, it is the first general purpose algorithm for using symmetry with search rearrangement backtracking.

There are many questions about whether better algorithms exist for the various calculations that arise in our procedures. *Symtest* can take exponential time, although it is usually much faster. Originally, we used Jerrum's algorithm repeatedly at a cost of $O(n^4)$ per variable set. Now we have an algorithm [5] that costs $O(n^2)$ per variable set. Even the algorithms that use $O(n^2)$ memory leave something to be desired when they are used on problems with $n \approx 1000$. (Of course, such problems can be solved only when the predicate is very efficient at rejecting solutions or the symmetry group is extremely large.) There are various ways to sometimes overcome these problems, but further study is needed to determine the best ways. The algorithm currently implemented has already proved efficient enough to be of great practical help in certain large searches. We are continuing to investigate ways of improving the efficiency of the group theory algorithms applied to backtracking.

References

1. James R. Bitner and Edward M. Reingold, "Backtrack Programming Techniques", *CACM* **18** (1975), pp. 121–136.
2. Roger W. Brockett and David Dobkin, "On the Number of Multiplications Required for Matrix Multiplication", *SIAM J. Comput.* **5** (1976), pp. 624–628.
3. C. A. Brown, L. A. Finkelstein, and P. W. Purdom, Jr., "Backtrack Searching in the Presence of Symmetry", NUTR NU-CCS-87-2 (1987).
4. C. A. Brown, L. A. Finkelstein, and P. W. Purdom, Jr., "Efficient Implementation of Jerrum's Algorithm for Permutation Groups", NUTR NU-CCS-87-19 (1987).
5. C. A. Brown, L. A. Finkelstein, and P. W. Purdom, Jr., "A New Base Change Algorithm for Permutation Groups", NUTR NU-CCS-87-30 (1987).
6. G. Butler, "Computing in Permutation and Matrix Groups II: Backtrack Algorithm", *Math. Comp.* **39** (1982), pp. 671–680.
7. G. Butler and C. W. H. Lam, "A General Backtrack Algorithm for the Isomorphism Problem of Combinatorial Objects", *Journal of Symbolic Computation* **1** (1985), pp. 363–382.
8. J. J. Cannon, "An Introduction to the Group Theory Language, Cayley", in *Computational Group Theory*, edited by M. D. Atkinson, Academic Press, 1984, pp. 145–184.
9. J. L. Carter, *On the Existence of a Projective Plane of Order 10*, Ph. D. Thesis, University of California at Berkeley (1974).
10. J. W. L. Glaisher, "On the Problem of the Eight Queens", *Philosophical Magazine* series 4, vol. 48 (1874), pp. 457–467.
11. Shafi Goldwasser and Michael Sipser, "Private Coins versus Public Coins in Interactive Proof Systems", *Proc. 18th Sym. on Theory of Computing* (1986) pp. 59–68.
12. J. Hopcroft and J. Musinski, "Duality Applied to the Complexity of Matrix Multiplication and Other Bilinear Forms", *SIAM J. Comput.* **2** (1973), pp. 159–173.
13. Mark Jerrum, "A Compact Representation for Permutation Groups", *Journal of Algorithms* **7** (1986), pp. 60–78.
14. Rodney W. Johnson and Aileen M. Mc Loughlin, "Noncommutative Bilinear Algorithms for 3×3 Matrix Multiplication", *SIAM J. Comput.* **15** (1986), pp. 595–603.
15. Julian D. Laderman, "A Noncommutative Algorithm for Multiplying 3×3 Matrices Using 23 Multiplications", *Bull. Amer. Math. Soc.* **82** (1976), pp. 126–128.
16. J. Leon, "Computing Automorphism Groups of Combinatorial Objects", in *Computational Group Theory*, edited by M. D. Atkinson, Academic Press (1984), pp. 321–337.
17. E. M. Luks, "Isomorphisms of Graphs of Bounded Valence Can Be Tested in Polynomial Time", *J. Comp. Sys. Sci.* **25** (1982), pp. 42–65.
18. Paul W. Purdom, "Tree Size by Partial Backtracking", *SIAM J. Comput.* **7** (1978), pp. 481–491.
19. David A. Seaman, *Fast Matrix Multiplication*, Indiana University Master's Thesis (1978).
20. C. C. Sims, "Computation with Permutation Groups", in *Proc. Second Sym. on Symbolic and Algebraic Manipulation*, edited by S. R. Petrick, ACM, New York, 1971.
21. Volker Strassen, "Gaussian Elimination is Not Optimal", *Numer. Math.* **13** (1969), pp. 354–356.

FAST SERIAL-PARALLEL MULTIPLIERS

Marco Bucci and Adina Di Porto
Fondazione Ugo Bordoni
I-00142 Roma, Italy

1. Introduction

Some cryptographic applications which make use of large numbers (hundreds bits) have made Serial Parallel (SP) multipliers [1] up to date again. Such multipliers, known since 1962, are actually the sequential version of Braun's multipliers [2]. Their cost (number of gates) is $O(n)$, where n is the number of bits of the multiplicand (coefficient). On the contrary, the cost of Braun's multipliers (or any combinatorial device) is $O(mn)$ where m is the number of bits of the multiplicator (factor).

In this paper, after analyzing SP multipliers in some detail and showing that they are very simple and efficient, a method to speed SP multipliers up is discussed.

Such a method, based upon a suitable application of the so-called *bit-scanning* technique [2], involves the use of AND gates, Full Adders and a single h–bit Carry Look–ahead Adder, h being the desired speed increasing factor.

The cost of this device is $O(hn)$.

2. Conventional serial–parallel multipliers

An n bit SP multiplier is depicted in fig.1 . In literature, such a multiplier is also referred to as Carry Save Shift Adder (CSSA) because it performs a sum in Carry Save (CS) representation and a right shift of one place (division by 2), each step.

Work carried out in the framework of the Agreement bettween the Italian PT Administration and the Fondazione Ugo Bordoni.

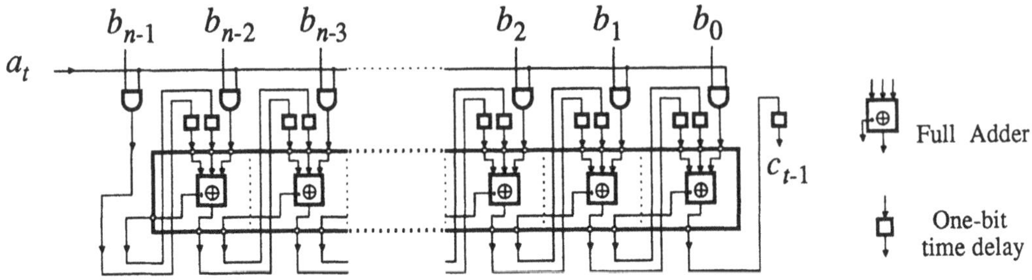

Fig.1 Carry Save Shift Adder (CSSA) with a one position right shift (conventional serial parallel multiplier).

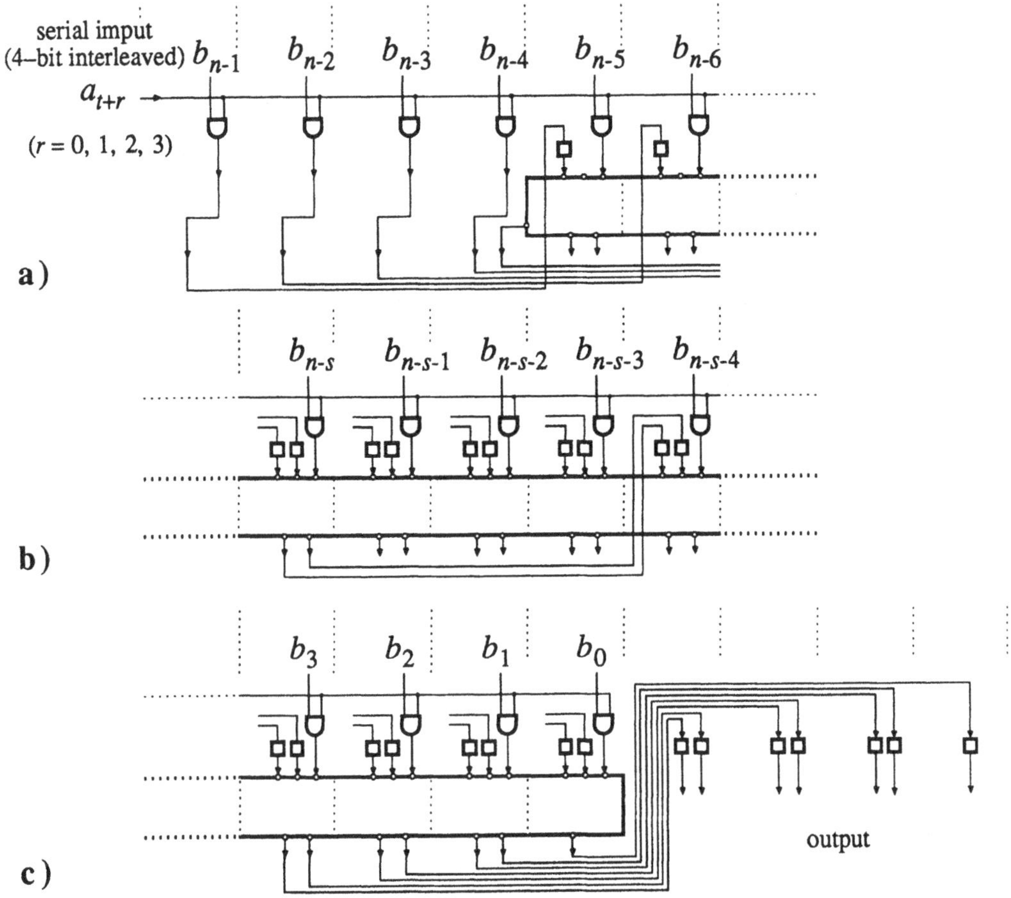

Fig.2 Example of Carry Save Shift Adder with a four position right shift (CSS^{4}A).
a) head cells, b) generic intermediate cells, c) tail (output) cells.

So, if

$$A = \sum_{i=0}^{m-1} a_i 2^i \qquad \text{and} \qquad B = \sum_{i=0}^{n-1} b_i 2^i \quad , \qquad a_i,\ b_i \in \{0, 1\} \qquad (1)$$

then the product is obtained as

$$A\,B = (1/2^{m+n}) \sum_{i=0}^{m-1} 2^{m+n+i} a_i B = \sum_{i=0}^{m-1} 2^i a_i B \qquad . \qquad (2)$$

The most interesting features of this kind of multipliers can be summarized in the following items:

- their cost depends only on the number of bits of the multiplicand, i.e. it is $O(n)$;
- SP multipliers are not affected by carry ripple; the carry is added to the result by the rightmost Full Adder (FA) of the device, automatically;
- if we suppose that the operands have approximately the same length, then each gate and latch is working for 3/4 the total calculation time, on the average;
- The FA's can be implemented using two levels of gating, i.e. using the minimun number of levels needed to implement a non–trivial combinatorial function(*). In this way, the speed of the device is limited only by the digital technology used.

In order to speed SP multipliers up, one could think of either (*i*) adding more than one term (see summation (2)) each step, or (*ii*) using a base greater than 2 . Consequently, the use of adders with more than 3 inputs (solutions (*i*) and (*ii*)) or the use of devices which implement the multiplicative table for bases greater than 2 (solution (*ii*)) seems quite natural. Unfortunately, the difference between the cost of these devices and the cost of FA's and AND gates is much greater than the obtainable speed increasing. So, the solution proposed in the next section involves only the use of AND gates and FA's .

As a concluding remark, we point out that the considerations we have done led us to believe that SP multipliers are *optimal* from several points of wiev, mainly as to the

(*) By the scheme of fig.1 it can be seen that the delay due to the AND gates can be easily eliminated by anticipating the phase of the serial input with respect to the clock of the latches.

bit–rate to cost ratio. In fact, all modifications to this scheme that have been proposed in literature (mainly addressed to obtain a device having both inputs serial and pipeline feeded [3,4]) imply a considerable worsening of the said bit–rate / cost ratio.

The lack of luck that SP multipliers have met in computer hardware is due, we think, to the fact that Braun's multipliers are more attractive for they do not require internal clock and their (quadratic) cost is not a problem when the operand are comparatively small.

On the contrary, in the case of operands exceeding one hundred bits, the one–dimensionality of the SP multipliers becomes necessary.

Now, the problem of increasing the bit–rate of these devices without worsening the bit–rate / cost ratio (so that no optimality feature gets lost, arises).

As usual, the simplest solution, among the large number of solutions taken into consideration, appeared to be the only solution fullfilling all requirements.

3. A method to speed SP multipliers up

Since the clock period is supposed to be as short as possible (depending on the digital technology), we thought of a multiplier which, instead of generating only one bit of the result each step, generates a bit–burst of length h . This can be easily obtained using a technique commonly referred to as *bit–scanning* .

In order to see how this can be done, let us refer to the SP multiplier shown in fig.1 that, from now onwards, will be called M1 . A suitable application of the bit–scanning principle consists into partitioning the sum (2) as follows:

$$A\,B = \sum_{r=0}^{h-1} \sum_{i=0}^{\lfloor \frac{m-1}{h} \rfloor} 2^{hi+r} a_{hi+r} B \qquad (3)$$

were $\lfloor x \rfloor$ denotes the integral part of x .

Let us consider the *meaning* of this operation from the circuitry point of wiev. Each of the h inner sums on the right–hand side of (3) implies a CSSA similar to the CSSA which constitutes M1 .

It must be noted that, in this case, the shift is of h positions to the right, instead of one. We denote this kind of CSSA by CSS^{h}A (fig.2). In this way, the result is generated as a sum of h addends CS . Each of these addends is computed by a

CSS^hA , h bits at a time each step, so that the addends are completely computed after $(\lfloor(m-1)/h\rfloor + 1) + (\lfloor(n-h-1)/h\rfloor + 1) + 1 \approx (m+n)/h$ steps. In fact, a number $\lfloor(m-1)/h\rfloor + 1$ of steps is necessary to perform the summations on the right–hand side of (3) and a number $(\lfloor(n-h-1)/h\rfloor + 1) + 1$ of steps is necessary to clear the CSS^hA's out by means of the introduction of an equivalent number of 0's at their serial inputs.

The Usual Binary Representation (UBR) of the result is simply obtained by adding the h CS addends up. In order to minimize the input–output delay introduced by this operation, the addends are processed by pipeline devices linked to the CSS^hA output.

The block diagram of such a multiplier is shown in fig.3 where four main blocks are sketched:

- a set of h CSS^hA ;
- a Rearranger;
- a pipeline adder with Progressive Reduction of the number of Addends (PRA) ;
- a h bit Carry Look–ahead Adder (CLA) .

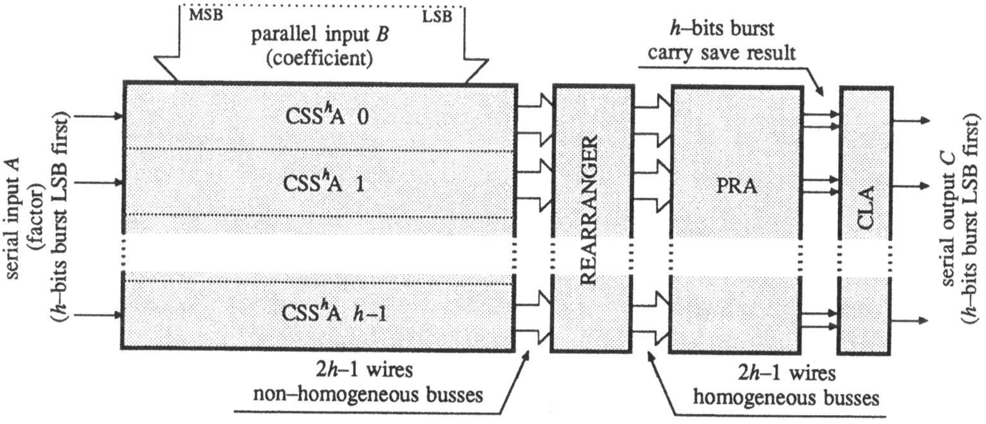

Fig.3 – Block diagram of a Mh multiplier

We shall denote this multiplier by Mh , because it has a bit–scanning coefficient equal to h , that is, its output format, and obviously the serial input format, is constituted by bursts of h consecutive bits each step.

As it can be seen, all the h CSS^hA's have the same parallel input which is, as usual, the multiplicand B (coefficient). The serial inputs are obtained by partitioning the bits of A in h interleaved strings. More precisely, the a_{th+r} bit ($r = 0, 1, \ldots, h-1$) enters the rth CSS^{h}A at the tth step.

It is worthwhile noting that, in this case, the number of outputs of each CSS^{h}A is $2h-1$. The weight of the bits at the output of the rth CSS^{h}A, at the tth step, lies within the interval $[2^{(t-1)h+r}, 2^{(t-1)h+r(h-1)}]$.

The total number P_i of bits having weight $2^{i+(t-1)h}$, generated at each step (by the set of all the CSS^{h}A's), is given by:

$$P_i = \begin{cases} 2i+1 & \text{if } i = 0, 1, \ldots, (h-1) \\ 2(2h-i-1) & \text{if } i = h, \ldots, 2(h-1) \end{cases} \quad . \qquad (4)$$

These binary variables are carried on wires which can be rearranged in such a way to get h groups (busses) of $2h-1$ wires carrying variables having the same weight (homogeneous busses). This is made by the rearranger.

The rearranger firstly groups the outputs of the CSS^{h}A's in order to create $2h-1$ homogeneous busses B_i having P_i wires. Then, it delays the variables on the busses B_i ($i = h, \ldots, 2(h-1)$) by one step and, finally, groups each delayed bus B_i with the bus B_{i-h}, thus obtaining h homogeneous busses having $2h-1$ wires.

The case $h = 4$ is shown in fig.4, where the outputs of the four CSS^{4}A's, the rearranger and the first row of the PRA are depicted. The number of wires P_i is also shown.

The PRA is a multistage non-conventional adder. All binary variables to be summed up are introduced into the first stage. The sum is performed stage by stage in such a way that the number of binary variables present at the input of each stage progressively decrease until the CS representation of the result is obtained at the output.

The CLA gives the UBR representation of the result.

In fig.5 the PRA and the CLA are shown for $h = 4$. As it can be seen, besides the CSS^{h}A's, FA's and latches only constitute the PRA.

Notice that the architecture of the PRA is somewhat similar to the Wallace and Dadda's trees [5].

For a better comprehension of the PRA architecture, let us think of it as partitioned in a certain number of cascade–connected stages each of which is constituted of h cells (intersections between stages and columns in fig.5). As it can be seen in the example of fig.6, if k is the number of inputs to a generic cell, then this cell is constituted of

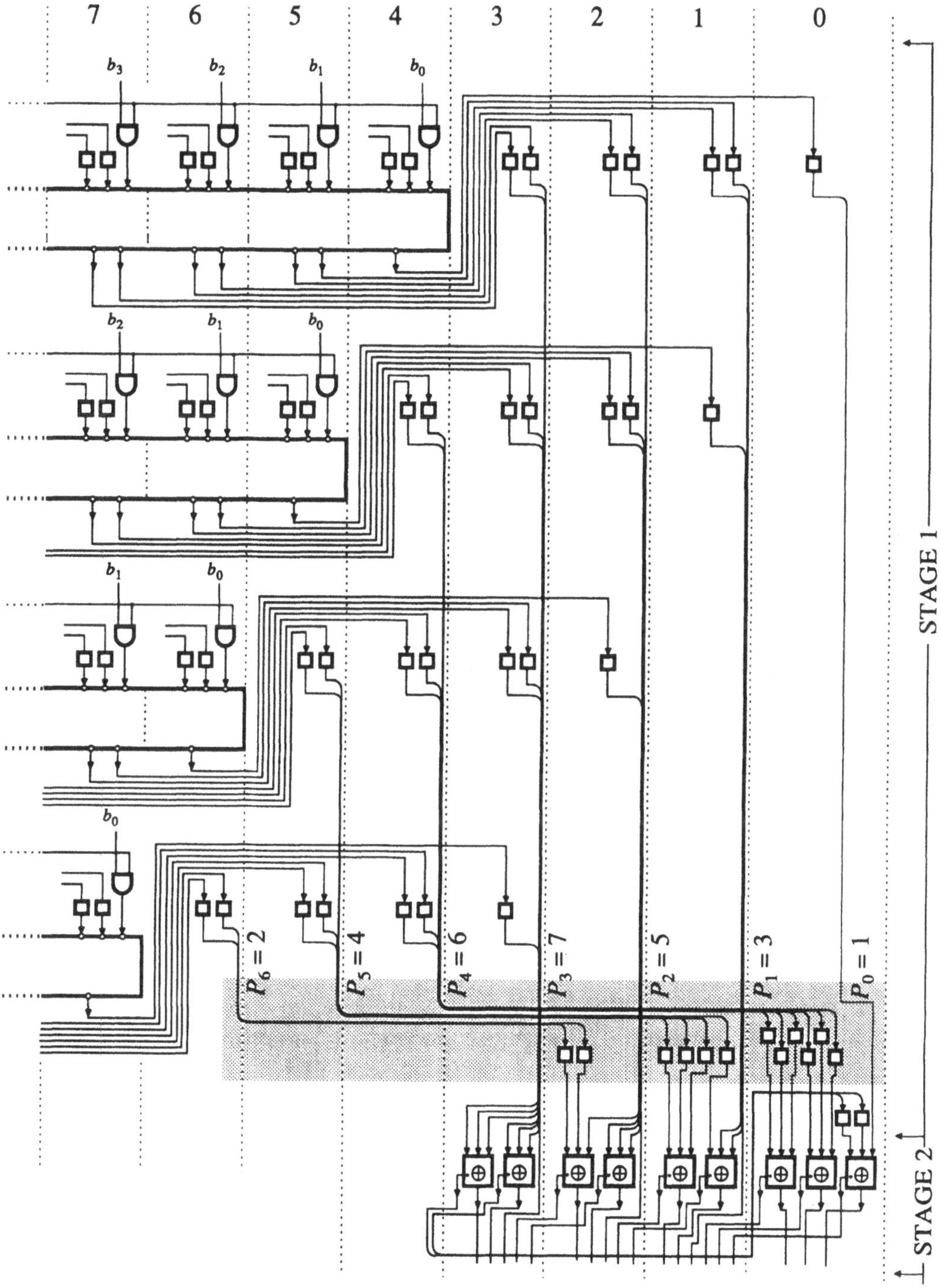

Fig.4 – Rearranger of an M4 multiplier

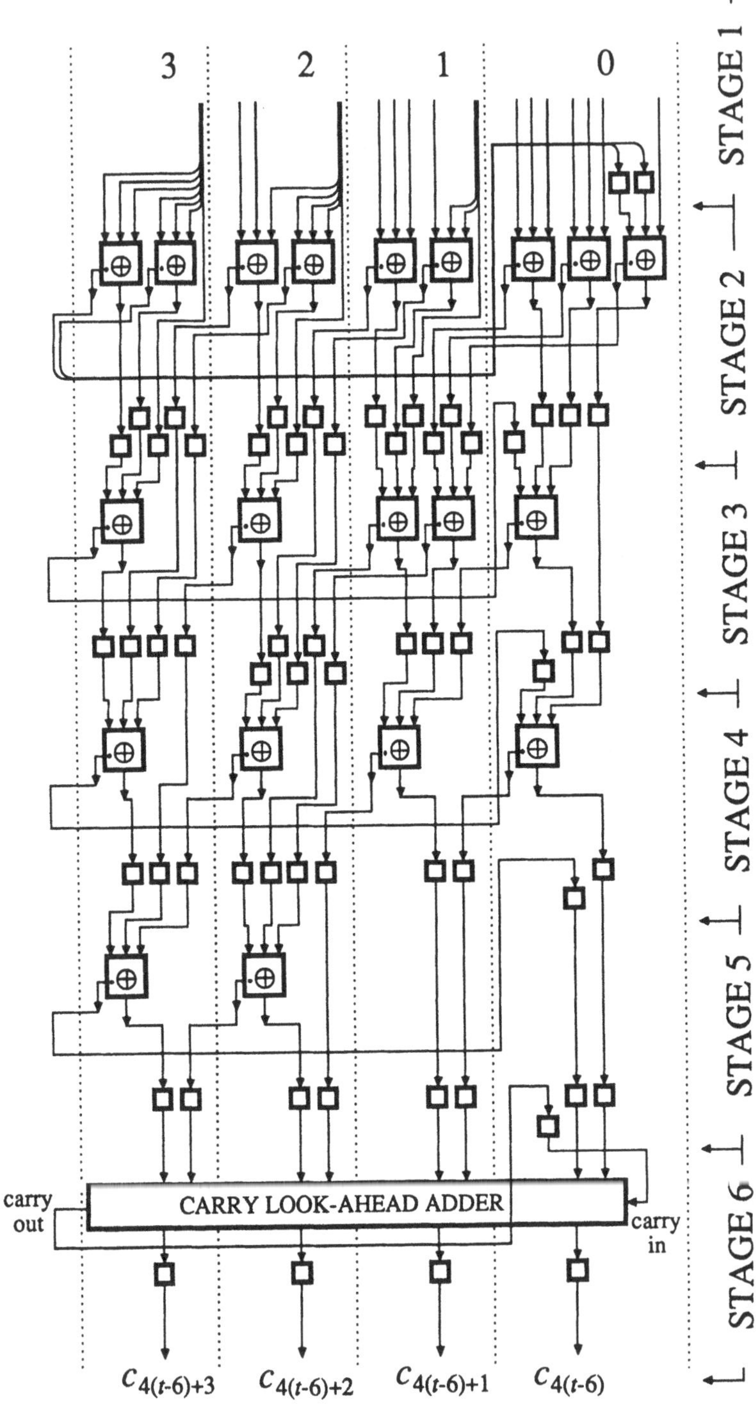

Fig.5 – PRA of an M4 multiplier

$\lfloor k/3 \rfloor$ FA's. The remaining $k - 3\lfloor k/3 \rfloor$ (possibly zero in number) inputs are directly tranferred to the output. From the example of fig.6 (concerning the case $k = 7$), it can be seen that the outputs are partitioned into two homogeneous busses, namely, the Sum bus (S) and the Carry bus (C). S carries variables having the same weight as that of the input variables, while C carries variables having double weight. Obviously, all outputs are sinchronized by means of latches in order to allow the subsequent stages to work pipeline–fashion.

Fig.7 shows the links among the cells. As it can be seen, the S busses proceed vertically through successive stages, while the C busses are left–shifted and then proceed diagonally, except for the leftmost C busses which enter the rightmost cell of its own stage.

The correct working of this scheme can be easily tested by checking up the correctness of the weights $W(t, i, j)$ of the processed variables. This check can be done with the aid of the expression

$$W(t, i, j) = 2^{4t+i-4j}$$

which denotes the weight of each binary variable within the latches of the cell with row (stage) index i and column index j, at the step t.

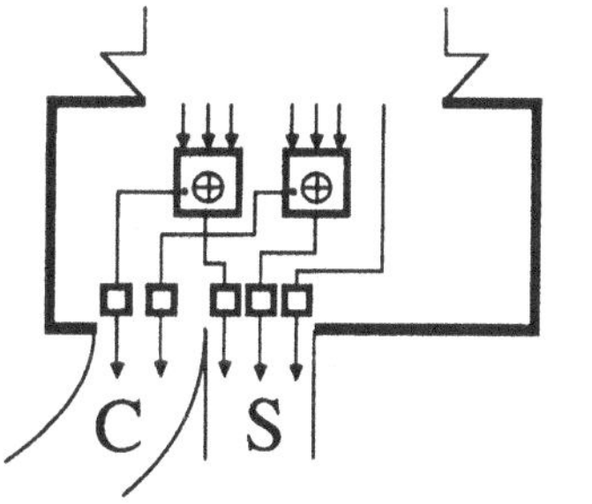

Fig.6 – Example of a 7 inputs PRA cell

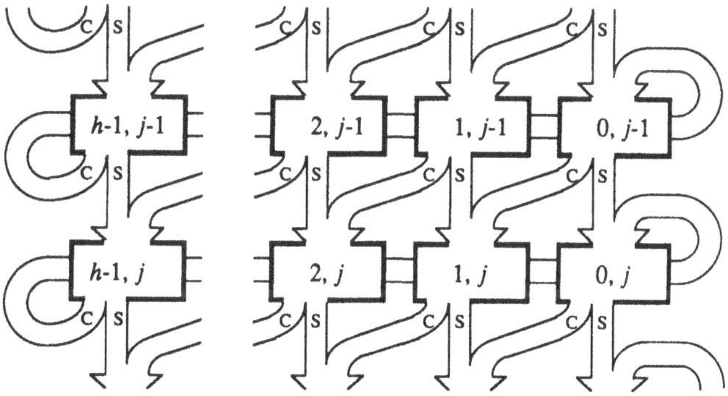

Fig.7 – PRA cell linking

3.1. Remarks

The number of stages required for the implementation of a PRA depends on the value of h and is roughly equal to $\log_{3/2}(2h{-}1)$, as we would expect, due to the fact that 3/2 is just the ratio of the number of inputs to the number of outputs of the FA's.

The behavior of the number of stages of a PRA vs h is shown in the following table, for $2 \le h \le 32$.

Number of input addends							
2	3	4÷5	6÷7	8÷10	11÷14	15÷21	22÷32
2	4	5	6	7	8	9	10
Number of PRA stages							

The logarithmic growth of the number of stages of the PRA as h increases is an important feature because it allows a short start–up, so that the multiplier can operate conveniently also over comparatively small operands.

As to the CLA , we observe that also its carry is delayed of one step and then enters the rightmost input of the CLA itself. This is the only part of the device where the carry is handled and, owing to this fact, it is the only bottleneck of the multiplier. As a matter of fact, since the CLA works cyclically on its carry output, it must be able to compute the carry at the same speed as that of the FA's . Otherwise, the clock frequency should be lowered.

Fortunately, for not too large h (say $h = 4 \div 6$) the CLA can still be implemented by using two levels of gating. For h greater than the said value, the CLA can be implemented by more than two levels of gating by using faster digital technologies (and a separate chip, if necessary) without an appreciable cost growth, owing to the fact that the CLA is a small portion of the multiplier hardware.

Another simple solution to implement large CLA's consists in using two or more pipelined stages constituted of smaller and simplified CLA's (fig.8).

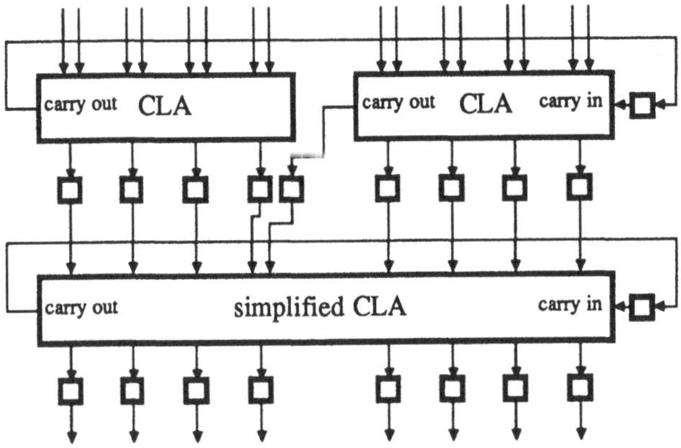

Fig.8 – Example of an 8 bit CLA implemented by means of two pipelined stages

Moreover, it must be noted that, for $h = 8 \div 12$, the multiplier is extremely fast and the use of these values of h makes sense only if the number of bits of the result is some hundreds.

Finally, we point out that, if h is much smaller than $m + n$, the cost of the whole device is $O(hn)$ and it is easily implementable, also in bit-slice technique.

References

[1] Y. Chu: *Digital computer design fundamentals*, New York: McGraw–Hill, 1962.

[2] S.Y.Kung, H.J.Whitehouse: *VLSI and modern signal processing*, Englewood Cliffs (NJ): Prentice–Hall, 1985.

[3] R.F.Lyon: "Two's Complement Pipeline Multipliers", *IEEE Trans on Comm*, ***COM–24***, No.4, Apr. 1976, pp 418–425.

[4] I-Ngo Chen, R.Willoner: "An $O(n)$ Parallel Multiplier with Bit–Sequential Input and Output", *IEEE Trans on Comp*, ***C–28***, No.10, Oct. 1976, pp 721–727.

[5] A.Habibi, P.A.Wintz: "Fast Multipliers", *IEEE Trans on Comp*, ***C–22***, No.2, Feb. 1970, pp 153–157.

On the complexity of computing class groups of algebraic number fields

J. Buchmann and M. Pohst
Mathematisches Institut der Universität Düsseldorf
4000 Düsseldorf, FRG

Abstract

Let n be a fixed natural number, $f \in \mathcal{Z}[x]$ a monic irreducible polynomial of degree n. Let $\mathcal{F} = \mathcal{Q}(\rho)$ be the algebraic number field which is generated by a root ρ of f and assume that $1, \rho, \rho^2, \ldots, \rho^{n-1}$ is a $\mathcal{Z}$-basis of the maximal order $\mathcal{O}$ of $\mathcal{F}$. In this paper we describe an algorithm by which the class group Cl of $\mathcal{F}$ can be computed in $\mathcal{D}^{1+\epsilon}$ binary operations where $\mathcal{D}$ denotes the discriminant of $\mathcal{F}$.

1 Introduction

Let $f \in \mathcal{Z}[x]$ be a monic irreducible polynomial of (fixed) degree n. Over the complex numbers $\mathcal{C}$ the polynomial f has s real and $2t$ complex zeros, $n = s + 2t$. Let $\mathcal{F} = \mathcal{Q}(\rho)$ be the algebraic number field which is generated by a root ρ of f and let $\mathcal{O}$ be its maximal order.

In the (multiplicative) abelian group

$$I = \{\frac{1}{d}\mathbf{a} : \ \mathbf{a} \neq (0) \text{ an ideal of } \mathcal{O}, d \in \mathcal{N}\}$$

of all **fractional ideals** of $\mathcal{O}$ the **principal ideals**

$$H = \{\frac{1}{d}(\alpha) : \ 0 \neq \alpha \in \mathcal{O}, d \in \mathcal{N}\}$$

form a subgroup of finite index. The factor group

$$Cl = I/H$$

is the **ideal class group** of $\mathcal{F}$. Its order is called the **class number** h of $\mathcal{F}$. In case $h > 1$ that group can be uniquely written in the form

$$Cl = \prod_{i=1}^{k} C_i$$

where the subgroups C_i are cyclic of order N_i subject to

$$1 < N_1 | N_2 | \dots | N_k.$$

In this paper we describe an algorithm which determines a generator $\mathbf{a}_i H$ and the order N_i for each cyclic factor C_i. Under the additional assumption that $1, \rho, \rho^2, \dots, \rho^{n-1}$ is a $\mathcal{Z}$-basis of the maximal order $\mathcal{O}$ of $\mathcal{F}$ we prove that our algorithm terminates after at most $\mathcal{D}^{1+\epsilon}$ binary operations where $\mathcal{D}$ is the discriminant of $\mathcal{F}$ and the O-constant only depends on the degree n of $\mathcal{F}$. In a forthcoming paper by J. Buchmann and H. W. Lenstra Jr. it will be shown that the assumption of a power integral basis is not necessary.

Complexity results for class group computations were known so far only for quadratic and complex cubic fields (see Williams [16]).

An efficient algorithm for computing class groups of arbitrary number fields was given by Pohst and Zassenhaus [13]. It was recently used to compute the class groups of all the totally real quartic fields of discriminant below 10^6 (see [4]). The complexity of this algorithm, however, has not yet been analysed.

2 The algorithm

We first compute the set P of all non-zero prime ideals of $\mathcal{O}$ whose norm is below the Minkowski bound

$$M = \frac{n!}{n^n}\left(\frac{4}{\pi}\right)^t \sqrt{|\mathcal{D}|}.$$

It is well known that

$$\{\mathbf{p}H : \mathbf{p} \in P\}$$

is a system of generators of the class group Cl. Each prime ideal $\mathbf{p}$ which divides a prime number p has a two element presentation

$$\mathbf{p} = p\mathcal{O} + \alpha\mathcal{O}$$

with an appropriate number $\alpha \in \mathbf{p}$ which, in turn, can be represented in the form

$$\alpha = \frac{1}{d}\sum_{i=1}^{n} a_i \rho^{i-1}$$

with rational integers a_i and $d > 0$. The denominator d is bounded by the index

$$I = [\mathcal{O} : \mathbb{Z}[\rho]]$$

and the coefficients a_i are bounded by dp.

P is computed by means of

Algorithm 2.1

- **Input** *The coefficients of the generating polynomial f and the Minkowski bound M as well as the prime numbers $p_1, \ldots, p_\ell$ below M.*
- **Output** *The set P of all prime ideals $\mathbf{p}$ with norm below M.*

1. *(Initialization) Set $i \leftarrow 1$, $P \leftarrow \emptyset$.*
2. *For $\gcd(I, p_i) = 1$ decompose f modulo p_i into irreducible factors:*

$$f(x) \equiv f_1^{e_1}(x) \cdots f_g^{e_g}(x) \bmod p_i\mathbb{Z}[x].$$

For each $j \in \{1,\ldots,g\}$ with $p_i^{\deg f_j} \leq M$ add the prime ideal

$$\mathbf{p} = p\mathcal{O} + f_j(\rho)\mathcal{O} \tag{1}$$

to the set P.

In case $\gcd(I, p_i) \neq 1$ apply the algorithm described in Pohst and Zassenhaus [14] to obtain the prime ideals dividing p_i.

3. Set $i \leftarrow i+1$. For $i > \ell$ terminate otherwise go to 2.

The decomposition of $f \bmod p_i$ in step 2 of Algorithm 2.1 is done by means of Berlekamp's algorithm (see [9]) or by the algorithm of Cantor and Zassenhaus [6].

Since each fractional ideal $\mathbf{a}$ of $\mathcal{F}$ is a free $\mathbb{Z}$-module of rank n it can be represented in terms of a $\mathbb{Z}$-basis $\alpha_1, \ldots, \alpha_n$:

$$\alpha_j = \frac{1}{d}\sum_{i=1}^{n} a_{i,j}\rho^{i-1} \quad (1 \leq j \leq n)\,.$$

There exists precisely one such basis such that the matrix $(a_{i,j})$ is in Hermite normal form and $\gcd(d, \{a_{i,j} : 1 \leq i,j \leq n\}) = 1$. This basis is called the **HNF-basis** of $\mathbf{a}$. Correspondingly, the matrix $(a_{i,j})$ is called the **HNF-matrix** of $\mathbf{a}$ and d the **denominator** of $\mathbf{a}$.

The prime ideals $\mathbf{p}$ computed in step 2 of Algorithm 2.1 are stored in terms of their HNF-matrices and their denominators. This representation can easily be computed from the 2-element representation (1) by applying the HNF-algorithm of Kannan and Bachem [8] or by using the modular HNF-algorithm of Domich, Kannan and Trotter [7] (with module p).

Once we have computed P we can determine the structure of Cl. For this purpose it is very important to have an efficient method for testing whether two fractional ideals $\mathbf{a}$ and $\mathbf{b}$ are eqivalent modulo H, i.e. whether $\mathbf{a}/\mathbf{b}$ is a principal ideal. This decision can me made by means of the reduction theory developed in Buchmann [1] and [2]. A fractional ideal $\mathbf{a}$ is called **reduced** if it contains no element $\alpha \neq 0$ all of whose conjugates are in absolute value less than 1. Each ideal class contains a reduced ideal. A reduced ideal $\mathbf{a}'$ in the class of a given ideal $\mathbf{a}$ can be computed by means of the algorithm of Buchmann and Williams [5], it is denoted by

$$\mathbf{a}' = \mathrm{red}\,(\mathbf{a})$$

in the sequel. Hence, we can decide whether two ideals $\mathbf{a}$ and $\mathbf{b}$ are equivalent by checking whether red$(\mathbf{a}/\mathbf{b})$ belongs to the (finite) set $\mathcal{R}$ of all reduced principal ideals which can be computed by means of the algorithm of Buchmann [1]. In order to be able to test ideals for principality, we therefore compute the set $\mathcal{R}$ of all reduced principal ideals of $\mathcal{O}$. Those ideals are stored in terms of their denominators and their HNF-matrices, i.e. as arrays

$$(d, a_{1,1}, a_{2,1}, a_{2,2}, \ldots, a_{n,1}, \ldots, a_{n,n}).$$

Those are ordered lexicographically.

Applying the results of Pohst and Zassenhaus [13] it is clear that the structure of Cl can be determined by

Algorithm 2.2

- **Input** *The set* $P = \{\mathbf{p}_1, \ldots, \mathbf{p}_l\}$ *of prime ideals.*
- **Output** *The orders* N_i *of* C_i *and reduced ideals* $\mathbf{a}_i$ *such that* $<\mathbf{a}_i H> = C_i$ $(1 \le i \le k)$.

  The (lexicographically ordered) set G *of all the reduced ideals of* $\mathcal{O}$ *which also contains for each element* $\mathbf{a}$ *the representation of* $\mathbf{a}H$ *in terms of the basis elements* $\mathbf{a}_i H$.

1. *(Initialization) Set* $k \leftarrow 0$, $i \leftarrow 0$, $G \leftarrow \{\mathcal{O}\}$.
2. *(Increase i) Set* $i \leftarrow i + 1$. *For* $i > l$ *terminate otherwise set* $\mathbf{a} \leftarrow red(\mathbf{p}_i)$.
3. *If* $\mathbf{a} \in G$ *go to 2.*
4.
   - *Compute the order* μ *of* $\mathbf{a}H$ *in* Cl *and factorize* μ.
   - *Determine the least positive integer* λ *such that* $red(\mathbf{a}^\lambda)$ *belongs to* G, *i.e.*

     $$(\mathbf{a}H)^\lambda \prod_{i=1}^{k} (\mathbf{a}_i H)^{\lambda_i} = \mathcal{O}\,.$$

   - *Compute the Smith normal form* $S = diag(\tilde{N}_0, \ldots, \tilde{N}_k)$ *of the matrix*

$$M = \begin{pmatrix} \lambda & 0 & 0 & \cdots & 0 \\ \lambda_1 & N_1 & 0 & \cdots & 0 \\ \vdots & \vdots & \ddots & & \vdots \\ \lambda_k & 0 & 0 & \cdots & N_k \end{pmatrix},$$

$S = UMV$ *where* $U, V \in Gl_{k+1}(\mathbb{Z})$. *Also compute the matrix* $U^{-1} = (\tilde{u}_{i,j})$. *Set* $\mathfrak{a}_0 \leftarrow \mathfrak{a}$.

- *For* $\tilde{N}_0 = 1$ *compute the new generators*

$$\mathfrak{a}_i \leftarrow red(\prod_{j=0}^{k} \mathfrak{a}_j^{\tilde{u}_{i,j}}) \quad (1 \leq i \leq k)$$

and set $N_i \leftarrow \tilde{N}_i$ *for* $1 \leq i \leq k$.

- *For* $\tilde{N}_0 \neq 1$ *compute the new generators*

$$\mathfrak{a}_{i+1} \leftarrow red(\prod_{j=0}^{k} \mathfrak{a}_j^{\tilde{u}_{i,j}}) \quad (0 \leq i \leq k)$$

and set $N_{i+1} \leftarrow \tilde{N}_i$ *for* $0 \leq i \leq k$. *Then increase* k *by* 1.

- *Update* G *and go to 2.*

Once we know the decomposition of *Cl* into its cyclic factors C_i, it is very easy to determine the p-part of *Cl* for each prime number p dividing the class number. We simply have to figure out the p-part of each C_i.

3 The analysis

In this section we analyze the complexity of the method for computing the class group described in the previous section for fixed degree n. All the O-constants used in this section only depend on n.

Since Berlekamps's algorithm factorizes the polynomial f modulo a prime number p in $O(p^{1+\epsilon})$ binary operations and since all the prime numbers under consideration are bounded by the Minkowski constant, the two element representations of all the prime ideals in the set P can be computed in $\mathcal{D}^{1+\epsilon}$ binary operations. Using the results of Kannan and Bachem [8] it is easy to see that the HNF-matrix of each ideal in P can be computed in $O(\mathcal{D}^{\epsilon})$

binary operations. Consequently , the HNF-matrices of the ideals of P can be computed in $\mathcal{D}^{1+\epsilon}$ binary operations and $O(\mathcal{D}^{1/2+\epsilon})$ bits are neccessary to store them all.

It was shown in Buchmann [3] that all the reduced principal ideals of $\mathcal{O}$ can be computed in $O(\mathcal{D}^{1/2+\epsilon})$ binary operations and that all those ideals can be stored by means of their HNF-matrices and their denominators in $O(\mathcal{D}^{1/2+\epsilon})$ bits. Moreover, it follows easily from the results of Buchmann and Williams [5] that each of the ideals which are considered in our algorithm can be reduced in $O(\mathcal{D}^{\epsilon})$ binary operations and that a reduced ideal can be stored in $O(\mathcal{D}^{\epsilon})$ bits.

In Algorithm 2.2 the full system G of representatives for the part of the class group which we know already is stored in lexicographical ordering. Therefore step 3 of Algorithm 2.2 can be executed in $O(\mathcal{D}^{\epsilon})$ binary operations. Moreover, this step is executed at most $O(\mathcal{D}^{1/2+\epsilon})$ times since the list P of prime ideals contains $O(\mathcal{D}^{1/2+\epsilon})$ elements.

Now we know from a theorem of Siegel [15] that the order of the class group is $O(\mathcal{D}^{1/2+\epsilon})$. In step 4 of Algorithm 2.2 the index of the part of the class group whose structure we have determined already in the full class group is diminished by a factor at least 2. Therefore this step is executed $O(\mathcal{D}^{\epsilon})$ times.

Now we analyze the complexity of the computations which have to be carried out in step 4. Again, by Siegel's Theorem the order μ can be computed in $O(\mathcal{D}^{1/2+\epsilon})$ binary operations and this order can be factored in $O(\mathcal{D}^{1/4+\epsilon})$ binary operations. Since the exponent λ of $\mathfrak{a}$ is neccessarily a divisor of the order μ it can be determined in $O(\mathcal{D}^{\epsilon})$ binary operations. The matrix M has $O(\mathcal{D}^{\epsilon})$ columns and rows, its entries are of size $O(\mathcal{D}^{1/2+\epsilon})$ and thus by means of the algorithm of Kannan and Bachem [8] the Smith normal form of this matrix can be computed in $O(\mathcal{D}^{\epsilon})$ binary operations. Finally, by using the sorting techniques described in Knuth [10] the set G can be updated in $O(\mathcal{D}^{1/2+\epsilon})$ binary operations. Thus we have proved

Theorem 3.1 *The structure of the class group of an algebraic number field with power integral basis can be computed in $O(\mathcal{D}^{1+\epsilon})$ binary operations. The storage requirement is $O(\mathcal{D}^{1/2+\epsilon})$ bits.*

References

[1] J. Buchmann, *On the computation of units and class numbers by a generalization of Lagranges's algorithm*, J. Number Theory **26** (1987), 8–30.

[2] J. Buchmann, *On the period length of the generalized Lagrange algorithm*, J. Number Theory **26** (1987), 31–37.

[3] J. Buchmann, *Zur Komplexität der Berechnung von Einheiten und Klassenzahlen algebraischer Zahlkörper*, Habilitationsschrift, Düsseldorf, 1988.

[4] J. Buchmann, J. v. Schmettow and M. Pohst, *On the computation of unit and class groups of totally real quartic fields*, to appear.

[5] J. Buchmann und H.C. Williams, *On principal ideal testing in algebraic number fields*, J. Symbolic Computation **4** (1987), 11–19.

[6] D. Cantor and H. Zassenhaus, *A new algorithm for factoring polynomials over finite fields*, Math. Comp. **36** (1981), 587–592.

[7] P.D. Domich, R. Kannan and L.E. Trotter Jr., *Hermite normal form computation using modulo determinant arithmetic*, Math. Oper. Research **12** (1987), 50–59.

[8] R. Kannan and A. Bachem, *Polynomial algorithms for computing Smith and Hermite normal forms of an integer matrix*, Siam J. Comput. **8** (1979), 499–507.

[9] D.E. Knuth, *The art of computer programming, Vol.2: Seminumerical algorithms*, Addison–Wesley, sec. ed., Reading, Mass., 1982.

[10] D.E. Knuth, *The art of computer programming, Vol.3: Sorting and searching*, Addison–Wesley, Reading, Mass., 1973.

[11] A.K. Lenstra, H.W. Lenstra Jr. und L. Lovasz, *Factoring polynomials with rational coefficients*, Math. Ann. **261** (1982), 515–534.

[12] H.W. Lenstra Jr., *On the computation of regulators and class numbers of quadratic fields*, Lond. Math Soc. Lect. Note Ser. **56** (1982), 123–150.

[13] M. Pohst und H. Zassenhaus, *Über die Berechnung von Klassenzahlen und Klassengruppen algebraischer Zahlkörper*, J. Reine Angew. Math. **361** (1985), 50–72.

[14] M. Pohst und H. Zassenhaus, *Algorithmic algebraic number theory*, Cambridge University Press, to appear 1988.

[15] C.L. Siegel, *Abschätzung von Einheiten*, Ges. Abh. IV , Berlin, Heidelberg, New York 1979, 66–81.

[16] H.C. Williams, *Continued fractions and number theoretic computations*, Rocky Mountain J. Math. **15** (1985), 621–655.

Some new effectivity bounds in computational geometry

Leandro Caniglia[1), André Galligo[2), Joos Heintz[1)

1) Working Group Noaï Fitchas, Instituto Argentino de Matemática, Viamonte 1636, 1er cuerpo, 1er piso, (1055) Buenos Aires, Argentina - mailing address

2) Institut de Mathématique et Sciences Physiques, Université de Nice, Parc Valrose, F - 06034 Nice, France
and INRIA, Sophia Antipolis, Valbonne, France

1. Introduction

The main purpose of this paper consists in giving satisfactory estimates for the worst case complexity of some well-known algorithms in computational geometry. Our upper bounds turn out to be the expected ones by the evidence of the existing literature in this field. Our complexity results are based on some geometrical considerations which include an effective version (indeed an almost optimal one) of Hilbert's Nullstellensatz for algebraically closed fields of *any* characteristic. To the existing literature, our work adds the following aspects which may be new:
First, a didactic one, because our proofs are almost elementary and moreover *independent of the characteristic of the ground field*. So they fill a gap left open by some of the already known results in the domain (see [5]). Secondly, a complexity theoretical one: for each of the geometrical problems we are considering we have algorithms which are essentially optimal, both in their sequential and parallel ("number of processors") complexity.

Let us now state our main results in more detail beginning with the geometrical ones: Let k be an *arbitrary* field with algebraic closure, say $\bar{k}$. Let $X_1,\dots,X_n$ be indeterminates over k and let $F,F_1,\dots,F_s \in k[X_1,\dots,X_n]$ with $d_1 := \deg F_1,\dots,d_s := \deg F_s$ and $d := \max\{d_1,\dots,d_s\}$. Then we have the following results :

(i) (effective Hilbert's Nullstellensatz) $1 \in (F_1,\dots,F_s)$ iff there exist $P_1,\dots,P_s \in k[X_1,\dots,X_n]$ with $\deg P_1,\dots,\deg P_s \leq d^{((n+1)(n+2))/2}$ such that $1 = P_1F_1 + \dots + P_sF_s$.
Closely related to this we have (ii) (effective Hentzelt's Nullstellensatz) $F \in \mathrm{rad}(F_1,\dots,F_s)$ iff $F^{d^{((n+1)(n+2))/2}} \in (F_1,\dots,F_s)$.

Moreover, if $s \leq n$ and $F_1,\dots,F_s$ is a regular sequence then $F \in \mathrm{rad}(F_1,\dots,F_s)$ iff $F^{d_1\cdots d_s} \in (F_1,\dots,F_s)$.

For char $k = 0$ a somewhat sharper version of these results with bounds of type $3n^2d^n$ is due to Brownawell ([5] and [6]). For char $k \neq 0$ our results seem to be new. By a well-known example of Masser-Philippon [5] one deduces that our bounds can't essentially be improved with respect to degree, they have to be at least of type $d^{O(n)}$.

Note added in proof: Based on our methods B. Shiffman [35] has recently extended and improved our results. A slightly improved bound with a direct proof will also appear in a Note by the authors in Comptes Rendus de l'Académie de Sciences, Paris [11]. A $d^{O(n)}$ upper bound for any characteristic has recently been achieved by J. Kollár[24] using non-elementary methods as local cohomology. His result is optimal in terms of the degrees of the generators of the ideal provided $d_1,\dots,d_s > 2$.

As a first consequence we obtain: Let Φ be a prenex first order formula involving m quantifier alternations and the polynomials $F_1,\dots,F_s$. Denote the length of Φ by $|\Phi|$. Then one can construct a quantifier free formula, which is equivalent to Φ in the elementary theory of $\bar{k}$, in sequential time $(sd)^{n^{O(m)}} + O(|\Phi|)$ and in parallel time $n^{O(m)}(\log sd)^{O(1)} + O(\log |\Phi|)$.

Note that the overall complexity of the problem of quantifier elimination over $\bar{k}$ is inherently doubly exponential in sequential time (see [22] and [36]) and simply exponential in parallel time (see [15] and [17]). So, essential improvements of this result (at least in the given parameters which measure the input Φ) can't be expected. A first bound of type $(sd)^{n^{O(n)}}$ for the sequential complexity of quantifier elimination in the elementary theory of $\bar{k}$ has been given in [23],[22] and [37]. A more precise result of type $(sd)^{n^{O(m)}}$ can be found in [13] and [21]. However these results were obtained by algorithms which are *not efficiently parallelizable*.
For char $k = 0$ our result (including parallel complexity) is shown in [15] and [17]. It is new for char k arbitrary.

As a further consequence of the effective Nullstellensatz we obtain fairly good complexity bounds for the following (affine) problems, which are new at least for their parallel aspect. (Compare also [28],[20] and [12]. These papers are dealing with similar problems from a projective point of view.)

Let $V := \{x \in \bar{k}^n;\ F_1(x) = 0,\dots, F_s(x) = 0\}$. We are considering the question to decide whether $V = \emptyset$ and whether $F(x) = 0$ for all $x \in V$. (For simplicity let us assume $\deg F \leq d$.) Moreover, we want to compute $\dim V$ and, in case that all irreducible components of V have equal dimension, $\deg V$.
From our geometrical results we can deduce that all these problems have sequential complexity $s^4 d^{O(n^3)}$ and parallel complexity $O(n^6 \log^2 sd)$.
Let us now suppose that V is finite (i.e. $\dim V \leq 0$). As an easy consequence of our proofs we obtain the inequality $\dim_k k[X_1,\dots,X_n]/(F_1,\dots,F_s) \leq d^n$.
In particular, if $s = n$ and $F_1,\dots,F_n$ is a regular sequence of polynomials we have $\dim_k k[X_1,\dots,X_n]/(F_1,\dots,F_n) \leq d_1 \cdots d_n \leq d^n$.

This implies that for any order of the monomials of $k[X_1,\dots,X_n]$ there exists a Gröbner (standard) basis of $(F_1,\dots,F_s)$ with total degree bounded by d^n. Together with the effective Nullstellensatz we obtain that any Gröbner base algorithm (improved in the sense of [27]) for $(F_1,\dots,F_s)$ has complexity $s^3 d^{O(n^3)}$. In particular, such an algorithm computes automatically a representation $1 = P_1F_1 + \dots + P_sF_s$ with

$\deg P_1, \ldots, \deg P_s = d^{O(n^2)}$ provided that $1 \in (F_1, \ldots, F_s)$. The stairs of $(F_1, \ldots, F_s)$ can be computed in sequential time $s^4 d^{O(n^3)}$ and in parallel time $O(n^6 \log^2 sd)$. In particular, we can compute in sequential time $s^3 d^{O(n^3)}$ a k-vector space base of $k[X_1, \ldots, X_n]/(F_1, \ldots, F_s)$.
Further results (with the same order of complexity) concern the computation of the image of a morphism of affine spaces (in collaboration with M. Chardin and M. Giusti, Ecole Polytechnique, Palaiseau Paris) and a quantitative and effective version of Suslin's Theorem which is related to Serre's Conjecture. This result, which relies on the effective Nullstellensatz, was obtained in collaboration with the working group Noaï Fitchas, Instituto Argentino de Matemática, Buenos Aires.

Now we are going to explain the motivations which led to our results. These are closely related to the existing literature in the field and based on the pioneer work of Buchberger [8], Lazard [28], Chistov-Grigor'ev [12] and Brownawell [5]. In general, problems of computational commutative algebra and geometry are inherently doubly exponential (see [31],[22] and [36]). However, very often, interesting *special* problems turn out to be simply exponential. The first result in this direction was obtained in [28] in the projective case and in [12] in the affine one. [5] resolves an important conjecture (see [4] and [1]) and throws a new light on earlier results.
We want to stress here that the results of [5] and ours give a geometrical interpretation for the improvements of Gröbner base computations introduced by Buchberger and others (see e.g. [27],[9]). In fact, if the variety defined by the input polynomials has some geometrical properties, for example, if it is zero dimensional, the *improved* Gröbner base algorithms do not have double exponential complexity as in the general case. This is known for the projective case [19], but seems to be new for the affine one. Now we can understand geometrically why in case $n = 2$ it is possible to construct a Gröbner base of any ideal using only polynomials of degree $\leq d^2$, where d bounds the degree of the input polynomials. (See [10] and [29].) This and the general upper bound in [18] seem to be the only complexity results with respect to *affine* Gröbner base computations previously known and not relying on projective hypotheses.

Let us also remark that the methods of [5],[7] and [2] give stronger results with respect to the growth of the coefficients of the polynomials involved in the Nullstellensatz. The reader interested only in the degree bounds of the effective Nullstellensatz may find a short and direct proof in [11].

Our methods are elementary. They are based on those used in [22] where the classical Bezout Theorem (which can't be applied without precaution) is replaced by a much simpler, multiplicity free affine version: the Bezout Inequality. Moreover, this Bezout Inequality can be shown in a completely elementary way. In the present paper multiplicity considerations enter in an implicit form, but the proofs remain elementary (Proposition 5 and Theorem 17).

2. Algebraic preliminaries

In this section we suppose that k is an algebraically closed field of *any* characteristic. Let $X_1,\dots,X_n$, $Y_1,\dots,Y_r$ be indeterminates over k. All rings to be considered are (commutative) k-algebras essentially of finite type [25].

Lemma 1 : Let A be a finitely generated $k[Y_1,\dots,Y_r]$-algebra and S a multiplicatively closed subset of A (with $1 \in S$). Suppose that for every minimal prime $p \in \operatorname{Spec} A$ the following holds :

$$p \cap S = \emptyset \quad \Rightarrow \quad \dim_{\text{Krull}} A/p = r .$$

Then $S^{-1}A \otimes_{k[Y_1,\dots,Y_r]} k(Y_1,\dots,Y_r)$ is a finite dimensional $k(Y_1,\dots,Y_r)$-algebra.

Proof : Let $B := k[Y_1,\dots,Y_r]$ and $K := k(Y_1,\dots,Y_r)$. First suppose that A is an integral domain. If the structural morphism $k[Y_1,\dots,Y_r] \longrightarrow A$ is not injective, then $A \otimes_B K = 0$, whence $S^{-1}A \otimes_B K = 0$. Therefore we may assume $k[Y_1,\dots,Y_r] \subset A$. Since $\dim_{\text{Krull}} A = r$, all elements of A are algebraic over $k[Y_1,\dots,Y_r]$. Therefore $A \otimes_B K$ is a finite dimensional K-algebra without zero divisors, hence a field. This implies $S^{-1}A \otimes_B K = A \otimes_B K$ is finite dimensional over K.
General case: Let $P := \{p \in \operatorname{Spec} A;\ p$ minimal with $p \cap S = \emptyset\}$. For $p \in P$ we have: A/p is an integral domain satisfying the hypothesis of the lemma. Hence, by our former argument, $S^{-1}A/S^{-1}p \otimes_B K$ is finite dimensional. Obviously P is finite and $0 \longrightarrow S^{-1}A/(\operatorname{nil}(S^{-1}A)) \otimes_B K \longrightarrow \bigoplus_{p \in P} S^{-1}A/S^{-1}p \otimes_B K$ is exact, where $\operatorname{nil}(S^{-1}A)$ denotes the ideal of nilpotent elements of $S^{-1}A$. Therefore $S^{-1}A/(\operatorname{nil}(S^{-1}A)) \otimes_B K$ is finite dimensional over K. Finally the finite dimensionality of $S^{-1}A \otimes_B K$ over K follows by induction on m considering the exact sequence
$0 \longrightarrow \operatorname{nil}^m(S^{-1}A)/(\operatorname{nil}^{m+1}(S^{-1}A)) \otimes_B K \longrightarrow S^{-1}A/(\operatorname{nil}^{m+1}(S^{-1}A)) \otimes_B K \longrightarrow S^{-1}A/(\operatorname{nil}^m(S^{-1}A)) \otimes_B K$. □

Lemma 2 : Let $y_1,\dots,y_r$ be a regular sequence of a (not necessarily finitely generated) k-algebra A. Consider the morphism $k[Y_1,\dots,Y_r] \longrightarrow A$ which sends $Y_1,\dots,Y_r$ into $y_1,\dots,y_r$ and suppose that $A \otimes_{k[Y_1,\dots,Y_r]} k(Y_1,\dots,Y_r)$ is finite dimensional over $k(Y_1,\dots,Y_r)$. Then $A/(y_1,\dots,y_r)$ is finite dimensional over k with

$$\dim_k A/(y_1,\dots,y_r) \;\le\; \dim_{k(Y_1,\dots,Y_r)} A \otimes_{k[Y_1,\dots,Y_r]} k(Y_1,\dots,Y_r) .$$

Proof : By induction on r, the case $r = 0$ being trivial.
Let $r > 0$. Put $\bar{A} := A/(Y_1)$, $B := k[Y_1,\dots,Y_r]$, $\bar{B} = k[Y_2,\dots,Y_r]$, denote the image of $a \in A$ in $\bar{A}$ by $\bar{a}$, and let $K := k(Y_1,\dots,Y_r)$, $\bar{K} := k(Y_2,\dots,Y_r)$. It suffices to show $\dim_{\bar{K}} \bar{A} \otimes_{\bar{B}} \bar{K} \le \dim_K A \otimes_B K$.
Let $a_1,\dots,a_m \in A$ such that $\bar{a}_1 \otimes 1,\dots,\bar{a}_m \otimes 1$ are $\bar{K}$-linearly independent in $\bar{A} \otimes_{\bar{B}} \bar{K}$. We show that $a_1 \otimes 1,\dots,a_m \otimes 1$ are K-linearly independent in $A \otimes_B K$, thus finishing the proof.
Suppose $a_1 \otimes 1,\dots,a_m \otimes 1$ K-linearly dependent in $A \otimes_B K$. Then there exist $F_1,\dots,F_m \in B$, not all 0, such that $0 = a_1 \otimes F_1 + \dots + a_m \otimes F_m \in A \otimes_B B \cong A$.

Let Y_1^{ℓ} be the highest power of Y_1 which divides all of the $F_1,\dots,F_m$ and let $F'_1 := F_1/Y_1^{\ell},\dots,F'_m := F_m/Y_1^{\ell}$. Then $0 = (a_1 \otimes F'_1 + \dots + a_m \otimes F'_m)\cdot Y_1^{\ell}$. Since y_1 is not a zero divisor in A, Y_1 is not a zero divisor in $A \otimes_B B$. From this we infer $0 = a_1 \otimes F'_1 + \dots + a_m \otimes F'_m \in A \otimes_B B$, whence for the images $\bar{F}'_1,\dots,\bar{F}'_m \in \bar{K}$ of $F'_1,\dots,F'_m \in K$ $0 = \bar{a}_1 \otimes \bar{F}'_1 + \dots + \bar{a}_m \otimes \bar{F}'_m \in \bar{A} \otimes_{\bar{B}} \bar{K}$. By construction not all of the $\bar{F}'_1,\dots,\bar{F}'_m$ are 0, which contradicts the linear independence of $\bar{a}_1 \otimes 1,\dots,\bar{a}_m \otimes 1$. □

Remark 3 : The statement of Lemma 2 is false without any hypothesis of regularity (in fact flatness). Consider the example $A := k[X,Y]/(X^2,XY) = k[x,y]$, x,y being the images of X,Y in A. Since $(X) \cap (X^2,Y)$ is a primary decomposition of (X^2,XY) in $k[X,Y]$, Ax is the unique minimal prime of (0) in A. We have $y \notin (x)$ and therefore $A/Ay \cong k[X]/(X^2)$ has dimension 2 over k.
On the other hand $\dim_{k(Y)} A \otimes_{k[Y]} k(Y) = 1$.

For later applications of Lemma 1 and Lemma 2 we need the following (in the sense of [26]) Bertini-type result :

Lemma 4 : Let $F \in k[X_1,\dots,X_n]$ and assume that F is not a p-th power of any polynomial if $\mathrm{char}\, k = p \neq 0$. Then there exists a nonempty Zariski-open set $O \subset k$ such that $F + t$ is square free for all $t \in O$.

Proof : By hypothesis we may assume $\frac{\partial F}{\partial X_1} \neq 0$. Let $G_1,\dots,G_N \in k[X_1,\dots,X_n]$ be all the factors of $\frac{\partial F}{\partial X_1}$. For $1 \leq j \leq N$ let $\Lambda_j := \{\lambda \in k \,;\, G_j^2 \text{ divides } F + \lambda\}$. Since $\#\Lambda_j \leq 1$, we may put $O := k \setminus \bigcup_{1\leq j\leq N} \Lambda_j$. □

3. A refined Bezout Inequality

Proposition 5 : Let $p \subset k[X_1,\dots,X_n]$ be a prime ideal and denote the degree of the variety V in k^n defined by p by $\deg p$. Let $A := k[X_1,\dots,X_n]/p$ and denote the quotient field of A by K. Let $F_1,\dots,F_r \in k[X_1,\dots,X_n]$ with images $f_1,\dots,f_r$ in A. Suppose that the following hypotheses are satisfied :

(i) $\dim_{\mathrm{Krull}} A = r$

(ii) $(f_1,\dots,f_r)$ is a proper ideal of A and $\dim_{\mathrm{Krull}} A/(f_1,\dots,f_r) = 0$.

Then $$[K : k(f_1,\dots,f_r)] \leq \deg p \prod_{1\leq i\leq r} \deg F_i . \qquad (*)$$

Before starting the proof let us consider the following special case :

Remark 6 : Let $\mathrm{char}\, k = 0$. Then things are rather simple: by the Theorem on Fibres ([34], Chapter I, § 6, Theorem 7) the hypotheses (i) and (ii) imply that the morphism of affine varieties $\chi : V \to k^r$ defined by $\chi(x) = (f_1(x),\dots,f_r(x))$ for $x \in V$ is dominating with generically finite fibres.

To χ there corresponds the finite field extension $k(f_1,\dots,f_r) \subset K$. Let h be the cardinality of a generic fibre of χ.
Then $h = [K : k(f_1,\dots,f_r)]_{sep} = [K : k(f_1,\dots,f_r)]$ by [22], Proposition 1, and the hypothesis char $k = 0$. On the other hand, for suitable $\lambda_1,\dots,\lambda_r \in k$ we have by the Bezout Inequality ([22], Theorem 1)
$h = \# V \cap \{F_1 - \lambda_1 = 0,\dots, F_r - \lambda_r = 0\} \leq \deg V \cdot \prod_{1\leq i\leq r} \deg F_i = \deg p \cdot \prod_{1\leq i\leq r} \deg F_i$.
This implies the assertion of Proposition 5 in case char $k = 0$.

Now let us consider the case of *arbitrary* characteristic.

Proof of Proposition 5 : Without loss of generality we may assume that $F_1,\dots,F_r$ are not p-th powers of any polynomial if char $k = p \neq 0$. Let us begin with some observations.

Observation 1 : For any $\lambda_1,\dots,\lambda_r \in k$ (*) is equivalent to
$[K : k(f_1,\dots,f_r)] \leq \deg p \prod_{1\leq i\leq r} \deg(F_i - \lambda_i)$.

Observation 2 : There exists $g \in k[f_1,\dots,f_r]$, $g \neq 0$, such that A_g is a free $k[f_1,\dots,f_r]_g$ -module of finite rank.

Proof : (i) and (ii) imply that K is a finite extension of $k(f_1,\dots,f_r)$. So there exists $g \in k[f_1,\dots,f_r]$, $g \neq 0$, such that A_g is a finite $k[f_1,\dots,f_r]_g$ -module. Moreover, since $k[f_1,\dots,f_r]$ is an integral domain, g can be chosen in such a way that A_g is a *free* $k[f_1,\dots,f_r]_g$ -module. □

Observation 3 : Let $g \in k[f_1,\dots,f_r]$ as in Observation 2.
There exist $\lambda_1,\dots,\lambda_r \in k$ such that
a) $F_1 - \lambda_1,\dots, F_r - \lambda_r \in k[X_1,\dots,X_n]$ are square free
b) $f_1 - \lambda_1,\dots, f_r - \lambda_r$ is a regular sequence in A
c) $g(\lambda_1,\dots,\lambda_r) \neq 0$. (Note: $f_1,\dots,f_r$ are algebraically independent over k.)

Proof : Let $G \in k[Y_1,\dots,Y_r]$ such that $g = G(f_1,\dots,f_r)$. Since $G \neq 0$, Lemma 4 implies that there exists $\lambda_1 \in k$ such that $F_1 - \lambda_1 \in k[X_1,\dots,X_n]$ is square free, $f_1 - \lambda_1 \neq 0$, and $G(\lambda_1, Y_2,\dots,Y_r) \neq 0$.
Suppose now that we have chosen $\lambda_1,\dots,\lambda_i \in k$, $i < r$, such that $F_1 - \lambda_1,\dots, F_i - \lambda_i \in k[X_1,\dots,X_n]$ are square free, $f_1 - \lambda_1,\dots, f_i - \lambda_i$ is a regular sequence in A, and $G(\lambda_1,\dots,\lambda_i, Y_{i+1},\dots,Y_r) \neq 0$ in $k[Y_{i+1},\dots,Y_r]$. First we observe that $(f_1 - \lambda_1,\dots, f_i - \lambda_i)$ is a proper ideal. To see this we choose $\lambda'_{i+1},\dots,\lambda'_r \in k$ such that $G(\lambda_1,\dots,\lambda_i,\lambda'_{i+1},\dots,\lambda'_r) \neq 0$. Since K is algebraic over $k(f_1,\dots,f_r)$ and $\dim_{Krull} A = \mathrm{transdeg}_k(K) = r$, $f_1,\dots,f_r$ are algebraically independent over k. We consider the morphism $\phi : k[f_1,\dots,f_r] \longrightarrow k$ which sends $f_1,\dots,f_i$ to $\lambda_1,\dots,\lambda_i$ and $f_{i+1},\dots,f_r$ to $\lambda'_{i+1},\dots,\lambda'_r$. Since $0 \neq G(\lambda_1,\dots,\lambda_i,\lambda'_{i+1},\dots,\lambda'_r) = \phi(g)$, ϕ can be extended to $\phi : k[f_1,\dots,f_r]_g \longrightarrow k$. But A_g is integral over $k[f_1,\dots,f_r]_g$. So by the Going-up-Theorem ϕ can be extended to a morphism $\phi : A_g \longrightarrow k$ which corresponds to a common zero of $F_1 - \lambda_1,\dots, F_i - \lambda_i$ in V.

With other words, $(f_1-\lambda_1,\dots,f_i-\lambda_i)$ is a proper ideal of A.
Let $\mathcal{Q}=\{q;\ q\subset A$ associated prime of $(f_1-\lambda_1,\dots,f_i-\lambda_i)\}$. For $q\in\mathcal{Q}$ let $\Lambda_q:=\{\lambda\in k;\ f_{i+1}-\lambda\in q\}$. Since $\#\Lambda_q\leq 1$, by Lemma 4 we find $\lambda_{i+1}\in k$ such that $F_{i+1}-\lambda_{i+1}$ is square free, $\lambda_{i+1}\notin\bigcup_{q\in\mathcal{Q}}\Lambda_q$, and $G(\lambda_1,\dots,\lambda_{i+1},Y_{i+2},\dots,Y_r)\neq 0$.
By the choice of λ_{i+1} we have $f_{i+1}-\lambda_{i+1}\notin q$ for all $q\in\mathcal{Q}$.
Therefore $f_1-\lambda_1,\dots,f_{i+1}-\lambda_{i+1}$ is regular. By the way, $(f_1-y_1,\dots,f_r-\lambda_r)$ is a maximal ideal of $k[f_1,\dots,f_r]_g$. □

Observations 1,2,3 imply that we may assume without loss of generality that there exists $G\in k[Y_1,\dots,Y_r]$ with $G(0,\dots,0)\neq 0$ such that for $g:=G(f_1,\dots,f_r)$ A_g is a free $k[f_1,\dots,f_r]_g$-module.
Therefore $[K:k(f_1,\dots,f_r)] = \mathrm{rk}_{k[f_1,\dots,f_r]_g} A_g = \dim_k A/(f_1,\dots,f_r)$. In particular $V\cap\{F_1=0,\dots,F_r=0\}$ is finite.
We are going to show

$$\dim_k A/(f_1,\dots,f_r)\leq \deg p\prod_{1\leq i\leq r}\deg F_i \qquad (**)$$

thus finishing the proof.

Let $x_1,\dots,x_n$ be the coordinate functions of A. We introduce rn new indeterminates over k:

$$X_{n+1},\dots,X_{2n};\ X_{2n+1},\dots\quad\dots\quad;\ X_{rn+1},\dots,X_{(r+1)n}.$$

We consider the following finitely generated reduced k-algebras defining equidimensional algebraic submanifolds of $k^{(r+1)n}$:

$$A'' := A\otimes_k k[X_{n+1},\dots,X_{2n}]\otimes_k\dots\otimes_k k[X_{rn+1},\dots,X_{(r+1)n}]$$
$$A' := A\otimes_k k[X_{n+1},\dots,X_{2n}]/(F_1')\otimes_k\dots\otimes_k k[X_{rn+1},\dots,X_{(r+1)n}]/(F_r'),$$

where $F_i':=F_i(X_{in+1},\dots,X_{(i+1)n})$, $1\leq i\leq r$. We denote the coordinate functions of A' by $x_1,\dots,x_n;\ x_{n+1},\dots,x_{2n};\ \dots;\ x_{rn+1},\dots,x_{(r+1)n}$.
Let A_{ij}, $n+1\leq i\leq(r+1)n$, $1\leq j\leq(r+1)n$, and Z_ℓ, $n+1\leq\ell\leq(r+1)n$ be new indeterminates. Consider the morphism $\phi: k[A_{ij},Z_\ell;\ i,j,\ell]\longrightarrow k[A_{ij};\ i,j]\otimes_k A'$ defined by $\phi(A_{ij}):=A_{ij}$, $\phi(Z_\ell):=\sum_{1\leq j\leq(r+1)n}A_{\ell j}x_j$. In fact ϕ induces a morphism of affine manifolds whose fibres can be interpreted as intersections of the rn-dimensional equidimensional submanifold $V\times\{F_1=0\}\times\dots\times\{F_r=0\}$ of $k^{(r+1)n}$ with rn affine hyperplanes (compare [22]). To ϕ there corresponds a separable $k(A_{ij},Z_\ell;\ i,j,\ell)$-algebra of dimension $\deg p\cdot\prod_{1\leq i\leq r}\deg F_i$. This follows easily from [22], Lemma 1, Proposition 2, and Remark 2(3), using the hypothesis that $F_1,\dots,F_r$ are square free.

Next we are going to identify all the k-algebras $k[X_{in+1},\dots,X_{(i+1)n}]$ with $k[X_1,\dots,X_n]$, which means that we intersect our product manifold $V\times\{F_1=0\}\times\dots\times\{F_r=0\}$ with the diagonal.

For $n+1\leq\ell\leq(r+1)n$ let $\bar{\ell}\in\{n,1,\dots,n-1\}$ be the unique number congruent ℓ modulo n.

Putting $\alpha_{\ell j} = \begin{cases} 1 & \text{if } j = \bar{\ell} \\ -1 & \text{if } j = \ell \\ 0 & \text{if } j \neq \ell, \bar{\ell} \end{cases}$ we obtain $\sum_j \alpha_{\ell j} X_j = X_{\bar{\ell}} - X_\ell$.

Specializing A_{ij} to α_{ij} and Z_ℓ to 0 , we pass from A'' to A and from A' to $A/(f_1,\ldots,f_r)$.

Having this in mind, we "change" ϕ into $\tilde{\Phi} : k[A_{ij}, Z_\ell ; i,j,\ell] \longrightarrow k[A_{ij} ; i,j] \otimes_k A''$, given by $\tilde{\Phi}(A_{ij}) = A_{ij} - \alpha_{ij}$, $\tilde{\Phi}(Z_\ell) = \sum_{1\leq j\leq n} A_{\ell j} x_j + \sum_{n+1\leq j'\leq (r+1)n} A_{\ell j'} X_{j'}$ and into $\tilde{\phi} : k[A_{ij}, Z_\ell ; i,j,\ell] \longrightarrow k[A_{ij} ; i,j] \otimes_k A'$ given by $\tilde{\phi}(A_{ij}) = A_{ij} - \alpha_{ij}$, $\tilde{\phi}(Z_\ell) = \sum_{1\leq j\leq (r+1)n} A_{\ell j} x_j$. $\tilde{\phi}$ is nothing else but a slight pertubation of ϕ .

Therefore to $\tilde{\phi}$ there corresponds a $k(A_{ij}, Z_\ell ; i,j,\ell)$ -algebra of dimension $\deg p \cdot \prod_{1\leq i\leq r} \deg F_i$. Moreover, for any ordering of the indices i, j and ℓ

$A_{ij} - \alpha_{ij}$, $n+1 \leq i \leq (r+1)n$, $1 \leq j \leq (r+1)n$; $\sum_{1\leq j\leq n} A_{\ell j} x_j + \sum_{n+1\leq j'\leq (r+1)n} A_{\ell j'} X_{j'}$, $n+1 \leq \ell \leq (r+1)n$, $F'_1,\ldots,F'_r$ is a regular sequence of $k[A_{ij} ; i,j] \otimes_k A''$.

This can be seen in the following way: dividing successively the $A_{ij} - \alpha_{ij}$ one passes from one integral domain to the other. The same holds for the successive divisions by the $\sum_{1\leq j\leq n} A_{\ell j} x_j + \sum_{n+1\leq j'\leq (r+1)n} A_{\ell j'} X_{j'}$. Finally the regularity of the divisions by $F'_1,\ldots,F'_r$ follows from the regularity of $f_1,\ldots,f_r$ in A .

The next step consists in changing the order of the regular sequence just constructed in such a way that $F'_1,\ldots,F'_r$ appear at the beginning of the new regular sequence. To make this possible we have to localize $k[A_{ij} ; i,j] \otimes_k A''$ suitably.

For each $P := (a_1,\ldots,a_n)$ of the finite set $V \cap \{F_1 = 0,\ldots,F_r = 0\}$ let $m_P \in \mathrm{Spec}_{max}(k[A_{ij} ; i,j] \otimes_k A'')$ be the maximal ideal generated by $A_{ij} - \alpha_{ij}$ for all i,j, by $x_\ell - a_\ell$ for $1 \leq \ell \leq n$, and by $X_\ell - a_{\bar{\ell}}$ for $n+1 \leq \ell \leq (r+1)n$.

Let S'' be the complement of $\bigcup_{P\in V\cap\{F_1=0,\ldots,F_r=0\}} m_P$. S'' is a multiplicatively closed subset of $k[A_{ij} ; i,j] \otimes_k A''$, $S''^{-1} k[A_{ij} ; i,j] \otimes_k A''$ is semi-local with Jacobson radical J , and

$A_{ij} - \alpha_{ij}$ for all i,j ; $\sum_{1\leq j\leq n} A_{\ell j} x_j + \sum_{n+1\leq j'\leq (r+1)n} A_{\ell j'} X_{j'}$ for all ℓ ; $F'_1,\ldots,F'_r$ is a J-regular sequence in this ring. Therefore, by [30], Theorem 28,

$F'_1,\ldots,F'_r$; $A_{ij} - \alpha_{ij}$ for all i,j ; $\sum_{1\leq j\leq n} A_{\ell j} x_j + \sum_{n+1\leq j'\leq (r+1)n} A_{\ell j'} X_{j'}$ for all ℓ is a J-regular sequence in this ring, too. Let S' be the image of S'' under the canonical morphism $k[A_{ij} ; i,j] \otimes_k A'' \longrightarrow k[A_{ij} ; i,j] \otimes_k A'$ whose kernel is generated by $F'_1,\ldots,F'_r$. Consequently, $A_{ij} - \alpha_{ij}$ for all i,j ; $\sum_{1\leq j\leq (r+1)n} A_{\ell j} x_j$ for all ℓ is a J'-regular sequence of the semi-local ring $S'^{-1} k[A_{ij} ; i,j] \otimes_k A'$ with Jacobson radical denoted by J' .

Therefore the morphism $\tilde{\phi} : k[A_{ij}, Z_\ell ; i,j,\ell] \longrightarrow S'^{-1} k[A_{ij} ; i,j] \otimes_k A'$ (with suitably extended domain) satisfies the hypotheses of Lemma 2 . This implies

$$\dim_k A/(f_1,\ldots,f_r) \leq \dim_K (S^{-1} k[A_{ij} ; i,j] \otimes_k A') \otimes_R K \leq \deg p \prod_{1\leq i\leq r} \deg F_i ,$$

where $R := k[A_{ij}, Z_\ell ; i,j,\ell]$ and K its fraction field. □

Corollary 7 : Let $\mathfrak{p} \subset k[X_1,\dots,X_n]$ be a prime ideal which defines an irreducible closed subset V of k^n, and let $A := k[X_1,\dots,X_n]/\mathfrak{p} = k[x_1,\dots,x_n]$, where $x_1,\dots,x_n$ are the images of $X_1,\dots,X_n$ in A. Let $r := \dim_{\text{Krull}} A = \dim V$ and let K be the fraction field of A. Moreover let $F_1,\dots,F_m \in k[X_1,\dots,X_n]$, $m \leq r$, with images $f_1,\dots,f_m \in A$, such that $\{f_1 = 0,\dots, f_m = 0\}$ is a (set-theoretical) complete (non-empty) intersection in V. With other words, assume that $\{f_1 = 0,\dots,f_m = 0\}$ is non-empty, every component being of dimension $r-m$. Choose $y_{m+1},\dots,y_r$ as "generic" linear combinations of $x_1,\dots,x_n$ (over k).
Then $[K : k(f_1,\dots,f_m, y_{m+1},\dots,y_r)] \leq \deg \mathfrak{p} \cdot \prod_{1\leq i\leq m} \deg F_i$.

Proof : Observe that by the hypotheses $f_1,\dots,f_m, y_{m+1},\dots,y_r$ induce a dominating morphism $V \longrightarrow k^r$ and that the fibre of $(0,\dots,0)$ is non-empty and finite. Therefore only finitely many maximal ideals contain the proper ideal $(f_1,\dots,f_m,y_{m+1},\dots,y_r)$ of A. Hence the hypotheses (i) and (ii) of Proposition 5 are satisfied. On the other hand $y_{m+1},\dots,y_r$ are images of linear polynomials of $k[X_1,\dots,X_n]$. The assertion follows now from the conclusion of Proposition 5. □

4. A Normalization Lemma

Proposition 8 : Let hypotheses and notations be as in Corollary 7. Let B be the integral closure of $k[f_1,\dots,f_m, y_{m+1},\dots,y_r]$.
Then there exists $t \in B$ with $t = a_1f_1 + \dots + a_mf_m + 1$, $a_1,\dots,a_m \in A$, such that $A_t = B_t$.

Proof : Let K be the fraction field of A. Choosing $y_{m+1},\dots,y_r$ as generic linear combinations of $x_1,\dots,x_n$, we may suppose that $y_{m+1},\dots,y_r$ satisfy the condition of Noether's Normalization Theorem for every component of $\{f_1 = 0,\dots,f_m = 0\}$ (compare [22], Lemma 1).
We consider the following diagram of ring extensions

$$\begin{array}{ccc} A & \subset & K \\ \cup & & \\ B & & \cup \\ \cup & & \\ k[f_1,\dots,f_m, y_{m+1},\dots,y_r] & \subset & k(f_1,\dots,f_m, y_{m+1},\dots,y_r) \end{array} ,$$

where B is the integral closure of $k[f_1,\dots,f_m,y_{m+1},\dots,y_r]$ in A.
By the geometrical hypothesis on $f_1,\dots,f_m$ and the choice of $y_{m+1},\dots,y_r$ we conclude that $f_1,\dots,f_m,y_{m+1},\dots,y_r$ are algebraically independent.
So $k[f_1,\dots,f_m,y_{m+1},\dots,y_r]$ is integrally closed (in its quotient field). Moreover, $A/(f_1,\dots,f_m)$ is a finite $k[y_{m+1},\dots,y_r]$-module by the choice of $y_{m+1},\dots,y_r$ and the proof of Lemma 1.
Let $S := \{a_1f_1 + \dots + a_mf_m + 1;\ a_1,\dots,a_m \in A\}$ and $T := S \cap B$. Using Zariski's Main Theorem ([25], Chapter IV,2, Theorem 2.1) we are going to show $T^{-1}B = T^{-1}A$. We consider $T^{-1}A$ as a $T^{-1}B$-module. In order to show $T^{-1}B = T^{-1}A$ by means of Zariski's

Main Theorem, it suffices to verify that each maximal ideal n of $T^{-1}B$ has a non-empty fibre in $T^{-1}A$ consisting only of maximal ideals of $T^{-1}A$.
Let n be a maximal ideal of $T^{-1}B$.

<u>Observation 1</u>: $(Af_1 + \dots + Af_m) \cap B \subset n \cap B$.

<u>Proof</u>: Let $b = a_1 f_1 + \dots + a_m f_m \in B \setminus n$ with $a_1, \dots, a_m \in A$. By the maximality of n in $T^{-1}B$ we have $b\,T^{-1}B + n = T^{-1}B$. Therefore there exist $t \in T$, $x \in n \cap B$, and $b' \in B$ such that $b'b + x = t$. Recalling the definition of T we see $x = t - b'b =$ $= t - (b'a_1 f_1 + \dots + b'a_m f_m) \in n \cap B \cap T = \emptyset$. Contradiction. □

<u>Observation 2</u>: $n \cap B$ is maximal in B.

<u>Proof</u>: By Observation 1 the image $\bar{T}$ of T in $B/(n \cap B)$ is $\{1\}$. Therefore $B/(n \cap B) \cong \bar{T}^{-1}B/(n \cap B) \cong T^{-1}B/n$ is a field. Now the assertion follows. □

Observations 1 and 2 imply that $n \cap B$ is the kernel of a morphism $\phi : B \to k$ with $\phi((Af_1 + \dots + Af_m) \cap B) = 0$.
ϕ induces therefore a morphism $\bar{\phi} : B/((Af_1 + \dots + Af_m) \cap B) \to k$. We consider now the following commutative diagram

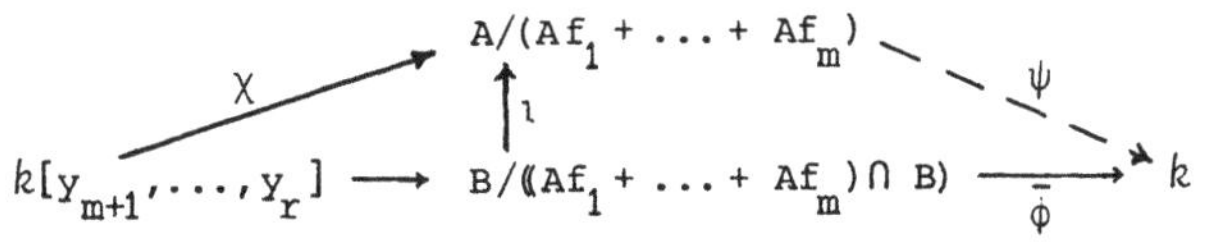

ι is injective and (with respect to χ) we can consider $A/(Af_1 + \dots + Af_m)$ as a finite $k[y_{m+1}, \dots, y_r]$-module. Therefore $A/(Af_1 + \dots + Af_m)$ is a finite $B/(Af_1 + \dots + Af_m)$-module. Since ι is injective, we can apply the Going-up-Theorem obtaining thus a morphism ψ extending $\bar{\phi}$. The kernel of ψ corresponds to a maximal ideal m of A with $Af_1 + \dots + Af_m \subset m$. Therefore $S \cap m = \emptyset$ which implies $T \cap m = \emptyset$. Moreover, since ψ extends $\bar{\phi}$, $n \cap B \subset m$. From this we deduce that $T^{-1}m$ is a maximal ideal of $T^{-1}A$ with $T^{-1}m \cap T^{-1}B = n$. Hence the fibre of n in $T^{-1}A$ is not empty.
Now suppose that $p \in \operatorname{Spec} T^{-1}A$ is in the fibre of n. We show that p is maximal. By Observation 2 we have $p \cap B = m \cap B$. To n there corresponds a maximal ideal $\bar{n}$ of $B/((Af_1 + \dots + Af_m) \cap B)$ by Observation 1. Since $f_1, \dots, f_m \in p$, to p there corresponds a prime ideal $\bar{p}$ of $A/(Af_1 + \dots + Af_m)$ lying over $\bar{n}$.
Since $A/(Af_1 + \dots + Af_m)$ is a finite $B/((Af_1 + \dots + Af_m) \cap B)$-module and ι is injective, $\bar{p}$ has to be maximal by the Going-up-Theorem. This implies p maximal.
$T^{-1}A = T^{-1}B$ follows now by Zariski's Main Theorem.
Since A is finitely generated over k there exists $t \in T$ such that $A_t = B_t$, $t \in B$, and t has the form $t = a_1 f_1 + \dots + a_m f_m + 1$ with $a_1, \dots, a_m \in A$. This finishes the proof of Proposition 8. □

The projective case is much simpler. This is the content of

Remark 9 : Let hypotheses and notations be as in Corollary 7. Suppose that p is homogeneous. Then A is a graded k-algebra. Suppose that $F_1,\dots,F_m \in k[X_1,\dots,X_n]$ are homogeneous. Choose $y_{m+1},\dots,y_r$ as "generic" homogeneous elements of degree 1. Then $k[f_1,\dots,f_m, y_{m+1},\dots,y_r] \subset A$ is an integral extension and

$$[K : k(f_1,\dots,f_m, y_{m+1},\dots,y_r)] \leq \deg p \cdot \prod_{1\leq i\leq m} \deg f_i .$$

We observe that in this case no localization is necessary. Moreover, the proof is completely elementary (no Zariski's Main Theorem and no Bezout Inequality).

Proof (Sketch): We proceed step by step. For simplicity assume $m=2$ and $d_1 := \deg f_1$, $d_2 := \deg f_2 > 0$. By Noether's Normalization Theorem there exist $y_1,\dots,y_r \in A$ homogeneous of degree 1 and algebraically independent over k such that $E := k[y_1,\dots,y_r] \subset A$ is an integral extension. Moreover E is graded, too, and the inclusion map is homogeneous of degree 0. We choose $y_1,\dots,y_r$ generically.
Let L be the fraction field of E. E is integrally closed in L. Therefore the minimal polynomial of f_1 over L has coefficients in E.
Let $q := [L(f_1):L]$. Since A is graded and f_1 homogeneous of degree d_1, the minimal polynomial of f_1 over L has the form $T^q + a_{q-1}T^{q-1} + \dots + a_0$, where $a_{q-1},\dots,a_0 \in E$ are homogeneous with $a_i = 0$ or $\deg a_i = (q-i)d_1$, $0 \leq i \leq q-1$, and T is a new indeterminate.
Since A is a domain, we have $a_0 \neq 0$. Hence $\deg a_0 = q\, d_1 > (q-i)d_1 \geq \deg a_i$, $1 \leq i \leq q-1$. a_0 is a homogeneous polynomial of degree qd_1 in the algebraically independent elements $y_1,\dots,y_r$. By the generic choice of $y_1,\dots,y_r$ we may assume $\deg_{y_1} a_0 = \deg a_0 = q\,d_1 > \deg a_i \geq \deg_{y_1} a_i$ for $i = 1,\dots,q-1$. Therefore

$$f_1^{\,q} + a_{q-1} f^{q-1} + \dots + a_0 = 0 \tag{*}$$

expresses an integral dependence of y_1 over $E' := k[f_1, y_2,\dots,y_r]$. Observing that $f_1, y_2,\dots,y_r$ are algebraically independent over k we see that E' is integrally closed in its fraction field L'. Reading (*) as an equation for y_1 over L' we see $[L'(y_1) : L'] \leq q\, d_1$. By the generic choice of $y_1,\dots,y_r$ we have $\deg p = [K : L]$. Moreover $[K:L] = [K:L(f_1)]\cdot[L(f_1):L] = [K:L(f_1)]\cdot q = [K:L'(y_1)]\cdot q$, since $L(f_1) = L'(y_1)$. Therefore $\deg p = [K:L'(y_1)]\cdot q$. Hence $[K:k(f_1,y_2,\dots,y_r)] = [K:L'] = [K:L'(y_1)]\cdot[L'(y_1):L'] \leq [K:L'(y_1)]\,q\,d_1 = \deg p \cdot d_1$.

Let N be the normal closure of K over L' and C the integral closure of E' (and hence of A) in N. Let $G := \mathrm{Aut}_{L'}(N)$. For $\sigma \in G$ we have $\sigma(C) = C$.
Let $p := [L'(f_2) : L']$ and $T^p + b_{p-1}T^{p-1} + \dots + b_0$ be the minimal polynomial of f_2 over L'. As before $b_{p-1},\dots,b_0 \in E'$ are homogeneous with $b_i = 0$ or $\deg b_i = (p-i)d_2$ for $i=1,\dots,p-1$ and $\deg b_0 = p\,d_2$.
We claim that f_1 doesn't divide b_0. Assume the contrary and let $q \subset C$ be any fixed minimal prime ideal over Cf_1. Thus $b_0 \in q$. By the Going-down-Theorem we have $q \cap E' = E'f_1$. Since q is prime and b_0 a product of conjugates $\sigma(f_2)$, $\sigma \in G$,

there exists some $\tau \in G$ such that $\tau(f_2) \in p$. But then $f_2 \in q' := \tau^{-1}(q)$. Since τ leaves f_1 fixed, we have $f_1 \in q'$. By the Going-down-Theorem q' is a minimal prime ideal over Cf_1. Hence $\text{height}(q') = 1$. This implies $\text{transdeg}_k\, C/q' = r-1$ whence for $\mathfrak{r} := A \cap q'$ $\text{transdeg}_k\, A/\mathfrak{r} = r-1$ and finally $\text{height}(\mathfrak{r}) = 1$. So $\mathfrak{r}$ is a minimal prime ideal over Af_1 and Af_2. Thus $V \cap \{f_1 = 0, f_2 = 0\}$ contains a component of codimension 1 which contradicts the hypothesis on f_1 and f_2. Consequently b_0 is not divisible by f_1. Hence b_0 must contain at least a monomial of degree pd_2 only in $y_2, \ldots, y_r$. By the generic choice of $y_2, \ldots, y_r$ we may assume $\deg_{y_2} b_0 = \deg b_0 = p\, d_2$. As before, we conclude that y_2 is integral over $k[f_1, f_2, y_3, \ldots, y_r]$.
Therefore $k[f_1, f_2, y_3, \ldots, y_r] \subset A$ is an integral extension. Moreover, repeating the former argument, we obtain $[K : k(f_1, f_2, y_3, \ldots, y_r)] \leq \deg p \cdot d_1 d_2$. □

5. Effective Nullstellensätze

In this section we suppose k algebraically closed if nothing else is said.

Proposition 10 : Let $p \subset k[X_1, \ldots, X_n]$ be a prime ideal which defines an irreducible closed subset V of k^n with $\dim V = r$. Let $m \leq r$ and $F_1, \ldots, F_m, G \in k[X_1, \ldots, X_n]$. Put $A := k[X_1, \ldots, X_n]/p$ and let $f_1, \ldots, f_m, g$ be the images of $F_1, \ldots, F_m, G$ in A. Write $\deg p := \deg V$. Assume that $\{f_1 = 0, \ldots, f_m = 0\}$ is a (non-empty) set-theoretical complete intersection in V.

Then $\quad g \in \text{rad}(f_1, \ldots, f_m) \quad$ iff $\quad g^{\deg p \prod_{1 \leq i \leq m} \deg F_i} \in (f_1, \ldots, f_m)$.

Proof : We consider exclusively the (non-trivial) only-if part of the assertion.
Let K be the fraction field of A and $x_1, \ldots, x_n$ the images of $X_1, \ldots, X_n$ in A.
Then $E := k[f_1, \ldots, f_m, y_{m+1}, \ldots, y_r]$ is integrally closed in its fraction field L.
Let B be the integral closure of E in A. By Proposition 8 there exists $t \in B$, $t = a_1 f_1 + \ldots + a_m f_m + 1$, $a_1, \ldots, a_m \in A$, such that $A_t = B_t$.
Let N be the normal closure of K over L and C the integral closure of E (and hence of B) in N. Put $G := \text{Aut}_L(N)$. We have $\sigma(C) = C$ and $\sigma(f_1) = f_1, \ldots, \sigma(f_m) = f_m$ for every $\sigma \in G$. Suppose $g \in \text{rad}(f_1, \ldots, f_m)$. This means that there exists $k \in \mathbb{N}$ such that $g^k \in Af_1 + \ldots + Af_m$. Since $A_t = B_t$ this implies that we may suppose $(tg)^k \in Bf_1 + \ldots + Bf_m \subset Cf_1 + \ldots Cf_m$.
Let q be any fixed prime ideal of C minimal over $Cf_1 + \ldots + Cf_m$.
Since $(tg)^k \in Cf_1 + \ldots + Cf_m = \sigma(C)\sigma(f_1) + \ldots + \sigma(C)\sigma(f_m) = \sigma(Cf_1 + \ldots + Cf_m) \subset \sigma(q)$, we have $tg \in \sigma(q)$ for any $\sigma \in G$. Therefore $\sigma(tg) \in q$ for any $\sigma \in G$.
Let U be a new indeterminate and $U^q + b_{q-1}U^{q-1} + \ldots + b_0 \in L[U]$ the minimal polynomial of tg over L, where $q := [L(tg) : L] \leq [K : L]$. We have $tg \in C$, i.e. tg is integral over E, whence $b_{q-1}, \ldots, b_0 \in E$, since E is integrally closed.
On the other hand $b_{q-1}, \ldots, b_0$ are homogeneous polynomials in the $\sigma(tg)$, $\sigma \in G$.
This implies $b_{q-1}, \ldots, b_0 \in E \cap p = Ef_1 + \ldots + Ef_m$ (apply the Going-down-Theorem and

use the fact that $Ef_1 + \ldots + Ef_m$ is prime). From $(tg)^q + b_{q-1}(tg)^{q-1} + \ldots + b_0 = 0$ we infer now $(tg)^q = - b_{q-1}(tg)^{q-1} - \ldots - b_0 \in Af_1 + \ldots + Af_m$.
Since $t = a_1 f_1 + \ldots + a_m f_m + 1$ with $a_1, \ldots, a_m \in A$, we have $(tg)^q - g^q \in Af_1 + \ldots + Af_m$.
Therefore $g^q \in Af_1 + \ldots + Af_m = (f_1, \ldots, f_m)$.
Furthermore we have $q \leq [K:L] = [K : k(f_1, \ldots, f_m, y_{m+1}, \ldots, y_r)] \leq \deg p \cdot \prod_{1 \leq i \leq m} \deg F_i$ by Corollary 7. □

We will pay attention to the following special case

<u>Corollary 11</u> : Let $p \in k[X_1, \ldots, X_n]$ be a prime ideal and $F, G \in k[X_1, \ldots, X_n]$. Then

$$G \in \operatorname{rad}(p, F) \quad \text{iff} \quad G^{\deg p \cdot \deg F} \in (p, F) .$$

<u>Proof</u> : Obvious by Proposition 10. □

<u>Corollary 12</u> : Let k be arbitrary. Let $F_1, \ldots, F_m \in k[X_1, \ldots, X_n]$, $m \leq n$, a regular sequence (or what comes to the same thing: let $\{F_1 = 0, \ldots, F_m = 0\}$ be a complete intersection of codimension m). Assume that $F_1, \ldots, F_m$ are irreducible.
Let $G \in k[X_1, \ldots, X_n]$.
Then

$$G \in \operatorname{rad}(F_1, \ldots, F_m) \quad \text{iff} \quad G^{\prod_{1 \leq i \leq m} \deg F_i} \in (F_1, \ldots, F_m) .$$

<u>Proof</u> : Put $p = (0)$ in Proposition 10 and pass to the algebraic closure $\bar{k}$ of k. The problem of membership of an ideal can be reduced to solving a system of inhomogeneous linear equations. Therefore the assertion for $\bar{k}[X_1, \ldots, X_n]$ implies the assertion for $k[X_1, \ldots, X_n]$. □

<u>Remark 13</u> : Using Remark 9, one can show Proposition 10, Corollary 11 and Corollary 12 in an absolutely elementary way if p, $F_1, \ldots, F_m$, and F are homogeneous.

<u>Theorem 14</u> (Effective Hentzelt's Nullstellensatz): Let k be an arbitrary field. Let $F_1, \ldots, F_s, G \in k[X_1, \ldots, X_n]$ and $d := \max_{1 \leq i \leq s} \deg F_i$.
Then

$$G \in \operatorname{rad}(F_1, \ldots, F_s) \quad \text{iff} \quad G^{d^{((n+1)(n+2))/2}} \in (F_1, \ldots, F_s) .$$

<u>Proof</u> : We only show the non-trivial only-if part of Theorem 14. As before, we may suppose k algebraically closed.
We consider only the non-trivial case : $s \geq 1$ and $F_i \neq 0$, $i = 1, \ldots, s$.
Put $M := \{F_1 = 0, \ldots, F_s = 0\}$.

In a preparatory step we show that we may assume $s = n+1$ and that for every $i = 1, \ldots, n+1$ each irreducible component C of $\{F_1 = 0, \ldots, F_i = 0\}$ not contained in M satisfies $\operatorname{codim} C = i$.
For $1 \leq j \leq n+1$ we define inductively $H_1, \ldots, H_j \in k[X_1, \ldots, X_n]$ with the following properties :

a) $H_1,\dots,H_j$ are k-linear combinations of $F_1,\dots,F_s$.

b) Each irreducible component C of $\{H_1=0,\dots,H_j=0\}$ not contained in M satisfies $\operatorname{codim} C = j$.

Case $j = 1$: We take $H_1 := F_1$.

Case $1 < j \leq n+1$: Let $\mathcal{C}$ be the set of all irreducible components of $\{H_1=0,\dots,H_{j-1}=0\}$ not contained in M. For $C \in \mathcal{C}$ let
$E_C := \{\lambda \in k^s;\ F_\lambda := \sum_{1\leq i\leq s} \lambda_i F_i \in$ ideal $I(C)$ of polynomials vanishing on $C\}$.
E_C is a *proper* linear subspace of k^s. We choose $H_j := F_{\lambda_0}$ for some (arbitrary) $\lambda_0 \in k^s \setminus \bigcup_{C\in\mathcal{C}} E_C$.

Since each irreducible component C' of $\{H_1=0,\dots,H_j=0\}$ not contained in M is an irreducible component of $\{H_j=0\} \cap C$ for some $C \in \mathcal{C}$, we have $\operatorname{codim} C' = \operatorname{codim} C + 1 = j$.
As a consequence of properties a) and b) we have :

1. $\max_{1\leq j\leq n+1} \deg H_j \leq d$
2. $(H_1,\dots,H_{n+1}) \subseteq (F_1,\dots,F_s)$
3. $\operatorname{rad}(H_1,\dots,H_{n+1}) = \operatorname{rad}(F_1,\dots,F_s)$

This finishes the preparatory step.

For $i = 0,\dots,n+1$ denote by $\mathcal{V}_i$ the set of all irreducible components of $\{F_1=0,\dots,F_i=0\}$. Note that $\mathcal{V}_0 = \{\mathbb{A}^n_k\}$.
Fix $G \in \operatorname{rad}(F_1,\dots,F_{n+1})$ (i.e. G vanishes on M).

In the next step we show inductively for $i = n,n-1,\dots,0$ that given $V \in \mathcal{V}_i$, there exist $A_V \in (F_{i+1},\dots,F_{n+1})$ and $d_V \in \mathbb{N}$ such that

1) $G^{d_V} + A_V$ vanishes on V

2) $\sum_{V\in\mathcal{V}_i} d_V \leq d^{n+1}\cdot d^n \cdot \ldots \cdot d^{i+1}$.

Note that for $i=0$ properties 1) and 2) imply $G^{d^{((n+1)(n+2))/2}} \in (F_1,\dots,F_{n+1})$.

Case $i = n$: Let $V \in \mathcal{V}_n$. If $V \subseteq M$, we take $d_V = 1$ and $A_V = 0$. If $V \nsubseteq M$, then $\operatorname{codim} V = n$. This means that V is a (closed) point with maximal ideal $I(V)$. Since $F_{n+1} \notin I(V)$, $1 \in I(V)+(F_{n+1})$. Therefore there exists $A_V \in (F_{n+1})$ such that $G + A_V$ vanishes on V. Again, we take $d_V = 1$. Then, by the Bezout Inequality, [22], Theorem 1,

$$\sum_{V\in\mathcal{V}_n} d_V \leq \sum_{V\in\mathcal{V}_n} \deg V \leq d^n \leq d^{n+1}.$$

Case $0 \leq i < n$: Let $V \in \mathcal{V}_i$. If $V \subseteq M$, we take $d_V = 1$ and $A_V = 0$. Suppose $V \nsubseteq M$. Let $W_1,\dots,W_{t_V}$ be the components of $V \cap \{F_{i+1}=0\}$ not contained in M (this includes the case $t_V = 0$). If $t_V = 0$ (i.e. $V\cap\{F_{i+1}=0\} = \emptyset$ or $V\cap\{F_{i+1}=0\}\subseteq M$), let $R_V := G$.

Let $t_V > 0$. Then we have $W_j \in \mathcal{V}_{i+1}$ (this follows from the fact that $\operatorname{codim} W_j = i+1$). From the induction hypothesis for each $1 \leq j \leq t_V$ there exist $A_{W_j} \in (F_{i+2},\dots,F_{n+1})$

and $d_{W_j} \in \mathbb{N}$ such that

$G^{d_{W_j}} + A_{W_j}$ vanishes on W_j and $\sum_{1 \leq j \leq t_V} d_{W_j} \leq \sum_{W \in \mathcal{V}_{i+1}} d_W \leq d^{n+1} \cdot d^n \cdot \ldots \cdot d^{i+2}$.

Now let $R_V := \prod_{1 \leq j \leq t_V} (G^{d_{W_j}} + A_{W_j}) = G^{\Sigma d_{W_j}} + B_V$ with $B_V \in (F_{i+2}, \ldots, F_{n+1})$.

Therefore, in any case (i.e. $t_V = 0$ or $t_V > 0$), R_V vanishes on $V \cap \{F_{i+1} = 0\}$. By Corollary 11 we have $R_V^{\deg V \cdot d} \in (I(V), F_{i+1})$. This implies that there exists $A_V \in (F_{i+1}, \ldots, F_{n+1})$ such that

$$G^{d_V} + A_V \quad \text{vanishes on } V ,$$

where $d_V := (\sum_{1 \leq j \leq t_V} d_{W_j}) \cdot \deg V \cdot d$ (with the convention $\sum_{1 \leq j \leq t_V} d_{W_j} = 1$ if $t_V = 0$).

Thus $d_V \leq d^{n+1} \cdot d^n \cdot \ldots \cdot d \cdot \deg V$, whence, by the Bezout Inequality,

$$\sum_{V \in \mathcal{V}_i} d_V \leq d^{n+1} \cdot \ldots \cdot d^{i+2} \cdot d^{i+1} .$$

□

<u>Remark 15</u> : (i) An elementary way to demonstrate Theorem 14 for homogeneous polynomials $F_1, \ldots, F_s, G \in k[X_1, \ldots, X_n]$ follows from Remark 13 adapting the proof of Theorem 14 (in particular the preparatory step) to homogeneous ideals.

(ii) A different bound for the problem treated in Theorem 14 is given in the classical paper: G. Hermann, Die Frage der endlich vielen Schritte in der Theorie der Polynomideale, Math.Ann. 95 (1926) 736-788. The bound there is doubly exponential in n and depends also on s .

More important from a practical point of view is the following

<u>Theorem 16</u> (Effective Hilbert's Nullstellensatz): Let k be arbitrary.
Let $F_1, \ldots, F_s \in k[X_1, \ldots, X_n]$ and $d := \max_{1 \leq i \leq s} \deg F_i$. Then the following holds :
$1 \in (F_1, \ldots, F_s)$ iff there exist polynomials $P_1, \ldots, P_s \in k[X_1, \ldots, X_n]$ such that for $i = 1, \ldots, s$ $\deg P_i \leq d^{((n+1)(n+2))/2}$ and $1 = \sum_{1 \leq i \leq s} P_i F_i$.

<u>Proof</u> : As in the proof of Corollary 12 we may suppose k algebraically closed.
Let X_0 be a new indeterminate and $\bar{F}_1, \ldots, \bar{F}_s \in k[X_0, \ldots, X_n]$ the homogenizations of $F_1, \ldots, F_s$. In particular, $F_i = \bar{F}_i(1, X_1, \ldots, X_n)$ for $i = 1, \ldots, s$.
Then $1 \in (F_1, \ldots, F_s)$ iff $X_0 \in \text{rad}(\bar{F}_1, \ldots, \bar{F}_s)$.
By Theorem 14 we have $X_0 \in \text{rad}(\bar{F}_1, \ldots, \bar{F}_s)$ iff $X_0^{d^{((n+1)(n+2))/2}} \in (\bar{F}_1, \ldots, \bar{F}_s)$ iff there exist $Q_1, \ldots, Q_s \in k[X_0, \ldots, X_n]$ homogeneous with $\deg Q_i = d^{((n+1)(n+2))/2} - \deg \bar{F}_i$ for $i = 1, \ldots, s$ and such that

$$X_0^{d^{((n+1)(n+2))/2}} = \sum_{1 \leq i \leq s} Q_i \bar{F}_i . \tag{$*$}$$

Put $P_i := Q_i(1, X_1, \ldots, X_n)$. We have $\deg P_i \leq d^{((n+1)(n+2))/2}$. Specializing $X_0 \to 1$ in $(*)$ we obtain $1 = \sum_{1 \leq i \leq s} P_i F_i$ as desired. □

Let us observe that for char $k = 0$ Theorem 16 implies Theorem 14 by the arguments used in [6].

We conclude this section with an affine Bezout Inequality.

Theorem 17 : Let k be an arbitrary field with algebraic closure $\bar{k}$.
Let $F_1,\dots,F_s \in k[X_1,\dots,X_n]$ and $d := \max_{1\le i\le s} \deg F_i$.
Suppose that $V := \{x \in \bar{k}^n ; F_1(x)=0,\dots,F_s(x)=0\}$ is finite.
Then $k[X_1,\dots,X_n]/(F_1,\dots,F_s)$ is finite dimensional over k and
$\dim_k k[X_1,\dots,X_n]/(F_1,\dots,F_s) \le d^n$.
Moreover, if $n=s$ (in particular if $F_1,\dots,F_n$ is a regular sequence), then

$$\dim_k k[X_1,\dots,X_n]/(F_1,\dots,F_n) \le \prod_{1\le i\le n} \deg F_i .$$

Proof : Without loss of generality assume k algebraically closed. Theorem 17 is trivial if $V=\emptyset$. Suppose $V\ne\emptyset$ and let $G_1,\dots,G_n$ be generic linear combinations of $F_1,\dots,F_s$. Then $G_1,\dots,G_n$ is a regular sequence with $(G_1,\dots,G_n)\subset(F_1,\dots,F_s)$ not defining the empty set. Furthermore $\deg G_i \le d$ for $1\le i\le n$. $G_1,\dots,G_n$ define finitely many points over $\bar{k}$.
Therefore $\dim_k k[X_1,\dots,X_n]/(G_1,\dots,G_n) < \infty$
and $\dim_k k[X_1,\dots,X_n]/(F_1,\dots,F_s) \le \dim_k k[X_1,\dots,X_n]/(G_1,\dots,G_n)$.
Then, by Lemma 1 and Lemma 2 and by Proposition 5, we have

$$\begin{aligned}&\dim_k k[X_1,\dots,X_n]/(G_1,\dots,G_n) \le \\ &\le \dim_{k(G_1,\dots,G_n)} k[X_1,\dots,X_n] \otimes_{k[G_1,\dots,G_n]} k(G_1,\dots,G_n) \\ &= [k(X_1,\dots,X_n) : k(G_1,\dots,G_n)] \le \prod_{1\le i\le n} \deg G_i \le d^n .\end{aligned}$$

If $s=n$ and $F_1,\dots,F_n$ is a regular sequence, this argument applies directly, i.e. without using $G_1,\dots,G_n$. □

6. Applications of the Effective Nullstellensatz

Theorem 18 : Let k be any field and $\bar{k}$ its algebraic closure. We consider first order formulae Φ in the elementary language over $\bar{k}$ involving only constants from k. Let $|\Phi|$ be the length of Φ. Suppose that Φ is prenex with m quantifier alternations. Furthermore assume that Φ is built up from polynomials $F_1,\dots,F_s \in k[X_1,\dots,X_n]$ with $d := \sum_{1\le i\le s} \deg F_i$.
Then there exists an algorithm which eliminates quantifiers modulo the elementary theory of $\bar{k}$ and works in

$$\text{sequential time} \quad (sd)^{n^{O(m)}} + O(|\Phi|)$$

and

$$\text{parallel time} \quad n^{O(m)} (\log sd)^{O(1)} + O(\log |\Phi|) .$$

The same bounds hold for the decision problem of the elementary theory of $\bar{k}$.

Proof : Substitute in the proof of [17], Theorem 3, or [15], Théorème 3, the application of [5], Theorem 1, by our Theorem 16 . □

Theorem 19 : Let k be arbitrary with algebraic closure $\bar{k}$.
Let $F_1,\ldots,F_s, F \in k[X_1,\ldots,X_n]$ with $d := \max_{1\le i\le s} \deg F_i$ and $d' := \max\{d, \deg F\}$.
Let $V := \{x \in \bar{k}^n \,;\, F_1(x) = 0,\ldots,F_s(x) = 0\}$.

Then the following problems can be resolved in sequential time $s^4\, d^{O(n^3)}$ and in parallel time $O(n^6 \log^2 sd)$:

(i) the decision whether $V = \emptyset$ (or equivalently $1 \in (F_1,\ldots,F_s)$);

(ii) the computation of $\dim V$;

(iii) the computation of $\deg V$, if all components of V have equal dimension.

(iv) In sequential time $s^4\, d'^{\,O(n^3)}$ and parallel time $O(n^6 \log^2 sd')$ it can be decided whether F vanishes on V (or equivalently: $F \in \operatorname{rad}(F_1,\ldots,F_s)$).

Proof (Sketch): (i) Using Theorem 16 translate the problem to decide whether $1 \in (F_1,\ldots,F_s)$ into the problem to decide the solvability of a $sd^{O(n^3)} \times sd^{O(n^3)}$-system of linear inhomogeneous equations. The bounds follow now by Theorem 16 and [33],[3] as in [15] or [17]. Instead of [33],[3] one can also use [14]. If one is interested only in sequential complexity we refer to [22], Lemma 7.

(ii) Apply test (i). If $V \ne \emptyset$ results, intersect V with one generic affine hyperplane (using indeterminates). If the intersection is again non-empty, intersect with two generic hyperplanes etc. until you get an empty intersection. To check the complexity of the tests, observe the following :
Let A_{ij}, A_i, $i = 1,\ldots,r$, $j = 1,\ldots,n$, be new indeterminates, moreover let $K = k(A_{ij}, A_i;\, i,j)$ and $\bar{K}$ its algebraic closure.
Then $\dim V < n$ iff $F_1,\ldots,F_s$ and $\sum_{1\le j\le n} A_{ij}X_j$, $i = 1,\ldots,r$, have no common zero in $\bar{K}$ iff $1 \in (F_1,\ldots,F_s, \sum_{1\le j\le n} A_{ij}X_j \,;\, i = 1,\ldots,r) \subset K[X_1,\ldots,X_n]$.

(iv) Just combining Theorem 16 with Rabinowitsch's Trick. □

We observe that one obtains Theorem 19 with complexity bounds of type $(sd)^{n^{O(1)}}$ (sequential complexity) and $(n \log sd)^{O(1)}$ (parallel complexity) just applying Theorem 18.

The following result has been obtained in collaboration with Silvia Jortack-Taich, La Plata.
For the notion of Gröbner base look at [8] or [16].

Theorem 20 : Let notations be as before. Suppose $\# V < \infty$. Let any computable order of the monomials of $k[X_1,\ldots,X_n]$ be given. Let $\mathfrak{a} := (F_1,\ldots,F_s)$.

(i) There exists a Gröbner base of $\mathfrak{a}$ with total degree bounded by d^n.

(ii) Any Gröbner base algorithm (improved in the sense of [27]) uses only polynomials of total degree $d^{O(n^2)}$. In particular, it has complexity $s^3\, d^{O(n^3)}$.
The stairs of $\mathfrak{a}$ can be computed in sequential time $s^4\, d^{O(n^3)}$ and in parallel time $O(n^6 \log^2 sd)$.

(iii) If $1 \in \mathfrak{a}$ such an algorithm computes automatically a representation $1 = P_1F_1 + \ldots + P_sF_s$ with $\deg P_1, \ldots, \deg P_s = d^{O(n^2)}$.

Proof (Sketch): (i) By Theorem 17 for each $1 \leq i \leq n$ there exists $G_i \in k[X_i]$ with $G_i \in \mathfrak{a}$ and $0 < \deg G_i \leq d^n$. Now one immediately sees that any Gröbner base for $\mathfrak{a}$ must contain an element with leading monomial $X_i^{d_i}$, for any i, where $d_i \leq d^n$. Therefore the leading monomials of every element of the Gröbner base have total degree bounded by nd^n. By standard arguments (compare [27]) one finds that the given Gröbner base can be changed into another one which has the property that *all* its monomials have degree bounded by nd^n.

(ii) Let Y be a new indeterminate. We have $1 \in (F_1, \ldots, F_s, 1 - YG_i) \subset k[X_1, \ldots, X_n, Y]$. By Theorem 17 there exist $P_1, \ldots, P_s, P \in k[X_1, \ldots, X_n, Y]$ with $\deg P_1, \ldots, \deg P_s$, $\deg P = d^{O(n^2)}$ such that $1 = P(1 - YG_i) + \sum_{1 \leq j \leq s} P_jF_j$.
Substituting Y by $1/G_i$ and multiplying by the denominator we obtain
$G_i^{d^{O(n^2)}} = \sum_{1 \leq j \leq s} Q_jF_j$ with $Q_j \in k[X_1, \ldots, X_n]$ and $\deg Q_j = d^{O(n^2)}$, $j = 1, \ldots, s$.
Let $G_i' := G_i^{d^{O(n^2)}}$ for $i = 1, \ldots, n$. We have $G_i' \in \mathfrak{a}$ and G_i' has a representation in terms of $F_1, \ldots, F_s$ with degree controlled by $d^{O(n^2)}$. Observe that $G_1', \ldots, G_n'$ is a Gröbner base for $(G_1', \ldots, G_n')$ in the given order.
Now let $H_1, \ldots, H_t \in k[X_1, \ldots, X_n]$ be a Gröbner base of $\mathfrak{a}$ with $\deg H_i \leq n\,d^n$, for $i = 1, \ldots, t$ ((i) in this theorem).
For $j = 1, \ldots, t$ write $H_j = \sum_{1 \leq \ell \leq s} A_\ell F_\ell$ with $A_1, \ldots, A_s \in k[X_1, \ldots, X_n]$. Apply Hironaka-Division to each $A_1, \ldots, A_s$: $A_\ell = \sum_{1 \leq k \leq n} Q_{\ell k} G_k' + R_\ell$, $\ell = 1, \ldots, s$. Each monomial of R_ℓ is situated below the stairs of $G_1', \ldots, G_n'$. Therefore $\deg R_\ell = d^{O(n^2)}$.
Now we have $H_j = \sum_{1 \leq \ell \leq s} R_\ell F_\ell + B_j$, where $B_j \in (G_1', \ldots, G_n')$.
Since $\deg H_j, \deg R_\ell F_\ell = d^{O(n^2)}$ for $1 \leq j \leq t$, $1 \leq \ell \leq s$, we have $\deg B_j = d^{O(n^2)}$.
But since $G_1', \ldots, G_n'$ is a Gröbner base of $(G_1', \ldots, G_n')$, $B_j = C_{j1}G_1 + \ldots + C_{jn}G_n$ with $\deg C_{j1}G_1, \ldots, \deg C_{jn}G_n = d^{O(n^2)}$. The assertions follow now easily.

(iii) immediate by the foregoing arguments. □

For similar ideas compare the proof of Lemma 3.4 of [2].

We also indicate the following applications of our results :

Proposition 21 (together with M. Chardin and M. Giusti, Ecole Polytechnique, Paris): Let k be arbitrary and $F_1, \ldots, F_s \in k[X_1, \ldots, X_n]$ with $d := \max_{1 \leq i \leq s} \deg F_i$. Let $T_1, \ldots, T_s$ be indeterminates. Then

(i) $\operatorname{transdeg}_k k(F_1, \ldots, F_s) < s$ iff there exists $P \in k[T_1, \ldots, T_s]$, $0 < \deg P \leq nd^n$, with $P(F_1, \ldots, F_s) = 0$.

(ii) $\operatorname{transdeg}_k k(F_1, \ldots, F_s)$ can be computed in sequential time $s\,d^{O(n^2)}$ and in parallel time $O(n^4 \log^2 sd)$.

The proof of Proposition 21 is based on Theorem 17. This result allows in particular to calculate "fast" the dimension of images of morphisms between affine spaces. *

We conclude the paper with a result connected with the classical solution of Serre's Conjecture. It is a quantitative and effective (and more general) version of Suslin's Theorem.

Theorem 22 (together with N. Fitchas, Buenos Aires): Let k be infinite and $A := k[X_1,\dots,X_n]$. Let $F = (F_{ij})_{1\leq i\leq r, 1\leq j\leq s} \in A^{r\times s}$ with $r \leq \min\{n,s\}$ a unimodular $r\times s$ matrix of polynomials $F_{ij} \in A$ (F unimodular means that the $r\times r$ -minors of F generate the unit ideal of A). Let $d := \deg F := \max_{1\leq i\leq r, 1\leq j\leq s} \deg F_{ij}$.

Then there exists a unimodular matrix $M \in A^{s\times s}$ such that

(i) $F \cdot M = \left(\begin{array}{cc|c} \underbrace{\begin{matrix} 1 & & 0 \\ & \ddots & \\ 0 & & 1 \end{matrix}}_{r} & & 0 \end{array}\right)\Big\}r$; (with s columns)

(ii) $\deg M \leq (sd)^{O(n^2)}$;

(iii) a matrix M with (i) and (ii) can be computed in sequential time $(sd)^{O(n^3)}$ and parallel time $O(n^6 \log^2 sd)$.

For $r = 1$ (i) corresponds to the classical Suslin Theorem, whereas (ii) gives a quantitative and (iii) an effective version of it.

* Note added in proof: J.P. Jouanolou communicated us recently that a slightly sharper result can be obtained using Resultant Theory.

References

[1] D. Bayer, M. Stillman: On the complexity of computing syzygies. Preprint 1985.

[2] C.A. Berenstein, A. Yger: Effective Bezout Identities in $\mathbb{Q}[z_1,\dots,z_n]$. Preprint University of Maryland 1987.

[3] S.J. Berkowitz: On computing the determinant in small parallel time using a small number of processors. Information Processing Letters 18 (1984) 147-150.

[4] J. Briançon: Sur le degré des relations entre polynômes. C.R. Acad.Sci. Paris 297, Série I (1982) 553-556.

[5] D. Brownawell: Bounds for the degrees in the Nullstellensatz. Ann.math. Second Series, Vol. 126 No 3 (1987) 577-591.

[6] W.D. Brownawell: Borne effective pour l'exponent dans le théorème des zéros. C.R. Acad.Sci. Paris 305, Série I (1987) 287-290.

[7] W.D. Brownawell: Local Diophantine Nullstellen Inequalities. Preprint Penn State University 1987.

[8] B. Buchberger: Ein algorithmisches Kriterium für die Lösbarkeit eines algebraischen Gleichungssystems. Aequat.math. 4 (1970) 374-383.

[9] B. Buchberger: A criterion for detecting unnecessary reductions in the construction of Gröbner bases. Sym. and Alg.Comp., Springer LN Comput.Sci. 72 (1979) 3-21.

[10] B. Buchberger: A note on the complexity of constructing Gröbner bases. Proc. Eurocal'83, Computer Algebra, ed. J.A. van Hulzen, Springer LN Comput.Sci. 162 (1983) 137-145.

[11] L. Caniglia, A. Galligo, J. Heintz: Borne simple exponentielle pour les degrés dans le théorème des zéros sur un corps de caractéristique quelconque. to appear in: C.R. Acad.Sci. Paris 1988.

[12] A.L. Chistov, D.Yu. Grigor'ev: Subexponential time solving systems of algebraic equations I,II. LOMI preprints E-9-83, E-10-83, Leningrad 1983.

[13] A.L. Chistov, D.Yu. Grigor'ev: Complexity of quantifier elimination in the theory of algebraically closed fields. Proc. 11th Symp. MFCS 1984, Springer LN Comput.Sci. 176 (1984) 17-31.

[14] A.L. Chistov: Fast parallel calculation of the rank of matrices over a field of arbitrary characteristic. Proc. Int.Conf. FCT 1985, Springer LN Comput.Sci. 199 (1985) 63-69.

[15] N. Fitchas, A. Galligo, J. Morgenstern: Algorithmes rapides en séquentiel et en parallel pour l'élimination de quantificateurs en géométrie élémentaire. to appear in: Séminaire Structures Algébriques Ordonnées, UER de Math., Université de Paris VII (1987); final version to appear in: Same Séminary, Publ.Univ. Paris VII.

[16] A. Galligo: Algorithmes de construction de bases standards. Preprint University of Nice 1985.

[17] A. Galligo, J. Heintz, J. Morgenstern: Parallelism and fast quantifier elimination over algebraically (and real) closed fields. Invited lecture Int.Conf. FCT'87 Kazan 1987.

[18] M. Giusti: Some effectivity problems in polynomial ideal theory. Proc. Eurosam 84, Springer LN Comput.Sci. 174 (1984) 159-171.

[19] M. Giusti: Complexity of standard bases in projective dimension zero. Preprint Ecole Polytechnique Paris 1987.

[20] M. Giusti: Combinatorial dimension theory of algebraic varieties. Preprint Ecole Polytechnique Paris 1988.

[21] D.Yu. Grigor'ev: The complexity of the decision problem for the first order theory of algebraically closed fields. Math. USSR Izvestija, Vol. 29, No 2 (1987) 459-475.

[22] J. Heintz: Definability and fast quantifier elimination in algebraically closed fields. Theoret.Comput.Sci. 24 (1983) 239-277; Russian transl.in: Kyberneticeskij Sbornik, Novaja Serija Vyp. 22, Mir Moscow (1985) 113-158.

[23] J. Heintz, R. Wüthrich: An efficient quantifier elimination algorithm for algebraically closed fields. SIGSAM Bull. 9(4) (1975) 11.

[24] J. Kollár: Sharp effective Nullstellensatz. Manuscript 1988.

[25] B. Iversen: Generic Local Structure in Commutative Algebra. Springer LN Math. 310 (1973).

[26] J.P. Jouanolou: Théorèmes de Bertini et applications. Birkhäuser PM 42 (1983).

[27] C. Kollreider, B. Buchberger: An improved algorithmic construction of Gröbner-bases for polynomial ideals. Bericht Nr. 170 (1978), Technical Report, Universität Linz.

[28] D. Lazard: Algèbre linéaire sur $K[X_1,...,X_n]$ et élimination. Bull.Soc.Math. France 105 (1977) 165-190.

[29] D. Lazard: Gröbner Bases, Gaussian Elimination and Resolution of Algebraic Equations.
Proc. Eurocal'83, Computer Algebra, ed. J.A. van Hulzen, Springer LN Comput.Sci. 162 (1983) 146-156.

[30] H. Matsumura: Commutative algebra.
W.A. Benjamin 1980 (first edition).

[31] E. Mayr, A. Meyer: The complexity of the word problem for commutative semigroups and polynomial ideals.
Advances in Math. 46 (1982) 305-329.

[32] H.M. Möller, F. Mora: New Constructive Methods in Classical Ideal Theory.
J. of Algebra, Vol. 100, No 1 (1986) 138-178.

[33] K. Mulmuley: A fast parallel algorithm to compute the rank of a matrix over an arbitrary field.
Proc. 18th Ann. ACM Symp. Theory of Computing (1986) 338-339.

[34] I.R. Shafarevich: Algebraic Geometry.
Springer Berlin 1974.

[35] B. Shiffman: New degree bounds for the Nullstellensatz in arbitrary characteristic.
Manuscript 1988.

[36] V. Weispfenning: The complexity of linear problems in fields.
J. on Symbolic Comput. Vol. 5, No 1-2 (1988) 3-27.

[37] R. Wüthrich: Ein schnelles Quantoreneliminationsverfahren für die Theorie der algebraisch abgeschlossenen Körper.
Ph.D.-Thesis, Univ. Zurich 1977.

Note added in proof :

In the meantime we obtained a $d^{O(n)}$ - bound for the effective Nullstellensatz combining our arguments with elementary properties of the functor Ext for modules.
We use the well-known fact that the existence of regular sequences for a module can be characterized by Ext. In this part of the proof our arguments are similar to Kollàr's who uses the more intricated cohomology with support for algebraic varieties.

Proofs will appear elsewhere.

SOME REMARKS ON THE DIFFERENTIAL DIMENSION*

GIUSEPPA CARRA' FERRO

Dipartimento di Matematica, Università di Catania
viale A.Doria, 6 - 95125 CATANIA - ITALY

It is shown that in the case of finite systems of ordinary algebraic differential equations with coefficients in a universal differential field K of characteristic zero it is possible to determine the differential dimension of the differential ideal I associated to the system without factoring polynomials. In fact in this case we can determine effectively an integer $s_0=2^{n-1}s+1$, depending on the number n of differential variables and the greatest order s of the ordinary differential polynomials in the system in such way as we can determine the differential dimension of I by looking only at the ideal $D(I_{s_0})$, which is an ideal in a ring of polynomials.

We suppose that every ring is commutative with identity element and every ring homomorphism preserves the identity.

Definition. A differential ring R is a ring with a finite set $\Delta=\{\partial_i,\ldots,\partial_m\}$ of derivation operators such that $\partial_i\partial_j=\partial_j\partial_i$ for all i and j. An ideal I of R is called a differential ideal if it is stable under each $\partial\in\Delta$. If $T\subseteq R$. [T] (respectively {T}) denotes the smallest differential ideal (respectively the smallest radical differential ideal) containing T.

Definition. Let R and S be differential rings. A differential ring homomorphism $\varphi:R\longrightarrow S$ is a ring homomorphism such that $f\partial=\partial f$ for all ∂.

If $\Delta=\{\partial_i,\ldots,\partial_m\}$, $\Theta=\{\partial_1^{e_1},\ldots,\partial_m^{e_m}:(e_1,\ldots,e_m)\in\mathbb{N}^n\}$ will denote the commutative semigroup of all derivative operators. If $\vartheta=\partial_1^{e_1},\ldots,\partial_m^{e_m}$, then $\mathrm{ord}(\vartheta)=\Sigma_{i=1,\ldots,m}e_i$. $\Theta_j=\{\vartheta\in\Theta:\mathrm{ord}(\vartheta)\leq j\}$.

For unspecified notations and definitions see [6].

---

* This paper was supported by Italian M.P.I. (40%-1985).

Let K be a differential field of char 0.
$R=K\{x_1,\dots x_n\}=K[\vartheta x_j:\ \vartheta\in\Theta$ and $i=1,\dots,n]$ is the differential ring of the differential polynomials in the differential indeterminates $x_1,\dots,x_n$. If $s\in\mathbb{N}$, then $R_s=K[\vartheta x_j:\ \vartheta\in\Theta_s,\ i=1,\dots,n]$

Definition. An orderly ranking O of $(x_1,\dots,x_n)$ is a total ordering of the set $\{\vartheta x_j:\ \vartheta\in\Theta$, and $j=1,\dots,n\}$ such that

(i) $\vartheta x_j <_o \vartheta' x_{j'}$ implies $\vartheta''\vartheta x_j <_o \vartheta''\vartheta' x_{j'}$ for all $\vartheta,\vartheta',\vartheta''\in\Theta$ and all j,j';

(ii) $\vartheta x_i < \vartheta' x_j$ whenever $\mathrm{ord}(\vartheta)<\mathrm{ord}(\vartheta')$.

If $R=K\{x_1,\dots x_n\}$, then T_Δ is the set of all mononials in R. By ([2], lemma 1 and remark 5, p.8-9) given an orderly ranking O of $(x_1,\dots x_n)$ there exists a term ordering $<_o$ on T_Δ, compatible with O, that induces a term ordering $<_s$ on R_s for all $s\in\mathbb{N}$ such that $<_{s+k}$ restricted to R_s is equal to $<_s$ for all $k\in\mathbb{N}$. Let M and N be monomials in T_Δ. If $s=\max\{\mathrm{ord}(M),\ \mathrm{ord}(N)\}$ and $k=n\binom{s+m}{m}$, then $R_s=K[y_1,\dots,y_k]$, $M=y_1^{a_1},\dots,y_k^{a_k}$ and $N=y_1^{b_1},\dots,y_k^{b_k}$. Now $M<_o N$ iff the first non zero difference b_j-a_j from the right is positive.

Let $I=[f_1,\dots,f_r]$ be a finetely differentially generated ideal in R. $I(s)=I\cap R_s$, while $I_s=(\vartheta f_i:\ \mathrm{ord}\ \vartheta f_i\le s,\ i=1,\dots,r)$ for all $s\in\mathbb{N}$. By ([2], p.13) given a orderly ranking O of $(x_1,\dots x_n)$ we can construct the ideals $D(I_s)$ in the following way. Let $\{h_{i,j(i)}:\ i=0,\dots,s\ j(i)=1,\dots,\mu(i)\}$ be the reduced Gröbner basis of I_s with respect to $<_o$ and suppose that $h_{i,j(i)}\in R_i-R_{i-1}$ for all $j(i)=1,\dots,\mu(i)$. Put $(I_s)_1=(I_s,\vartheta h_{i,j(i)}:\ i\le s-1,\ \vartheta h_{i,j(i)}\in R_s)$. If $I_s=(I_s)_1$, then $D(I_s)=I_s$. If $I_s\subset(I_s)_1$, let $(I_s)_2=((I_s)_1)_1$. If $(I_s)_2=(I_s)_1$, then $D(I_s)=(I_s)_1$. If not, we can construct inductively the ascending chain of ideals $I_s\subset(I_s)_1\subset\dots\subset(I_s)_k\subset\dots$ in R_s. Since R_s is noetherian, then there exists $k\in\mathbb{N}$ such that $(I_s)_k=(I_s)_{k+1}$. $D(I_s)=(I_s)_k$.

Let $D^1(I_s)=D(I_{s+1})\cap R_s$. Of course $D(I_s)\subseteq D^1(I_s)$.

Lemma 1. Let $I=[f_1,\dots,f_r]$ be a finetely differentially generated ideal in R and let O be an orderly ranking of $(x_1,\dots,x_n)$. For any $s\in\mathbb{N}$ we have the following statements:

(i) if $f\in D^1(I_{s-1})$, then $\vartheta f\in D(I_s)$ whenever $\vartheta f\in R_s$;
(ii) if $f\in rad(D^1(I_{s-1}))$ then $\vartheta f\in rad(D^1(I_s))$ whenever $\vartheta f\in R_s$;
(iii) if $f\in rad(D^1(I_s))\cap R_{s-1}$, then $\vartheta f\in rad(D^1(I_s))$ whenever $\vartheta f\in R_s$;
(iv) if p is a minimal prime ideal over $rad(D^1(I_s))$ and $f\in P\cap R_{s-1}$, then $\vartheta f\in P$ whenever $\vartheta f\in R_s$;

Proof. (i) Let $\{h_{i,j(i)}$: $i=0,\dots,s$ $j(i)=1,\dots,\mu(i)\}$ be the reduced Gröbner basis of $D(I_s)$ with respect to $<_o$ and suppose that $h_{i,j(i)}\in R_i-R_{i-1}$ for all $i=1,\dots,s$ and $j(i)=1,\dots,\mu(i)$. Let $f\in D(I_s)\cap R_{s-k}=(h_{i,j(i)}$: $i=0,\dots,s-k$, $j(i)=1,\dots,\mu(i))$ by definition of $<_o$. So $f=\Sigma_{i=0,\dots,s-k}(\Sigma_{j(i)=1,\dots\mu(i)}\ A_{i,j(i)}h_{i,j(i)})$ with $A_{i,j(i)}\in R_{s-k}$ for all i and j(i). If $\partial\in\Delta$, then $\partial f=\Sigma_{i=0,\dots,s-k}$. $(\Sigma_{j(i)=1,\dots\mu(i)}\ \partial A_{i,j(i)}h_{i,j(i)}+A_{i,j(i)}\partial h_{i,j(i)})\in D(I_s)\cap R_{s-k+1}$ by definition of $D(I_s)$.
(ii) Let $f\in rad(D^1(I_{s-1}))$. There exists $n\in\mathbb{N}$ such that $f^n\in(D^1(I_{s-1}))$. By (i) $f^{n-1}\partial f\in D(I_s)$ and $(n-1)f^{n-2}\partial f^2+f^{n-1}\partial^2 f)\in D(I_{s+1})$, being $f\in R_{s-1}$. So $f^{n-3}\partial f^2\in D^1(I_s)$. By (i) $\partial(f^{n-2}\partial f^3)\in D(I_{s+1})$ and then $f^{n-3}\partial f^5\in D^1(I_s)$. After a finite number of derivations we have $\partial f^{2n-1}\in D^1(I_s)$.
(iii) Let $f\in rad(D^1(I_s))\cap R_{s-1}$. There exists $n\in\mathbb{N}$ such that $f^n\in D^1(I_s)$. So $f^{n-1}\partial f\in D^1(I_s)$. By repeating the proof as in (ii) we have $\partial f^{2n-1}\in D^1(I_s)$.
(iv) By hypothesis on P there exists $x\in R_s$, $x\notin P$ such that $P=(rad(D^1(I_s)):x)$. Let $f\in P\cap R_{s-1}$. $fx\in rad(D^1(I_s))$ and $\partial fx+f\partial x$ is in $rad(D^1(I_s))$. $\partial fx^2+fx\partial x$ $\in$ $rad(D(I_{s+1}))$ and then ∂fx^2 is in $rad(D(I_{s+1}))\cap R_s=rad(D^1(I_s))$. So ∂fx is in $rad(D^1(I_s))$ and $\partial f\in P$.

Let P be a prime differential ideal in R and let S=R/P. Let F=q·f(S). F is a differential field with the derivation operators induced by those on R ([6], p.64). The canonical homomorphism of $K\{x_1,\dots,x_n\}$ into F maps $(x_1,\dots,x_n)$ onto a family $\bar{x}=(\bar{x}_1,\dots,\bar{x}_n)$ and maps K isomorphically onto a differential field that we can identify with K. After this identification F can be written as $K\langle\bar{x}\rangle=K\langle\bar{x}_1,\dots,\bar{x}_n\rangle=K(\vartheta\bar{x}_i$: $i=1,\dots,n$ $\vartheta\in\Theta)$

Definition. Let F and G be differential fields such that $F\subset G$. If

$a\in G$, a is called differentially algebraic over F if there exists $f\in F\{x\}-\{0\}$ such that $f(a)=0$. If $a\in G$ and a is not differentially algebraic over F, then a is differentially trascendental over F ([6], p.69]).

Remark. If $a\in G$ and a is differentially algebraic over F then $\vartheta(a)$ is differentially algebraic over F for all $\vartheta\in\Theta$.

Definition. Let F and G be differential fields such that $F\subset G$ and let $a_1,\dots,a_d\in G$. $a_1,\dots,a_d$ are called differentially algebraic dependent over F if there exists $f\in F\{x_1,\dots,x_d\}-\{0\}$ such that $f(a_1,\dots,a_d)=0$ ([6], prop.7, p.100). If char $F=0$ then the differential trascendence degree of G over F is the largest number of elements in G such that no derivates of them of any order are differentially algebraic dependent over F. The set B of all such elements is called the differential trascendence basis of G over F. ([6], p.104- 105-108).

Remark. If $B=\{a_h: h\in H\}$ and $F\langle a_h: h\in H\rangle=F(\vartheta a_h: h\in H, \vartheta\in\Theta)=H$ denotes the differential extension of F by the a_h's, then H is a differential field, $F\subset H\subseteq G$ and G is differentially separable over H([6], prop.10, p.104). The differential trascendence degree of G over F is equal to the differential trascendence degree of H over F. Furthermore G as field can have an infinite trascendence basis over H.

Definition. Let P be a prime differential ideal in $R=K\{x_1,\dots,x_n\}$. The differential dimension of P is the differential trascendence degree of $F=q{\cdot}f\ (R/P)$ over K and it is denoted by diff $\dim_K P$ ([6], p.129).

Remark. It is well know that if P is a prime differential ideal in R then diff $\dim_K P\le n$ and diff $\dim_K P=n$ iff $P=(0)$.

Definition. Let I be a differential ideal in R. diff $\dim_K I=$ $=\max\{$diff $\dim_K P$: P minimal prime differential ideal over I$\}$.

It is well known in algebraic geometry that if I is an ideal in $K[x_1,\dots x_n]$ then $\dim_K I=\max\{h: I\cap K[x_{i_1},\dots,x_{i_h}]=(0)$ with $1\le i_1<\dots<i_h\le n\}$, whenever K is algebraically closed. We shall extend this property to

the case of the differential dimension.

Proposition 1. Let I be a finitely differentially generated ideal in R. diff $\dim_K I=\max(h: I\cap K(x_{i_1},\dots,x_{i_h})(0)$ with $1\le i_1<\dots<i_h\le n)$, whenever K is a universal differential field.

Proof. First of all suppose that I is a prime differential ideal $F=q\cdot f\ (R/I)=(f/g: f,g\in R,\ g\notin I)$. So F can be identified with the differential extension $K(a_1,\dots,a_n)$ of K where $f(a_1,\dots,a_n)=0$ for all $f\in I$. Let d=diff $\dim_K I$. Without loss of generality we can suppose that $a_1,\dots,a_d$ are differentially algebraic independent over K. So for each $f\in K(x_1,\dots,x_d)-(0)$, $f(a_1,\dots,a_d)\neq 0$ while for each $i=d+1,\dots,n$ there exists $f_i\in K(x_1,\dots,x_d,x_i)$ such that $f_i(a_1,\dots,a_d,a_i)=0$ by definition of differential trascendence degree. $f_i\in I$ by definition of F for all $i=d+1,\dots,n$. Furthermore for each subset $(i_1,\dots,i_{d+1})$ of $(1,\dots,n)$ with $1\le i_i<\dots<i_{d+1}\le n$ there exists $f\in K(x_{i_1},\dots,x_{i_d},x_{i_{d+1}})$ such that $f(a_{i_1},\dots,a_{i_{d+1}})=0$ by definition of differential trascendence degree. Of course such $f\in I$. So d=diff $\dim_K I$ implies $I\cap K(x_1,\dots,x_d)=(0)$ while $I\cap K(x_1,\dots,x_{i_{d+1}})\neq($ for all subsets $(i_i,\dots,i_{d+1})$ of $(1,\dots,n)$ with $1\le i_1<\dots<i_{d+1}<n$. Conversely suppose that $d=\max(h: I\cap K(x_{i_1},\dots,x_{i_h}=(0),\ 1\le i_1<\dots<i_h\le n)$. So $a_{i_1},\dots,a_{i_d}$ are differentially algebraic independent over K for some subset $(i_1,\dots,i_d)$ of $(1,\dots,n)$ with $1\le i_1,<\dots<i_d\le n$ while $a_{i_1},\dots,a_{i_{d+1}}$ are differentially algebraic dependent over K for all subsets $(i_1,\dots,i_{d+1})$ of $(1,\dots,n)$ with $1\le i_1<\dots<i_{i_{d+1}}\le n$. So diff $\dim_K F=d=$ diff $\dim_K I$. Now suppose that I is not prime. By ([6], p.147) $(I)=\mathrm{rad}I=\cap(P_j: j=1,\dots,k)$ where P_j is a prime differential ideal in R for all i. Let d_j=diff $\dim_K P_j$ and let $d=\max(d_j: j=1,\dots,k)$ diff $\dim_K I$. Let $d'=\max(h: I\cap K(x_{i_1},\dots,x_{i_h})=(0)$, $1\le i_i<\dots i_h\le n)$. Since $I\subset P_j$ for all $i=1,\dots,k$, then $P_j\cap K(x_{i_1},\dots,x_{i_r})==(0)$ implies $I\cap K(x_{i_1},\dots,x_{i_r})=(0)$, so $d\le d'$. If $d<d'$ then $I\cap K(x_{i_1},\dots,x_{i_{d'}})=(0)$. On the other hand $P_j\cap K(x_{i_1},\dots,x_{i_{d'}})\neq(0)$

for all $j=1,\dots,k$. Let $f_j \in P_j \cap K\{x_{i_1},\dots,x_{i_{d'}}\}$ for all $j=1,\dots,k$. $f=\Pi_{j=1,k} f_j \in \Pi_{j=1,k} P_j \subseteq \{I\}$. So there exists $n \in N$ such that $f^n \in I$, being $\{I\}=\mathrm{rad} I$ and we have a contradiction.

Remark. The computation of the differential dimension of a differential ideal I in $K\{x_1,\dots,x_n\}$ is possible when K is an arbitrary differential field of char. 0. In fact the hypothesis that the ground field K is universal differential is necessary only in order to make the algebraic and the geometric definitions of differential dimension coincide.

Definition. Suppose that we fix a ranking of $(x_1,\dots,x_n)$ and let $f \in R=K\{x_1,\dots,x_n\}$. The highest ranking derivative ϑx_j in f is called the leader of f and it is denoted by u_f. If $f,g \in R$, g is called reduced with respect to f if g is free of every proper derivative of u_f and $\deg_{u_f} g < \deg_{u_f} f$. A subset A of R is called autoreduced if $A \cap K = \emptyset$ and each element of A is reduced with respect to all the others.

Remark. Each autoreduced set is finite by ([6], lemma 15 a, p.49). Autoreduced sets were introduced by Ritt, who called them chains, as a tool in the algorithm of reduction of differential polynomials ([9], p.2-7) and ([9], p.163-166). Autoreduced sets can be also defined in the case $R=K[x_1,\dots,x_n]$ (i.e. $|\Delta|=0$), in fact if this is the case we have only to consider derivatives of order zero.

Definition. Let A be an autoreduced subset of $R=K\{x_1,\dots,x_n\}$ with respect to some ranking of $(x_1,\dots,x_n)$. A is called coherent iff whenever $a,a' \in A$ and $v=\vartheta u_a = \vartheta' u_{a'}$, then $S_{a'}\vartheta a - S_a \vartheta' a'$ is in $(A_v):H_A^{\infty}$, where $A_v=\{\vartheta a: a \in A,\ \vartheta \in \Theta \text{ and } \vartheta u_a < v\}$, $S_a = \partial a/\partial u_a$ and H_a is as in ([6], p.136).

Given a differential ideal $I=[f_1,\dots,f_r]$ in $R=K\{x_1,\dots,x_n\}$ and a ranking of $(x_1,\dots,x_n)$ in ([9], p.5) and ([9], p.95) it is described an algorithm in order to find the corresponding autoreduced set associated to I in the case $|\Delta|=0,1$. This algorithm is extended in ([6], p.168) to the case $|\Delta|=m$ and for every ranking of $(x_1,\dots,x_n)$ in order to find the corresponding autoreduced and coherent set associated to I. Since each autoreduced set is coherent in the case $|\Delta|=0,1$, then fixed a ranking O of $(x_1,\dots,x_n)$ the algorithm of Ritt-Kolchin associates to a differential ideal $I=[f_1,\dots,f_r]$ in R the corresponding autoreduced and coherent set A that we call "the characteristic set" of I and it can be different from the one defineted in ([6], p.81-82). $A=\{f_{i,\rho(i)}: i=h,h+1,\dots,n,\ \rho(i)=1,\dots,\lambda(i)\}$, where (in ([6], p.81-82)) $x_1<\dots<x_n$ with respect to O, $h\geq 1$, $u_{f_{j,\rho(j)}}=\theta_{\rho(j)}x_j$ for all $j=h,\dots n$ $i(j)=1,\dots,\lambda(j)$ and $f_{j,\rho(j)}\in K(x_i,\dots,x_{j-1})\{x_j\}$. If $|\Delta|=0,1$, then $\lambda(j)=1$ for all j. Furthermore $A\subseteq I\subseteq [A]:H_A^\infty$.

In ([9], chp. V, p.107-121) when $|\Delta|=0,1$ and in [6], p.166-169) when $|\Delta|=m$ it is described an algorithm in order to find the differential dimension of a finitely differentially generated ideal in R. This algorithm of Ritt-Kolchin is based on the following facts. First of all it is possible to find after a finite number of derivations and factorizations of polynomials characteristic sets of all prime differential ideals minimal over I. Furthermore, given the characteristic set of a prime differential ideal P in R it is possible to know the differential dimension of P.

Proposition 2. Let $I=[f_1,\dots,f_r]$ be a finetely differentially generated ideal in $K(x_1,\dots,x_n)$, with K universal differential field of char. O. Let s_0 be the smallest integer for which every characteristic set of I is contained in $D(I_{s_0})$. If $d=\max\{h: D(I_{s_0+1})\cap \cap K(x_{i_1},\dots,x_{i_h})=(0)\}$, then $d=\text{diff dim}_K I$.

Proof. Let $d'=\text{diff dim}_K I$. By proposition 1 $d'\leq d$, being $D(I_{s_0+1})\subset I$ Every orderly ranking O of $(x_1,\dots,x_n)$ determines a permutation of $(1,\dots,n)$ in such way as $x_{i_1}<\dots<x_{i_n}$. So O determines a term ordering on R_{s_0} by taking $\theta x_i<\theta' x_j$ iff either $x_i<x_j$ or $i=j$ and $\theta x_i<\theta' x_i$ with respect to O. So $D(I_{s_0+1})$ has a reduced Gröbner basis with respect to this term ordering $\{K_{l,j(l)}: l=1,\dots,n,$

(l)=1, ..., v(l)) with $K_{l,j(l)} \in K\{x_{i_1}, ..., x_{i_l}\} - K\{x_{i_1}, ..., x_{i_{l-1}}\}$ for all l and i(l). Let $H_l = (K_{l,j(l)}: i=1, ..., l,\ j(i)=1, ..., v(i))$ for all l. $D(H_l)=H_l$ by its own definition and by (i) of lemma 1. Without loss of generality we can suppose $I \cap K\{x_1, ..., x_{d'}\} = (0)$ and $D(I_{s_o+1}) \cap K\{x_1, ..., x_d\} = (0)$. Let L be the smallest universal extension of the differential field $K\langle x_1, ..., x_d\rangle$ and let $\varphi: K\{x_1, ..., x_d\}\{x_{d+1}, ..., x_n\} \longrightarrow L\{x_{d+1}, ..., x_n\}$ be the corresponding differential ring homomorphism. $\varphi(I)^e = [\varphi(I)] = [f_1, ..., f_r]$, where $f_1, ..., f_r$ are considered as differential polynomials in $L\{x_{d+1}, ..., x_n\}$. So $D(\varphi(I)^e_{s_o+1}) = \varphi(D(I_{s_o+1}))^e = D(I_{s_o+1})$ as ideal in $L\{x_{d+1}, ..., x_n\}$ because $D(I_{s_o+1}) \cap K\{x_1, ..., x_d\} = (0)$. Suppose that $d' < d$. diff $\dim_L \varphi(I)^e = -1$, while $o = \max(h: D(I_{s_o+1}) \cap L\{x_{i_1}, ..., x_{i_h}\} = (0)$, with $d+1 \le i_1 < ... < i_h \le n)$. First of all suppose that $d+1=n$. So $D(I_{s_o+1}) = H_n$ and $H_n \neq (0)$ by hypothesis on d.

Let $V=V(H_n)$ be the affine algebraic variety defined by rad $D^1(I_{s_o})$ in $L^{t(1)}$ being $t(r) = r\binom{s_o+m}{m}$. By (iii) and (iv) of lemma 1 there exists $\mathbf{a}_{s_o} = (\vartheta a: \vartheta \in \Theta_{s_o}) \in L^{t(1)}$ in V. Since the characteristic set of $\varphi(I)^e$ is contained in $\varphi(D(I_{s_o}))^e = D(I_{s_o})$ and $D(I_{s_o}) \subset D^1(I_{s_o})$. Then a satisfies all integrability conditions in order to be a solution of the system $f_i=0$ i=1, ..., r. So diff $\dim_L \varphi(I)^e \ge 0$ and we have a contradiction. Now suppose that $d+1<n$ and the system $f_i=0$ i=1, ..., r has a solution when $n \le d+k$, $k \ge 1$. Now let $n=d+k+1$. Let $V_1 = V(H_{d+k})$ be the affine algebraic variety defined by rad $(H_{d+k} \cap R'_{s_o}) \subset$ rad $D^1(I_{s_o})$ in $L^{t(k)}$, where $R' = L\{x_{d+1}, ..., x_n\}$. By induction hypothesis there exists $\mathbf{a}_{s_o} = (\vartheta a_i: i=1, ..., k\ \vartheta \in \Theta_{s_o}) \in V_1$. Let $L_1 = L\langle \mathbf{a}(k)\rangle$, where $\mathbf{a}(k) = (\vartheta a_i: i=1, ..., k,\ \ \vartheta \in \Theta)$ and let G be the smallest universal differential field containing L_1. Let $R'' = G\{x_n\}$ and let J_{s_o+1} be the ideal in R″ generated all $g \in D(I_{s_o+1}) - (D(I_{s_o+1}) \cap L\{x_{d+1}, ..., x_{n-1}\})$. Since $n=d+k+1$ by repeating the proof as in the

case $n=d+1$, there exists $a_{s_0}(1)=(\vartheta a:\vartheta\in\Theta_{s_0})$, that is a solution of the system $g=0$ for all $g\in I_{s_0+1}\cap R''_{s_0}$. So $(a_{s_0}(k),a_{s_0}(1))$ is a solution of the system $g=0$ for all $g\in D(I_{s_0+1})\cap R_{s_0}$ and then the system $f_i=0$ $i=1,\dots,r$ has a solution in G^n. On the other hand diff $\dim_L \varphi(I)^{*}=-1$ implies diff $\dim_G\varphi(I)^{*}=-1$, because $L\subseteq G$. Now we have a contradiction, because diff $\dim_G \varphi(I)^{*}\geq 0$.

Remark. The number s_0 depends on s, n and $m=|\Delta|$ and it can be very big. An estimation of s_0 can be given by looking at the particular case of systems of homogeneous linear differential equations. If this is the case, let $S=K[\partial_1,\dots,\partial_m]$ with $\partial_i\partial_j=\partial_j\partial_i$ and $\partial_i a\partial_j=\partial_i(a)\partial_j+a\partial_i\partial_j$ for $a\in K$, $i,j=1,\dots,m$ and let $F=Sx_1\oplus\dots\oplus Sx_n$. There is a one to one correspondence between submodules of F and such systems, moreover the problem of finding s_0 is equivalent to find a Gröbner basis of a such submodule, that there exists by ([4],[5],[7],[8]).

By extending a result in [1] for the dimension of an ideal in a ring of polynomials with coefficients in a field K, we have the following lemma, on the dimension (respectively differential dimension) in terms of the variables (respectively the variables with their derivatives) that appear in the polynomials (respectively differential polynomials) of the ideal.

Proposition 3. Let $I=[f_1,\dots,f_r]$ be a finetely differentially generated ideal in R. Let s_0 as in proposition 2 and let $(g_1,\dots,g_{\lambda(s_0+1)})$ be the reduced Gröbner basis of $D(I_{s_0+1})$. Let $A=\|a_{ij}\|$ $i=1,\dots,\lambda(s_0+1)$, $j=1,\dots,n$ be the matrix where $a_{ij}=0$ if no power of some ϑx_j with $\vartheta\in\Theta_{s_0}$, appears in the maximal monomial in g_i and $a_{ij}=1$ if some power of ϑx_j, with $\vartheta\in\Theta_{s_0}$ appears in the maximal monomial in g_i. diff $\dim_K I=n-h$, where $h=\min\{r$: there exists a subset $\{i_1,\dots,j_r\}$ of $\{1,\dots,n\}$ with $1\leq i_1<\dots<j_r\leq n$ such that for some i $a_{i,j_e}\neq 0$, $e=1,\dots,r\}$.

Proof. Since $|\Theta_{s_0+1}|=\binom{s_0+1+m}{m}$ we can consider the matrix

$B=\|B_1B_2,\dots,B_n\|$, where $B_j=\|b_{i,l(j)}\|$ where $i=1,\dots,\lambda(s_o+1)$, $l(j)=1,\dots,\binom{s_o+1+m}{m}$ and $j=1,\dots,n$ with the condition $b_{i,l(j)}=1$ if some powewr of ϑx_j appears in the maximal monomial of g_i and 0 otherwise. Let $d=\mathrm{diff\ dim}_k I$. By propositions 1 and 2 there exist $x_{i_1},\dots,x_{i_d}$ such that $D(I_{s_o+1})\cap K(x_{i_1},\dots,x_{i_d})=0$ and $D(I_{s_o+1})\cap$ $\cap K(x_{i_1},\dots,x_{i_d},x_j)\neq(0)$ for all $j=d+1,\dots,n$. So $\dim_K D(I_{s_o+1})=$ $=d\binom{s_o+1+m}{m}+a$ with $a\geq 0$. By ([1], p.2630) and [3]) for each set of $r\leq(n-d)\binom{s_o+1+m}{m}-a-1$ columns in B there is a row in B in which the corresponding elements are zero $D(I_{s_o+1})\cap K(x_{i_1},\dots,x_{i_d},x_{i_j})\neq(0)$ for all $j=d+1,\dots,n$ implies that for each such j there exists a g_i such that its maximal monomial is in $K(x_{i_1},\dots,x_{i_d},x_{i_j})$. So $(n-d)\cdot$ $\cdot\binom{s_o+1+m}{m}-a\geq(n-d-1)\binom{s_o+1+m}{m}+1$ by maximality of d. Let A be the matrix as in the hypothesis $a_{ij}=0$ iff $b_{i,\lambda(j)}=0$ for all $l(j)=1,\dots,\binom{s_o+1+m}{m}$.

By definition of h we have $(h-1)\binom{s_o+1+m}{m}\leq(n-d)\binom{s_o+1+m}{m}-a-1\leq$ $\leq(n-d)\binom{s_o+1+m}{m}-1$ i.e. $h\leq n-d$. On the other hand $(n-d-1)\binom{s_o+1+m}{m}\leq$ $\leq(h-1)\binom{s_o+1+m}{m}$ by maximality of h-1, i.e. $n-d\leq h$.

Proposition 4. Let $I=[f_1,\dots,f_r]$ be a finetely differentially generated ideal in $R=\{x_1,\dots,x_n\}$ and let $s=\max\{\mathrm{ord}(f_i): i=1,\dots,r\}$. If $|\Delta|=1$, then each characteristic set of I is contained in $D(I_{2^{n-1}s})$.

Proof. First of all suppose that $s=1$. In this case we have only one orderly ranking of $\{x_1\}$. So we have only one characteristic set with only one element f. By Ritt algorithm ([9], p.5 and p.15) and by lemma 1 in order to find f we need only differential polynomials that are in $D(I_s)$. Now suppose that the proposition is true for differential ideals in $K\{x_1,\dots,x_{n-1}\}$. Let $S=K\langle x_1,\dots,x_{n-1}\rangle\{x_n\}$ and let $\varphi:K\{x_1,\dots,x_n\}\longrightarrow S$ be the canonical differential ring homomorphism. $[\varphi(I)]$ is again a differential ideal. If $f_j\in K\{x_1,\dots,x_{n-1}\}$ for all $i=1,\dots,r$ then the proposition is true by induction hypothesis, being $D(I_{2^{n-2}s})\subset D(I_{2^{n-1}s})$. Suppose that there is some $j=1,\dots,r$ such that $f_j\notin K\{x_1,\dots,x_{n-1}\}$. Without loss of generality we can suppose that $f_1,\dots,f_h\in K\{x_1,\dots,x_{n-1}\}$ and $f_{h+1},\dots,f_r$ are in $K\{x_1,\dots,x_n\}-K\{x_1,\dots,x_{n-1}\}$. Let $J=[f_{h+1},\dots,f_r]$ $[\varphi(J)]=[f_{h+1},\dots,f_r]$ where $f_{h+1},\dots,f_r$ are considered as differential polynomials in S. Let $r_j=\text{ord}(f_j)$ as polynomials in S. Let $s_n=\max\{r_j:\ j=h+1,\dots,r\}$ $s_n\le s$. Without loss of generality we can suppose that $r_{h+1}\le r_{h+2}\le\dots\le r_r$. If $r_{h+1}=0$ and $r_j>0$ for all $j=h+2,\dots,r$ then by Ritt's algorithm we need derivates of f_{h+1} up to order $r_r\le s$. The maximal order of the derivates of the x_i, $i=1,\dots,n-1$ that appear in these derivates of f_{k+1} is $s+r_r\le 2s$. The partial remainder R_j of each f_j with respect to f_{h+1} has order not greater then 2s and $R_j\in D(I_{2s})$ when it is considered as differential polynomial in $K\{x_1,\dots,x_n\}$. If $r_{h+1}>0$ by Ritt's algorithm the remainder R_j has order not greater than $2s-r_{h+1}$. If we have $R_j\in K\{x_1,\dots,x_{n-1}\}$ for all $j=h+1,\dots,r$, then f_{h+1} is the characteristic set of [9]) in S and we can use the induction hypothesis on the ideal $[f_1,\dots,f_{h-1},f_h,R_{h+2},\dots,R_r]\subset K\{x_1,\dots,x_{n-1}\}$. If not we can repeat the algorithm with a new $s'_n=r_{h+2}$. At each step the maximal order of each remainder is not greater than 2s and it is in $D(I_{2s})$ by its own definition. In order to find the characteristic set of I now we need to find the characteristic set of $I'=[f_1,\dots,f_h,R_{h+2},R_r]$. By induction hypothesis such set is contained in $D(I'_{2^{n-2}(2s)})=D(I'_{2^{n-1}s})$. Since $I'\subseteq I$, we have the proof.

Example 1. Let $[\partial^s(x_2)-x_1,\dots,\partial^s(x_{n-1})-x_{n-2},\ \partial^s(x_n)-x_{n-1},\partial^s x_1-x_n]$ in $K\{x_1,\dots,x_n\}$ and let O be an orderly ranking of $(x_1,\dots,x_n)$ with

$x_1<\ldots<x_n$ $\{\partial^{ns}x_1 - x_1,\ x_2-\partial^{(n-1)s}x_1, \ldots, x_{n-1}-\partial^{2s}x_1, x_n-\partial^{s}x_1\}$ is the characteristic set of I.

Example 2. Let $I=[x_2\partial x_1-1,\ x_1\partial x_2-1]\subset K\{x_1,x_2\}$. For any orderly ranking of (x_1,x_2) $\{x_2\partial x_1-1,\ x_1\partial x_2-1\}$ is the reduced Gröbner basis of $D(I_1)=I_1$. $D(I_s)=(x_2\partial x_1-1,\quad x_1\partial x_2-1,\quad \partial^k(\partial^2 x_1+\partial x_1^2\partial x_2),$ $\partial^k(\partial^2 x_2+\partial x_1\partial x_2^2)$: $k=0,\ldots,s-2)$ for all $s\geq 2$ $2^{n-1}s=2$ and diff $\dim_K I=0$.

REFERENCES

[1] **G. CARRA' FERRO**, *"Some properties of the lattice points and their application to differential algebra"*, Communications in Algebra, vol.15 (1987) p.2625-2632.

[2] **G. CARRA' FERRO**, *"Gröbner bases and differential Algebra"* (1987) Preprint submitted to AAECC5.

[3] **G. CARRA' FERRO**, *"Gröbner bases and Hilbert schemes I"*, (1986), To appear in Journal of Symbolic Computation.

[4] **A. GALLIGO**, *"Some algorithmic questions on ideals of differential operators"*, Proc. Eurocal 1985 II, Lecture Notes in Computer Science, 204 (1985), p.413-421.

[5] **A. KANDRI-RODY, V. WEISPFENNING**, *"Non commutative Gröbner base in Algebras of solvable type"*, (1986) to appear in Journal of Symbolic in Computation.

[6] **E. R. KOLCHIN**, *"Differential algebra and algebraic groups"* Academic Press, New York (1973).

[7] **F. MORA**, *"Seven variations on standard bases"*, Preprint (1987).

[8] **F. MORA**, *"Gröbner bases in non commutative Algebra"*, Preprint (1988).

[9] **J.F. RITT**, *"Differential Algebra"* AMS publications (1950).

ON THE ASYMPTOTIC BADNESS OF CYCLIC CODES WITH BLOCK-LENGTHS COMPOSED FROM A FIXED SET OF PRIME FACTORS

G. Castagnoli

Institute for Signal and Information Processing
Swiss Federal Institute of Technology
ETH-Zentrum, 8092 Zurich, Switzerland

Abstract: q-ary cyclic codes of rate $r \geq R$ and with blocklength $n = p_1^{\alpha_1} \dots p_s^{\alpha_s}$ composed from a fixed finite set of primes $S = \{p_1, \dots, p_s\}$ are shown to have a minimum distance d_{min} which is upper bounded by a function $d(S, R)$. Hence, cyclic codes with blocklengths such that all prime factors are in S and which have rate $r \geq R$ are asymptotically bad in the sense that for these codes d_{min}/n tends to zero as n increases to infinity.

1. INTRODUCTION

The question of whether long cyclic codes are good is an unsolved problem in algebraic coding theory. To date several classes of cyclic codes are known which are asymptotically bad in the sense that sequences $(C_i)_{i \in \mathbf{N}}$ of codes in these classes with rates $r_i \geq R$ and blocklengths n_i yield $\lim_{n_i \to \infty} \frac{d_{min}(C_i)}{n_i} = 0$.

We consider here q-ary cyclic codes that have blocklengths n with prime factors in a fixed finite set $S = \{p_1, \dots, p_s\}$ of primes. S.D. Berman proved that when $q = p^m$ and $p \notin S$ or when $S = \{p\}$, q-ary cyclic codes of rate $r \geq R$ and blocklength $n = p_1^{\alpha_1} \dots p_s^{\alpha_s}$ have a minimum distance d_{min} not greater than some value $d(S, R)$ which is independent of $\{\alpha_1, \dots, \alpha_s\}$. Thus for these codes the ratio d_{min}/n goes to zero in a particularly strong way as n increases to infinity and so these codes are asymptotically bad. We will give a simple derivation of Berman's result, which we will also extend to allow S to be an arbitrary finite set of primes.

2. NOTATION AND PRELIMINARIES

In this paper we use the following notation:

Let $S = \{p_1, \dots, p_s\}$ be a finite set of primes. We then define:

$N_S \stackrel{\text{def}}{=} \{n = p_1^{\alpha_1} \dots p_s^{\alpha_s} \mid \alpha_i \in \mathbf{N} \text{ for } i = 1, \dots, s\}$, and $N_S \stackrel{\text{def}}{=} \{1\}$ when $S = \emptyset$.

$P(S,q) \stackrel{\text{def}}{=} \{f(x) \in GF(q)[x] \mid f(x)$ is irreducible, $ord\ f(x) \in N_S$ and there exists no

$g(x) \in GF(q)[x]$ and $u > 1$ such that $f(x) = g(x^u)\}$.
Observe that $P(S \cup \{p\}, p^m) = P(S, p^m)$ for all S and all prime powers p^m, since the order of an irreducible polynomial cannot contain the characteristic of the coefficient field $GF(p^m)$ as a factor.

At several places we make use of the notation $x \vee y$ for $max(x, y)$, were x, y are integers. For $n \in N_S$ and $t > 0$, we define $g_{n,t}(x)$ as the product of all irreducible factors $f(x)$ of $x^n - 1$ for which $ord\ f(x) > n/t$ and $d_{n,t} \stackrel{\text{def}}{=} deg\ g_{n,t}(x)$, i.e., $d_{n,t} = \left|\{\text{ n-th roots of unity over } GF[q], \text{ of order } > n/t\}\right|$. Finally, $w_H(v(x))$ denotes the Hamming weight of $v(x)$, i.e., the number of nonzero coefficients of $v(x)$.

Example: Let $S = \{3\}$ and $q = 2$. Then $N_S = \{3^u | u \geq 0\}$ and $P(S, q) = \{x^2+x+1,\ x+1\}$. The irreducible polynomials $f(x) \in GF(2)[x]$ for which $ord\ f \in N_S$, i.e., for which $ord\ f = 3^u$ for some $u \in \mathbf{N}$ are $x + 1$ for $u = 0$ and $x^{2 \cdot 3^{u-1}} + x^{3^{u-1}} + 1$ for $u = 1, 2, \ldots$

The next three theorems form the basis of our approach.

Theorem 1 implies that the irreducible polynomials over $GF(q)$ with orders equal to products of powers of primes from a fixed set S can be obtained by forming polynomials $f(x^u)$ in all possible ways, whereby u is a product of powers of primes contained in S and $f(x)$ is from the *finite* set $P(S, q)$ of irreducible polynomials.

Theorem 2 asserts that the product $g_{n,t}(x)$ of irreducible factors $f(x)$ of $x^n - 1$ with large orders, i.e., for which $ord\ f > n/t$, can be made to have an arbitrarily large degree $d_{n,t}$ relative to n, for all $n \in N_S$, by choosing t sufficiently large. Theorem 2 does not hold when the set S of possible prime factors of n is infinite.

Theorem 3 is required to extend Berman's theorem from the case of simple-root cyclic codes to repeated-root cyclic codes, i.e., to q-ary cyclic codes whose the blocklength contains a power of the characteristic p of $GF(q)$.

Theorem 1: $P(S, q)$ is finite for any prime power q and any finite set S of primes.

Outline of Proof: Let $p_1, \ldots, p_u$ be primes such that $(p_i, q) = 1\ \forall i$. Then consider $n = p_1 \ldots p_u$ and let m be the smallest integer for which $n \mid q^m - 1$.
For $j = 1, 2, \ldots, u$ let t_j be the multiplicity of p_j in $q^m - 1$. Consider now an irreducible polynomial $f(x) \in GF(q)[x]$ of order $ord\ f = p_1^{\gamma_1} \ldots p_u^{\gamma_u}$ with $\gamma_j \geq 1\ \ \forall j$; then we have

$$deg\ f = m \cdot p_1^{(\gamma_1 - t_1) \vee 0} \ldots p_u^{(\gamma_u - t_u) \vee 0},$$

(except when for one of the primes $p_i \in S$ we have $p_i = 2$ with $t_i = 1$, $\gamma_i > 1$ and $2^2 \mid q+1$). Let $f(\sigma) = 0$. Then $\sigma^{p_1^{(\gamma_1 - t_1) \vee 0} \ldots p_u^{(\gamma_u - t_u) \vee 0}}$ has the order $ord\ f / p_1^{(\gamma_1 - t_1) \vee 0} \ldots p_u^{(\gamma_u - t_u) \vee 0}$ which is a divisor of $q^m - 1$. Hence $f(x)$ must equal $\tilde{f}(x^{p_1^{(\gamma_1 - t_1) \vee 0} \ldots p_u^{(\gamma_u - t_u) \vee 0}})$ where

$\tilde{f}(x) \in GF(q)[x]$ is one of finitely many irreducible polynomials with $ord\ \tilde{f}(x) \mid q^m - 1$ and $deg\ \tilde{f}(x) = m$.

Theorem 2: Let S be a finite set of primes which does not contain the characteristic of $GF(q)$ and let $d_{n,t}$ denote the number of n-th roots of unity over $GF[q]$, of order $> n/t$. We then have

$$\lim_{t \to \infty} \left[\inf_{n \in N_S} \frac{d_{n,t}}{n} \right] = 1.$$

Outline of Proof: Consider $e_{n,t} \stackrel{\text{def}}{=} |\{$ n-th roots of unity over $GF[q]$, of order $\leq n/t\}| = n - d_{n,t}$. When $S = \{p\}$, Theorem 2 follows from $e_{p^\alpha,t} = p^{\alpha - \lceil \log_p t \rceil}$. When S contains several primes, the statement can be proved by induction on $\mid S \mid$. Consider $E(n,t) \stackrel{\text{def}}{=} \{d \in \mathbf{N} \mid d \mid n,\ d \leq n/t\}$, let p^α be a prime power not in N_S and $\bar{n} \in N_S$. Then

$$e_{\bar{n}p^\alpha,t} = \sum_{d \in E(\bar{n}p^\alpha,t)} \phi(d) = \sum_{j=0}^{\alpha} \Big(\sum_{d \in E(\bar{n},tp^{j-\alpha})} \phi(p^j)\phi(d) \Big) \leq \sum_{j=0}^{\alpha} p^j e_{\bar{n},tp^{j-\alpha}} = \sum_{j=0}^{\alpha} p^{\alpha-j} e_{\bar{n},t/p^j}$$

From this inequality we can derive an upper bound on $e_{\bar{n}p^\alpha,t}/\bar{n}p^\alpha$ which is independent of $\bar{n}p^\alpha$

$$\frac{e_{\bar{n}p^\alpha,t}}{\bar{n}p^\alpha} \leq \sum_{j=0}^{\alpha} \frac{e_{\bar{n},t/p^j}}{\bar{n}p^j} \leq \sum_{j=0}^{\infty} \frac{e_{\bar{n},t/p^j}}{\bar{n}p^j} \leq \inf_{j \in \mathbf{N}} \Big(j \frac{e_{\bar{n},t/p^j}}{\bar{n}} + \frac{2}{p^j}\Big) \leq \inf_{j \in \mathbf{N}} \Big(j \cdot \sup_{\bar{n} \in N_S} \frac{e_{\bar{n},t/p^j}}{\bar{n}} + \frac{2}{p^j}\Big).$$

The right side of this inequality can be made arbitrarily small by choosing t sufficiently large: given $\epsilon > 0$, choose j such that $2/p^j < \epsilon/2$ and then choose t large enough such that $e_{\bar{n},t/p^j}/\bar{n} < \epsilon/2j$ for all $\bar{n} \in N_S$.

Theorem 3 (see Massey et al.[4] for $q = 2$): Let $q = p^m$; then a q-ary repeated-root cyclic code of blocklength $n = p^\alpha \bar{n}$ $(p \nmid \bar{n})$ generated by $g(x)$, for which at least one irreducible factor $f(x)$ of $x^{\bar{n}} - 1$ occurs in $g(x)$ with multiplicity $m_o \leq \bar{t} \stackrel{\text{def}}{=} p^\alpha - p^{\alpha-\bar{\alpha}} = (p-1)p^{\alpha-1} + \ldots + (p-1)p^{\alpha-\bar{\alpha}}$ for some $\bar{\alpha} \leq \alpha$, has $d_{min} \leq p^{\bar{\alpha}} d_s$, where d_s is the minimum distance of the simple-root cyclic code generated by a certain $\bar{g}(x)$ with $\bar{g}(x) \mid g(x)/f^{m_o}(x)$.

Outline of Proof: Let $\bar{g}(x)$ be the product of those irreducible factors (taken with simple multiplicity) which occur in $g(x)$ with a multiplicity greater than $\bar{t}$. Then, under the conditions of the theorem there exist nonzero codewords in the repeated-root cyclic code generated by $g(x)$ which have the following form: $c(x) = (x^{\bar{n}} - 1)^{\bar{t}} \bar{g}(x)^{p^\alpha} u(x)^{p^\alpha}$ mod $(x^{\bar{n}p^\alpha} - 1)$. It then follows that

$$w_H(c(x)) = w_H((x^{\bar{n}} - 1)^{\bar{t}}) \cdot w_H(\bar{g}(x)^{p^\alpha} u(x)^{p^\alpha}) = w_H((x-1)^{\bar{t}}) \cdot w_H(\bar{g}(x)u(x)).$$

According to Massey et al. [3] and Berman [1], $w_H((x-1)^{\bar{t}}) = p^{\bar{\alpha}}$ and, by adequately choosing $u(x)$, $w_H(\bar{g}(x)u(x))$ can be made to equal the minimum distance of the simple-root cyclic code generated by $\bar{g}(x)$.

3. PROOF AND GENERALIZATION OF BERMAN'S THEOREM

Let S be a finite set of primes not containing the characteristic of $GF(q)$ and let C be a q-ary cyclic code of rate $r \geq R$ and blocklength $n \in N_S$. We now deduce the existence of a codeword of bounded Hamming weight as follows. Let $g(x)$ be the generator polynomial of C. Then $deg\ g(x) \leq n(1-R)$. According to Theorem 2 we can choose t large enough, so that $d_{\tilde{n},t}/\tilde{n} > 1-R$ for all $\tilde{n} \in N_S$, in particular for $\tilde{n} = n$. This means that there exists at least one irreducible factor $f(x)$ of $g_{n,t}(x)$ which is not contained in $g(x)$. Hence $c(x) = (x^n - 1)/f(x)$ is the polynomial of a codeword $c \in C$. Since $f(x) = \tilde{f}(x^{p_1^{\beta_1} \cdots p_s^{\beta_s}})$ for some $\tilde{f}(x) \in P(S,q)$ and certain integers $\beta_1, \ldots, \beta_s$, we obtain:

$$c(x) = (x^n - 1)/\tilde{f}(x^{p_1^{\beta_1} \cdots p_s^{\beta_s}}) = (x^{up_1^{\beta_1} \cdots p_s^{\beta_s}} - 1)/\tilde{f}(x^{p_1^{\beta_1} \cdots p_s^{\beta_s}}) = \tilde{c}(x^{p_1^{\beta_1} \cdots p_s^{\beta_s}})$$

with $\tilde{c}(x) = (x^u - 1)/\tilde{f}(x)$. Since $f(x)$ as a factor of $g_{n,t}(x)$ satisfies $ord\ f(x) > n/t$, we have $n/ord\ f(x) = u/ord\ \tilde{f}(x) < t$, which implies $u < t \cdot ord\ \tilde{f}(x)$. Since $P(S,q)$ is finite, $ord\ \tilde{f}(x)$ and hence also u cannot exceed certain values determined by S, q and t. Therefore $w_H(c(x)) = w_H(\tilde{c}(x)) \leq u$ and so there exists an upper bound $d(S,R)$ on d_{min} which is independent of $n \in N_S$ in the case where the characteristic of $GF(q)$ is not in S.

Now let the characteristic p of $GF(q)$ be contained in S. Then $n \in N_S$ can be written as $n = p^\alpha \tilde{n}$ with $p \nmid \tilde{n}$. According to Theorem 2, for any $\delta < R$ we can choose t large enough so that $d_{\tilde{n},t}/\tilde{n} > 1 - R + \delta$ holds for all $\tilde{n} \in N_{S \setminus \{p\}}$ (note: if $S \setminus \{p\} = \emptyset$, we have $N_{S\setminus\{p\}} = \{1\}$).

Let $u \cdot d_{\tilde{n},t} = deg\ g(x)$; then we can infer that at least one irreducible factor $f(x)$ of $g_{\tilde{n},t}(x)$ does not occur in $g(x)$ with multiplicity greater than u, where

$$u = \frac{deg\ g(x)}{d_{\tilde{n},t}} < \frac{(1-R)p^\alpha \tilde{n}}{(1-R+\delta)\tilde{n}} = p^\alpha \frac{1-R}{1-R+\delta} = p^\alpha - p^\alpha \frac{\delta}{(1-R+\delta)}.$$

Choose an $\tilde{\alpha} \in \mathbf{N}$ such that $p^{-\tilde{\alpha}} \leq \frac{\delta}{(1-R+\delta)}$. Then we get $\lfloor u \rfloor \leq p^\alpha - p^{(\alpha-\tilde{\alpha})\vee 0}$ so, by Theorem 3, it now follows that $d_{min} \leq p^{min(\tilde{\alpha},\ \alpha)} D \leq p^{\tilde{\alpha}} D$, where D is the minimum distance of the length $\tilde{n}$ (simple-root) cyclic code generated by $\bar{g}(x) \stackrel{\text{def}}{=} (x^{\tilde{n}} - 1)/f(x)$. By the same argument as in the proof of Berman's theorem in the case of simple-root cyclic codes, we can now conclude that D is bounded above by a value which is determined by S, q and t and, in particular, is independent of $\tilde{n}$. Therefore there exists a $d(S,R)$ which is an upper bound on the minimum distance d_{min} of q-ary cyclic codes of rates $r > R$ and blocklengths $n \in N_S$ with S containing the characteristic of $GF(q)$.

References:

[1] S.D. Berman, "On the Theory of Group Codes", *Cybernetics*, 3(1), pp.25-31, 1967.

[2] S.D. Berman, "Semisimple Cyclic and Abelian Codes", *Cybernetics*, 3(3), pp.17-23, 1967.

[3] J.L. Massey, D.J. Costello, and J. Justesen, "Polynomial Weights and Code Constructions", *IEEE Trans. Inform. Th.*, IT-19, pp.101-110, 1974.

[4] J.L. Massey, N. von Seeman, and Ph. Schöller, "Hasse Derivatives and Repeated-Root Cyclic Codes", *Abstracts of papers*, IEEE Int. Symp. Info. Th., Ann Arbor, MI, USA, p.39, 1986.

BIG NUMBERS P-ADIC ARITHMETIC:

A PARALLEL APPROACH

Attilio Colagrossi

Istituto di Analisi dei Sistemi
ed Informatica
Consiglio Nazionale delle Ricerche
Viale Manzoni 30
00185 Roma, Italy

Carla Limongelli

Dipartimento di Informatica
e Sistemistica
Università di Roma "La Sapienza"
Via Buonarroti 12
00185 Roma, Italy

In the past years, great attention has been payed to approximated p-adic arithmetic expressed in the form of Hensel codes and several contributions have made this p-adic arithmetic really effective, just according to this new consideration of it. Howewer it has been shown that the appliability of p-adic arithmetic is strongly constrained by the size of rational numbers which constitutes the output of the given computation. The key idea we like to present here, it is based on the intention of getting any advantage out the variation of the prime p, choosing it as a very large number. This choise suggests a definition of a new algorithm which will benefit from a parallel execution. Thus our aim is to perform "big" rational numbers arithmetic applying the so called g-adic approach, based on the theory of g-adic numbers. This paper describes a general schema of g-adic computation and then presents two algorithms to perform the inverse mapping together with the related complexity analysis.

This work has been partially supported by M.P.I.: Progetti di Manipolazione Algebrica, Calcolo Algebrico; and by C.N.R.: Progetto Strategico Matematica computazionale.

1. Introduction

Approximated p-adic arithmetic, expressed in the form of Hensel codes, has received great attention in the past years; the initial proposal by Gregory [GRE78] based on the look-up table approach, the contribution by Miola [MIO84] developing the inverse mapping algorithm, and further refinements by Gregory, Dittenberger, Colagrossi & Miola [G&K84,DIT85,C&M87], have made this p-adic arithmetic really effective.

The effectiveness of this approximated arithmetic derives from a basic property, and namely from the assumption that for a given prime p and an approximation r (number of digits represented), the exact result over the rationals can be detected if and only if it belongs to the Farey's fractions set defined by p and r.

In order to enlarge the appliability of the approximated p-adic arithmetic, in this paper we like to get a real and complete advantage out of the variation of the prime p, choosing it as a very large number; this choice suggests the definition of a new algorithm particulary suitable for a parallel execution. Our aim is to perform "big" rational numbers arithmetic applying the so called g-adic approach.

The theory of g-adic numbers has been already treated by Mahler [MAH73]. In substance such theory enables us to construct a result of a given computation over the rationals by working with the Farey's fractions set defined by $g = p_1 p_2 \cdots p_k$, with p_i prime, and by the approximation r, starting from the k p-adic computations defined by the different p_i and by the same approximation r.

In the following sessions we will present a general schema of g-adic computation and two algorithms to perform the inverse mapping together with the related complexity analysis.

2. The general schema of g–adic computation

In the sequel we will assume as given the concepts about p-adic and Hensel code's representation as described in [KOB77] and in [G&K84]. The following notations will be used:

$H(p,r,a/b)$: is the Hensel code representation of the rational number a/b with respect to the prime p and the approximation r; this is formed by the ordered pair (μ, exp_μ) (i.e. mantissa, exponent).

$\mathbf{H}(p,r)$: is the Hensel's codes set with respect to the prime p and approximation r (i.e. the number of digits of each code);

$\mathbf{F}(p,r)$: is the Farey's fraction set, i.e. the set of rational numbers a/b such that:

$$a, b \in \mathbf{N}, \quad 0 \leq a \leq N, \quad 0 < b \leq N \quad with \quad N = \left\lfloor \sqrt{\frac{p^r - 1}{2}} \right\rfloor$$

Let us assume also that k distinct primes $p_1, \ldots, p_k$ are given together with a positive integer r and let be $g = p_1 p_2 \cdots p_k$.

Let p denote the maximum of p_i, with $i = 1 \ldots k$. Then $g < p^k$.

For sake of simplicity and without loss of generality, let us consider the simple problem of a single binary operation op between two given rational numbers:

$$\frac{c_1}{c_2} = \frac{a_1}{a_2} \; op \; \frac{b_1}{b_2}$$

We choose g and r such that $c_1/c_2 \in \mathbf{F}(g, r)$. As described in detail in [G&K84] and [MIO84], this choice guarantees that the exact result (i.e. c_1/c_2) will be detected.

Furthermore let us suppose that c_1/c_2 doesn't belong to any of the Farey's fractions sets $\mathbf{F}(p_i, r)$, for $i = 1, \ldots, k$.

The computation will proceed according to the following three steps:

1. Both the rational numbers a_1/a_2, b_1/b_2 are mapped into k Hensel codes and each of these codes belongs to one of the following sets: $\mathbf{H}(p_1, r), \ldots, \mathbf{H}(p_k, r)$. Let $(\alpha_i, exp_{\alpha_i})$ and (β_i, exp_{β_i}) be the images of a_1/a_2 and of b_1/b_2 respectively in the i–th Hensel code set.
2. In each of the sets $\mathbf{H}(p_i, r)$ the following expression is computed

$$(\gamma_i, exp_{\gamma_i}) = (\alpha_i, exp_{\alpha_i}) \quad op' \quad (\beta_i, exp_{\beta_i})$$

   in order to obtain the k Hensel codes of the expected result. We note that op' is the corresponding operator of op in the direct homomorphism.
3. From the k results obtained in the previous step **2.**, the unique rational result c_1/c_2 is then reconstructed.

The two initial steps can be easily carried out by usual p–adic methods and operations as described in [G&K84]. In order to perform the third step, a specific inverse mapping algorithm must be devised. In the next section we are presenting two inverse mapping algorithms, and in the following section **4.** we will compare them from a computational point of view.

3. Inverse mapping algorithms

The inverse mapping algorithms start from the k given Hensel codes

$$(\gamma_i, exp_{\gamma_i}) = (\gamma_{0,i} \ldots \gamma_{j,i} \ldots \gamma_{r-1,i}, exp_{\gamma_i})$$

where $\gamma_{j,i} \in Z_{p_i}$ represents the j–th digits of the mantissa of the i–th Hensel code.

Let us note that the inverse mapping algorithms handle with Hensel codes whose exponents are equal to zero. Thus if the above mentioned k Hensel codes have the exponent exp_{γ_i} different from zero, we must first perform same operations in order to transform the codes into the corresponding ones.

The first inverse mapping algorithm to be considered, it is a very natural extension of the algorithm used to map a single Hensel code into the rationals; it uses the Chinese Remainder Algorithm (CRA), and it proceeds according to the following steps.

Algorithm C: CRA based inverse mapping algorithm.

C1. From each Hensel code $(\gamma_i, exp_{\gamma_i})$ compute in $Z_{p_i^r}$

$$d_i = \sum_{j=0}^{r-1} \gamma_{j,i} p_i^j = \gamma_{0,i} + \gamma_{1,i} p_i + \ldots + \gamma_{r-1,i} p_i^{r-1}$$

C2. compute the moduli $m_i = p_i^r$;

C3. apply CRA to the values d_i with the moduli m_i; to obtain the unique $d \in Z_{g^r}$;

C4. apply the Modified version of the Extended Euclidean Algorithm (MEEA) (the same algorithm used in [MIO84]) to d, g and r, in order to map d into the unique rational number c_1/c_2 belonging to $\mathbf{F}(g,r)$.

The second method is based on a "Hensel like" lifting approach. Essentially it uses the first j-1 digits of each Hensel code γ_i to obtain the j–*th* digit δ_j of the g–adic expansion of the result:

$$d = \delta_0 + \delta_1 g + \cdots + \delta_{r-1} g^{r-1} \qquad (1)$$

according to the following steps.

Algorithm L: Hensel lifting based inverse mapping algorithm

L1. apply CRA to the Hensel codes' digit $\gamma_{0,i}$ and moduli p_i, for $i = 1, \ldots, k$, to obtain δ_0; set j to 1;

L2. apply the lifting algorithm **LA** (as described later) to δ_{j-1}, $\gamma_{j,i}$ and moduli p_i to obtain the values $\overline{\gamma}_{j,i}$, for $i = 1, \ldots, k$;

L3. apply CRA to the images $\overline{\gamma}_{j,i}$ and moduli p_i to obtain δ_j;

L4. if $j < r - 1$ then set j to $j + 1$ and go to **L2.** else perform next step;

L5. transform the digits of the p–adic expansion already obtained, according to the relation (1);

L6. like step **C4.**.

Let us now outline the lifting algorithm **LA**.

Let $(\gamma_i, exp_{\gamma_i})$ be the i–*th* Hensel code result, with

$$\gamma_i = \gamma_{0,i}\gamma_{1,i} \cdots \gamma_{r-1,i}.$$

Because

$$\delta_0 + \delta_1 g + \cdots + \delta_{r-1} g^{r-1} \equiv \gamma_{0,i} + \gamma_{1,i} p_i + \cdots + \gamma_{r-1,i} p_i^{r-1} \pmod{p_i^r}$$

with $i = 1 \ldots k$, the following system of k congruences holds:

$$\delta_0 \equiv \gamma_{0,i} \pmod{p_i}$$
$$\delta_0 - \delta_1 g \equiv \gamma_{0,i} + \gamma_{1,i} p_i \pmod{p_i^2}$$
$$\cdots\cdots\cdots$$
$$\delta_0 - \delta_1 g + \cdots + \delta_j g^j \equiv \gamma_{0,i} + \gamma_{1,i} p_i + \cdots + \gamma_{j,i} {p_i}^j \pmod{p_i^{j+1}}$$

with $j = 0, \ldots, r-1$ and $i = 1, \ldots, k$.

By applying CRA to the first congruence we obtain δ_0

The j-*th* digit δ_j is computed by **LA** algorithm which is now shortly described.

Algorithm LA: Lifting Algorithm:

Let: $\overline{\gamma}_{0,i} = s_{0,i} = \gamma_{0,i}$;

$$s_{1,i} \equiv \frac{s_{0,i} - \delta_0}{p_i} + \gamma_{1,i} \pmod{p_i};$$

$$\overline{\gamma}_{1,i} \equiv \frac{\overline{s}_{1,i}}{g_i} \pmod{p_i} = \overline{s}_{1,i}\overline{g}_i \pmod{p_i}$$

where

$$g_i = g/p_i, \qquad \overline{g}_i \equiv g_i^{-1} \pmod{p_i}$$

and

$$\overline{s}_{1,i} \equiv s_{1,i} \pmod{p_i}$$

for $i = 1, \ldots, k$;

(LA1):

$$s_{j,i} \equiv \frac{(\overline{s}_{j-1,i} - \delta_{j-1} g_i^{j-1})}{p_i} + \gamma_{j,i} \pmod{p_i};$$

(LA2):

$$\overline{\gamma}_{j,i} \equiv (\overline{s}_{j,i}\overline{g}_i^j) \pmod{p_i}$$

where

$$\overline{s}_{j,i} = s_{j,i} \pmod{p_i}$$

for $i = 1, \ldots, k$.

Let us underline the fact that we have performed two different methods only to recover the g-adic Hensel code that will be the input for the final step of the transformation: the usual MEEA.

Now let us consider the case in where the exponent of the codes is different from zero. In such a case we operate to transform each pair $(\gamma_i, exp_{\gamma_i})$, into the code $(\gamma_i', 0)$.

We must distinguish two cases:

a) $exp_{\gamma_i} > 0$:

in this case we can simply operate a right shift of the mantissa, exactly of exp_{γ_i} positions. So we have:

$$\gamma_i' = 0 \ldots 0 \gamma_{0,i} \cdots \gamma_{r-1-exp_{\gamma_i}}.$$

b) $exp_{\gamma_i} < 0$:

in this case we multiply each of the k codes by $H(p_i, r, -exp_{\gamma_i})$. We must pay attention when we obtain the rational result, since we must multiply it by $p_i^{exp_{\gamma_i}}$ to obtain the exact result of the computation.

In the Figure 1. it is shown an example of the application of both algorithms (the behaviour of C algorithm is expressed by the deshed lines, the plain full lines describe the **L** algorithm).

Example: $r=4$; $p_1=3$; $p_2=5$; $p_3=7$; $g=105$;

$$\frac{137}{43}+\frac{29}{13}=\frac{3973}{559} \in F(g,r)$$

	137/43	29/13	3973/559
H(3,4)	(2212,0)	(2122,0)	(1111,0)
H(5,4)	(4101,0)	(3143,0)	(2432,0)
H(7,4)	(4216,0)	(6012,0)	(3155,0)

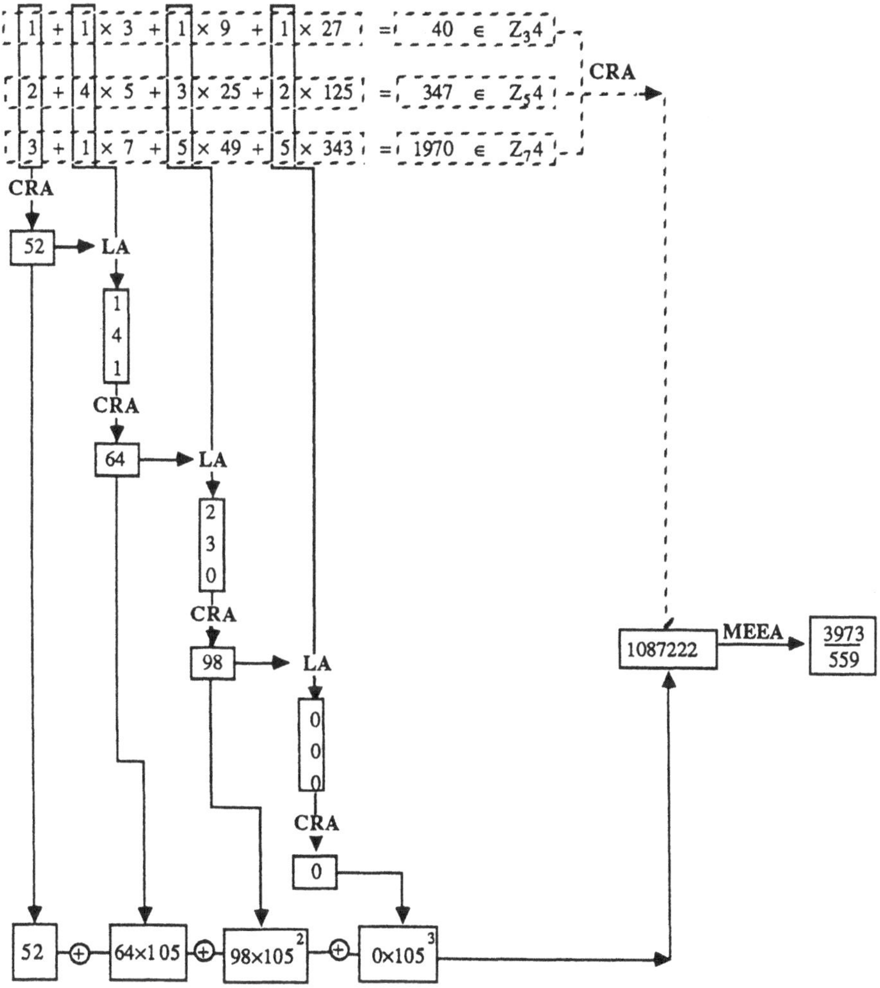

Fig.1: example of application.

4. Complexity analysis of the two inverse mapping algorithms

As mentioned in the introduction, the two algorithms can benefit of a parallel execution. Therefore the comparative complexity analysis will be carried out assuming this parallel execution.

In particular the complexity analysis of the two inverse mapping algorithms considers three main components, respectively related to:

a) the transformation of the Hensel codes into $Z_{p_i^r}$ (which is performed only by the C algorithm: step **C1.**).

b) CRA applications (steps **C3.**, **L3.**);

c) LA algorithm application (only for the L algorithm: step **L2.**).

In order to allow a complete complexity analysis and comparison of the proposed algorithms, a schema which takes into account the parallelism, is given (Figures 2 and 3).

Let us note that a linear (respect to k) number of processors are required for the parallel execution of both the algorithms.

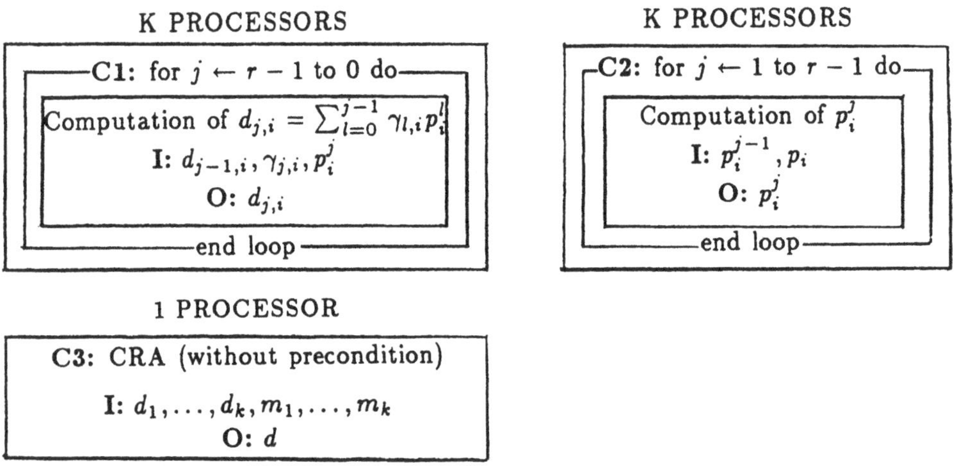

Fig.2: C Algorithm.

Note: In these figures **I** and **O** represent the input and output, respectively, of the functions described in each block. The number of processors depends on the number of p_i, for $i = 1 \ldots k$.

K PROCESSORS	K PROCESSORS	1 PROCESSOR
	Computation of g_i I: g, p_i O: g_i	CRA (without precondition) I: $\gamma_{0,1}, \ldots, \gamma_{0,k}, p_1, \ldots p_k$ O: δ_0
for $j \leftarrow 1$ to $r-1$ do		
LA1 I: $\bar{s}_{j-1}, g_i^{j-1}, p_i, \gamma_{j,i}$ O: $s_{j,i}$	Computation of $\bar{g}_1$ I: g_1, p_i O: $\bar{g}_1$	Computation of $d_l = \sum_{l=1}^{j} \delta_l g^l$ I: d_{l-1}, δ_j, g^j O: d_l
LA2 I: $\bar{s}_{j,i}, \bar{g}_i^j, p_i$ O: $\bar{\gamma}_{j,i}$	Computation of g_i^j I: $g_i^{j-1} \pmod{p_i}$ O: $g_i^j \pmod{p_i}$	Computation of g^{j+1} I: g^j O: g^{j+1}
		CRA (with precondition) I: $\bar{\gamma}_{j,1}, \ldots, \bar{\gamma}_{j,k}, p_1, \ldots, p_k$ O: δ_j
end loop		

Fig.3: L Algorithm (including LA algorithm)

The comparative analysis will be based on asymptotic valuation, so low costs operations will be discarded. As usual the O_B notation will indicate the order of magnitude under the bitwise computational model.

We also note that it is sufficient only to compare the C algorithm to the loop of the L because the remaining part of the L algorithm is not relevant to the complexity analysis.

-Complexity analysis of C algorithm.

The complexity of each cycle j of the steps C1. and C2., corresponds to that one of a multiplication of two numbers of size $r \log p$. Step C3. corresponds to a CRA (without precondition) application to primes of size $kr \log p$. So, the total cost of C algorithm is:

$$O_B\left(\sum_{j=1}^{r-1} (j \log p)^{1.59} + (kr \log p)^{1.59} \log k + k(r \log p)^{1.59} log(r \log p)\right) \qquad (2)$$

-Complexity analysis of L algorithm.

For each cycle j, the complexity of steps LA1. and LA2., corresponds to that of a multiplication of numbers of size $jk \log p$. Step L3. corresponds to a CRA (with precondition [AHU75]) application to values and moduli of size $\log p$. Therefore the total cost of L algorithm is:

$$O_B(\sum_{j=1}^{r-1}(kj \log p)^{1.59} + r(k \log p)^{1.59} \log k) \quad (3)$$

In order to compare (2) and (3), let us note that for practical values of r, the following approximation holds:

$$\sum_{j=1}^{r-1} j^{1.59} \sim r^2$$

Let us also note that when the cost of C algorithm is greater than the cost of L algorithm, the following inequality should hold:

$$r + k^{1.59} r^{1.59-1} log\ k + kr^{1.59-1} log(r\ log\ p) > rk^{1.59} + k^{1.59} log\ k \qquad (4)$$

Actually the inequality (4) is always satisfied for every choice of p, r and k, so the L algorithm turns out to be preferable to the C one. The experimental results, are in accordance with the above theoretical results as shown in the following Figure 4., where the behaviour of the costs of C and L is depicted for $k = 8$ and $p = 23$ (the execution time is computing in 100-*ths* of seconds). Howewer the same behaviour has been detected for several choice of k and p.

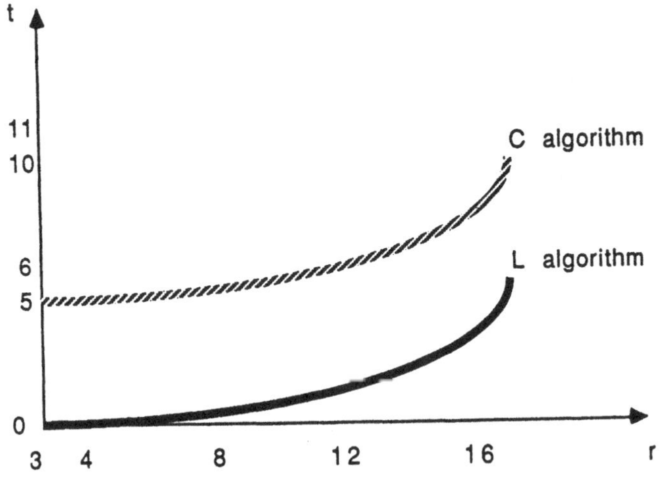

Fig. 4.

5. Conclusions

In this paper it has been faced the problem of big rational numbers computation. The recently achieved results on p-adic arithmetic have been applied to this problem with a generalization to the g-adic arithmetic.

Following the classical methodological approaches to the realization of inverse mappings, two algorithms have been presented: the first one, based on a CRA application (C algorithm), the other based on a Hensel-like construction (L algorithm).

Despite to its simple formulation, the C algorithm requires the manipulation of big integers. The L algorithm, on the contrary, requires the manipulation of small integers. From a qualitative point of view the L algorithm, when compared with the C one, shows a more attractive behaviour due essentially to the digit-by-digit generation of the g-adic code of the result. It has to be noted, in fact, that the C algorithm furnishes only the usual fractional representation of the result, because the g-adic expansion is not generated; so, in some sense, it is not completely correct to speak of g-adic approach when the C algorithm is applyed.

Moreover, during the practical computations, it is possible that the Farey fractions set must be enlarged by increasing either p or r. If we increase p, we must restart all the computation in both algorithms, and it is not a suitable situation. If we increase r (for example of h digits), in C algorithm we must restart all the CRA computation, whereas in L algorithm we must apply only h further times the LA algorithm and the CRA computations, in order to obtain the h remaining digits of the g-adic code result.

Further development can be carried out by the parallelization of the whole schema of computation, in particular by the parallelization of the MEEA algorithm for the inverse mapping.

Other algorithms, which take into account a mixed approach CRA-Hensel, deserve to be studied for further developments.

References

[AHU75] Aho, Hopcroft, Ullmann: *The Design and Analysis of Computer Algorithms.* Addison Wesley-Publishing Company 1975.

[C&M87] A. Colagrossi, A. Miola: A normalization algorithm for truncated p--adic arithmetic. Eight Symposium on Computer Arithmetic, IEEE; Como 1987

[DIT85] K. Dittenberger: An Efficient Method for Exact Numerical Computation. Diploma Thesis University of Linz 1985.

[GRE78] R.T. Gregory: The use of Finite Segment P-adic Arithmetic for Exact Computation, Bit 18, 1978.

[G&K84] R.T.Gregory, E.V. Krishnamurthy: *Methods and Applications of Error-Free Computation.* Springer-Verlag 1984.

[HOR78] E. Horowitz: *Fundamentals of Computer Algorithms.* PITMAN 1978.

[KOB77] N. Koblitz: *P-Adic Numbers, P-Adic Analysis and Zeta Functions.* Springer Verlag, New York, 1977.

[MAH73] K. Mahler: *Introduction to P-adic Numbers and their Functions.* Cambridge University Press, 1973

[MIO84] A. Miola: *Algebraic Approach to P-Adic Conversion of Rational Numbers.* IPL 18, 1984.

MONOMIAL IDEALS, GROUP ALGEBRAS AND ERROR CORRECTING CODES

Vesselin Drensky *)
Institute of Mathematics
Bulgarian Academy of Sciences
1090 Sofia, P.O.Box 373
Bulgaria

Piroska Lakatos
Institute of Mathematics
University of Debrecen
4010 Debrecen, Pf. 12
Hungary

0. Introduction

We fix a finite field K of characteristic p and a finite p-group $G=\{g_1,\ldots,g_n\}$, $n=p^m$. The group algebra $K[G]$ is a K-algebra with a basis $g_1,\ldots,g_n$ and a multiplication defined by the multiplication of G. The mapping $\varphi: K[G]\rightarrow K^n$, $\varphi(a_1g_1+\ldots+a_ng_n)=(a_1,\ldots,a_n)\in K^n$, $a_i\in K$, is an isomorphism of vector spaces. By φ we identify $K[G]$ with K^n and consider the subspaces of $K[G]$ as linear codes in K^n. In particular, the Hamming weight $w(a_1g_1+\ldots+a_ng_n)$ equals the number of the non-zero a_i's.

For G abelian, Berman [1] has initiated the investigation of the powers of the Jacobson radical $J^{\ell}(K[G])$ of $K[G]$ as codes. He has obtained that for the elementary abelian group C_2^m, $J^{\ell}(GF(2)[C_2^m])$ coinsides with the Reed-Muller (RM) code of order $m-\ell$. Charpin [2] has proved a similar result for an arbitrary prime integer p and $G=C_p^m$, when the powers of the radical are generalized Reed-Muller (GRM) codes. (This class of codes has been introduced by Kasami, Lin and Peterson [8].) Landrock and Manz [10] have established the connection between the theorems of Berman and Charpin and the classical result of Jennings [7] on the radical of a modular group algebra. A short survey on the subject is given in [6].

The main purpose of this paper is to continue the investigation of the radical of modular group algebras in the spirit of [10]. It turns out that it is convenient to consider "monomial" ideals in the ring $K[x_1,\ldots,x_m]/(x_1^p-1,\ldots,x_m^p-1)$ (such ideals have been studied by Poli, see e.g. [15]). They are a generalization of the radical codes and enjoy some good properties.

1. Monomial Ideals

1.1. Definition. Let $R=K[x_1,\ldots,x_m]/(x_1^p-1,\ldots,x_m^p-1)\cong K[C_p^m]$ be the group algebra of the elementary abelian p-group C_p^m as a vector space with a basis $x_1^{a_1}\ldots x_m^{a_m}$, $0\leq a_i<p$. In what follows we use R for this

*) Partially supported by Ministry of Culture, Science and Education, Bulgaria, Contract No. 876/1988.

object only. We call the ideal C of R a monomial code, if C is generated by a set of "monomials"

$$\{(x_1-1)^{b_1}\dots(x_m-1)^{b_m} \mid b=(b_1,\dots,b_m)\in B\}.$$

The following proposition allows to calculate easily the minimum distance if the weights w(u) of the generating monomials u are known. Actually, the proof is contained in Lemma 1.9 [1].

1.2. Proposition. Let $C=(u_b=(x_1-1)^{b_1}\dots(x_m-1)^{b_m} \mid b\in B)$. Then

$$d(C)=\min\{w(u_b) \mid b\in B\}=\min\{\prod(b_i+1) \mid b\in B\}.$$

Proof. We use an induction on m. For m=1 the assertion is well known and is an easy consequence of Theorem 1.1 [1]. For m>1, let $x\in C$ and let

$$x=\sum\lambda_i(x_1-1)^{i_1}\dots(x_m-1)^{i_m}=(x_m-1)^s(c_s+c_{s+1}(x_m-1)+\dots+c_{s+r}(x_m-1)^r),$$

where $\lambda_i\in K$, $c_j\in K[x_1,\dots,x_{m-1}]/(x_1^p-1,\dots,x_{m-1}^p-1)$ and

$$c_s=\sum\lambda_i(x_1-1)^{i_1}\dots(x_{m-1}-1)^{i_{m-1}}\neq 0,\ i=(i_1,\dots,i_{m-1},s).$$

Inductively, there exists an $i=(i_1,\dots,i_{m-1},s)$ such that $\lambda_i\neq 0$ and

$$w(c_s)\geq w((x_1-1)^{i_1}\dots(x_{m-1}-1)^{i_{m-1}})=d_1.$$

Let $c_s=\sum\mu_j x_1^{j_1}\dots x_{m-1}^{j_{m-1}}$. Hence in this expression of c_s there are at least d_1 non-zero coefficients μ_j. Rewriting x in the form

$$x=(x_m-1)^s\sum f_j(x_m)x_1^{j_1}\dots x_{m-1}^{j_{m-1}},\ f_j(x_m)=\mu_j+\mu_{j1}(x_m-1)+\dots+\mu_{jr}(x_m-1)^r,$$

we have at least d_1 non-zero summands $f_j(x_m)$. Bearing in mind the validity of the assertion for m=1 we obtain that $w(x)\geq(s+1)d_1$ and this completes the proof.

1.3. Definition. Let $h_1,\dots,h_m$ be positive integers. Then it is defined a filtration by degree of R by $R=R^0(h)\supset R^1(h)\supset\dots$, where

$$R^\ell(h)=(\prod(x_i-1)^{b_i} \mid \sum b_ih_i\geq\ell).$$

It turns out that the powers of the radical $J^\ell(K[G])$ for any p-group G are equivalent to $R^\ell(h)$ for a suitable h. In particular [1,2], for $h_1=\dots=h_m=1$, the GRM-codes coinsidc with $R^\ell=R^\ell(1,\dots,1)$.

1.4. Definition. Let G be a finite p-group. Then the Brauer-Jennings-Zassenhaus M-series $M_1>M_2>\dots$ is defined by $M_1=M_1(G)=G$, $M_s=M_s(G)=\langle(M_{s-1},G),\ M^{(p)}_{\lceil s/p\rceil}\rangle$ for s>1, where $(g_1,g_2)=g_1^{-1}g_2^{-1}g_1g_2$, $M_i^{(p)}=\{g^p \mid g\in M_i\}$.

It is known [7] that $M_s=\{g\in G \mid g-1\in J^s(K[G])$ and the factors M_s/M_{s+1} are elementary abelian groups.

1.5. Proposition. [7] Let $M_s/M_{s+1}=\langle g_{s1}M_{s+1}\rangle\times\dots\times\langle g_{sk_s}M_{s+1}\rangle$ be a decomposition of M_s/M_{s+1} into a direct product of cyclic groups. Then:

(i) The vector space $J^\ell(K[G])$ has a basis $\prod_{s\geq 1}\prod_{i\geq 1}(g_{si}-1)^{a_{si}}$, where $\sum s a_{si}\geq \ell$.

(ii) Let $f(z)=\prod(1+z^s+z^{2s}+\ldots+z^{(p-1)s})^{k_s}=1+f_1z+\ldots+f_{t-1}z^{t-1}$. Then $\dim J^\ell(K[G])=f_\ell+f_{\ell+1}+\ldots+f_{t-1}$.

<u>1.6. Theorem</u>. Let G be a finite p-group and let $\{g_{si}\}$ be the elements of Proposition 1.5. Then, as a code the ℓ-th power of the Jacobson radical $J^\ell(K[G])$ is equivalent to $R^\ell(h)$, where $h_1=\ldots=h_{k_1}=1$, $h_{k_1+1}=\ldots=h_{k_1+k_2}=2,\ldots$ and the equivalence is given by the identification of $g_{11},\ldots,g_{1k_1},g_{21},\ldots,g_{2k_2},\ldots$ respectively with $x_1,\ldots,x_{k_1}$, $x_{k_1+1},\ldots,x_{k_1+k_2},\ldots$.

<u>Proof</u>. Let $\{g_{si}\}\subset G$ be the elements of Proposition 1.5. We define a mapping $\Psi:G\to C_p^m=\langle x_1\rangle\times\ldots\times\langle x_m\rangle$ by $\Psi(g_{11})=x_1,\ldots,\Psi(g_{1k_1})=x_{k_1},\Psi(g_{21})=x_{k_1+1},\ldots,\Psi(g_{2k_2})=x_{k_1+k_2},\ldots$ and $\Psi(\prod_{s\geq 1}\prod_{i\geq 1}g_{si}^{a_{si}})=\prod_{s\geq 1}\prod_{i\geq 1}\Psi(g_{si})^{a_{si}}$.
Then the linear mapping $\tilde{\Psi}:K[G]\to R$ determined by $\tilde{\Psi}(g)=\Psi(g)$, $g\in G$, is an isometry of the vector spaces $K[G]$ and R. By Proposition 1.5 (i) $\tilde{\Psi}(J^\ell(K[G]))=R^\ell(h)$ which completes the proof.

The following assertion is an immediate consequence of Theorem 1.6.

<u>1.7. Corollary</u>. Let G and H be p-groups such that the factors $M_s(G)/M_{s+1}(G)$ and $M_s(H)/M_{s+1}(H)$ are isomorphic for all s. Then the codes $J^\ell(K[G])$ and $J^\ell(K[H])$ are equivalent for $\ell=1,2,\ldots$. Hence for the investigation of $J^\ell(K[G])$ it suffices to know only the factors of the M-series of G.

Clearly, the monomial codes have some advantages, first of all they can be easily defined and the minimum distance is determined immediately. On the other hand, the code parameters of some of them are not sufficiently good.

<u>1.8. Proposition</u>. (i) Let $h=(h_1,\ldots,h_m)$ define a filtration on R. Then $R^s(h)$ is embedded into a GRM-code $R^\ell=R^\ell(1,\ldots,1)$ and $d(R^s(h))=d(R^\ell)$.

(ii) For p=2 every monomial code is a subcode of a Reed-Muller code with the same minimum distance.

<u>Proof</u>. (i) Without loss of generality we assume that $h_1\geq\ldots\geq h_m$. Let ℓ be the maximal integer such that $R^s(h)\subset R^\ell=R^\ell(1,\ldots,1)$. Then $d(R^s(h))\geq d(R^\ell)$ and there exists $u_b=(x_1-1)^{b_1}\ldots(x_m-1)^{b_m}\in R^s(h)\cap(R^\ell\setminus R^{\ell+1})$, i.e. $\sum b_i=\ell$. It is known [1] that $d(R^\ell)=d(J^\ell(K[C_p^m])=p^v(w+1)$, where $\ell=v(p-1)+w$, $0\leq w<p-1$. Clearly, for the chosen u_b, $\sum b_ih_i\leq(p-1)(h_1+\ldots+h_v)$

$+wh_{v+1}$, hence $u'=(x_1-1)^{p-1}\dots(x_v-1)^{p-1}(x_{v+1}-1)^w\in R^s(h)$ and $d(R^s(h))\leq w(u')=p^v(w+1)=d(R^\ell)$. This inequality gives immediately that $d(R^s(h))=d(R^\ell)$.

(ii) The case is trivial because for p=2 the polynomials generating the monomial code C are of the form $u_i=(x_{i_1}-1)\dots(x_{i_s}-1)$, $i_1<\dots<i_s$ and $w(u_i)=2^s$. Therefore, for the maximal ℓ with $C\subset R^\ell$ we establish $d(C)=d(R^\ell)$.

Fortunately, for p>2 there exist monomial codes with parameters better than those of the GRM-codes. The proof of the following assertion is an immediate consequence of Proposition 1.2.

<u>1.9. Proposition</u>. There exists a unique maximal monomial code of length $n=p^m$ and with minimum distance $\geq d$ and this is

$I_d=(\prod(x_i-1)^{b_i} \mid \prod(b_i+1)\geq d)$.

For p>2 some of the codes I_d do not coinside with the GRM-codes and have better parameters.

<u>1.10. Examples</u>. Let p=5, m=2. Then I_d and R have the following parameters: $I_2=R^1$ - (25,24,2); $I_3=R^2$ - (25,22,3); I_4 - (25,20,4) and R^3 - (25,19,4); I_5 - (25,17,5) and R^4 - (25,15,5) etc. For p=5, m=3 C_{20} - (125,63,20) and R^7 - (125,53,20). Other examples of good codes for different I_d are given in [9].

<u>2. Dual Codes</u>

It is known [11] (see also [5]) that the orthogonal (or dual) code $C^\perp$ of a left ideal C of K[G] coinsides with the left annihilator $\mathrm{Ann}_{K[G]}C^*$ of C^*, where $*$ is the involution of K[G] defined by $g^*=g^{-1}$, $g\in G$.

<u>2.1. Proposition</u>. Let C be a monomial code in R. Then:

(i) $C^\perp$ is also a monomial code and $C^\perp=\mathrm{Ann}_R C$.

(ii) If $C=R^\ell(h)$, $R^{t-1}(h)\neq 0$, $R^t(h)=0$, then $C^\perp=R^{t-\ell}(h)$; in particular, if $J=J(K[G])$, $J^{t-1}\neq 0$, $J^t=0$, then $(J^\ell)^\perp=J^{t-\ell}$.

<u>Proof</u>. (i) As a consequence of Proposition 1.5 (i), we obtain easily that the "monomials" $(x-1)^b$ and $((x-1)^b)^*=(x^{-1}-1)^b$ generate the same ideal in $K[x]/(x^p-1)=K[C_p]$ which coinsides with $J^b(K[C_p])$. Hence for any monomial code C in R, $C^\perp=\mathrm{Ann}_R C^*=\mathrm{Ann}_R C$. Bearing in mind that the sum and the intersection of monomial codes are monomial again and $(C_1+\dots+C_r)^\perp=C_1^\perp\cap\dots\cap C_r^\perp$, it suffices to consider only the case when C is a principal ideal and $C=(u_b=(x_1-1)^{b_1}\dots(x_s-1)^{b_s})$, $b_i>0$. Since the monomials $(x_1-1)^{c_1}\dots(x_m-1)^{c_m}$, $0\leq c_i<p$, form a basis for R, it is easy to see that $\mathrm{Ann}_R(u_b)=((x_1-1)^{p-b_1},\dots,(x_s-1)^{p-b_s})$ is also a monomial

code and this completes the proof.

(ii) Let us consider the polynomial

$$f(z)=\prod(1+z^{h_i}+z^{2h_i}+\ldots+z^{(p-1)h_i})=f_0+f_1z+\ldots+f_{t-1}z^{t-1},$$

$t-1=(p-1)(h_1+\ldots+h_m)$. As in Proposition 1.5 (ii), $\dim R^{\ell}(h)=f_{\ell}+\ldots+f_{t-1}$, $f(1)=p^m$ and $f(z)=z^{t-1}f(z^{-1})$. Hence $\dim R^{t-\ell}(h)=f_{t-\ell}+\ldots+f_{t-1}=f_0+\ldots+f_{\ell-1}$ and $\dim R^{\ell}(h)+\dim R^{t-\ell}(h)=p^m=\dim R$. Since the group algebras are Frobenius algebras (see [4] for details) $\dim R^{\ell}(h)+\dim \operatorname{Ann}_R R^{\ell}(h)=\dim R$. But $R^{\ell}(h)R^{t-\ell}(h)=0$ and this yields $(R^{\ell}(h))^{\perp}=R^{t-\ell}(h)$.

<u>2.2. Corollary</u>. Let $p=2$ and let $h=(h_1,\ldots,h_m)$ be a filtration by degree of R such that $R^{t-1}(h)\neq 0$, $R^t(h)=0$. Then $R^{\ell}(h)$ is a self-dual code if and only if t is even and $\ell=t/2$.

<u>2.3. Corollary</u>. Let $p=2$ and let G be a finite 2-group. Then a suitable power of either $J(K[G])$ or $J(K[G\times C_2])$ is a self-dual code.

<u>Proof</u>. Let $J=J(K[G])$ and let $J^{t-1}\neq 0$, $J^t=0$. Clearly, $K[G\times C_2]\cong K[G]\otimes_K K[C_2]$. Then, for $J_1=J(K[G\times C_2])$ it holds $J_1^t\neq 0$, $J_1^{t+1}=0$. The proposition follows immediately from Corollary 2.2 because one of the integers t and $t+1$ is even and the other is odd.

<u>2.4. Examples</u>. Let $p=2$. For every m the ideal of R generated by x_1-1 is a self-dual $(2^m,2^{m-1},2)$-code, which is a direct sum of trivial $(2,1,2)$-codes. For m odd, $R^{\lceil m/2\rceil}$ is a self-dual $(2^m,2^{m-1},2^{\lceil m/2\rceil})$-RM-code. For $m\leq 5$ the other self-dual codes (up to equivalence) are the following: $C_0=R^3(2,1,1,1)=J^3(K[C_4\times C_2^2])$ and $C_0'=R^4(2,2,2,1)$ with parameters $(16,8,4)$, $C_1=R^4(2,2,1,1,1)=J^4(K[C_4^2\times C_2])$, $C_2=R^5(3,2,2,1,1)$, $C_3=R^6(4,2,2,2,1)$, $C_4=R^6(3,3,3,1,1)$, $C_5=R^5(4,2,1,1,1)=J^5(K[C_8\times C_2^2])$ with parameters $(32,16,4)$. Using the system LINCOR for investigation of linear codes due to Manev [13] we have computed the weight distribution of these codes and have identified them with the known self-dual codes [3]. The results are given in the following table:

	A_0	A_4	A_8	A_{12}	A_{16}	A_{20}	A_{24}	A_{28}	A_{32}
C_0, C_0'	1	28	198	28	1				
C_1	1	8	700	13 496	37 126	13 496	700	8	1
C_2	1	24	860	12 712	38 342	12 712	860	24	1
C_3, C_4	1	56	1 180	11 144	40 774	11 144	1 180	56	1
C_5	1	120	1 820	8 008	45 638	8 008	1 820	120	1

Although C_0 and C_0', C_3 and C_4 have the same weight distribution, some additional arguments show that these codes are non-equivalent and, in the notation of [14], $C_0=E_{16}$, $C_0'=A_8\oplus A_8$, $C_3=E_{16}\oplus E_{16}$, $C_4=A_8\oplus A_8\oplus A_8\oplus A_8$. For the covering radius of C_0 and C_0' we have obtained $r(C_0)=r(C_0')=4$, $r(C_3)=r(C_4)=8$. These four codes together with the RM-code $R^3(1,1,1,1,1)$ (with r=6, see [12]) reach the Delsarte bound for covering radius, i.e. they are uniformly packed.

For m>5 there exist self-dual monomial codes which cannot be obtained by a suitable filtration. For example, it is easy to see that the following code has such a property: $C=J^4(R)+(X_1X_2X_3, X_1X_2X_4, X_1X_2X_6, X_1X_4X_5, X_1X_4X_6, X_2X_3X_4, X_2X_4X_5, X_2X_4X_6, X_2X_5X_6, X_3X_4X_6)$, where $X_i=x_i-1$, $R=GF(2)[x_1,\ldots,x_6]/(x_1^2-1,\ldots,x_6^2-1)$.

2.5. Problem. We call two monomial codes C and C' of $R=K[x_1,\ldots,x_m]/(x_1^p-1,\ldots,x_m^p-1)$ monomially equivalent if they can be obtained by a permutation of the variables $x_1,\ldots,x_m$. Obviously, the monomially equivalent codes are equivalent. Does the opposit statement hold: If C and C' are equivalent monomial codes, are they monomially equivalent? In particular, is this true for self-dual monomial codes?

2.6. Problem. Explicit examples show that for small m and $d\leq\lceil m/2\rceil$ there exists an abelian 2-group G_d of order 2^m such that a power of $J(GF(2)[G_d])$ is a self-dual code with parameters $(2^m,2^{m-1},2^d)$. Does this phenomenon appear for all m? This has been verified by computer for $m\leq 30$ [9].

3. Wreath Products

Let G and H be groups and let $Fun(G,H)=H^G$ be the Cartesian product of $|G|$ copies of H, i.e. Fun(G,H) is the group of functions from G into H with the usual multiplication. For $g\in G$ and $f\in Fun(G,H)$ we define a function f^g by $f^g(u)=f(gu)$, $u\in G$. By definition, the wreath product HwrG is the split extension $G\ltimes Fun(G,H)$ with multiplication $g_1f_1\cdot gf=g_1gf_1^gf$. In the sequel we fix $H=C_p\cong GF(p)$, $G=\{g_1,\ldots,g_n\}$ and write C_p additively. Then $Fun(G,C_p)$ has a canonical structure of a GF(p)-vector space given by $f\rightarrow(f(g_1),\ldots,f(g_n))$. The mapping $\theta: Fun(G,C_p)\rightarrow GF(p)[G]$ defined by $\theta(f)=\sum f(g_i)g_i^{-1}$ is an isomorphism of vector spaces and θ is an isometry.

3.1. Proposition. The mapping θ is an isometry of $Fun(G,C_p)$ and GF(p)[G] which defines a one-to-one correspondence between the normal subgroups of $C_p$wrG which are in $Fun(G,C_p)$ and the right ideals of GF(p)[G].

Proof. Clearly, θ maps the class of subgroups of $Fun(G,C_p)$ on the class of subspaces of GF(p)[G]. Since $Fun(G,C_p)$ is an abelian group, the subgroup L of $Fun(G,C_p)$ is a normal subgroup of $C_p$wrG when $L^g=L$ for all

$g\in G$. Obviously $\theta(f^g)=\Sigma f^g(g_i)g_i^{-1}=\Sigma f(gg_i)g_i^{-1}=(\Sigma f(g_i)g_i^{-1})g$. Hence the equality $L^g=L$ holds if and only if $\theta(L)g=\theta(L)$, i.e. when $\theta(L)$ is a right ideal of $GF(p)[G]$.

It turns out that the powers of the radical of a modular abelian group algebra have a natural description in the language of wreath products.

3.2. Theorem. Let G be a finite abelian p-group and let $U=U_1>U_2>\dots>U_{t+2}=\langle 0\rangle$ be the lower central series of $U=C_p\mathrm{wr}G$. Then the codes $U_{\ell+1}\subset \mathrm{Fun}(G,C_p)$ and $J^{\ell}(GF(p)[G])$, $\ell=1,2,\dots,t$, are equivalent and the isometry is given by the mapping θ.

Proof. Since G is an abelian group, the only non-trivial commutators of length $\ell+1$ in U are $(f,h_1,\dots,h_\ell)=(\dots(f,h_1),\dots,h_\ell)$, where $f\in \mathrm{Fun}(G,C_p)$ and $h_i\in G$. Bearing in mind that $\mathrm{Fun}(G,C_p)$ is written additively, we obtain $(f,h_1)=-f+f^{h_1}$ and $\theta(f,h_1)=\theta(f)(h_1-1)$. Hence $\theta(f,h_1,\dots,h_\ell)=\theta(f)(h_1-1)\dots(h_\ell-1)$ and $\theta(U_{\ell+1})\subset J^{\ell}$, $J=J(GF(p)[G])$. Since every element of J^{ℓ} is a linear combination of elements $1(h_1-1)\dots(h_\ell-1)$ and θ is a one-to-one correspondence, we establish that $\theta(U_{\ell+1})=J^{\ell}$.

In particular, as an immediate consequence of Theorem 3.2 and the results of [1,2] we obtain a new description of the GRM-codes over GF(p).

3.3. Theorem. Let $U=U_1>U_2>\dots>U_{m+2}=\langle 0\rangle$ be the lower central series of $U=C_p\mathrm{wr}C_p^m$. Then the subcode U_ℓ of $\mathrm{Fun}(C_p^m,C_p)$, $\ell>1$, is equivalent to the GRM-code of order $m+1-\ell$.

Acknowledgments

The authors are very grateful to Stefan Dodunekov for general discussions and useful suggestions and to Krasimir Manev for the possibility to use the system LINCOR for computing the code parameters in 2.4.

References

1. S.D.Berman, On the theory of group codes, Kibernetika, 3, 1967, 1, 31-39.
2. P.Charpin, A new description of some polynomial codes: the primitive generalized Reed-Muller codes, Publ. du L.I.T.P., Paris, 1985, No.14.
3. J.H.Conway, V.Pless, On the enumeration of self-dual codes, J. Comb. Theory, Ser.A, 28, 1980, No.1, 26-53.
4. C.W.Curtis, I.Reiner, Representation Theory of Finite Groups and Associative Algebras, Wiley-Interscience, New York, London, 1962.
5. I.Damgård, P.Landrock, Ideals and codes in group algebras, Math. Inst., Aarhus Univ., Preprint No.12, 1986/87.
6. V.Drensky, Modular group algebras and error correcting codes, Proc. International Workshop "Algebraic and Combinatorial Coding Theory", Varna, Bulgaria, 18-24 Sept.1988, Sofia, 1988, 41-48.

7. S.A.Jennings, The structure of the group ring of a p-group over a modular field, Trans. Amer. Math. Soc., 50, 1941, 175-185.
8. T.Kasami, S.Lin, W.W.Peterson, New generalizations of the Reed-Muller codes. Part I. Primitive codes, IEEE Trans. Inform. Theory, 14, 1968, 189-198.
9. P.Lakatos, To modular abelian group codes, Proc. International Workshop "Algebraic and Combinatorial Coding Theory", Varna, Bulgaria, 18-24 Sept.1988, Sofia, 1988, 126.
10. P.Landrock, O.Manz, Classical codes as ideals in group algebras, Math. Inst., Aarhus Univ., Preprint No.18, 1986/87.
11. F.J.MacWilliams, Codes and ideals in group algebras, in R.C.Bose and T.A.Dowling, eds., "Combinatorial Mathematics and its Applications", Univ. of North Carolina Press, Chapel Hill, 1969.
12. A.Mc Loughlin, The covering radius of the (m-3)-rd order Reed-Muller codes and a lower bound on the (m-4)-th order Reed-Muller codes, SIAM J. Appl. Math., 37, 1979, 419-422.
13. K.N.Manev, LINCOR - a system for linear codes researches, in "Mathematics and Education in Mathematics", Sofia, 1987, 500-503.
14. V.Pless, A classification of self-orthogonal codes over GF(2), Discr. Math., 3, 1972, 209-246.
15. A.Poli, Important algebraic calculations for n-variable polynomial codes, Discr. Math., 56, 1985, 255-263.

FAST INTEGER MULTIPLICATION BY BLOCK RECURSIVE, NUMBER THEORETICAL TRANSFORMS

Ricardo Ferré
Department of Computer Sciences
Lund University, Sweden
Box 118
221 00 Lund, Sweden

ABSTRACT

The use of multidimensional techniques permits to increase the sequence length to powers of the word length, when Discrete Fourier Transforms are performed on congruences modulo a Fermat or a Mersenne Number. The number of dimensions is equivalent to the number of recursion levels in block-recursive techniques, so that we can interchange the concepts multidimensional techniques and block-recursive techniques.

Sequences of length $2^{2M+\delta}$, with $\delta \in Z_2$ can be convolved (via DFT) with increased multiplicative efficiency, as it has been shown in earlier papers[Fer1,Fer2], with the use of Fermat, Pseudo Fermat, Mersenne and Pseudo Mersenne numbers of the form $(2^{(2.4-1)\nu} \pm 1)/D$.

In this paper, the sequence length is generalized to $2^{kM+\delta}$, where M represents the number of recursion levels and k the number of blocks for each level, with $\delta \in Z_k$ and $k = 2^{\lambda}$.

After a natural generalization of the concept of byte, it is shown in this paper that Pseudo Mersenne Numbers of the form $(2^{(2k-1)\nu} - 1)/D$ permits to replace the additions by permutations of bytes, in performing DFT in congruences with these numbers as bases.

An algorithm is proposed in this paper, in which the additions are replaced by permutations of bytes for multiplication of large integer numbers, with appropriate values of the constants k and M, and with careful byte organization of the elements of the sequences representing the large integers.

When the number of blocks is arbitrarily large, the complexity of the corresponding algorithm is shown to be $n^{1+\varepsilon}$, where ε is arbitrarily small.

INTRODUCTION

When Cooley and Tukey[Co&T] published their article containing their n log n algorithm to implement Discrete Fourier Transforms, they gave an enormous impulse to the search of algorithmic solutions to a series of problems.

Such a reduction from n^2 to (nlog n) in one step should be equivalent to an improvement in the computers' physical base that could take years. Among the different implementations of the Discrete Fourier Transforms, those based on finite numerical rings are remarkable. As the base of these rings appear Fermat or Mersenne numbers in whose case they are called Fermat Number Transforms respectively Mersenne Number Transforms. Multidimensional techniques and the equivalent block recursive techniques of these types permit increasing the length of the sequences(polynomials) to be transformed. Another advantage of the extension to more dimensions is that we can use the simplicity of the numerical rings to obtain the consequent simple algorithms.

This simplicity of the algorithms gives considerable advantages over more sophisticated algorithms up to certain limits, which can be sufficiently large for most applications i. e. in signal processing, where the algorithms must be built-in in digital filters.

This approach is both efficient and easy to implement, and it can effectively compete with the algorithm by Singleton[Sing], Split-Radix of Duhammel[Duh], the fixed-radix multidimensional approach proposed by Agarwal and Burrus[Ag&B] and other algorithms as it is shown in a comparison table for multiplicative efficiency. The additive efficiency can also be improved in the case of integer multiplication, when we can choose appropriate bit and byte organizations, and in the case when sequence elements are stored in many appropriate word(byte) length.

THE DISCRETE FOURIER TRANSFORM OVER A COMMUTATIVE RING WITH UNITY

Given a commutative ring with unity, and an element ω, called a Nth. root of the unity in this ring with the following properties:

1. $\omega^N=1$
2. $\omega \neq 1$
3. $\sum_{k \in Z_N} \omega^{kp} = 0$, for $0<p<N$

we can define Discrete Fourier Transforms for sequences in this ring.

A sequence of elements in this ring is a function of the natural numbers into the ring. We obtain:

Definition Convolution of two sequences x[n] and h[n] of length N is a sequence y[n] given by the formula:

$$y[n] = \sum_{q \in \Theta} x[q]h[n-q] \text{ , for } n \in \Lambda$$

The convolution is cyclic if the indices of x and h are evaluated modulo N and $\Lambda=\Theta=Z_N$. It is linear if the values of x and h are defined to be zero for indices outside the closed interval [0,N-1] and $\Lambda= Z_{2N-1}$.

The linear or acyclic convolution corresponds to the usual multiplication of polynomials.

The cyclic convolution is also called "wrapped"(see Aho, Hopcroft, Ullman[Ah,H&U]).

Definition Fourier Transform of a sequence x of length N is the sequence X of length N given by the formula [0]:

$$X[k] = \sum_{p \in Z_N} x[p]\, \omega^{pk} \quad \text{for } k \in Z_N \qquad [0]$$

It has been shown by Aho, Hopcroft and Ullman[Ah,H&U] that cyclic convolutions in a commutative ring with unity can be obtained by the formula [1]

$$y = F^{-1}(F(x)*F(h)) \qquad [1]$$

where the symbol * represents pointwise multiplication of the vectors F(x) and F(h) and F respectively F^{-1} represents Discrete Fourier Transform and its inverse.

It is usual to consider the elements of the sequences x[n] and h[n] in the field of the reals or in the field of the complex numbers, in the domain of the integers or in the ring of the integers modulo an integer, in most cases a Fermat or a Mersenne number. In the latter case they are called respectively the Fermat Number Transforms and the Mersenne Number Transforms. In these transforms there is a useful root of the unity ω which is a power of 2, which implies that the multiplications by powers of ω are reduced to be shifts in a digital computer.

These transforms are very useful, principally for two reasons:

a) The Fermat Number Transforms or the Mersenne Number Transforms are computationally faster, result in no round-off errors and involve no other multiplications than the pointwise multiplications of the transformed sequences.

b) Sometimes only a partial output from the convolution is necessary, either a partial sequence or a sampled sequence every L-th value, which can be obtained efficiently by means of this technique.

These transforms have the limitation that the length of the sequences to be convolved is proportional to the number of bits taken by the base(the word length).

Therefore it is sometimes more useful to consider the elements of these sequences x and h as being themselves sequences.

In these cases we can consider multidimensional techniques for the convolution.

Bidimensional convolution can be implemented by a bidimensional transform with the advantages mentioned above and with the length of the convolved sequences proportional to the square of the word length.

MULTIDIMENSIONAL TECHNIQUES

Bidimensional techniques for the computation of linear convolution arise naturally by considering sequences(polynomials) whose elements are themselves sequences (polynomials).

We must define formally the operations of addition and multiplication for sequences, the same as additive and multiplicative unities, which we do in the natural way, generalizing the same operations in the ring, and using convolution as the multiplicative operation for sequences.

In this case we can apply the result to elements of the commutative ring (or the integral domain, when they are polynomials) with unity of the sequences. Observe that the rings of sequences respectively polynomials are formally different having cyclic and acyclic convolutions as the ring product.

Suppose that we have two such sequences X and H of length M each, whose elements being themselves sequences of length L. We first perform the DFT with, e.g. a FFT-like technique for length M. We obtain M transformed sequences of length L in the order of Mlog M sequence operations, each such operation involving L element operations. We obtain the total of LMlog M element operations. Let us now multiply each of the transformed sequences of X with the corresponding transformed sequences of H again by means of a FFT-technique, this time in the original ring of the elements. This takes Llog L + L arithmetic operations for each of the M transformed sequences; in total M(Llog L +L) element operations. The transformation process then takes in total MLlog M + MLlog L = Nlog N element operations. The inverse transform takes again Nlog N element operations. This gives 2Nlog N + N element operations. This is asymptotically worse than algorithms implemented on FFT. The advantage of this method compared with the usual FFT is that if we operate in a finite ring in which w , the root of the unity of order N, is a power of 2, all the multiplications become shifts and we have only to perform additions. The computer universe is always bounded and discrete, so operating with finite elements is not a real limitation.

In the case of polynomials, when linear convolution is to be performed, we often want to transform a one dimensional convolution into a bidimensional one to be able to increase sequence length without increasing the word length. If the length of the polynomials to be convolved is N = LM then we can perform bidimensional convolution by defining L sequences as each having M elements. We can always obtain a favorable sequence length by adding zeros to the sequences to be convolved. If we perform acyclic convolutions then we must partially overlap and add the resulting sequences in order to obtain the final sequence. This can easily be noticed i.e. by comparing the exponents of the terms of the resulting partial polynomials or the total length of the product polynomial, given as the final sequence.

<u>Generalization</u>

This procedure can be generalized for more dimensions in the same way. It is possible to obtain an algorithm for a short sequence length K and apply M-dimensional technique for sequences of the order of K^M. This has the advantage over radix K algorithms that we do not need more than a root of the unity of the order K, or of the order of 2K-1 if linear convolution is to be performed.

Example:

Let us suppose that $N = 2^P$. It is always possible to obtain a decomposition of $N = 4^M.2^\delta = 2^{2M+\delta}$. The interesting cases appear when M>0, and δ is the Kronecker's symbol, taking the value 1 if P is odd and 0 if P is even . In this case we can apply M-dimensional convolution of length 4, eventually followed by one (one-dimensional) convolution of length 2.

Acyclic convolution for length 4 sequences

We add 3 zeros to the sequences and compute cyclic convolution of the order 7. We need a ring modulo B such that $\omega^7=1$ mod B and w is a power of 2. The number $2^{21}+1$ fulfills these demands, but it has 3 as a divisor(its prime factor decomposition is actually $3^2.43.5419$), which restricts the length of the sequences to be convolved to 2. To be able to convolve length 7 sequences we can use the Pseudo Fermat Number $(2^{21}+1)/9$ which has the required properties, but permits a maximum sequence length of 42, as asserted by Nussbaumer[Nus77] (check above its prime factorization). The word length of this base is approximately 18, which prevents the risk of overflow, when operating in 32 bits machines. The arithmetic modulo B = $(2^{21}+1)/9$ is obviously more complex than the one modulo B' = $2^{21}+1$. This difficulty can be avoided by computing the convolution modulo B' and then obtaining the final result performing a last operation modulo B on the convolution evaluated modulo B', because z mod B = (z mod B') mod B. Obviously, all operations modulo B' are performed on word length longer than that of the final result. It is easy to see, however that the increase in word length is only about 3 bits. **In all this paper we assume the same technique to be applied.**

To perform the convolution we apply the idea first suggested by Rader[Rad], of rearranging the indices of the DFT for prime numbers thus obtaining the schema of fig. 1.

The right side of the second equality (fig. 2) is the cyclic convolution of the sequences $[\omega^1\ \omega^3\ \omega^2\ \omega^6\ \omega^4\ \omega^5] = \Omega$ and [x[1] x[3] x[2] x[6] x[4] x[5]] = X, of length 6, which can be computed using the algorithm of Winograd[Win] or directly. If we operate directly we can perform this multiplication matrix-vector by means of 3 shifts and 2 additions per line, because the three last elements of the vector x are zero, in total 18 shifts and 12 additions. The element X[0] is calculated with three additions, giving 15 additions. Adding x[0] gives 6 more additions totalling 21 additions.

After transforming the two length 7 sequences we multiply pointwise the transformed sequences by means of 7 element multiplications. These are the only ones to be performed because all other multiplications involve only shifts. To obtain the partial sequences of the product we perform the inverse Pseudo Fermat Number Transform in the same way with only shifts, because all the involved multiplications are again with powers of 2. We shall see later that careful bit organization can radically reduce the number of additions.

fig. 1

$$
\begin{array}{l}
X[0]\\ X[1]\\ X[3]\\ X[2]\\ X[6]\\ X[4]\\ X[5]
\end{array}
=
\begin{array}{ccccccc}
1 & 1 & 1 & 1 & 1 & 1 & 1\\
1 & \omega^1 & \omega^3 & \omega^2 & \omega^6 & \omega^4 & \omega^5\\
1 & \omega^3 & \omega^2 & \omega^6 & \omega^4 & \omega^5 & \omega^1\\
1 & \omega^2 & \omega^6 & \omega^4 & \omega^5 & \omega^1 & \omega^3\\
1 & \omega^6 & \omega^4 & \omega^5 & \omega^1 & \omega^3 & \omega^2\\
1 & \omega^4 & \omega^5 & \omega^1 & \omega^3 & \omega^2 & \omega^6\\
1 & \omega^5 & \omega^1 & \omega^3 & \omega^2 & \omega^6 & \omega^4
\end{array}
\quad
\begin{array}{l}
x[0]\\ x[1]\\ x[3]\\ x[2]\\ x[6]\\ x[4]\\ x[5]
\end{array}
$$

$x[4]=x[5]=x[6]=0$

Then

$$
\begin{array}{l}
X[1]-x[0]\\ X[3]-x[0]\\ X[2]-x[0]\\ X[6]-x[0]\\ X[4]-x[0]\\ X[5]-x[0]
\end{array}
=
\begin{array}{cccccc}
\omega^1 & \omega^3 & \omega^2 & \omega^6 & \omega^4 & \omega^5\\
\omega^3 & \omega^2 & \omega^6 & \omega^4 & \omega^5 & \omega^1\\
\omega^2 & \omega^6 & \omega^4 & \omega^5 & \omega^1 & \omega^3\\
\omega^6 & \omega^4 & \omega^5 & \omega^1 & \omega^3 & \omega^2\\
\omega^4 & \omega^5 & \omega^1 & \omega^3 & \omega^2 & \omega^6\\
\omega^5 & \omega^1 & \omega^3 & \omega^2 & \omega^6 & \omega^4
\end{array}
\cdot
\begin{array}{l}
x[1]\\ x[3]\\ x[2]\\ x[6]\\ x[4]\\ x[5]
\end{array}
$$

Observe that the last 3 columns of the matrices do not need to be considered because the last 3 elements of the vector x are zero.

In this example, $\omega = 2^6$, which means that multiplying by ω implies shifting 6 bits. Then, computing the first row of the multiplication above involves the addition of x[1] shifted 6 bits, plus x[3] shifted 18 bits, plus x[2] shifted 12 bits.

If we organize our coefficients in 6 bits long bytes then it is sufficient to perform the addition of 4 such bytes, because the other bytes do not overlap. Remember that the words are 18 bits long in this example.

In other words, to perform the multiplication of the first row of the matrix by the vector x it is sufficient adding 24 bits, somewhat less than a usual word of 32 bits.

To perform the complete matrix-vector multiplication is then sufficient to add 14 such bytes, that is, 84 bit-additions, or about 3 additions of 32 bit words. To calculate the element X[0] demands 3 more additions, so we can perform the transform with 6 additions, provided that we organize the coefficients in the suggested way. Each row demands also a modulo operation.

We must finally overlap 3 positions and add the partial sequences obtained.

The same procedure could be applied for other appropriate bases of the ring of integers in which a root of the unity of the order of 7 is a power of 2. These alternative bases are Fermat, Pseudo Fermat , Mersenne and Pseudo Mersenne numbers of the form $(2^{(2.4-1)v} \pm 1)/D$, where D must be any divisor of the number $2^{(2k-1)v} \pm 1$ less than 7, if it exist, otherwise it may be some other divisor of this number. Examples of useful Pseudo Fermat numbers are: $(2^{21}+1)/17$, $(2^{35}+1)/33$ with approximate word length 24 respectively 30, which could be used in 32 bits devices and $(2^{49}+1)/129$ with approximate word length 41 in 64 bits machines(perhaps abstract machines).

The Pseudo Mersenne numbers were studied by Nussbaumer[Nus76]. Among these $(2^{21}-1)/49$, $(2^{35}-1)/127$, $(2^{35}-1)/31$ and $(2^{49}-1)/127$ are also appropriate bases of the finite numerical rings that hold the same essential properties.

Linear convolution for length 2 sequences

The simple scheme:

$Y[0] = x[0]h[0]$

$Y[1] = (x[0]+x[1]).(h[0]+h[1]) - Y[0] - Y[2]$

$Y[2] = x[1]h[1]$

gives the result with 3 element multiplications, and 4 additions.

Theorem **The general complexity** of this method follows the equation:

$T(n) = 3^{\delta}.[7T(n/4) + 6S(n/4)]$, where S(n) represents addition of n-length vectors which gives formula [2]

$$M(n) = 3^{\delta}.7^{1/2\log n} = 3^{\delta}.n^{1/2\log 7}, \qquad [2]$$

where M(n) represents the total number of multiplications of the algorithm.

The above formula gives $3^{\delta}.n^{1/2\log 7}$ multiplications and $2.3^{\delta}.n^{1/2\log 7}$ additions.

As usual in these cases log represent logarithm in base 2.

In Table I the approximate multiplicative efficiency of some methods is given. The other methods are the one described by Singleton[Sing], Split-Radix by Duhammel[Duh], the proposed by Agarwal and Burrus[Ag&B] and lastly the method proposed in this paper which we have called Multidimensional Mixed 4-2 Radix(MDM4-2). We can check in the table that my proposed method has a better behavior for sequences up to at least 1024 elements.

Table I

N	Singleton	Split-Radix	Ag.-Burr	MDM4-2
8	66	36	27	21
16	162	100	81	49
32	386	260	243	147
64	1037	644	729	343
128	2304	1540	2187	1372
256	5120	3588	6561	2401
512	12288	8196	19683	7203
1024	26624	18436	59049	16807

We find at this point that if we use M-dimensional techniques for $N = 2^P$, using only the linear convolution scheme for length 2 sequences, that this is the same result as given by Knuth[Knu]("How Fast Can We Multiply?") using recursive block techniques.

Multidimensional techniques used for short bases like 2, 3, etc. are equivalent to the divide and conquer methods developed by Knuth[Knu](Op. Cit.).

In practice, however, this method by Knuth for length-4 sequences(polynomials) is both more complicated to implement and less effective.

The difference is made by the use of Pseudo Theoretical Number Transforms as defined by Nussbaumer[Nus76,Nus77].

Otherwise, everything I have told about multidimensional techniques can be applied to block recursive techniques.

Example

Perform the acyclic convolution of two sequences of length 8 by means of a bidimensional linear convolution.

$y = x.h \quad x = x_0\, x_1\, x_2\, x_3\, x_4\, x_5\, x_6\, x_7$

$h = h_0\, h_1\, h_2\, h_3\, h_4\, h_5\, h_6\, h_7$

length of y = 15

We first divide x in two sequences of 4 elements each x_0 and x_1 and in the same way h in h_0 and h_1

$x_0 = x_0\, x_1\, x_2\, x_3 \qquad x_1 = x_4\, x_5\, x_6\, x_7$

$h_0 = h_0\, h_1\, h_2\, h_3 \qquad h_1 = h_4\, h_5\, h_6\, h_7$

Compute now

$y_0 = x_0 \cdot h_0$

$y_1 = (x_0 + x_1) \cdot (h_0 + h_1) - y_0 - y_2$

$y_2 = x_1 \cdot h_1$

y_0, y_1 and y_2 are subsequences of the product sequence y of length 7 each. Now we can compute y_0, y_1 and y_2 following the algorithm described in the preceding pages.

Observe that the total length of the three sequences is 21.

$y_0 = \text{¥}_{00}\, \text{¥}_{01}\, \text{¥}_{02}\, \text{¥}_{03}\, \text{¥}_{04}\text{¥}_{05}\, \text{¥}_{06}$

$y_1 = \text{¥}_{10}\, \text{¥}_{11}\, \text{¥}_{12}\, \text{¥}_{13}\, \text{¥}_{14}\text{¥}_{15}\, \text{¥}_{16}$

$y_2 = \text{¥}_{20}\, \text{¥}_{21}\, \text{¥}_{22}\, \text{¥}_{23}\, \text{¥}_{24}\text{¥}_{25}\, \text{¥}_{26}$

Now we must overlap 3 positions and add to obtain the product sequence y. This means that we add at most 2 elements of the ¥s in order to obtain an element of the product sequence y Observe that we always overlap 3 places, that is the second dimension minus one and that y has the desired length, 15 places.

fig. 2

y_0		y_1		y_2		y
¥$_{00}$					=	y_0
¥$_{02}$					=	y_1
¥$_{02}$					=	y_2
¥$_{03}$					=	y_3
¥$_{04}$	+	¥$_{10}$			=	y_4
¥$_{05}$	+	¥$_{11}$			=	y_5
¥$_{06}$	+	¥$_{12}$			=	y_6
		¥$_{13}$			=	y_7
		¥$_{14}$	+	¥$_{20}$	=	y_8
		¥$_{15}$	+	¥$_{21}$	=	y_9
		¥$_{16}$	+	¥$_{22}$	=	y_{10}
				¥$_{23}$	=	y_{11}
				¥$_{24}$	=	y_{12}
				¥$_{25}$	=	y_{13}
				¥$_{26}$	=	y_{14}

We get a general result.

Theorem When the length of the sequences is 2M-1 the overlapping is M-1. This can be easily shown by considering that the total length $2N-1 = 2ML-1 = (2L-2)(2M-1-(M-1))+2M-1$. Observe that the overlapping is independent of the first dimension.

MULTIPLICATION OF LARGE INTEGERS

Definition A **byte** is an operational unit of the machine memory, consisting of b bits.

Definition A **word** is a machine memory unit consisting of B bytes.

Division in 2^λ blocks

Theorem DFT can be performed without additions, operating in an integer ring with pseudo Mersenne base.

Proof:

Let us assume the pseudo Mersenne number being $(2^{(2k-1)\nu} - 1)/D$.

Then

$\omega = 2^\nu = 1.2^\nu = 2^{(2k-1)\nu}.2^\nu = 2^{2k\nu}$ and multiplications by ω^q imply shifting $2k\nu.q$ bits. If the number of bits $b = (2k-1)\nu$, then these multiplications can be substituted by placing the coefficient from place number $2k\nu.q$, that is qbytes and ν bits from the beginning of a long word with all the other bits = 0. Under these conditions, adding coefficients multiplied by different powers of ω implies placing these coefficients in different parts of this long word.The elements of the matrix form of the DFT are then $a_{ij} = \omega^{i.j}$, where $i,j \in Z_{2k-1}$ and the product i.j is calculated modulo the base. If 2k -1 is a prime then the rows 2,3, ... of the matrix consist of 1 followed by permutations of the powers of ω : $\omega^1\ \omega^2 \dots \omega^{2k-2} = \Omega$.

Then multiplying each of these rows of the matrix by the sequence x consists only of permutations of the bytes of x[0], x[1], ... x[2k -2] (the same permutations as above), and we do not need to perform additions. We get the result in words of $B = (2k-1)\,b + \nu.(2k-1)$ bits each. To continue with the process it is sufficient to perform one modulo operation per row, that is subtract a multiple of the base, **no matter how large k is.** The first row, composed only by 1's, can be multiplied by x in the same way, observing that

$1 = 2^{(2k-1)\nu} = 2^{q(2k-1)\nu}$. With different values of q we obtain a permutation of the coefficients in the long word.

We have shown the following

Theorem **The complexity** of this method(see formula [2]) is $T(n) = K.n^{1/\lambda \log(2k-1)} = K.n^{1+\varepsilon}$ because $2k = 2.2^\lambda$.

When k is arbitrarily large, ε is arbitrarily small.

Let us suppose that we want to multiply two large integers. We store these integers x and h in N bytes each, that is

$$x = \sum_{j=0}^{N-1} x_j 2^{bj} \text{ and } h = \sum_{j=0}^{N-1} h_j 2^{bj}$$

We want to perform this multiplication by means of block recursive techniques, operating with $k = 2^\lambda$ blocks.

The result consists of 2k -1 blocks, which means that we need a base number of the form $(2^{(2k-1)\nu} - 1)/D$, where ν is an integer constant, as described above. This assures the root of the unity of the order 2k - 1 to be a power of 2. The multiplication of 2^δ blocks must be implemented in the same way , for all $\delta \in Z_k$. An algorithm for performing the multiplication of b-bits numbers is also necessary, but it may be constructed in the same way proposed by this paper.

Example:

$b = 7$, $k = 4$, $2k - 1 = 7$, base number $(2^7 - 1)/D$, $D = 1$

DFT-matrix:

$$
\begin{pmatrix} X[0] \\ X[1] \\ X[3] \\ X[2] \\ X[6] \\ X[4] \\ X[5] \end{pmatrix} = \begin{pmatrix} 1 & 1 & 1 & 1 \\ 1 & \omega^1 & \omega^3 & \omega^2 \\ 1 & \omega^3 & \omega^2 & \omega^6 \\ 1 & \omega^2 & \omega^6 & \omega^4 \\ 1 & \omega^6 & \omega^4 & \omega^5 \\ 1 & \omega^4 & \omega^5 & \omega^1 \\ 1 & \omega^5 & \omega^1 & \omega^3 \end{pmatrix} \cdot \begin{pmatrix} x[0] \\ x[1] \\ x[3] \\ x[2] \end{pmatrix}
$$

Then

$$
\begin{pmatrix} X[1]-x[0] \\ X[3]-x[0] \\ X[2]-x[0] \\ X[6]-x[0] \\ X[4]-x[0] \\ X[5]-x[0] \end{pmatrix} = \begin{pmatrix} \omega^1 & \omega^3 & \omega^2 \\ \omega^3 & \omega^2 & \omega^6 \\ \omega^2 & \omega^6 & \omega^4 \\ \omega^6 & \omega^4 & \omega^5 \\ \omega^4 & \omega^5 & \omega^1 \\ \omega^5 & \omega^1 & \omega^3 \end{pmatrix} \cdot \begin{pmatrix} x[1] \\ x[3] \\ x[2] \end{pmatrix}
$$

X[0] can also be computed by placing the bytes x[0],x[1],x[2],x[3] in a 4-bytes word because $1 = 2^{h7}$, where h takes the values 1,2,3.

That is:

$$X[0] = x[0] + 2^7 x[1] + x[2]2^{2.7} + x[3]2^{3.7}$$

One modulo operation must follow at last.

Performing the multiplication of integers stored in 2^P words, as usual in DFT, implies for $P = kM + \delta$ that we must solve in the same way as before the multiplications of 2^δ blocks for all $\delta \in Z_k$.

Different values of b and k give algorithms for performing DFT of large integer numbers **with only shifts and byte-permutations** .

To complete these algorithms in practice, we just need an algorithm for performing effectively the multiplication of two b-bits numbers.

REFERENCES

Ag&B. Agarwal, R. C. and Burrus C. S. *Fast One-Dimensional Digital Convolution by Multidimensional Techniques* IEEE Transactions on Acoustics, Speech and Signal Processing Vol. ASSP-22, No. 1, Feb. 1974,.

Ah,H&U. Aho, Hopcroft, Ullman *The Design and Analysis of Computer Algorithms*, Addison Wesley, Reading, MA, 1974.

Co&T. Cooley, J. W. and J. W. Tukey *An Algorithm for Machine Calculation of Complex Fourier Series* , Math. Comp. Vol. 19, No. 297, 1966.

Duh. Duhammel, P. *Implementation of Split-Radix" FFT Algorithms for Complex, Real and Real-Symmetric Data.* IEEE Transactions on Acoustics, Speech and Signal Processing, Vol. ASSP-34, No. 2, 1986.

Fer1. Ferré, R. E. *An Algebraic Approach tto Multidimensional Convolution.* ACM Computer Science Conference 1987. Proceedings.

Fer2. Ferré, R. E. *Linear Convolution by Multidimensional Techniques based on Mersenne Number Transforms.* International Conference on Industrial and Applied Mathematics, 1987.

Knu. Knuth, D. E.*The Art of Computer Programming* Vol. 2, Semi-Numerical Algorithms, Addison Wesley, Reading, MA, 1981.

Nus76. Nussbaumer, H. J. *Digital Filtering Using Complex Mersenne Transforms*, IBM Journal of Research and Development, Sept. 1976.

Nus77. Nussbaumer, H.J. *Digital Filtering Using Pseudo Fermat Number Transforms* IEEE Trans. on Acoustics, Speech and Signal Processing, Vol. ASSP-25 No. 1, Feb 1977.

Rad. Rader, C. M. *Discrete Fourier Transforms When the Number of Data Samples Is Prime* Proceedings of the IEEE, Vol. 56, No. 6, 1968

Sing. Singleton, R. C. *An Algorithm for Computing the Mixed Radix Fast Fourier Transform*, IEEE Trans. on Audio and Electroac., Vol. AU-17, No. 2, 1969.

Win. Winograd, S. *On Computing the Discrete Fourier Transform* Mathematics of Computation, Vol. 32, No. 141, 1978.

LAMBDA-UPSILON-OMEGA: AN ASSISTANT ALGORITHMS ANALYZER

PHILIPPE FLAJOLET
INRIA, Rocquencourt
78150 Le Chesnay (France)

BRUNO SALVY
INRIA and Ecole Polytechnique
91405 Palaiseau (France)

PAUL ZIMMERMANN
INRIA, Rocquencourt

Abstract. Lambda-Upsilon-Omega, $\Lambda\Upsilon\Omega$, is a system designed to perform automatic analysis of well-defined classes of algorithms operating over "decomposable" data structures.

It consists of an 'Algebraic Analyzer' System that compiles algorithms specifications into generating functions of average costs, and an 'Analytic Analyzer' System that extracts asymptotic informations on coefficients of generating functions. The algebraic part relies on recent methodologies in combinatorial analysis based on systematic correspondences between structural type definitions and counting generating functions. The analytic part makes use of partly classical and partly new correspondences between singularities of analytic functions and the growth of their Taylor coefficients.

The current version $\Lambda\Upsilon\Omega_0$ of $\Lambda\Upsilon\Omega$ implements as basic data types, term trees as encountered in symbolic algebra systems. The analytic analyzer can treat large classes of functions with explicit expressions. In this way, $\Lambda\Upsilon\Omega_0$ can generate in the current stage about a dozen non-trivial average case analyses of algorithms like: formal differentiation, some algebraic simplification and matching algorithms. Its analytic analyzer can determine asymptotic expansions for large classes of generating functions arising in the analysis of algorithms.

The outline of a design for a full system is also discussed here. The long term goal is to include a fairly rich set of data structuring mechanisms including some general recursive type definitions, and have the analytic analyzer treat wide classes of functional equations as may be encountered in combinatorial analysis and the analysis of algorithms.

1. Introduction

Ideally, a system for automatic program analysis should take as input a procedure or function specification

I1. `procedure Quicksort {instructions} end;`

I2. `procedure Diff {instructions} end;`

for sorting or computing symbolic derivatives, and produce an "analysis" of the program. We concern ourselves here with *average–case* analysis and optimization of programs, and we would like the system to output something like

O1. `Time for Quicksort on random inputs of size n is`
`11.67 n ln(n) - 1.74 n + O(ln(n))`

O2. `Time for Diff on random inputs of size n is`

$$8\,\frac{n\,(-\,240 + 37\;42^{1/2})}{(-\,13 + 2\;42^{1/2})\;42^{1/2}} + O(1)$$

These two analyses will naturally depend on *type specifications* that are companion to (I1) and (I2), a description of a *complexity model* (e.g. an '`if`' takes 5 units of time), and a description of a (random input) *statistical model*. For Quicksort, we could have some way of specifying that all permutations of n are taken equally likely, while for Diff we could decide that all expression trees of size n with the proper type are equally likely.

We shall describe here a system whose current state performs automatic analysis of a whole *class* of algorithms in the realm of symbolic manipulation algorithms and contains a good deal of what is needed in order to analyze permutation algorithms like (I1). Result (O1) is taken directly from [Knuth 1973], but (O2) was literally produced by our system.

Our system is called $\Lambda\Upsilon\Omega$. The name $\Lambda\Upsilon\Omega$ (Lambda-Upsilon-Omega) comes from the Greek word λύω which means (amongst other things) 'I solve', and it is from this verb that "analysis" derives. Implementation was started in mid 1987. We shall describe here its overall design principles as well as the state of the current implementation $\Lambda\Upsilon\Omega_0$. There are two major components in $\Lambda\Upsilon\Omega$:

- An *Algebraic Analyser System*, ALAS, that accepts algorithms specifications in a suitably restricted programming language. That part produces *type descriptors* and *complexity descriptors* in the form of *generating functions*.
- An *Analytic Analyser System*, ANANAS, that accepts generating functions (for type descriptors and complexity descriptors) and tries to determine automatically an asymptotic expansion of its Taylor coefficients.

The algebraic component is currently implemented in Lisp, though ML is also considered for later implementations. In its present form, it permits to analyze a class of symbolic term (tree) manipulation programs and comprises about 500 Lisp instructions (in the Le_Lisp dialect). The analytic component is already a fairly large set of symbolic "algebra" routines written in Maple and comprising about 3000 instructions.

Both components encapsulate a fair amount of mathematical expertise at a rather abstract level.

- The algebraic system is based on research in combinatorial analysis developed mostly during the 1970's regarding correspondence between structural definitions of combinatorial objects and generating functions, together with some new extensions to program schemes.
- The analytic system is based on some recent developments of late 19th and early 20th century complex asymptotic analysis concerning the correspondence between singularities or saddle points of functions and the asymptotic order of coefficients in Taylor expansions.

The $\Lambda\Upsilon\Omega$ system has of course no claim of being universal, since program termination is in general undecidable. Its interest lies in consideration of a restricted class of purely "functional" procedures that operate through recursive descent over a large class of "decomposable" structures defined by powerful type structuring mechanisms. Such a class contains algorithms and data structures, like binary search trees, unbalanced heaps for priority queues, quicksort, digital tries and radix exchange sort, merge sort, several versions of hashing, pattern matching, recursive parsing. Our long term objective is to have a system that will perform automatically analysis of a non-negligible fractions of these algorithms as well as many other of the same style. The current system implements a complexity calculus on term trees along the lines of [Flajolet, Steyaert 1987] and the analytic analyzer is already appreciably more general.

For an interesting alternative approach to automatic complexity analysis, the reader is referred to [Hickey, Cohen 1988] and references therein.

2. A Sample Session

A typical $\Lambda\Upsilon\Omega$ session† starts with by calling a script, which (using Unix virtual tty's) initiates a joint Lisp and Maple session. We then load ALAS, apply it to the program to be analyzed. This generates a set of equations over generating functions, that are passed to Maple initialized with ANANAS.

The example considered is a program that computes symbolic derivatives (without simplification) of expressions (terms, trees) built from the operator set

$$1^{(0)},\ x^{(0)},\ \exp^{(1)},\ +^{(2)},\ *^{(2)},\ \div^{(2)}$$

with superscripts denoting arities. The key steps are

1. The recursive definition of the type 'term' is reflected by a quadratic equation for its generating function $t(z)$.
2. The recursive structure of the Diff procedure is reflected by a linear equation for its complexity descriptor $\tau\,diff$ (generating function of average costs).

These two steps are completed automatically by ALAS, which also uses a small Maple procedure to derive explicit expressions. At the next stage, ANANAS is used on those generating functions:

3. Both $t(z)$ and $\tau\,diff(z)$ are recognized as having singularities at a finite distance of a so-called 'algebraico-logarithmic' type. Local singular expansions are then determined (through Maple's Taylor capability).
4. Using general theorems from complex asymptotic analysis, singular expansions can be transformed automatically into asymptotic expansions of the coefficients. This is achieved by means of the versatile '`equivalent`' command of ANANAS.

Dividing the asymptotic form of the coefficients of $[z^n]$ in $t(z)$ and $\tau\,diff(z)$, we obtain the *asymptotic* average complexity of symbolic differentiation in an either algebraic or floating point form. The same device may be used to analyse the variant DiffCp of Diff that proceeds by copying subexpressions instead of sharing them.

In this way we obtain average case analyses which we summarized here, compared to the obvious best case and worst case results.

Algorithm	**Best Case**	**Average Case**	**Worst Case**
`Diff` [sharing]	$O(n)$	$c.n + O(1)$	$O(n)$
`DiffCp` [copy]	$O(n)$	$c.n^{3/2} + O(n)$	$O(n^2)$

The *order* of the cost for Diff was to expected. The $O(n^{3/2})$ result for DiffCp is harder to guess, and it is related to the behaviour of the average path length in trees as discussed in [Knuth 1968] or [Meir, Moon 1978].

† The necessary concepts will be developed in Sections 3 (Algebraic System) and 4 (Analytic System). The script that follows has been slightly edited and a few commands have been decomposed for the sake of readability.

```
Script started on Fri Jan 29 12:08:24 1988
% maplelisp     [Initialize Lisp and Maple]
; Le-Lisp (by INRIA) version 15.21  (25/Dec/87)  [sun]
    |\^/|
._|\|   |/|_. INRIA - Rocquencourt
 \  MAPLE  /  Version 4.1 --- May 1987
 <____ ____>  For on-line help, type  help();
      |

[Load Algebraic Analyzer, Alas]
? (load luo)
Function to analyse : (diff copy1 distrib size simpl count1 filter
copy evalf diffcp)
? diff [Diff with subexpression sharing]
(de diff (x)
   (caseroot   *    ('+    (maketree    '+
   [etc...]
Analysis of diff on trees formed with ...
   symbol x of arity 0 etc...
Calling analyse ...
? (maple)
*** Initial system ***    [Equations Produced by Alas]
                   2                    2
 phi = 2 + 3 z  + z, t = 2 + 3 z t  + z t,
                               2
   tau_diff = 21 z t  + 6 z t tau_diff + 4 z t + z tau_diff + 2 z
*** Explicit expression(s) ***    [etc...]

[Start Analytic Analyzer, Ananas]
> tn:=equivalent(t); [Asymptotic number of inputs: 16 sec on sun3/60]
                             1/4              1/2 1/2
                           42     (13 - 2 42    )
tn := - 1/6 ------------------------------------------------
                             1/2   3/2   1/2              1/2 n
             (- 13 + 2 42    ) n     Pi     (13 - 2 42    )
                      1/2 (- n)
        (13 - 2 42    )
   + O(---------------------)
                   5/2
                  n
>  [etc...]     [average diff time]
                                                    1/2
                               n (- 240 + 37 42    )
               av_diffn := 8 ------------------------ + O(1)
                                          1/2    1/2
                               (- 13 + 2 42    ) 42
> evalf(av_diffn);      [Floating point eval]
                               6.82942 n + O(1)
> suspend_ maple;
? (analyse 'diffcp)     [Diff with copy of arguments]
> [etc ...]
> evalf(av_diffcpn);
                                        3/2
                             16.6226 n     + O(n)
> suspend_maple;
? (end)     [The whole thing!]
Que Le-Lisp soit avec vous.
% ^D
script done on Fri Jan 29 12:44:51 1988
```

Figure 1. A ΛΥΩ session showing the automatic analysis of symbolic differentiation.

3. The Algebraic Analyzer System

3.1 Combinatorial Principles

The algebraic part of our system – ALAS – relies on recent research in combinatorial analysis. Till the mid twentieth century, the field of combinatorial enumerations was mostly conceived as an art of obtaining *recurrences* for the counting of combinatorial structures, with generating functions entering as an *ad hoc* solution device in more complex cases. The books by Riordan, and many of the analyses in Knuth's *magnum opus* are witnesses of this approach.

From research conducted by Rota, Foata, Schützenberger and their schools, there has emerged a general principle:

A rich collection of combinatorial constructions have direct translation into generating functions.

More precisely, let $\mathcal{A}$ be a class of combinatorial objects, with $\mathcal{A}_n$ the subclass consisting of objects of size n, and $A_n = \mathrm{card}(\mathcal{A}_n)$. We define the *ordinary generating function* (OGF) and *exponential generating function* (EGF) of $\mathcal{A}$ by

$$A(z) = \sum_{n \geq 0} A_n z^n \qquad \text{and} \qquad \hat{A}(z) = \sum_{n \geq 0} A_n \frac{z^n}{n!}. \tag{3.1}$$

A combinatorial construction, say $\mathcal{C} = \Phi[\mathcal{A}, \mathcal{B}]$ is said to be *admissible* if the counting sequence $\{C_n\}_{n \geq 0}$ of the result depends only on the counting sequences $\{A_n\}$ and $\{B_n\}$ of the arguments. An admissible construction then defines an *operator* (or a *functional*) on corresponding generating functions:

$$C(z) = \Psi[A(z), B(z)] \qquad \text{and} \qquad \hat{C}(z) = \hat{\Psi}[\hat{A}(z), \hat{B}(z)].$$

For instance the cartesian product construction is admissible since

$$\mathcal{C} = \mathcal{A} \times \mathcal{B} \qquad \Longrightarrow \qquad C_n = \sum_{k=0}^{n} A_k B_{n-k} \text{ and } C(z) = A(z) \cdot B(z).$$

Combinatorial enumerations are developed systematically within a comparable framework in the book by Goulden and Jackson [1983]. The tables in Figure 2 summarize a collection of admissible constructions borrowed from [Flajolet 1985, 1988].

In this context, the primary object for combinatorial enumerations is no longer *integer sequences* but rather *generating functions.* Furthermore, that approach fits nicely with asymptotic analysis, the main tool for asymptotic analysis being analytic function theory rather than explicit integer sequences. It is on the conjunction of these two principles that our system is built.

The task of enumerating a class of combinatorial structures is then reduced to *specifying* it (up to isomorphism) by means of admissible constructions. Once this is done, the task of computing a set of generating function equations reduces to performing simply a purely formal translation. In the context of the analysis of algorithms, data structure declarations are thus converted to generating functions (GF's), each data type having its own GF also called its type *descriptor*. An interesting approach similar to ours and based on an extension of context–free grammars is presented in Greene's thesis [1983].

OGF:		
Disj. Union	$\mathcal{C} = \mathcal{A} \uplus \mathcal{B}$	$c(z) = a(z) + b(z)$
Cart. Product	$\mathcal{C} = \mathcal{A} \times \mathcal{B}$	$c(z) = a(z) \cdot b(z)$
Diagonal	$\mathcal{C} = \Delta(\mathcal{A} \times \mathcal{A})$	$c(z) = a(z^2)$
Sequence	$\mathcal{C} = \mathcal{A}^*$	$c(z) = (1 - a(z))^{-1}$
Marking	$\mathcal{C} = \mu\mathcal{A}$	$c(z) = z\frac{d}{dz}a(z)$
Substitution	$\mathcal{C} = \mathcal{A}[\mathcal{B}]$	$c(z) = a(b(z))$
PowerSet	$\mathcal{C} = 2^{\mathcal{A}}$	$c(z) = \exp\left(a(z) - \frac{1}{2}a(z^2) + \frac{1}{3}a(z^3) - \cdots\right)$
MultiSet	$\mathcal{C} = \mathbf{M}\{\mathcal{A}\}$	$c(z) = \exp\left(a(z) + \frac{1}{2}a(z^2) + \frac{1}{3}a(z^3) + \cdots\right)$
EGF:		
Disj. Union	$\mathcal{C} = \mathcal{A} \uplus \mathcal{B}$	$\hat{c}(z) = \hat{a}(z) + \hat{b}(z)$
Label. Product	$\mathcal{C} = \mathcal{A} * \mathcal{B}$	$\hat{c}(z) = \hat{a}(z) \cdot \hat{b}(z)$
Label. Sequence	$\mathcal{C} = \mathcal{A}^{(*)}$	$\hat{c}(z) = (1 - \hat{a}(z))^{-1}$
Marking	$\mathcal{C} = \mu\mathcal{A}$	$\hat{c}(z) = z\frac{d}{dz}\hat{a}(z)$
Label. Subst.	$\mathcal{C} = \mathcal{A}[\mathcal{B}]$	$\hat{c}(z) = \hat{a}(\hat{b}(z))$
Label. Set	$\mathcal{C} = \mathcal{A}^{[*]}$	$\hat{c}(z) = \exp(\hat{a}(z))$

Figure 2. A catalog of admissible constructions and their translation to ordinary or exponential generating functions. The OGF constructions are relative to unlabelled structures, the EGF constructions are relative to labelled structures.

3.2. Analysis of Algorithms

Let Γ be an algorithm that takes its inputs from a data type $\mathcal{I}$ and produces some output of type $\mathcal{O}$. We consider exclusively *additive complexity measures*, thereby restricting ourselves to *time complexity* analyses. Let $\tau\Gamma[e]$ denote the complexity of an execution of Γ on input e. By the additive character:

$$\Gamma = (\Gamma^{(1)}; \Gamma^{(2)}) \qquad \Longrightarrow \qquad \tau\Gamma = \tau\Gamma^{(1)} + \tau\Gamma^{(2)}.$$

The purpose of average case analysis is to determine the expectation of $\tau\Gamma[e]$ when e is a random element of $\mathcal{I}$ with size n. Thus, assuming $\mathcal{I}_n$ is a finite set, that quantity is a quotient

$$\frac{\tau\Gamma_n}{I_n} \qquad \text{where} \qquad \tau\Gamma_n = \sum_{e \in \mathcal{I}_n} \tau\Gamma[e],$$

$\tau\Gamma_n$ being thus a cumulated value of $\tau\Gamma$ over $\mathcal{I}_n$.

The ordinary *complexity descriptor* (OCD) of algorithm Γ is defined as the generating function

$$\tau\Gamma(z) = \sum_{n \geq 0} \tau\Gamma_n z^n.$$

(There is an obvious analogue for exponential descriptors.)

A program construction $\Gamma = \Phi[\Gamma^{(1)}, \Gamma^{(2)}]$ is said to be *admissible* if the cost sequence $\{\tau\Gamma_n\}$ of Γ depends only on the cost sequences of the arguments Γ_1, Γ_2 and the counting sequences of intervening data structures. An admissible construction again defines an operator over corresponding generating functions.

Assume for instance that $P(z : \mathcal{C})$ is a procedure that operates on inputs $z = (a, b)$ of type $\mathcal{C} = \mathcal{A} \times \mathcal{B}$ and is defined by

```
P(z :  C) := Q(a);
```

where Q is of type $Q(x : \mathcal{A})$. Then, is is easy to see that

$$\tau P(z) = \tau Q(z) \cdot b(z).$$

If in addition, we make use of the additivity of τ, the scheme

```
P(x :  C) := Q(a); R(b)
```

translates into

$$\tau P(z) = \tau Q(z) \cdot b(z) + a(z) \tau R(z).$$

It turns out that there is a collection of program schemes naturally associated to constructions described above that are admissible. Corresponding to $\mathcal{C} = \mathcal{A} \uplus \mathcal{B}$, $\mathcal{C} = \mathcal{A} \times \mathcal{B}$, $\mathcal{C} = \mathcal{A}^*$, $\mathcal{C} = 2^{\mathcal{A}}$, $\mathcal{C} = \mathbf{M}\{\mathcal{A}\}$, we find

$P(c) = \textbf{if } c \in \mathcal{A} \textbf{ then } Q(c) \textbf{ else } R(c)$	$\tau P(z) = \tau Q(z) + \tau R(z)$
$P((a,b)) = Q(a)$	$\tau P(z) = \tau Q(z) b(z)$
$P((a_1, \ldots, a_k)) = Q(a_1); \ldots; Q(a_k)$	$\tau P(z) = \tau Q(z)/(1 - a(z))^2$
$P(\{a_1, \ldots, a_k\}) = Q(a_1); \ldots; Q(a_k)$	$\tau P(z) = c(z)(\tau Q(z) - \tau Q(z^2) + \tau Q(z^3) - \cdots)$
$P(\{a_1, \ldots, a_k\}) = Q(a_1); \ldots; Q(a_k)$	$\tau P(z) = c(z)(\tau Q(z) + \tau Q(z^2) + \tau Q(z^3) + \cdots)$

For instance, a recursive type definition for trees

$T \ = \ \{a\} \times T^*;$

together with a recursive procedure specification sheme

$Q(x : T) \ := \ R(x);$ **for** y **root_subtree_of** x **do** $Q(y)$;

will result in the system of equations

$$\begin{cases} T(z) = \dfrac{z}{1 - T(z)} \\ \tau Q(z) = \tau R(z) + z \dfrac{\tau Q(z)}{(1 - T(z))^2}. \end{cases}$$

Observe that $T(z)$ is an algebraic function of degree 2, and owing to the structure of the algorithm, $\tau Q(z)$ is expressed *linearly* in terms of itself. This is roughly the situation that we encounter when analyzing symbolic differentiation as well as many similar algorithms [Steyaert 1984].

The algebraic analyzer of $\Lambda\Upsilon\Omega_0$ implements a calculus based on previously exposed principles, but restricted to trees. Nonetheless (cf Fig. 1), it can produce automatic analyses of versions of matching, simplification, or various types of evaluations etc.

4. The Analytic Analyzer System

At this stage, our task is to take a generating function, defined either explicitly (for non recursive data types) or implicitly via a functional equation (for most recursive data types). The current version of $\Lambda\Upsilon\Omega$ treats only functions that lead to explicit expressions after a possible usage of the 'solve' routine of Maple. We shall therefore limit ourselves to this case.

4.1. Analytic Principles

Let $f(z)$ be a function analytic at the origin. We assume further that $f(z)$ is explicitly given by an *expression*, a blend of sums, products, powers, exponential and logarithms. Most explicit generating functions constructed by the combinatorial tools of Section 3 are of this type.

The starting point is Cauchy's coefficient formula

$$[z^n]f(z) = \frac{1}{2i\pi}\int_\Gamma f(z)\,\frac{dz}{z^{n+1}}, \tag{4.1}$$

where $[z^n]f(z)$ is the usual notation for the coefficient of z^n in the Taylor expansion of $f(z)$. Two major classes of methods are applicable to determine the Taylor coefficients of those functions:

- For functions with singularities at a finite distance, the local behaviour of the function near its dominant singularities (the ones of smallest modulus) determines the growth order of the Taylor coefficients of the function. Asymptotic information is obtained by taking Γ to be a contour that comes close to the dominant singularities.
- For entire functions, saddle point contours Γ are usually applicable.

Several observations are useful here. First, functions defined by expressions are analytically continuable – except for possible isolated singularities – to the whole of the complex plane (though they may be multivalued). Second, by Pringsheim theorem, functions with positive coefficients (such is the case for our generating functions) always have a positive dominant singularity, a fact that eases considerably the search for singularities.

Though a complete algorithm covering all elementary functions is not (yet) available since the classification of singularities, even for such functions, is not fully complete, a good deal of functions arising in practice can be treated by the following algorithm.

Procedure equivalent(f : expression) : expression;
{determines an asymptotic equivalent of $[z^n]f(z)$}

1. Determine whether $f(z)$ is entire or $f(z)$ has singularities at a finite distance.

2. If $f(z)$ has finite singularities, let ρ be the modulus of a dominant singularity. We know at least that

$$f_n = [z^n]f(z) \approx \rho^{-n}.$$

Compute a local expansion of $f(z)$ around its dominant singularity (-ies). This is called a singular expansion.

2a. If a singular expansion is of an 'algebraico-logarithmic' type, namely

$$f(z) \sim (1 - z/\rho)^\alpha \log^\beta (1 - z/\rho)^{-1} \qquad \text{as} \qquad x \to \rho \tag{4.2}$$

then apply methods of the Darboux-Pólya type to transfer singular expansions to coefficients

$$f_n = [z^n]f(z) \sim \rho^{-n}\frac{n^{-\alpha-1}}{\Gamma(-\alpha)}\log^\beta n \tag{4.3}.$$

This applies generally to functions that are "not too large" near a singularity.

2b If the function is large near its singularity, for instance

$$f(z) \approx \exp\Big(\frac{1}{(1 - z/\rho)^\alpha}\Big),$$

then apply saddle point methods like (3) below.

3. If $f(z)$ is entire, then use a saddle point integral. If this succeeds, we get

$$f_n = [z^n]f(z) \sim \frac{e^{h_n(R_n)}}{\sqrt{2\pi h_n(R_n)}}$$

where $h_n(z) = \log f(z) - (n+1)\log z$, and R_n is such that

$$\left[\frac{dh_n(z)}{dz}\right]_{z=R_n} = 0.$$

This is the outline of the algorithm that we have implemented in Maple, with the minor exception of step (2b) (saddle point at a finite distance) and with the current limitation that singularities and saddle points should be within reach of Maple's 'solve' routine.

It is important to note that a few *theorems*, whose conditions can be automatically tested, are used to support this algorithm.

Singularity Analysis. The classical form of the Darboux-Pólya method requires differentiability conditions on error terms. However, from [Flajolet, Odlyzko 1987], we now know that analytic continuation is enough to ensure the transition from (4.2) to (4.3), and by our earlier discussion, these conditions are always fulfilled for functions defined by expressions. Thus, the use of (2a) is guaranteed to be sound. Furthermore, that approach makes it possible to cope with singularities involving iterated logarithms as well (not yet implemented).

Saddle Point Integrals. There has been considerable interest for those methods, due to their recognized importance in mathematical physics and combinatorial enumerations. We thus know, from works by Hayman, Harris and Schoenfeld, or Odlyzko and Richmond, classes of functions *defined by closure properties* for which saddle point estimates are valid. Such conditions, that are extremely adequate for combinatorial generating functions, can be checked inductively on the expression.

4.2. Some Applications

Let us take here the occasion of a few examples to discuss some further features of ANANAS. The next three examples are all taken from combinatorial enumerations: (E1) Trees of cycles of cycles of beads; (E2) Involutive permutations; (E3) Children's Rounds of [Stanley 1978]; (E4) Bell numbers counting partitions of n.

```
[E1]> equivalent(1/2*(1-sqrt(1-4*log(1/(1-log(1/(1-z)))))));
                              1/2
-1/2   exp(exp(-1/4))     exp(-3/8)
           3/2                          1/2
     1 / (n     (- exp(exp(-1/4)) exp(-1) + 1)    )
                                          n   1/2
        / ((- exp(exp(-1/4)) exp(-1) + 1)   Pi    ) + etc ...
                                     (1/2 - n)
        (- exp(exp(-1/4) - 1) + 1)
   + O(------------------------------------)
                              7/2
                             n
```

```
[E2]> equivalent(exp(z+z^2/2));
                               1/2                                   1/2 2
exp(- 1/2 - 1/2 (1 + 4 n)         + 1/2 (- 1/2 - 1/2 (1 + 4 n)     ) )

                              1/2 n       1/2     1/2                           1/2 1/2
   / ((- 1/2 - 1/2 (1 + 4 n)      )  (-2)    Pi      (- 1/2 - 1/2 (1 + 4 n)     )    )

                   1/4
   / (1 + 4 n)
```

```
[E3]> equivalent((1-z)^(-z),5);
            ln(n)     gamma     1      ln(n)          1      gamma        ln(n)
1 - 1/n - ------- - ------- + ---- - ------- + 5/2 ---- - ------- + O(--------)
              3         3       3       4             4       4             5
             n         n       n       n             n       n             n
```

```
[E4]> equivalent(exp(exp(z)-1));
                                 1/2
1/2    exp(exp(W(n + 1)) - 1) 2      + etc...  [W(z)exp(W(z))=z]
```

Example 1 demonstrates the processing of functions with singularities at a finite distance. The singularity has an explicit form and the function behaves locally like a square root, whence the final result of the form:

$$Cn^{-3/2}\left(1-e^{e^{-1/4}-1}\right)^{-n}.$$

Example 2 shows the asymptotic analysis of the Involution numbers (Knuth [1973, p.65-67] does it by the Laplace method). It is treated here by "Hayman admissibility" (a classical notion bearing no direct relation to our previous usage of this word). Hayman's Theorem provides a class of admissible functions for which a saddle point argument (Step 3 of procedure 'equivalent') can be applied: *If f and g are admissible, h is an entire function and P is a real polynomial with a positive leading coefficient, then*

$$\exp(f),\ f+P,\ P(f),\ \textit{and under suitable conditions},\ f+h$$

are admissible. These conditions can be checked syntactically here.

Example 3 is a further illustration of singularity analysis in a non trivial case. Example 4, the classical asymptotics of Bell numbers [De Bruijn 1981] resembles Example 2. It is treated here by Harris–Schoenfeld admissibility (which also provides complete expansions), and the corresponding step in the algorithm implements a theorem of Odlyzko and Richmond relating Hayman admissibility to Harris–Schoenfeld admissibility.

5. Conclusions

We have presented here some preliminary design considerations for a system that would assist *research* in the analysis of algorithms. There are two benefits to be expected from such a research.

The first and most obvious benefit is to help an analyst explore some statistical phenomena that seem "tractable" in principle, but too intricate to be done by hand.

The second and most important one in our view is that the design of such a system creates *needs* of a new nature in algorithmic analysis methodology. (Thus, our approach departs radically from "Artificial Intelligence").

a. There is a need for extracting *general* program schemes that can be analyzed by these methods. In this way, we wish to attack the analysis of elementary but structurally complex programs, and we can hope to find general theorems relating complexity and structure of algorithms (cf [Steyaert 1984]).

b. When making algorithmic some parts of complex asymptotics, we naturally discover "gaps" that have never been revealed before. For instance, nobody seems to have considered such a simple asymptotic problem as

$$[z^n]\exp\left(\log^2\left(\frac{1}{1-z}\right)\right)$$

In summary, all we hope is that the development of ΛϒΩ will bring even more questions than answers!

6. Bibliography

N. G. DE BRUIJN [1981]. *Asymptotic Methods in Analysis.* Dover, New York, 1981.

L. COMTET [1974]. *Advanced Combinatorics.* Reidel, Dordrecht, 1974.

P. FLAJOLET [1985]. "Elements of a general theory of combinatorial structures", in *Proc. FCT Conf., Lecture Notes in Comp. Sc*, Springer Verlag, 1985, 112-127.

PH. FLAJOLET [1988]. "Mathematical Methods in the Analysis of Algorithms and Data Structures," in *Trends in Theoretical Computer Science*, E Börger Editor, Computer Science Press, 1988.

P. FLAJOLET AND A. M. ODLYZKO [1987]. "Singularity Analysis of Generating Functions", preprint, 1987.

P. FLAJOLET AND J-M. STEYAERT [1987]. "A Complexity Calculus for Recursive Tree Algorihms", *J. of Computer and System Sciences* **19**, 1987, 301-331.

I. GOULDEN AND D. JACKSON [1983]. *Combinatorial Enumerations.* Wiley, New York, 1983.

D. H. GREENE [1983]. "Labelled Formal Languages and Their Uses," Stanford University, Technical Report STAN-CS-83-982, 1983.

B. HARRIS AND L. SCHOENFELD [1968]. "Asymptotic Expansions for the Coefficients of Analytic Functions", *Illinois J. Math.* **12**, 1968, 264-277.

W. K. HAYMAN [1956]. "A Generalization of Stirling's Formula", *J. Reine und Angewandte Mathematik* **196**, 1956, 67-95.

P. HENRICI [1977]. *Applied and Computational Complex Analysis.* Three Volumes. Wiley, New York, 1977.

T. HICKEY AND J. COHEN [1988]. "Automatic Program Analysis", *J.A.C.M.* **35**, 1988, 185-220

D. E. KNUTH [1973a]. *The Art of Computer Programming.* Volume 1: *Fundamental Algorithms.* Addison-Wesley, Reading, MA, second edition 1973.

D. E. KNUTH [1973b]. *The Art of Computer Programming.* Volume 3: *Sorting and Searching.* Addison-Wesley, Reading, MA, 1973.

A. MEIR AND J. W. MOON [1978]. "On the Altitude of Nodes in Random Trees," *Canadian Journal of Mathematics* **30**, 1978, 997–1015.

G. PÓLYA [1937]. "Kombinatorische Anzahlbestimmungen für Gruppen, Graphen und chemische Verbindungen", *Acta Mathematica* **68**, 1937, 145–254. Translated in: G. Pólya and R. C. Read, *Combinatorial Enumeration of Groups, Graphs and Chemical Compounds*, Springer, New-York, 1987.

V. N. SACHKOV [1978]. *Verojatnostnie Metody v Kombinatornom Analize,* Nauka, Moscow, 1978.

R. SEDGEWICK [1983]. *Algorithms.* Addison-Wesley, Reading, 1983.

R. P. STANLEY [1978]. "Generating Functions," in *Studies in Combinatorics,* edited by G-C. Rota, M. A. A. Monographs, 1978.

R. P. STANLEY [1986]. *Enumerative Combinatorics,* Wadsworth and Brooks/Cole, Monterey, 1986.

J-M. STEYAERT [1984]. "Complexité et Structure des Algorithmes", These de Doctorat ès-Sciences, Université Paris 7, 1984.

GLOBAL DIMENSION OF ASSOCIATIVE ALGEBRAS

Tatiana Gateva-Ivanova
Institute of Mathematics
Bulgarian Academy of Sciences
P.O. Box 373, Sofia 1090, Bulgaria

ABSTRACT Let G be a connected graded s.f.p. (standard finitely presented) associative algebra over a field K. We show that the global dimension of G is effectively computable in the following cases: 1) G is a finitely presented monomial algebra; 2) G is a connected graded s.f.p. algebra and the associated monomial algebra A(G) has finite global dimension. The situation is considerably simpler when G has polynomial growth of degree d and gl.dim A(G) $< \infty$. We show that in this case gl.dim G = gl.dim A(G) = d.

1. INTRODUCTION

In [2] Anick studied a number of fundamental properties of monomial algebras G, among which were the global dimension and the Hilbert series. These results depend only on the monomial generators of a defining ideal for G. Of central importance is Anick's notion of an n-chain.

The purpose of this paper is to consider these properties from the standpoint of effective computation. In theorem I we find a linear function N of the degree and number of generators which bounds the global dimension (if finite) from above. Moreover, if there are no (N+1)-chains in the sense of Anick, the global dimension is indeed finite. This, combined with [2,Thm. 4] gives the effective computability of the global dimension.

We show that if G is a connected graded s.f.p. algebra such that the associated monomial algebra A(G) has finite global dimension then gl.dim G $\leq$ gl.dim A(G) and therefore, the global dimension of G is also effectively computable.

In [2, p.301] Anick conjectured that if G is a connected graded K-algebra with polynomial growth and global dimension d $< \infty$, then the Hilbert series of G is of the form

Partially supported by Contract No.62/1988, Committee of Science, Bulgaria.

$$H_G(z) = \sum_{i=1}^{d}(1 - z^{e_i})^{-1}$$

for some positive integers e_i. He noted there that the conjecture was known for G commutative, an enveloping algebra, or a monomial ring and (in an unpublished result of M. Lorenz), for Noetherian PI-rings. Here we establish Anick's conjecture for an algebra G whose associated monomial algebra A(G) is of finite global dimension. In this case gl.dim G = gl.dim A(G). One sees immediately that this case contains the commutative, monomial, and enveloping algebras.

We would like to thank Luchezar Avramov for drawing our attention to Anick's paper as well as for several useful discussions.

2. DEFINITIONS AND RESULTS

In this paper K denotes a fixed field and the term K-algebra is used to denote an associative algebra with unit over K. A K-algebra G is connected graded if there are K-vector spaces $\{G_n\}_{n \geq 0}$, such that: $G = \bigoplus_{n \geq 0} G_n$, $G_0 = K$, and $G_m.G_n \subseteq G_{m+n}$ for all m, n.

Given a non-empty set $X = \{x_1,\dots,x_s\}$, $\langle X\rangle$ will denote the free monoid with 1 generated by X, and $K\langle X\rangle$ will denote the free associative K-algebra (with 1) generated by X. As a K-vector space $K\langle X\rangle \approx \mathrm{Span}\langle X\rangle$. If the set X is graded, i.e. if there is a function $e: X \to Z_+$, then e extends uniquely to a map of monoids $X \to Z_+ \cup \{0\}$ $(e(1) = 0)$ and makes $H = K\langle X\rangle$ into a connected graded algebra ($H_n = \mathrm{Span}\{h \in \langle X\rangle, e(h) = n\}$). We shall assume that the $x_1,\dots,x_s$ are indexed so that $e(x_i) \leq e(x_j)$ if $i < j$. (When no grading is specified e will be taken to be the grading by length, i.e. $e(x_i) = 1$, $1 \leq i \leq s$.) A polynomial $f \in K\langle X\rangle$ is called homogeneous if $f \in H_n$, for some n.

With the goal of determining a standard basis for an ideal of $K\langle X\rangle$ we fix the so-called mixed order on $\langle X\rangle$. We recall the definition: define a total order $<_T$ on the monoid $\langle X\rangle$ by setting a) $x_1 <_T x_2 <_T \dots <_T x_s$, b) $u <_T v$ if $e(u) < e(v)$ or $e(u) = e(v)$ and $u < v$ in the lexicographic order. It is clear that $(\langle X\rangle, <_T)$ is well ordered. For any $g = \sum_{i=1}^{m} \alpha_i u_i$, $\alpha_i \in K$, $u_i \in \langle X\rangle$ we shall denote by HT(g) the highest term of g, i.e. $HT(g) = u_i$ if $u_j <_T u_i$ for all $j \neq i$.

(2.1) Given a set of polynomials $F \subseteq K\langle X\rangle$ the monomial u is called normal modulo F if it does not contain any of the monomials HT(f), $f \in F$, as a subword. A polynomial $g = \sum \alpha_i u_i$, $\alpha_i \in K$, $u_i \in \langle X \rangle$ is in normal form modulo F if all the monomials u_i are normal modulo F. NM(F) will denote the set of all normal modulo F monomials in $\langle X \rangle$, and (F) will denote the two-sided ideal in $K\langle X \rangle$ generated by F. If all polynomials in F are homogeneous, then the ideal I = (F) is graded, i.e. $I = \bigoplus I_n$, with $I_n = I \cap H_n$, $n \geq 0$, and the algebra $G = K\langle X \rangle/(F)$ is a connected graded algebra.

We give here one of the equivalent definitions of a standard basis of an ideal I (with respect to $<_T$). (We shall always assume that $X \cap I = \emptyset$.) For more details cf. [3], [4], [5], [6].

(2.2) A set of polynomials $F \subseteq K\langle X \rangle$ generating I as a two-sided ideal is a standard (or Groebner) basis for I if the set NM(F) projects to a K-basis of $K\langle X \rangle/I$. Clearly any ideal has a (possibly infinite) standard basis.

The following facts are well-known, but for convenience we formulate them explicitly.

On the monoid $\langle X \rangle$ consider the relation $\prec$: for $u,v \in \langle X \rangle$, $u \prec v$ holds if and only if u is a segment (subword) of v (u = v is also possible). Clearly $\prec$ is a partial order on $\langle X \rangle$.

(2.3) For a given ideal I the set of obstructions V(I) is defined as:

$$V(I) = \left\{ u \in \langle X \rangle \;\middle|\; \begin{array}{l} \exists\, g \in I : u = HT(g); \\ \text{if } h \in I \text{ has } HT(h) \preceq u, \\ \text{then } HT(h) = u \end{array} \right\},$$

i.e., V(I) is the set of all monomials in the set $\{ HT(g) \mid g \in I \}$ which are minimal relative to $\prec$. Note that V(I) depends on the ideal I as well as on the order $<_T$ on $\langle X \rangle$.

For any $u \in V(I)$ we take an element $f_u \in I$ such that $f_u = u - h_u$, $h_u \in K\langle X \rangle$, $HT(h_u) <_T u$. Note that if the ideal I is graded, f_u can always be taken to be homogeneous. Consider the set

$$F = F(I) = \{ f_u \mid u \in V(I) \}.$$

It is clear that NM(F) = NM(V(I) = NM(I) , so that F is a standard basis of I. Note that F is minimal in the sense that for any $f \in F$, $F \setminus \{f\}$ is not a standard basis of I. If, moreover, any $f \in F$ is in normal form modulo $F \setminus \{f\}$, the set F is called a reduced standard basis of I. It is known that for any ideal I of $K\langle X\rangle$ there exists a uniquely determined (possibly infinite) reduced standard basis F(I) of I. Obviously, $V(I) = \{ HT(f) \mid f \in F(I) \}$. If moreover, the algebra $G = K\langle X\rangle/I$ is connected graded (i.e. the ideal I is graded) then the reduced basis F(I) consists of homogeneous polynomials. It is clear that for any standard basis Φ of I one has $V(I) \subseteq \{ HT(\varphi) \mid \varphi \in \Phi \}$ and therefore:

(2.4) REMARK. An ideal I of $K\langle X\rangle$ has a finite standard basis if and only if the set of obstructions V(I) is finite.

(2.5) DEFINITION. An algebra $G \approx K\langle X\rangle/I$ is called s.f.p. (standard finitely presented) algebra if the ideal I has a finite standard basis.

(2.6) Recall that if $W \neq \emptyset$ is a set of monomials then the algebra $A = K\langle X\rangle/(W)$ is called a monomial algebra. We shall always assume that $W \cap X = \emptyset$, and no u is a segment of v for some $u,v \in W$, $u \neq v$, i.e. W is a nontrivial antichain of monomials (with respect to $\prec$). It is clear that if W is finite then A is a connected graded s.f.p. algebra with W as a reduced basis.

(2.7) Let $G = K\langle X\rangle/I$ and $V = V(I)$, as in (2.3). The algebra $A(G) = K\langle X\rangle/(V)$ is called the monomial algebra associated with G.

It is clear that the set NM(V) projects onto a K-basis for G and for A(G) as well.

Next we recall some definitions and results of Anick.

(2.8) DEFINITION [2, p.295] . Let V be a nontrivial antichain in $\langle X\rangle$. Define the set of n-chains recursively as follows. A (-1)-chain is the monomial 1, a 0-chain is any element of X, and a 1-chain is a word in V. An (n+1)-prechain is a word $w \in \langle X\rangle$ which has two factorizations $w = uvt = usq$ in $\langle X\rangle$, where u is an (n-1)-chain, s is a proper left segment of v, uv is an n-chain, and $q \in V$. The right factor t is called the tail of the prechain. An (n+1)-prechain w is an (n+1)-chain if no proper left segment of w is also an (n+1)-prechain.

(2.9) THEOREM (Anick, [2, Thm. 4]). Let V be a nontrivial antichain of monomials in the free monoid on the graded set X. If $A = K\langle X\rangle/(V)$ and $i \geq 0$, then $\mathrm{Tor}_i^A(K,K)$ is isomorphic as a graded K-module to the span of the set of (i-1)-chains of V. In particular, gl.dim $A \leq d$ if and only if there are no d-chains.

(2.10) Recall that a graded algebra G has polynomial growth if there is a polynomial $p(x) \in Z[x]$ such that $\dim_K G_n \leq p(n)$ for n large.

(2.11) CONJECTURE (Anick [2, p.301]). Suppose G is a connected graded K-algebra with polynomial growth and with global dimension $d < \infty$. Then for the Hilbert series of G, $H_G(z) = \sum_{n\geq 0} \dim_K(G_n)\, z^n$, one has

$$H_G(z) = \prod_{i=1}^{d} (1 - z^{e_i})^{-1}$$

for some positive integers $\{e_i\}$.

As noted in the Introduction, we deal here with Anick's results from the point of view of effective computation. In Section 3 we treat finitely presented monomial algebras and there establish the following result:

(2.12) THEOREM I. Let $V = \{w_1,\ldots,w_p\}$ be a nontrivial antichain of monomials in the free monoid of the graded set $X = \{x_1,\ldots,x_n\}$, let $N = \sum_{i=1}^{p} \deg w_i - p$. The monomial algebra $A = K\langle X\rangle/(V)$ has finite global dimension if and only if there are no (N+1)-chains on V. In particular,

$$N + 1 \geq \text{gl.dim } A = 1 + \max\{n \mid 1 \leq n \leq N, \text{ there exists an n-chain}\}$$

In Section 4 we consider the case when G is a connected graded s.f.p. algebra whose associated monomial algebra A(G) is of finite global dimension. It follows from routine observations that in this case gl.dim G is effectively computable. We also give a positive answer to the conjecture (2.11) in the case in which the associated monomial algebra A(G) has finite global dimension:

(2.13) THEOREM II . Let $G = K\langle X\rangle/I$ be a connected graded algebra such that: 1) G has polynomial growth of degree d and 2) the associated monomial algebra A(G) has finite global dimension. Then:

a) G is a standard finitely presented algebra ,

b) gl.dim G = gl.dim A(G) = d , and

c) for the Hilbert series of G, $H_G(z) = \sum_{n\geq 0} \dim_K G_n\, z^n$ one has $H(z) = \prod_{i=1}^{d} (1 - z^{e_i})^{-1}$, for some positive integers $\{e_i\}$.

3. GLOBAL DIMENSION OF FINITELY PRESENTED MONOMIAL ALGEBRAS

Throughout this section A denotes a graded finitely presented monomial algebra $A = K\langle X\rangle/(V)$, where the set of generators $X = \{x_1,\ldots,$

$x_s\}$ and the set of monomials $V = \{w_1,\dots,w_p\}$ are fixed. Furthermore, as in the last section, we assume that V is a nontrivial antichain of monomials.

(3.1) <u>REMARK</u>. It follows from (2.8) and [1, Lemma 1.3] that if c is an n-prechain on V, then for any $i < n$ there is a uniquely determined i-chain which is a proper left segment of c. We shall denote it by c(i). In particular, for $n \geq 1$, one has $c = c(n-1).t$, where t is the tail of c . By t(c) we shall denote the tail t of c.

Let us consider the set V defined as

$$\tilde{V} = \left\{ \begin{array}{llll} (u_{11},v_{11}), & (u_{12},v_{12}), & \dots, & (u_{1m_1},v_{1m_1}) \\ (u_{21},v_{21}), & (u_{22},v_{22}), & \dots, & (u_{2m_2},v_{2m_2}), \\ \dots & \dots & \dots & \dots \\ (u_{p1},v_{p1}), & (u_{p2},v_{p2}), & \dots, & (u_{pm_p},v_{pm_p}) \end{array} \right\}, \qquad (3.2)$$

where $m_i = \deg w_i - 1$, $\deg u_{ij} = j$, and $u_{ij}.v_{ij} = w_i$ for $1 \leq i \leq p$, $1 \leq j \leq m_i$. Thus

$$\# \tilde{V} = \sum_{i=1}^{p} m_i = \sum_{i=1}^{p} \deg w_i - p \quad . \qquad (3.3)$$

(3.4) <u>DEFINITION</u>. For the monomials $u, v, w \in \langle X \rangle \setminus \{1\}$ we say that v <u>overlaps with</u> w (<u>on</u> u) if $v = au$, $w = ub$ for some $a, b \in \langle X \rangle$.

Note that v and w can overlap in different ways. For example if $v = x_1x_2^2$, $w = x_2^3x_1$, then

$$v = a_1u_1,\ w = u_1b_1, \text{ where } a_1 = x_1,\ b_1 = x_2x_1 ,\ u_1 = x_2^2$$

and

$$v = a_2u_2,\ w = u_2b_2, \text{ where } a_2 = x_1x_2,\ b_2 = x_2^2x_1 ,\ u_2 = x_2$$

are two different ways of overlapping.

(3.5) <u>REMARK</u>. With any n-prechain c, $n \geq 2$, one can associate the uniquely determined sequence, called an n-prechain

$$\tilde{c} = ((a_1,b_1),(a_2,b_2),\dots,(a_n,b_n)),$$

where $a_i = a_i(c)$, $b_i = b_i(c)$ depend on c and

i) $(a_i,b_i) \in \tilde{V}$, $i = 1,\dots n$;

ii) $a_1 \in X$;

iii) a_j is a right segment of b_{j-1}, $j = 2,\dots,n$, i.e. $a_{j-1}b_{j-1}$

overlaps with a_jb_j on a_j;

iv) for $i = 1,\ldots,n$, and $c(i)$ as in (3.1) the equalities $\widetilde{c(i)} = ((a_1,b_1),\ldots,(a_i,b_i))$, $c(i) = a_1b_1b_2\ldots b_i$ hold and b_i is the tail of $c(i)$.

(3.6) <u>LEMMA</u>. Let c and f be (n+1)-prechains and let f be a proper left segment of c. Then $c(n) = f(n)$ and there exist monomials a, b, u, v, such that a is a proper right segment of u, v is a proper left segment of b, (a,b), $(u,v) \in \tilde{V}$, $\tilde{c} = (\widetilde{c(n)},(a,b))$, $f = (\widetilde{c(n)},(u,v))$, b and v are the tails of c and f respectively.

<u>PROOF</u>. Note that f being a left segment of c, three cases are possible: i) $f(n)$ is a proper left segment of $c(n)$; ii) $c(n)$ is a proper left segment of $f(n)$; iii) $c(n) = f(n)$. Cases i) and ii) lead to a contradiction with the definition of an n-chain (cf. (2.8)). Hence, $c(n) = f(n)$. It follows by (3.5) that there exist monomials a, b, u, v, such that a, u are right segments of $c(n)$, (a,b), $(u,v) \in \tilde{V}$, $\tilde{c} = (\widetilde{c(n)},(a,b))$, $\tilde{f} = (\widetilde{c(n)},(u,v))$. The monomial $f = c(n)v$ being a proper left segment of $c = c(n)b$, one obtains that v is a proper left segment of b. From the fact that V is an antichain of monomials it follows that uv is not a segment of ab, hence, a is a proper right segment of u.

Applying (2.8), (3.5) and lemma (3.6) one obtains:

(3.7) <u>LEMMA</u>. Let $c = a_1b_1\ldots b_n$ be an n-chain, let $\tilde{c} = ((a_1,b_1),\ldots,(a_n,b_n))$ and let $(a,b) \in \tilde{V}$. The monomial $a_1b_1\ldots b_nb$ is an (n+1)-chain ($\widetilde{\tilde{c},(a,b)}$ is an (n+1)-chain) if and only if a is a right segment of b_n and there is no element (u,v) of $\tilde{V}$ such that u is a right segment of b_n, v is a proper left segment of b.

(3.8) <u>LEMMA</u>. Let c be a q-chain, $q \geq 2$, $\tilde{c} = ((a_1,b_1),\ldots,(a_r,b_r),\ldots,(a_q,b_q))$ as in (3.5). For some r in the range $1 \leq r < q$, let $b_r = b_q$. Then i) The monomial cb_{r+1} is a (q+1)-chain, with $\widetilde{cb_{r+1}} = ((a_1,b_1),\ldots,(a_r,b_r),\ldots,(a_q,b_q),(a_{r+1},b_{r+1}))$; ii) For any $n \geq -1$ there exists an n-chain on V.

<u>PROOF</u>. By (3.5), $c(r) = a_1b_1\ldots b_r$ is an r-chain with b_r as a tail, and b_r overlaps with $a_{r+1}b_{r+1}$ on a_{r+1}. From the equality $b_r = b_q$ it follows that the tail of c, b_q, overlaps with $a_{r+1}b_{r+1}$, on a_{r+1}, so that cb_{r+1} is a (q+1)-prechain with $\widetilde{cb_{r+1}} = (\tilde{c}, (a_{r+1},b_{r+1}))$. Assume

cb_{r+1} is not a (q+1)-chain. Then by (2.8) there exists a (q+1)-prechain f which is a proper left segment of cb_{r+1}. It follows from (3.6) that $f(q) = (cb_{r+1})(q) = c$ and $\tilde{f} = (\tilde{c},(u,v))$, where $(u,v) \in \tilde{V}$, $b_q = b_r$ overlaps with uv on u, and v is a proper left segment of b_{r+1}. We obtained that c(r)v is an (r+1)-prechain which is a proper left segment of the (r+1)-chain , c(r+1) - a contradiction. Hence, cb_{r+1} is a (q+1)-chain. This proves(i). Let $q = r+k$, $k \geqslant 1$. In order to prove(ii) consider the following sequence of monomials: $c_{q+1}, c_{q+2}, \ldots, c_n, \ldots$, where for $n = ik + j$, $0 \leqslant j \leqslant k-1$, one has $c_n = a_1 \ldots b_{r-1}(b_r \ldots b_{r+k-1})^i b_r \ldots b_{r+j}$. Using induction on n it follows from (i) that c_n is an (r+n)-chain for any n, $n \geqslant 0$.

(3.9) <u>COROLLARY</u>. Let A be the monomial algebra $K\langle X\rangle/(w_1,\ldots,w_p)$, $N = \sum_{i=1}^{p} \deg w_i - p$, and assume that there exists an (N+1)-chain. Then there exists an n-chain for any $n \geqslant -1$. In particular, gl.dim $A = \infty$.

<u>PROOF</u>. Assume that c is an (N+1)-chain. Consider $\tilde{c} = ((a_1,b_1), \ldots,(a_{N+1},b_{N+1}))$, $(a_i,b_i) \in \tilde{V}$, $1 \leqslant i \leqslant N+1$. Since V is a set with N elements, there are r,p such that $1 \leqslant r < p \leqslant N+1$ and $(a_r,b_r) = (a_p,b_p)$. By Lemma (3.8) it follows that there exists an n-chain for any $n \geqslant -1$. Applying (2.9) we obtain gl.dim $A = \infty$.

Now Theorem I follows from (3.9) and Anick's Theorem(2.9) .

Note that the global dimension of A does not depend on the grading function $e: X \longrightarrow Z_+$ determining the grading, but only on the set V.

(3.10) $R = R(V)$ denotes the set of all different proper right segments of length $\geqslant 2$ of the monomials $w_i \in V$, $i = 1,\ldots,p$. Obviously,

$$\#(R \cup X) \leqslant \#V = N \, . \qquad (3.11)$$

In (3.12) we give a graphic interpretation of the observations above which was proposed to the author by Victor Ufnarovsky.

(3.12) Consider the <u>oriented graph</u> $\overline{C} = \overline{C}(V)$ defined as follows:

The <u>set of vertices</u> of C is $\{1\} \cup X \cup R$, and

(i) for any $x \in X$ there is an <u>edge</u> $1 \longrightarrow x$;

(ii) for $x \in X$, $r \in X \cup R$ there is an <u>edge</u> $x \longrightarrow r$ if and only if $xr \in V$ (i.e.,$(x,r) \in \tilde{V}$).

(iii) for $r,t \in R$ there is an <u>edge</u> $r \longrightarrow t$ if and only if there is a right segment a of r (a = r is also possible), such that $(a,t) \in \tilde{V}$ and there is no element $(u,v) \in \tilde{V}$, u is a right segment of r, v is a proper left segment of t.

After canceling all vertices $v \neq 1$ in $\bar{C}$, such that there is no route from 1 to v, one obtains the finite oriented graph $C = C(V)$, called the <u>graph of chains</u> on V.

(3.13) <u>REMARK</u>. For any $n \geq 0$ there is a one to one correspondence between the set of all routes of length n + 1 in C, beginning with 1, and the set of all n-chains, $V^{(n)}$

Indeed, if $1 \longrightarrow t_0 \longrightarrow t_1 \longrightarrow \dots \longrightarrow t_{n-1} \longrightarrow t_n$ is a route of length n+1 in C, then by the definition of C, cf.(3.12), and Lemma (3.8) it follows that the monomial $t_0 t_1 \dots t_{n-1} t_n$ is an n-chain. Conversely, if c is an n-chain, $\tilde{c} = ((a_1,b_1),(a_2,b_2),\dots,(a_n,b_n))$, then applying Lemma (3.8) again one obtains a route $1 \longrightarrow a_1 \longrightarrow b_1 \longrightarrow b_2 \longrightarrow \dots \longrightarrow b_n$ in the graph C.

By (3.13) and (2.9) one obtains

(3.14) <u>REMARK</u>. The monomial algebra $A = K\langle X\rangle/(V)$ has finite global dimension if and only if there are no cycles in the graph C(V). If this is the case for A then gl.dim A is the maximal length of a route in C(V). It follows from (3.11) that the number of vertices in C(V) is at most N, hence, we obtained again that gl.dim $A \leq N + 1$.

We are ready now to solve the following

(3.15) <u>PROBLEM</u>: Given: X, V, $A = K\langle X\rangle/(V)$.
Find gl.dim A.

(3.16) <u>PROCEDURE</u>.

1) Find R(V) cf. (3.10)
2) Construct the graph $\bar{C}(V)$. cf. (3.12)
3) Construct the graph C(V).
4) If there is a cycle in C(V) then gl.dim A := ∞, else gl.dim A := the maximal length of a route in C(V).

(3.17) <u>EXAMPLE</u>. $X = \{x_1, x_2\}$, $V = \{x_2^4, x_2^2 x_1\}$, $A = K\langle X\rangle/(V)$.

Obviously, $R(V) = \{x_2^3, x_2^2, x_2 x_1\}$.

The graphs $\bar{C}(V)$ avd C(V) are given in Figures 1 and 2 respectively.

Figure 1: 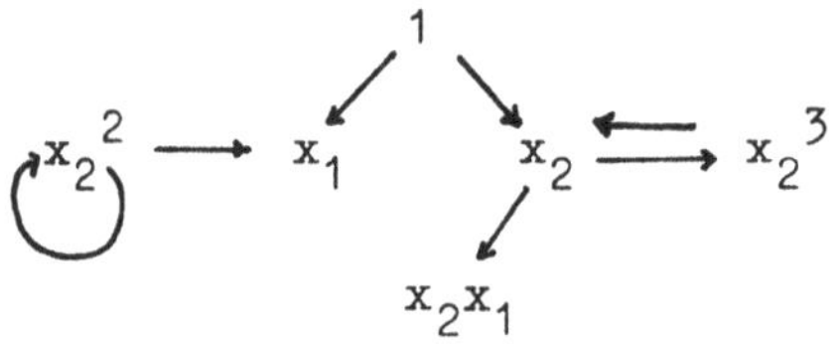

Figure 2: 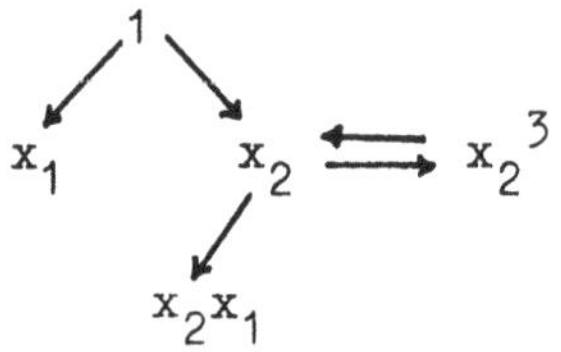

The graph C(V) contains a cycle, hence, by (3.14) gl.dim A $= \infty$.

4. GLOBAL DIMENSION OF CONNECTED GRADED S.F.P. ALGEBRAS

We keep the preceding notation.

(4.1) By $V^{(n)}$ we shall denote the set of all n-chains on V, $V^{(n)}K = \mathrm{Span}_K V^{(n)}$.

(4.2) Let $\alpha_i \in K$, $a_i \in V^{(n)}$, $b_i \in M(V)$, $1 \leq i \leq q$,

$$\omega = \sum_{i=1}^{q} \alpha_i a_i \otimes b_i \in V^{(n)}K \otimes_K G .$$

We assume that $a_i \otimes b_i \neq a_j \otimes b_j$ for $i \neq j$. By HT(ω) we shall denote the element $a_i \otimes b_i$ if and only if $a_j b_j <_T a_i b_i$ for all $j \neq i$. (By the definition of n-chains it follows that no equality $a_i b_i = a_j b_j$ is possible for $i \neq j$.)

It is shown in [1] Theorems 1.4, 1.5 that there is a homogeneous free G-resolution of K:

$$0 \leftarrow K \xleftarrow{\varepsilon} G \xleftarrow{\delta_0} XK \otimes_K G \xleftarrow{\delta_1} VK \otimes_K G \xleftarrow{\delta_2} V^{(2)}K \otimes_K G \xleftarrow{\delta_3} \cdots \qquad (4.3)$$

in which $\delta_0(x \otimes 1) = x$, $x \in X$, $\varepsilon(\bigoplus_{i\geq 1} G_i) = 0$, $\varepsilon|_K = \mathrm{id}_K$ and for $n \geq 1$ δ_n are homomorphisms of graded G-modules, which are inductively defined so that for any n-chain v :

$$\delta_n(v \otimes 1) = v(n-1) \otimes t(v) + \omega \quad , \text{ where } HT(\omega) <_T v \quad , \qquad (4.4)$$

v(n-1), t(v) as in (3.1) .

Now we shall make some routine observations. It is known that

$$gl.dim\ G = pd_G\ K \ . \tag{4.5}$$

(4.6) REMARK. There is an inequality:

$$gl.dim\ G \leq gl.dim\ A(G) \quad . \tag{4.7}$$

PROOF. It suffices to assume that gl.dim $A(G) = d < \infty$. It follows from (2.9) that $V^{(d)} = \emptyset$, $V^{(d-1)} \neq \emptyset$, hence the resolution (4.3) stops at $V^{(d-1)}K \otimes G$. It follows that $pd_G\ K \leq d$ which together with (4.5) gives (4.7).

We shall further assume that gl.dim $A(G) = d < \infty$. It is known that

$$pd_G\ K = \max \left\{ n \,\middle|\, Tor_n^G(K,K) \neq 0 \right\}. \tag{4.8}$$

In order to compute $Tor^G(K,K)$ we apply the functor $- \otimes_G K$ to (4.3) and obtain the complex of K-vector spaces

$$0 \leftarrow K \leftarrow G \otimes_G K \xleftarrow{\overline{\delta}_0} (XK \otimes_K G) \otimes_G K \leftarrow \ldots \xleftarrow{\overline{\delta}_{d-1}} (V^{(d-1)}K \otimes_K G) \otimes_G K \leftarrow 0 \tag{4.9}$$

where $\overline{\delta}_n = \delta_n \otimes 1$, $0 \leq n \leq d-1$. Note that since V is a finite set of monomials $V^{(n)}$ also is finite for any $n \geq 1$. Let $V^{(n)} = \{w_1^n, \ldots, w_{p_n}^n\}$,

(Obviously $p_1 = p$.) By the equality

$$a \otimes_K b \otimes_G 1 = 0,\ a \in V^{(n)},\ b \in M(V),\ b \neq 1, \tag{4.10}$$

it follows that the set $\{ w_j^n \otimes 1 \otimes 1 \mid 1 \leq j \leq p_n \}$ forms a K-basis of $V^{(n)}K \otimes G \otimes K \approx V^{(n)}K$ as a K-space. The images $\delta_n(w_j^n \otimes 1)$ can be found as in the proof of Thm. 1.4, [1]. Applying (4.10) again one can find $\overline{\delta}_n(w_j^n \otimes_K 1 \otimes_G 1) = (\delta_n(w_j^n \otimes_K 1)) \otimes_G 1$ and hence $\dim_K \operatorname{Im} \overline{\delta}_n$ can also be found. Because of:

$$\dim_K \operatorname{Ker}_n = \dim V^{(n)}K - \dim_K \operatorname{Im} \overline{\delta}_n = p_n - \dim_K \operatorname{Im} \overline{\delta}_n \tag{4.11}$$

one sees that for k defined by

$$k = \max \left\{ n \,\middle|\, 1 \leq n \leq d,\ \dim \operatorname{Ker} \overline{\delta}_{n-1} > \dim \operatorname{Im} \overline{\delta}_n \right\}, \tag{4.12}$$

one has $k = pd_G\ K = gl.dim\ G$. It can easily be shown that $\operatorname{Ker} \overline{\delta}_1 = VK \otimes_K G \otimes_G K \supsetneq \operatorname{Im} \overline{\delta}_2$, so that $gl.dim\ G \geq 2$.

We proved the following

(4.13) PROPOSITION. In the notation above, let G be a connected graded s.f.p. algebra and let A(G) be the associated monomial algebra of G. If gl.dim A(G) = d < ∞ then

$$d \geq \text{gl.dim}\, G = \max\{n \mid 1 \leq n \leq d,\ \dim \operatorname{Ker} \bar{\delta}_{n-1} > \dim \operatorname{Im} \bar{\delta}_n\}.$$

(4.14) COROLLARY. In case that gl.dim A(G) < ∞ , the global dimension of the graded s.f.p. algebra G is effectively computable.

(4.15) EXAMPLE. In the notation above, let $F = \{f_1 = x_3x_2 - x_1x_2,\ f_2 = x_2x_1 - x_1^2,\ f_3 = x_3x_1^2 - x_1^3\}$, $G = K\langle x_1,x_2,x_3\rangle/(F)$, graded by length. We shall find gl.dim G. Note first that G is s.f.p., $V = \{w_1 = x_3x_2,\ w_2 = x_2x_1,\ w_3 = x_3x_1^2\}$, $V^{(1)} = V$, $V^{(2)} = \{x_3x_2x_1\}$, $V^{(3)} = \emptyset$, hence gl.dim A(G) = 3. $\delta_1(x_3x_2 \otimes 1) = x_3 \otimes x_2 - x_1 \otimes x_2$, $\delta_1(x_2x_1 \otimes 1) = x_2 \otimes x_1 - x_1 \otimes x_1$, $(x_3x_1^2 \otimes 1) = x_3 \otimes x_1^2 - x_1 \otimes x_1^2$, $\bar{\delta}_1(w \otimes 1 \otimes 1) = 0$, for i = 1,2,3. $\operatorname{Ker} \bar{\delta}_1 = VK \otimes G \otimes K \approx VK$, $\delta_2(x_3x_2x_1 \otimes 1) = x_3x_2 \otimes x_1 - x_3x_1^2 \otimes 1$, $\bar{\delta}_2(x_3x_2x_1 \otimes 1 \otimes 1) = x_3x_1^2 \otimes 1 \otimes 1 \neq 0$.

Obviously $\operatorname{Im} \bar{\delta}_2 \subsetneq \operatorname{Ker} \bar{\delta}_1$, hence gl.dim G ≥ 2 , which together with the equality $\operatorname{Ker} \bar{\delta}_2 = 0$, gives gl.dim G = 2 < gl.dim A(G) .

We shall prove that if G is a connected graded s.f.p. algebra with polynomial growth of degree d (cf. (2.10)) and the associated monomial algebra A(G) has finite global dimension, then gl.dim G = gl.dim A(G) = d.

Till the end of the section A denotes a monomial algebra with a fixed presentation $A = K\langle X\rangle/(V)$.

The following facts can easily be abstracted from [2, Theorems 5 and 6].

(4.16) Suppose, following the notation above, that the monomial algebra A has global dimension d < ∞ and polynomial growth. Then there are normal monomials (called atoms) $y_1,\dots,y_d$ in $\langle X\rangle \setminus \{1\}$, $y_i \neq y_j$ for $i \neq j$, such that $X \subseteq \{y_1,\dots,y_d\}$ and such that the following properties hold:

i) If the monomial y_py_q is normal modulo V , $1 \leq q < p \leq d$, there exists a t , $1 \leq t \leq d$, such that $y_py_q = y_t$;

ii) For any p,q, $1 \leq p,q \leq d$, if y_p is a proper left segment of y_q , then p < q ; if y_p is a proper right segment of y_q then q < p.

iii) Any normal modulo V monomial w may be uniquely factored in the form $w = y_1^{\alpha_1} y_2^{\alpha_2} \dots y_d^{\alpha_d}$, $\alpha_i \geq 0$.

iv) Any monomial $w \in V$ can be factored as $w = y_p y_q$ for some p,q, $1 \leq q < p \leq d$.

v) The monomial algebra A is finitely presented: $\# V \leq d(d-1)/2$.

We shall now assume that the monomial algebra A is as in (4.16) with global dimension $d < \infty$ and polynomial growth.

(4.17) LEMMA. If y_p, y_q are atoms, $p \neq q$ and y_p overlaps with y_q (cf. (3.4)), then $p > q$.

PROOF. By the condition of the Lemma, $y_p = au$, $y_q = ub$ for some normal monomials a, b, u, $u \neq 1$. Since $p \neq q$, a and b cannot both be 1. If a = 1 or b = 1, then the assertion of the Lemma follows directly from (4.16 ii). Assume now that $a \neq 1$, $b \neq 1$. Since the monomial u is normal, there is a unique presentation

$$u = y_{k_1}^{\alpha_1} \dots y_{k_s}^{\alpha_s} , \quad k_1 < \dots < k_s , \quad \alpha_i \geq 1. \tag{4.18}$$

(Note that it is possible that u is an atom and in that case s = 1, $\alpha_1 = 1$.) Applying (4.16 ii) again we obtain:

$$p > k_s \quad \text{(since } y_{k_s} \text{ is a proper right segment of } y_p\text{)}, \tag{4.19}$$

$$k_1 > q \quad \text{(since } y_{k_1} \text{ is a proper left segment of } y_q\text{)}. \tag{4.20}$$

From (4.18), (4.19), and (4.20) it follows that $p > q$.

(4.21) LEMMA 2 . The monomial $y_{i+k} y_{i+k-1} \dots y_i$ is a k-chain with y_i as a tail for any i,k, $1 \leq i < d$, $1 \leq k \leq d - i$.

Since no proof is given in [2], we provide an argument here. Again we argue by cases.

k = 1. One must check that $y_{i+1} y_i \in V$. Suppose that this is not true. Then a) $y_{i+1} y_i$ is normal or b) $y_{i+1} y_i$ is not normal but there exists a monomial $w \in V$ which is a proper segment of $y_{i+1} y_i$. In case (a), (4.16. i) applies to give the existence of a t, $q \leq t \leq d$, such that $y_{i+1} y_i = y_t$. Then (4.16.ii) applies once again to show that $i + 1 > t > i$, which is impossible. Assume now that (b) holds. By (4.16.iv), $w = y_p y_q$ for some atoms y_p, y_q, $p > q$. Four cases can occur:

b1) y_{i+1} overlaps with y_q, y_q overlaps with y_i;

b2) y_{i+1} overlaps with y_p, y_p overlaps with y_i;

b3) y_p is a proper right segment of y_{i+1}, y_q is a left segment of y_i (here $q = i$ is admissible);

b4) $p = i + 1$, y_q is a proper left segment of y_i.

By applying (4.16.ii) again it can be easily shown that any of the subcases b1), b2), b3) and b4) leads to a contradiction. Hence (b) is impossible, and so $y_{i+1}y_i \in V$.

<u>k = 2</u>. It is clear that $u = y_{i+2}y_{i+1}y_i$ is a 2-prechain. Assume there exists a 2-prechain v which is a proper left segment of u. Since the set V is an antichain, v can be presented as $v = l(y_{i+2})y_py_q$, where $l(y_{i+2})$ is a proper left segment of y_{i+2}, $y_py_q \in V$. Depending on how the monomials $y_{i+2}y_{i+1}$ overlaps with y_py_q in the 2-prechain v , five cases are possible:

a) y_{i+2} overlaps with y_q, y_q overlaps with y_i;

b) y_{i+1} is a proper left segment of y_q, y overlaps with y_i;

c) y_{i+2} overlaps with y_p, y_p overlaps with y_{i+1}

d) y_{i+2} overlaps with y_p, y_{i+1} is a proper right segment of y_p;

e) y_{i+2} overlaps with y_p, y_p overlaps with y_i , y_{i+1} is a proper segment of y_p.

By applying (4.16.ii) one can show that each one of the cases (a) through (e) leads to a contradiction, so that $u = y_{i+2}y_{i+1}y_i$ is always a 2-chain It is obvious that its tail equals y_i.

Assume now that we have shown that if $i < d$, $2 \leqslant k \leqslant d - i$, then $y_{i+k}\cdots y_{i+1}$ is a $(k - 1)$-chain with y_{i+1} as a tail. Then we shall show that $u = y_{i+k}\cdots y_{i+1}y_i$ is a k-chain whose tail is y_i. Obviously u is a k-prechain. From the existence of a k-prechain v which is a proper left segment of u, it follows easily that there is a monomial $w \in V$ such that v is a proper segment of $y_{i+1}y_i \in V$. This contradicts the condition that V is an antichain.

(4.22) <u>LEMMA</u>. If k is an integer, $1 \leqslant k \leqslant d - 1$, then any k-chain can be presented as $u = y_{i_1}r(y_{i_2})r(y_{i_3})\dots r(y_{i_s})$, where $d > i_1 > i_2 \dots > i_s > 1$, and $r(y_{i_j}) \neq 1$ is a right segment of y_{i_j} .(It is also possible that $r(y_{i_j}) = y_{i_j}$ for some i_j). In the presentation above the tail of u is a right segment of $r(y_{i_{s-1}})r(y_{i_s})$.

PROOF. We use induction on k. For $k = 1$ we know that any 1-chain w is an element of V, hence by (4.16.iv) $w = y_p y_q$ for some p,q with $p > q$, and this is the presentation we need.

Assume that the assertion is true for $k - 1 < d - 1$ and let u be a k-chain, $u = u(k-1)t$, where $u(k-1)$ is as in (3.1) and t is the tail of u. By our assumption, $u(k-1) = y_{i_1} r(y_{i_2}) \ldots r(y_{i_{s-1}}) r(y_{i_s})$, $i_1 > i_2 > \ldots > i_s$, with the tail of $u(k-1)$ as a right segment of $r(y_{i_{s-1}})r(y_{i_s})$. By the definition of a k-chain the tail of $u(k-1)$ and therefore $r(y_{i_{s-1}})r(y_{i_s})$ itself overlaps with an element $w \in V$. By (4.16.iv) one has $w = y_p y_q$, $p > q$. The following cases can occur:

a) $r(y_{i_{s-1}})r(y_{i_s})$ overlaps with y_p, y_p is not a right segment of $r(y_{i_{s-1}})r(y_{i_s})$;

b) y_p is a right segment of $r(y_{i_{s-1}})r(y_{i_s})$;

c) $r(y_{i_s})$ overlaps with y_q, $r(y_{i_s})$ is not a left segment of y_q;

d) $r(y_{i_s})$ is a left segment of y_q;

e) $r(y_{i_s})$ overlaps with y_q.

Assume that (a) holds. Then the following subcases can arise:

a1) $r(y_{i_s})$ overlaps with y_p. Then $i_s > p > q$, and $u = u(k-1).$ $.r(y_p)y_q$ is the desired presentation of u. It is clear that the tail in this case is $r(y_p)y_q$.

a2) y_{i_s} is a proper left segment of y_p. Then $i_s > p > q$, $u = y_{i_1} r(y_{i_2}) \ldots r(y_{i_s}) r(y_p) y_q$, and the tail of u is $r(y_p)y_q$.

a3) $y_{i_{s-1}}$ overlaps with y_p. Here $i_{s-1} > p > q$, y_{i_s} is a segment of $y_p y_q$, and $u = y_{i_1} r(y_{i_2}) \ldots r(y_{i_{s-1}}) r(y_p) y_q$ with $r(y_p)y_q$ as a tail.

The proof of the assertion in cases (b) through (e) is analogous to that of (a) and so we leave it to the reader.

(4.23) COROLLARY. Under the hypotheses and with the assumption of (4.22), if u is a (d-2)-chain, then $e(u) < e(y_d y_{d-1} \ldots y_2 y_1)$.

PROOF. From the definition of a k-chain it follows that if u is a k-chain it is not an n-chain for $n < k$. In particular $y_d \dots y_2 y_1$ is not a (d-2)-chain. Now the assertion follows immediately from (4.22).

(4.24) PROOF OF THEOREM II. The set of normal modulo V monomials, NM(V) is a homogeneous K-basis both for G and A(G), so the monomial algebra A(G) also has polynomial growth of degree d.

(4.25) Applying Anick's theorem [2, Thm.6] one obtains that

i) $A(G) = K\langle X\rangle/(V)$ is finitely presented ($\# V \leqslant d(d-1)/2$);

ii) gl.dim A(G) = d;

iii) $H_{A(G)}(z) = \sum_{i=1}^{d} (1 - z^{e_i})^{-1}$;

iv) For A = A(G), $y_1, \dots, y_d$ as in (4.16) the assertions (4.16) through (4.23) hold.

Now the assertion (a) of the theorem follows from (2.4) and (4.25.i). By the obvious equality $H_G(z) = H_{A(G)}(z)$ and by (4.25.iii) one obtains (c). We shall prove (b). In order to compute gl.dim G we argue as in the proof of (4.13). Consider the free G-resolution (4.3). It follows from (2.9) and (4.25.ii) that $V^{(d-1)} \neq \emptyset$, $V^{(d)} = \emptyset$, so that the resolution stops at $V^{(d-1)}K \otimes G$. By [1, Thm.1.5] the right-hand side of (4.4) is homogeneous with $|\omega| = e(v)$. In particular, for the (d-1)-chain $y_d y_{d-1} \dots y_1$ one has

$$\delta_{d-1}(y_d \dots y_2 y_1 \otimes 1) = y_d \dots y_2 \otimes y_1 + \sum_{i=1}^{r} \alpha_i u_i \otimes v_i \qquad (4.26)$$

where for $i = 1, \dots, r$, u_i is a (d-2)-chain, $v_i \in NM(V)$ and $e(u_i v_i) = e(y_d \dots y_2 y_1)$. By (4.23) the inequality $e(u_i) < e(y_d \dots y_2 y_1)$ holds for $1 \leqslant i \leqslant r$, hence

$$v_i \neq 1, \text{ for } 1 \leqslant i \leqslant r . \qquad (4.27)$$

It follows from (4.26), (4.27) and (4.10) that $y_d \dots y_1 \otimes 1 \otimes 1$ is a non-zero element of Ker $\bar{\delta}_{d-1}$. Thus by (4.13) one obtains gl.dim G = d = gl.dim A(G).

We shall show that the properties 1) and 2) in the condition of Theorem II can be established algorithmically.

(4.28) REMARK [2, Lemma 1]. Let $A = K\langle X\rangle/(V)$ be an f.p. monomial algebra with the set function $e: X \to Z_+$ determining its grading. Let $\tilde{A}$ be an algebra with the same presentation as A but let $\tilde{A}$ be graded by length, i.e. set each $|x_i| = 1$. Then A has polynomial growth of degree d if and only if $\tilde{A}$ has polynomial growth of degree d.

(4.29) $\widetilde{A(G)}$ will denote the algebra $K\langle X\rangle/(V)$ graded by length. From (4.28) we immediately have:

(4.30) The algebra G has polynomial growth of degree d if and only if $\widetilde{A(G)}$ has polynomial growth of degree d.

(4.31) REMARK. For an s.f.p. algebra G the properties 1) and 2) in the condition of Theorem II (cf. (2.13) can be established algorithmically. More precisely, an algorithm is given in [5] for finding the type of growth of a given monomial algebra A (graded by length) and in case that 1) holds, by (4.30) and by (2.9) it follows that 2) is equivalent to:

(4.32) There are no d-chains on V .

REFERENCES

1. D. Anick, On the homology of associative algebras, Trans. AMS 296 (1986) 641-659.
2. ————, On monomial algebras of finite global dimension, Trans. AMS 291 (1985) 291-310.
3. G. Bergman, The diamond lemma for ring theory, Adv. in Math. 29 (1979) 178-218.
4. B. Buchberger, An algorithm for finding a basis for the residue class ring of a zero-dimensional polynomial ideal, Ph.D. Thesis, Univ. of Insbruck (1965).
5. T. Gateva-Ivanova, V. Latyshev, On recognizable properties of associative algebras, to appear in J.Symb. Comp.
6. T. Mora, Groebner basis for non-commutative polynomial rings, Proc. AAECC 3, L.N.C.S. 229 (1986).

Symmetry and Duality in Normal Basis Multiplication

Willi Geiselmann
Dieter Gollmann

Institut für Algorithmen und Kognitive Systeme
Universität Karlsruhe
Germany

Abstract.

This paper is concerned with efficient hardware architectures for normal basis multipliers in GF(2^n). New serial input / parallel output architectures are derived. The complexity of these multipliers is equivalent to the complexity of the Massey-Omura multiplier. We also combine dual basis and normal basis techniques. The duality of normal bases is shown to be equivalent to the symmetry of the logic array of the serial input / parallel output architectures proposed in this paper.

1. Introduction

Applications in cryptography require efficient (hardware) implementations of exponentiation in GF(2^n). Square-and-multiply decomposes exponentiation in two subroutines. Squaring is a linear operation in GF(2^n) but this does not necessarily imply a "cheap" hardware implementation. However, in a normal basis representation squaring becomes particularly simple. This has to be matched by efficient normal basis architectures for multiplication in GF(2^n). Massey and Omura have proposed a parallel input / serial output normal basis multiplier. A potential drawback in their design is the fact that it requires in general O(n^2) logic gates. This is a severe problem in particular in cryptographic applications where $n \geq 1000$ is envisaged.

We will consider alternative architectures to the Massey-Omura multiplier. We will derive serial input / parallel output architectures which are analogous to standard polynomial basis multipliers. Circuits of this type have also been described in [3]. We will also examine architectures that use different basis representations. This approach offers the designer a further degree of freedom in choosing the most convenient multiplier. Empirical results, however, do not suggest that this will in general allow for a substantial reduction in complexity. Finally we combine dual basis and normal basis techniques. We will show that the duality of normal bases is equivalent to the symmetry of the logic array of our serial input / parallel output architecture.

2. Normal bases

We start with an informal definition of normal bases. For a detailed mathematical treatment of this subject the reader is referred to [1][4]. Let α be an element of $GF(2^n)$ so that $< \alpha, \alpha^2, \alpha^4, ..., \alpha^{2^{n-1}} >$ is a basis of $GF(2^n)$. This basis is called a normal basis. Squaring an element u in a normal basis is a cyclic shift of the coordinates $(u_0, ..., u_{n-1})$ because of

$$\left(\sum_{i=0}^{n-1} u_i \alpha^{2^i} \right)^2 = \sum_{i=0}^{n-1} u_i \alpha^{2^{i+1}} \quad \text{and} \quad \alpha^{2^n} = \alpha .$$

Let w be the product of u and v. Let w_{n-1} denote the highest significant bit of w, i.e.

$$w := u \cdot v \ , \ w_{n-1} =: f(u,v) .$$

Massey and Omura have observed that the function f that computes w_{n-1} can also be used to compute the remaining coefficients of w. As a matter of fact we have

$$w_{n-1-i} = (w^{2^i})_{n-1} = f(u^{2^i}, v^{2^i}) \quad \text{for } i=0,..,n-1.$$

Thus we obtain a parallel input / serial output multiplier as described in Fig.1 .

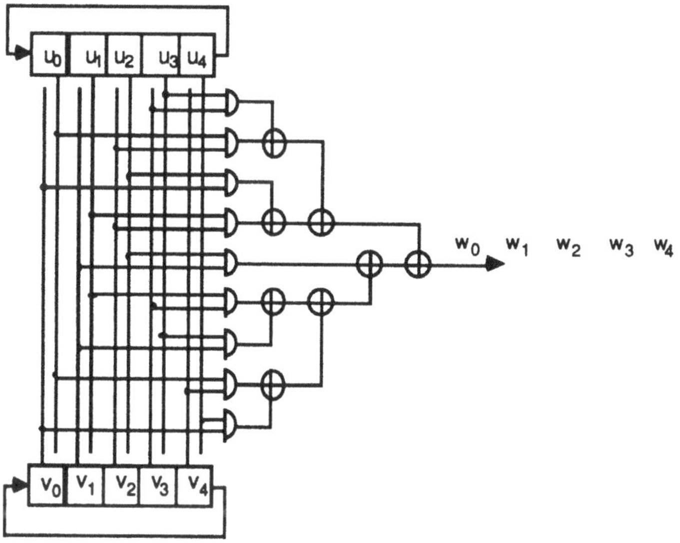

Fig.1. The Massey-Omura multiplier for $GF(2^5)$ with $p(x)=x^5+x^4+x^2+x+1$

3. Serial input / parallel output normal basis multipliers

In the Massey-Omura multiplier both inputs have to be entered at the start of the multiplication and the output is produced serially. In analogy to polynomial basis multipliers (Fig.2) we propose an architecture where one of the inputs is entered serially and the output bits are only available at the end of the computation. To this purpose the computation of u·v is decomposed so that we may enter the factor v either with its lowest significant bit first or with its highest significant bit first. Square and square root can of course be obtained by a cyclic shift. The second decomposition corresponds to Horner's rule for polynomial basis multiplication. The resulting circuit is given in Fig.3.

$$u \cdot v = u \cdot \sum_{i=0}^{n-1} v_i \alpha^{2^i} = v_0 u\alpha + v_1 (u^{1/2}\alpha)^2 + v_2 (u^{1/4}\alpha)^4 + \ldots + v_{n-1} (u^{1/2^{n-1}}\alpha)^{2^{n-1}}$$

$$= (\ldots((v_{n-1} u^{1/2^{n-1}} \alpha)^2 + v_{n-2} u^{1/2^{n-2}} \alpha)^2 \ldots + v_1 u^{1/2} \alpha)^2 + v_0 u\alpha \, .$$

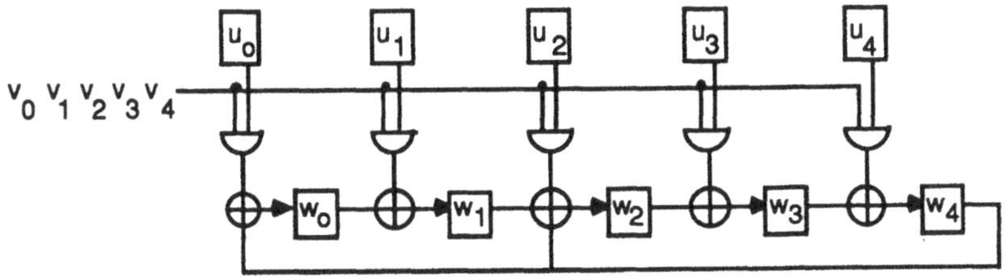

Fig.2. A polynomial basis multiplier for GF(2^5) with $p(x)=x^5+x^2+1$

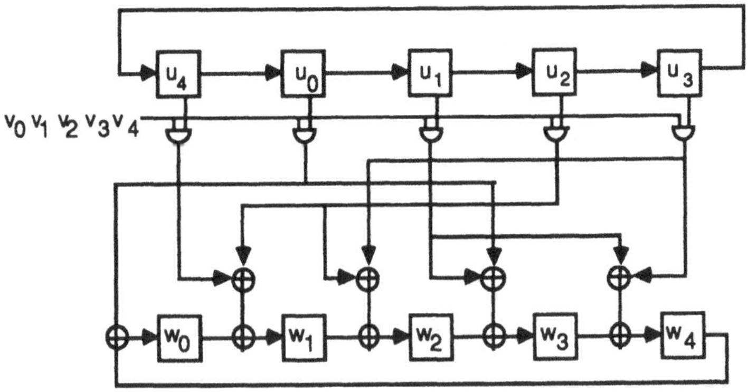

Fig.3. A serial input / parallel output normal basis multiplier for GF(2^5) with $p(x)=x^5+x^4+x^2+x+1$

In analogy to dual basis multipliers [1][3][6] we may also consider to use different basis representations for u and v. Multiplication by α can be represented as multiplication by a matrix $A := (a_{ij})$ defined by

$$\alpha^{2^j+1} = \sum_{i=0}^{n-1} a_{ij}\,\alpha^{2^i}\text{ . This yields } w = \sum_{i=0}^{n-1} v_i \cdot A^{2^i} \cdot u\,.$$

We transform u and w by multiplication with a regular matrix T into a representation in the basis $(\beta,\beta^2,\beta^4,\ldots,\beta^{2^{n-1}})$, i.e. $(\beta,\beta^2,\beta^4,\ldots,\beta^{2^{n-1}})T = (\alpha,\alpha^2,\alpha^4,\ldots,\alpha^{2^{n-1}})$. We get

$$Tw = \sum_{i=0}^{n-1} v_i \cdot (T \cdot A \cdot T^{-1})^{2^i} Tu\;.$$

Note that we have for TAT^{-1}

$$(\alpha\beta,\alpha\beta^2,\alpha\beta^4,\ldots,\alpha\beta^{2^{n-1}}) = (\beta,\beta^2,\beta^4,\ldots,\beta^{2^{n-1}})TAT^{-1}\,.$$

Alternatively we could represent v and w in $<\beta,\beta^2,\beta^4,\ldots,\beta^{2^{n-1}}>$ and u in $<\alpha,\alpha^2,\alpha^4,\ldots,\alpha^{2^{n-1}}>$. We define a new matrix A' by

$$(\beta\alpha,\beta\alpha^2,\beta\alpha^4,\ldots,\beta\alpha^{2^{n-1}}) = (\beta,\beta^2,\beta^4,\ldots,\beta^{2^{n-1}})A'\,.$$

We informally define the complexity of a normal basis multiplier to be the number of AND-gates in its logic array. The following theorem shows that a simple change from parallel input / serial output to serial input / parallel output architecture will not change this complexity. It also shows that there will be no difference between the two "two bases" multipliers we have proposed. We may, however, hope that a good choice of these two bases will reduce the complexity of the logic array.

Theorem 1: Let A, A', and T be defined as above and write the function f(u,v) of the Massey-Omura multiplier as $f(u,v) = u^t \cdot F \cdot v$. The following relations hold.

(1) $(A')_{ij} = (TAT^{-1})_{i-j,n-j}$
(2) $F_{ij} = A_{n-1-j,i-j}$

Indices are computed mod n.

Proof: Defining $TAT^{-1} =: (x_{ij})$ and $A' =: (z_{ij})$ we get

$$\beta\alpha^{2^j} = \sum_{i=0}^{n-1} z_{ij}\,\beta^{2^i} \quad \text{and} \quad \alpha\beta^{2^k} = \sum_{i=0}^{n-1} x_{ik}\,\beta^{2^i}.$$

From $\beta\alpha^{2^j} = (\alpha\beta^{2^{n-j}})^{2^j}$ we conclude $z_{ij} = x_{i-j,n-j}$.

To show (2) consider

$$u\cdot v = \left[\sum_{i=0}^{n-1} u_i\alpha^{2^i}\right]\left[\sum_{j=0}^{n-1} v_j\alpha^{2^j}\right] = \sum_{i=0}^{n-1}\sum_{j=0}^{n-1} u_i v_j\,\alpha^{2^i+2^j} = \sum_{i=0}^{n-1}\sum_{j=0}^{n-1} u_i v_j\left[\alpha^{2^{i-j}+1}\right]^{2^j} =$$

$$= \sum_{i=0}^{n-1}\sum_{j=0}^{n-1} u_i v_j \cdot \sum_{k=0}^{n-1} a_{k,i-j}\,\alpha^{2^{k+j}},$$

$$w_{n-1} = f(u,v) = \sum_{i=0}^{n-1}\sum_{j=0}^{n-1} u_i v_j a_{n-1-j,i-j} = \sum_{r=0}^{n-1}\sum_{s=0}^{n-1} u_{s+n-1-r}\,a_{rs}\,v_{n-1-r} = \sum_{r=0}^{n-1}\left[\sum_{s=0}^{n-1} a_{rs}u_{s+n-1-r}\right]v_{n-1-r} =$$

$$= \sum_{j=0}^{n-1}\left[\sum_{s=0}^{n-1} a_{n-1-j,s}u_{s+j}\right]v_j = \sum_{j=0}^{n-1}\left[\sum_{i=0}^{n-1} a_{n-1-j,i-j}u_i\right]v_j .$$

q.e.d.

A closer analysis reveals an intrinsic symmetry in the column structure of A. Let $\underline{a}_j$ denote the jth column of A, denote by $\underline{a}_j^{(i)}$ the cyclic shift of $\underline{a}_j$ so that the first component of $\underline{a}_j$ becomes the ith component of $\underline{a}_j^{(i)}$. We have

$$\alpha^{2^j+1} = \sum_{i=0}^{n-1} a_{ij}\alpha^{2^i} \quad \text{and} \quad \alpha^{2^j+1} = (\alpha^{2^{n-j}+1})^{2^j} = \sum_{i=0}^{n-1} a_{i,n-j}\alpha^{2^{i+j}} = \sum_{i=0}^{n-1} a_{i-j,n-j}\alpha^{2^i}$$

and thus $\underline{a}_j = \underline{a}_{n-j}^{(j)}$.

The multiplier of Fig.4 makes use of this regularity. Only the first n/2 columns of A have to be implemented, the inputs v_i have to be stored as they will be used again at later stages of the multiplication. The detailed algorithm is derived below.
Set t:= n/2 for n even or t:= (n+1)/2 for n odd. Let u·v also stand for the corresponding coefficient vector. We get

$$u\cdot v = \sum_{i=0}^{n-1} v_i (Au^{2^{-i}})^{2^i} = \sum_{i=0}^{n-1} v_i \Big(\sum_{j=0}^{n-1} u_{i+j}\,\underline{a}_j\Big)^{(i)} = \sum_{i=0}^{n-1} v_i \left[\Big(\sum_{j=0}^{t} u_{i+j}\,\underline{a}_j\Big)^{(i)} + \Big(\sum_{j=t+1}^{n-1} u_{i+j}\,\underline{a}_j\Big)^{(i)} \right]$$

With $\Big(\sum_{j=t+1}^{n-1} u_{i+j}\,\underline{a}_j\Big)^{(i)} = \Big(\sum_{j=t+1}^{n-1} u_{i+j}\,\underline{a}_{n-j}^{(j)}\Big)^{(i)} = \sum_{j=1}^{n-t-1} u_{i+n-j}\,\underline{a}_j^{(n-j+i)}$ we arrive at

$$u\cdot v = \sum_{k=0}^{n-1} \left[v_k \Big(\sum_{j=0}^{t} u_{k+j}\,\underline{a}_j\Big) + \sum_{j=1}^{n-t-1} v_{k+j} u_k \underline{a}_j \right]^{(k)} =$$

$$= \sum_{k=0}^{n-1} \left[v_k u_k \underline{a}_0 + \sum_{j=1}^{n-t-1} (v_k u_{k+j} + v_{k+j} u_k)\underline{a}_j + v_k u_{k+n/2}\,\underline{a}_{n/2} \right]^{(k)} .$$

The last term in the above sum will occur only for n even.

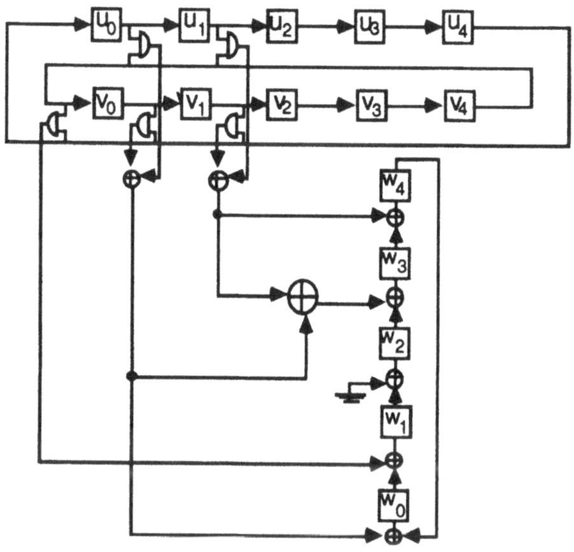

Fig.4. A modified normal basis multiplier for $GF(2^5)$ with $p(x)=x^5+x^4+x^2+x+1$

4. Symmetry and Duality

Results from our search through different normal bases and different combinations of normal bases for selected fields $GF(2^n)$ can be summarized as follows.

- The number of entries in A has its minimum when the normal basis is defined by a root of the polynomial

$$\sum_{i=0}^{n} x^i .$$

- In general between 40% and 60% of the entries in the matrices A and TAT^{-1} are ones.

- For some bases the matrix A or A´ or TAT^{-1} is symmetric.

The first result is already known [7], the second not too encouraging but the third deserves some closer attention. Before we come to our main result we have to define the notion of dual bases. For a detailed treatment the reader is again refered to [1][4]. Self dual normal bases have been studied in [2].

Let $< \alpha_0,\ldots,\alpha_{n-1} >$ be a basis of $GF(2^n)$. The dual basis $< \beta_0,\ldots,\beta_{n-1} >$ is defined by $tr(\alpha_i \cdot \beta_j) = \delta_{ij}$ (Kronecker´s delta), where tr is the trace function

$$tr(u) := \sum_{i=0}^{n-1} u^{2^i} .$$

A basis is called self dual when $tr(\alpha_i \cdot \alpha_j) = \delta_{ij}$ holds. We have for $u \in GF(2)$

$$u = \sum_{i=0}^{n-1} tr\,(u\alpha_i)\beta_i .$$

Theorem 2: Let $< \alpha,\alpha^2,\ldots,\alpha^{2^{n-1}} >$ and $< \beta,\beta^2,\ldots,\beta^{2^{n-1}} >$ be two normal bases. Let A , A´ and the transformation matrix T be defined as above. We have

(1) A´ is symmetric if and only if $< \alpha,\alpha^2,\ldots,\alpha^{2^{n-1}} >$ and $< \beta,\beta^2,\ldots,\beta^{2^{n-1}} >$ are dual.
(2) A is symmetric if and only if $< \alpha,\alpha^2,\ldots,\alpha^{2^{n-1}} >$ is self dual.
(3) TAT^{-1} is symmetric if and only if $< \beta,\beta^2,\ldots,\beta^{2^{n-1}} >$ is self dual.

Proof: Let $< \alpha,\alpha^2,\ldots,\alpha^{2^{n-1}}>$ and $< \beta,\beta^2,\ldots,\beta^{2^{n-1}}>$ be two dual normal bases. Define

$$\mathcal{A} := (\alpha^{2^{i+j}})_{i,j=0,..,n-1}\,,\ \mathcal{B} := (\beta^{2^{i+j}})_{i,j=0,..,n-1}\,.$$

For some $\gamma \in GF(2^n)$ let $\mathfrak{D}(\gamma)$ be a diagonal matrix with $(\mathfrak{D}(\gamma))_{ii} := \gamma^{2^i}$. We will make use of the following facts about the matrices $\mathcal{A}$, $\mathcal{B}$, A, A′, and TAT^{-1}.

- $(\mathcal{A}\mathcal{B})_{ij} = tr(\alpha^{2^i}\beta^{2^j}) \in GF(2)$
- $(\mathcal{A}\cdot\mathfrak{D}(\gamma)\cdot\mathcal{B})_{ij} = tr(\gamma\alpha^{2^i}\beta^{2^j})$
- $(\mathcal{A}^2)_{ii} = 1$
- $\mathcal{A}$ and $\mathcal{B}$ are nonsingular symmetric matrices, $\mathcal{A}^{-1}$ and $\mathcal{B}^{-1}$ are again symmetric.
- $\mathcal{A}$ and $\mathcal{B}$ are dual if and only if $\mathcal{A}\mathcal{B} = I$. I is the n×n identity matrix.
- $A = \mathcal{A}^{-1}\cdot\mathfrak{D}(\alpha)\cdot\mathcal{A}$
- $A' = \mathcal{B}^{-1}\cdot\mathfrak{D}(\beta)\cdot\mathcal{A}$
- $T^{-1}AT = \mathcal{B}^{-1}\cdot\mathfrak{D}(\alpha)\cdot\mathcal{B}$

If $< \alpha,\alpha^2,\ldots,\alpha^{2^{n-1}}>$ and $< \beta,\beta^2,\ldots,\beta^{2^{n-1}}>$ are dual we have $\mathcal{B}^{-1} = \mathcal{A}$ and thereby

$$(A')^t = (\mathcal{B}^{-1}\cdot\mathfrak{D}(\beta)\cdot\mathcal{A})^t = \mathcal{A}^t\cdot\mathfrak{D}(\beta)^t\cdot\mathcal{A}^t = \mathcal{B}^{-1}\cdot\mathfrak{D}(\beta)\cdot\mathcal{A} = A'.$$

Thus A′ is symmetric. In the same way we can prove that A and TAT^{-1} are symmetric when the bases $< \alpha,\alpha^2,\ldots,\alpha^{2^{n-1}}>$ and $< \beta,\beta^2,\ldots,\beta^{2^{n-1}}>$ respectively are self dual. Conversely assume that A′ is symmetric. We get

$$A' = \mathcal{B}^{-1}\cdot\mathfrak{D}(\beta)\cdot\mathcal{A} = \mathcal{A}\cdot\mathfrak{D}(\beta)\cdot\mathcal{B}^{-1} = (A')^t.$$

From $(\mathcal{A}\mathcal{B})^t = \mathcal{B}^t\cdot\mathcal{A}^t = \mathcal{B}\mathcal{A}$ we derive $(\mathcal{A}\mathcal{B})^t\cdot\mathfrak{D}(\beta) = \mathfrak{D}(\beta)\cdot\mathcal{A}\mathcal{B}$. Because of $(\mathcal{A}\mathcal{B})_{ij} = tr(\alpha^{2^i}\beta^{2^j}) \in GF(2)$ and $\beta^{2^i} \neq \beta^{2^i}$ the equation

$$(\mathcal{A}\mathcal{B})_{ji}\beta^{2^j} = (\mathcal{A}\mathcal{B})_{ij}\beta^{2^i}$$

implies that $\mathcal{A}\mathcal{B}$ is a diagonal matrix. $\mathfrak{D}(\beta) = (\mathcal{B}\mathcal{A})^{-1}\mathfrak{D}(\beta)\mathcal{A}\mathcal{B}$ demands that $\mathcal{A}\mathcal{B}$ is regular. Thereby we get $\mathcal{A}\mathcal{B} = I$ and $< \alpha,\alpha^2,\ldots,\alpha^{2^{n-1}}>$ and $< \beta,\beta^2,\ldots,\beta^{2^{n-1}}>$ are dual. This concludes the proof of part (1) of our theorem. Parts (2) and (3) follow in a similar way from the symmetry of A and TAT^{-1}.

q.e.d.

5.Conclusion

We have explored the feasability of efficient normal basis multipliers. Serial input / parallel output architectures are as efficient or inefficient as the Massey-Omura multiplier. Architectures that use different basis representations give the designer a further degree of freedom in choosing the most convenient multiplier. Empirical results, however, do not suggest that this will in general allow for a substantial reduction in complexity. The combination of dual basis and normal basis techniques does not yield the same efficient designs as in the case of the dual of a polynomial basis. Duality is mirrored in the symmetry of the logic array of our architectures.

Acknowledgement

This work was supported by DFG grant Be 887/3-1.

Literature

[1] Th.Beth, *On the Arithmetics of Galoisfields and the Like*, in Proc. of AAECC3, Springer LNCS 229, 1986

[2] Th.Beth, W.Fumy, R.Mühlfeld, *Zur algebraischen Diskreten Fourier-Transformation*, Archiv der Mathematik, Vol.40, pp.238-244, 1983

[3] W.Fumy, *Über orthogonale Transformationen und fehlerkorrigierende Codes*, Dissertation, Universität Erlangen, 1986

[4] R.Lidl, H.Niederreiter, *Finite Fields*, Cambridge University Press, 1984

[5] J.L.Massey, J.K.Omura, *Computational method and apparatus for finite field arithmetic*, U.S Patent application, submitted 1981

[6] R.McEliece, *Finite Fields for Copmputer Scientits and Engineers*, Kluwer, 1987

[7] C.C.Wang, et al.,*VLSI Architectures for Computing Multiplications and Inverses in GF(2^m)*, IEEE Trans. on Computers, Vol. C-34, No.8, pp.709-717, 1985

Multiple Error Correction with Analog Codes

Werner Henkel
Institut für Netzwerk- und Signaltheorie
Technische Hochschule Darmstadt
Merckstr. 25, D-6100 Darmstadt

1 Introduction

In 1983 J.K. Wolf wrote a paper titled "Redundancy, the Discrete Fourier Transform, and Impulse Noise Cancellation" [1], where he published some ideas on using RS- or BCH-codes based on complex numbers, later introduced as *"Analog Codes"* or *"DFT-Codes"* [2]. The paper treated a simple example for correction of single errors in noise, where the single error, called 'burst', had a much higher amplitude than the noise samples.

In this contribution, results on correcting multiple errors (bursts) are given, which analyse the influence of additive equally distributed or Gaussian noise on the error (burst) correction process. But firstly the expected benefits of coding with complex numbers are explained using a new idea of defining a syndrome by applying the divided difference scheme of the Newton method of interpolation. After giving some simulation results on the effect of additive noise, the sensitivity of the 'error search' is studied, which means, that a relation between relative errors (noise) in the time-domain codeword and relative errors in the inverse transform of the error locator polynomial is derived. Furthermore, it is shown, that the solutions of the Toeplitz subsystems which are generated during the execution of the recursive Berlekamp-Massey-Algorithm yield bounds for the condition numbers of the corresponding submatrices. These bounds are determined using the fact that the BMA leads to a triangular 'square-root' factorization of the inverse Toeplitz submatrices similar to that achieved by the Cholesky method for symmetric matrices.

2 Benefits of coding with complex numbers

In [1] a proof is given, that (almost always) *Analog Codes* are capable of correcting a number of errors equal to $M-1$, where M equals the number of parity samples.

Here, another proof is given based on the Newton-method of interpolation, also called interpolation by divided differences. This proof leads to a new way of defining a syndrome that reveals the special feature that an error-free range of the codeword can be detected from the syndrome without any further operations.

An analog RS-codeword can be defined as samples x_i of a polynomial $X(\xi)$ of the form

$$x_i = X(z^{-i}) = \frac{1}{N}\sum_{k=0}^{K-1} X_k \cdot (z^{-i})^k\,, \quad z = e^{j\frac{2\pi}{N}}\,, \quad i = 0,\dots,N-1\,, \tag{1}$$

which has a degree of $K-1$.

The divided differences of the Newton-method are given by

$$\Delta^1 x_i = \frac{x_{i+1} - x_i}{z^{-(i+1)} - z^{-i}} \tag{2}$$

$$\Delta^k x_i = \frac{\Delta^{k-1}x_{i+1} - \Delta^{k-1}x_i}{z^{-(i+k)} - z^{-i}}\,. \tag{3}$$

Analogous to the derivative of polynomials they have the property that

$$\Delta^K x_i = 0\,, \tag{4}$$

if $K-1$ is the degree of the polynomial (see e.g. [3]). With random errors of continuous value, it is nearly impossible that an erroneous codeword would be given according to a polynomial of degree $K-1$. It follows that a zero K'th divided difference is equivalent to a region of $K+1$ correct samples (see figure 1). Applying $\binom{N}{K+1}$ permutations at most, one can be sure that a zero appears at a special location in the syndrome of K'th divided differences if any $K+1 = N-M+1$ samples are error-free. This proves the error correction capability of analog codes and presents an interesting way of defining a syndrome. In the case of maximum number of errors $e_{max} = M-1$ a portion of $Ne_{max}!(N-e_{max})!/N! = N/\binom{N}{e_{max}}$ of all permutations yield the desired zero K'th divided difference.

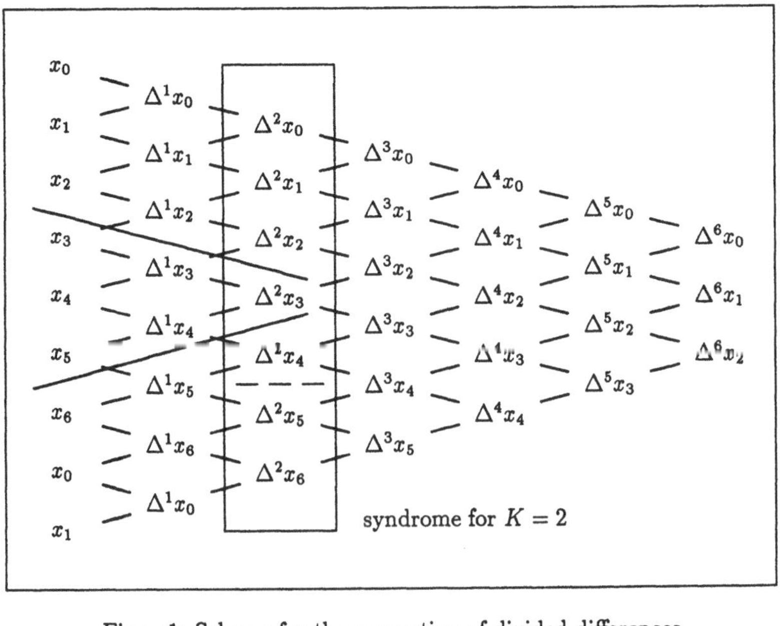

Figure1: Scheme for the generation of divided differences

Unfortunately, up to now, no algorithm of practical significance has been found to profit from that theoretically higher error correction capability. The 'new' syndrome may be a step into the right direction. Besides the fact that more errors (bursts) may be correctable, *"Analog Codes"* have the property to be tolerant of background noise, if its amplitude is considerably lower than that of the 'bursts' to be corrected. There is no analogue for codes over Galois fields, because of the lack of magnitude relations in such finite fields.

This paper is especially devoted to the examination of the influence of additional noise in determining the error locations (solution of the key equation) by usual, algebraic methods. A treatment of the calculation of the error magnitudes has been omitted, because it is only an application of well-known interpolation or approximation methods, e.g. Lagrange-Interpolation, Newton-Method, Forney-Algorithm, recursion using the key equation, or least-squares approximation.

To start, computer simulations are studied to give a first impression.

3 Some simulation results

Simultion results are given, showing the influence of noise on the solution of the key-equation or the corresponding Toeplitz system. Firstly, we quote some results for the case of additive background noise with equally distributed amplitude (in an interval) and phase. Gaussian noise will be treated afterwards. The 'bursts' are restricted to be of constant amplitude and random, equally distributed phase.

Ideally, the 'bursts' are to be located by zeros in the coefficients of the inverse transform $c(\xi)$ of the error locator polynomial $C(\xi)$. With additional noise, no exact zeros would appear in that time-domain error locator. The minima of the samples will then represent the 'burst'-locations, which is far more practical than determining the exact complex zeros. So the quotient

$$Q = \frac{Min\{|c_i|_{\text{error-free locations}}\}}{Max\{|c_i|_{\text{error locations}}\}} \tag{5}$$

represents a measure for the quality of error detection. Figure 2 shows the dependence of this measure on the maximal amplitude of the noise (divided by the 'burst' amplitude) and the number of errors e. A hyperbolic relation becomes obvious. Note that for $Q > 1$ all e errors can be found and no error-free samples are taken for being erroneous.

Results for additive normally distributed noise are presented in figure 3. The portion of correctable error pattern (by solving the key equation) is shown as function of the standard deviation. Upper bounds are added, representing the probability of being able to distinguish 'bursts' from the noise. This probability is given by

$$P_d = P(\min\{\mathbf{z}_i | i \in U\} > \max\{\mathbf{z}_i | i \in \overline{U}\}), \tag{6}$$

where U symbolizes the set of indices of the 'burst'-locations, $\overline{U}$ the one without 'bursts', and $\mathbf{z}_i$ the random variables at location i. For illustration see figure 4. Omitting some steps of the

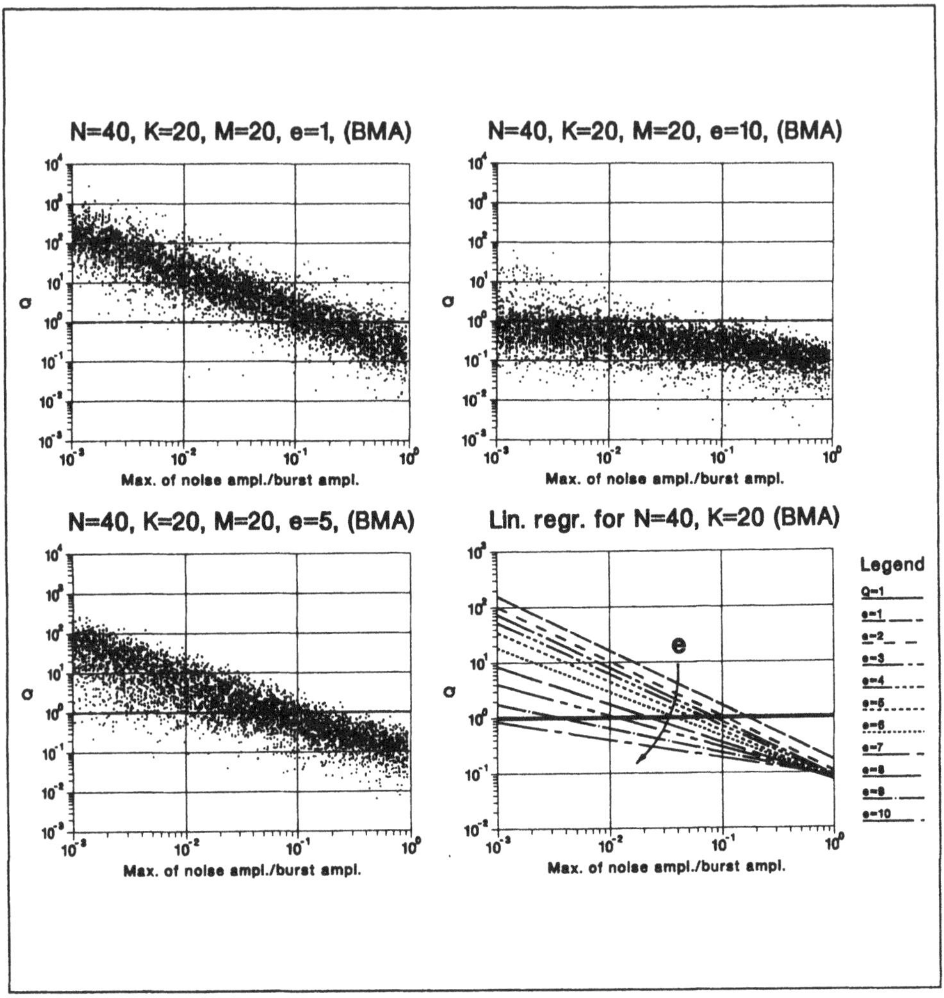

Figure2: The influence of equally distributed noise on determining the error locations

derivation, the probability follows as

$$P_d = \int_{u=-\infty}^{\infty} \left(F_{z_{\overline{U}}}(u)\right)^{N-e} \cdot e \cdot (1 - F_{z_U}(u))^{e-1} \cdot f_{z_U}(u)\, du \tag{7}$$

$$F_{z_U}(u) = \frac{1}{2}\left(1 + \mathrm{erf}\left(\frac{u-1}{\sqrt{2}\sigma}\right)\right) \tag{8}$$

$$F_{z_{\overline{U}}}(u) = \frac{1}{2}\left(1 + \mathrm{erf}\left(\frac{u}{\sqrt{2}\sigma}\right)\right) \tag{9}$$

$$f_{z_U}(u) = \frac{1}{\sqrt{2\pi}\sigma} e^{-(u-1)^2/(2\sigma^2)} \tag{10}$$

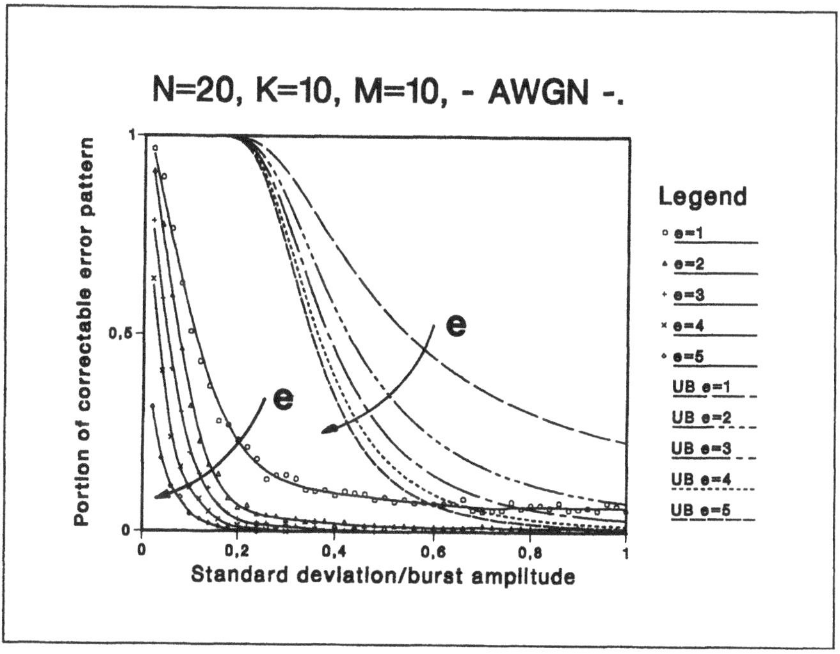

Figure3: Portion of correctable error pattern under Gaussian noise

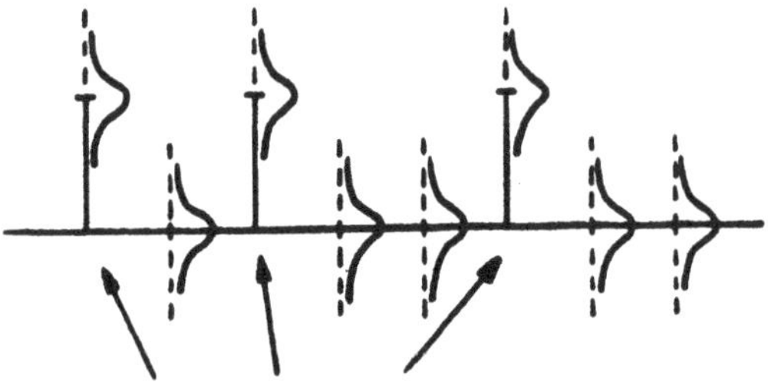

Figure4: Illustration for derivation of upper bounds in (6)

4 The sensitivity of the 'error search'

The 'error search' consists of three main operations:

I Transform of the time-domain codeword ($\vec{f}$) into the frequency domain ($\vec{F}$)
- *DFT* -

$$\vec{F} = \underline{Z}_{N\times N}\vec{f} \qquad \underline{Z}_{N\times N} = \begin{pmatrix} 1 & 1 & 1 & \cdots \\ 1 & z^1 & z^2 & \\ 1 & z^2 & z^4 & \\ \vdots & & & \end{pmatrix} \tag{11}$$

II Extraction of the syndrome $(S_0, S_1, \ldots, S_{2e-1})$ from $\vec{F}$ and solution of the Hankel (Toeplitz) system

$$\begin{pmatrix} S_{2e-2} & S_{2e-3} & \cdots & S_{e-1} \\ S_{2e-3} & & & \vdots \\ \vdots & & & \vdots \\ S_{e-1} & \cdots & \cdots & S_0 \end{pmatrix} \begin{pmatrix} C_1 \\ C_2 \\ \vdots \\ C_e \end{pmatrix} = - \begin{pmatrix} S_{2e-1} \\ S_{2e-2} \\ \vdots \\ S_e \end{pmatrix} . \tag{12}$$

In abbreviated form

$$\underline{S} \cdot \vec{C} = -\vec{s}$$

III Inverse transform of the error locator
- DFT^{-1} -

$$\vec{c} = \underline{Z}^{-1}_{N\times e}\vec{C} \qquad \underline{Z}^{-1}_{N\times e} = \frac{1}{N} \left.\overbrace{\begin{pmatrix} 1 & 1 & 1 & \cdots & 1 \\ 1 & z^{-1} & z^{-2} & & \\ 1 & z^{-2} & z^{-4} & & \\ \vdots & & & & \\ \vdots & & & & \end{pmatrix}}^{e}\right\} N \tag{13}$$

These three steps will now be studied to achieve an approximate relationship for the relative error $\rho_{\vec{c}}$ of the error locator $\vec{c}$ as function of the relative error $\rho_{\vec{f}}$ of the time-domain codeword $\vec{f}$ ($\rho_{\vec{c}} = \| \Delta\vec{c} \|_\infty / \| \vec{c} \|_\infty$, $\rho_{\vec{f}} = \| \Delta\vec{f} \|_\infty / \| \vec{f} \|_\infty$).

For the DFT-Matrices we obtain

$$\| \underline{Z}_{N\times N} \|_\infty = N\,, \quad \| \underline{Z}^{-1}_{N\times N} \|_\infty = \frac{N}{N} = 1\,,$$

$$\| \underline{Z}^{-1}_{N\times e} \|_\infty = \frac{e}{N}\,, \quad \| \underline{Z}_{e\times N} \|_\infty = N\,. \tag{14}$$

First step

If $\epsilon\vec{e}$ represents the noise vector and the index i stands for the 'bursty' vectors without background noise, it follows that

$$\vec{F} = \underline{Z}_{N\times N}\vec{f} = \underline{Z}_{N\times N}(\vec{f}_i + \epsilon\vec{e}) \tag{15}$$

$$\Longrightarrow \Delta\vec{F} = \underline{Z}_{N\times N} \cdot \epsilon\vec{e} \tag{16}$$

$$\Longrightarrow \| \Delta\vec{F} \| \leq \| \underline{Z}_{N\times N} \| \cdot \epsilon \cdot \| \vec{e} \|\,, \tag{17}$$

Furthermore,

$$\vec{f}_i = \underline{Z}^{-1}_{N\times N}\vec{F}_i \tag{18}$$

$$\Longrightarrow \| \vec{f}_i \| \leq \| \underline{Z}^{-1}_{N\times N} \| \cdot \| \vec{F}_i \| \quad \Longrightarrow \quad \frac{\| \underline{Z}^{-1}_{N\times N} \| \cdot \| \vec{F}_i \|}{\| \vec{f}_i \|} \geq 1 \tag{19}$$

Combining relations (17) and (19) we get

$$\| \Delta\vec{F}_i \| \leq \| \underline{Z}_{N\times N} \| \cdot \epsilon \cdot \| \vec{e} \| \frac{\| \underline{Z}^{-1}_{N\times N} \| \cdot \| \vec{F}_i \|}{\| \vec{f}_i \|} \tag{20}$$

$$\Longrightarrow \frac{\| \Delta\vec{F} \|}{\| \vec{F}_i \|} \leq \underbrace{\| \underline{Z}_{N\times N} \| \cdot \| \underline{Z}^{-1}_{N\times N} \|}_{\kappa\{\underline{Z}_{N\times N}\}} \cdot \frac{\epsilon \| \vec{e} \|}{\| \vec{f}_i \|}\,, \tag{21}$$

which means

$$\rho_{\vec{F}} \leq \kappa\{\underline{Z}_{N\times N}\} \cdot \rho_{\vec{f}}\,, \tag{22}$$

where $\kappa\{\underline{Z}_{N\times N}\}$ is the condition number of $\underline{Z}_{N\times N}$.

Second step

Regarding the Hankel (Toeplitz) system

$$\underline{S} \cdot \vec{C} = -\vec{s}\,, \tag{23}$$

we obtain according to [4]

$$\frac{\| \Delta\vec{C} \|}{\| \vec{C}_i \|} \leq \underbrace{\| \underline{S} \| \cdot \| \underline{S}^{-1} \|}_{\kappa\{\underline{S}\}} \cdot (\rho_A + \rho_B) \ \left(+O(\epsilon^2)\right) \ , \tag{24}$$

with

$$\rho_A = \frac{\| \Delta\underline{S} \|}{\| \underline{S}_i \|} \quad \text{and} \quad \rho_B = \frac{\| \Delta\vec{s} \|}{\| \vec{s}_i \|}$$

Using the L_∞-Norm and taking into account that $\vec{s}$ is just a section of $\vec{F}$ we obtain

$$\rho_B = \frac{\| \Delta\vec{s} \|_\infty}{\| \vec{s}_i \|_\infty} = \frac{\| \Delta\vec{s} \|_\infty}{\| \vec{F}_i \|_\infty} \cdot \frac{\| \vec{F}_i \|_\infty}{\| \vec{s}_i \|_\infty} \leq \rho_{\vec{F}} \cdot \frac{\| \vec{F}_i \|_\infty}{\| \vec{s}_i \|_\infty} . \tag{25}$$

Similar considerations for ρ_A yield

$$\rho_A = \frac{\| \Delta\underline{S} \|_\infty}{\| \underline{S}_i \|_\infty} \leq e \cdot \rho_{\vec{F}} \cdot \frac{\| \vec{F}_i \|_\infty}{\| \underline{S}_i \|_\infty} . \tag{26}$$

Third step

Analogous to the first step, the inverse Transform of the error locator

$$\vec{c} = \underline{Z}^{-1}_{N\times e}\vec{C} \tag{27}$$

leads to

$$\frac{\| \Delta\vec{c} \|}{\| \vec{c}_i \|} \leq \| \underline{Z}^{-1}_{N\times e} \| \cdot \| \underline{Z}_{e\times N} \| \cdot \frac{\| \Delta\vec{C} \|}{\| \vec{C}_i \|} . \tag{28}$$

Combining the results already obtained, the relative error $\rho_{\vec{c}} = \frac{\|\Delta\vec{c}\|_\infty}{\|\vec{c}_i\|_\infty}$ of the error locator in the 'time-domain' is given by (all norms are L_∞ norms)

$$\frac{\| \Delta\vec{c} \|}{\| \vec{c}_i \|} \leq \| \underline{Z}^{-1}_{N\times e} \| \cdot \| \underline{Z}_{e\times N} \| \cdot \| \underline{S} \| \cdot \| \underline{S}^{-1} \| \cdot \left(\frac{\| \vec{F}_i \|}{\| \vec{s}_i \|} + e\frac{\| \vec{F}_i \|}{\| \underline{S}_i \|} \right) \cdot \tag{29}$$

$$\cdot \| \underline{Z}_{N\times N} \| \cdot \| \underline{Z}^{-1}_{N\times N} \| \cdot \frac{\epsilon \| \vec{e} \|}{\| \vec{f}_i \|} =$$

$$= \frac{e}{N} \cdot N \cdot \kappa\{\underline{S}\} \cdot (\ldots) \cdot N \cdot 1 \cdot \frac{\epsilon \| \vec{e} \|}{\| \vec{f}_i \|} .$$

$$\Longrightarrow \rho_{\vec{c}} \leq e \cdot N \cdot \kappa\{\underline{S}\} \cdot \left(\frac{\| \vec{F}_i \|_\infty}{\| \vec{s}_i \|_\infty} + e\frac{\| \vec{F}_i \|_\infty}{\| \underline{S}_i \|_\infty} \right) \cdot \rho_{\vec{f}} \tag{30}$$

Knowing that bounds obtained from condition considerations are gross overestimations in general, (30) could be approximated by

$$\boxed{\rho_{\vec{c}} \leq 2eN \cdot \kappa\{\underline{S}\} \cdot \rho_{\vec{f}}} \tag{31}$$

(Simulations showed this approximated bound to be a gross one, too.)

5 Bounds for the condition numbers of Hankel (Toeplitz) submatrices used during execution of Berlekamp's algorithm (BMA)

The last section showed that the bound for the relative error $\rho_{\vec{c}}$ of the time-domain error locator is especially dependent on the condition number $\kappa\{\underline{S}\}$ of the syndrome matrix. Now bounds for $\kappa\{\underline{S}\}$ are derived using parameters that are directly available during execution of Berlekamp's recursive algorithm ([5]). For this, the fact is used that BMA leads to a triangular 'square root'-factorization.

For *"Analog Codes"* the usual BMA has to be modified slightly. The check on discrepancy equal or not equal to zero has to be replaced by a threshold decision like '**IF d < d₀ THEN**' with an appropriately chosen d_0. Another possibility is to skip that operation totally, which means setting d unequal to zero always. In that case, every second recursion, the BMA yields a system of linear equations (before each length change) of the form

$$(1, C_1, C_2, \ldots, C_l)\begin{pmatrix} S_l & & S_{2l} \\ & \ddots & \\ S_0 & & S_l \end{pmatrix} = (0, \ldots, 0, d_{2l}) \tag{32}$$

$$\Longleftrightarrow \underbrace{\begin{pmatrix} S_{2l} & S_l \\ & \\ S_l & S_0 \end{pmatrix}}_{=:\underline{S}_l}\begin{pmatrix} 1 \\ C_1 \\ \vdots \\ C_l \end{pmatrix} = \begin{pmatrix} d_{2l} \\ 0 \\ \vdots \\ 0 \end{pmatrix}, \tag{33}$$

where d_{2l} is a discrepancy before a length change and the vector $(1, C_1, \ldots, C_l)$ consists of the corresponding coefficients of the error locator polynomial, we obtain

$$(S_e)\cdot \underbrace{\begin{pmatrix} 1 & 0 & 0 & 0 \\ C_{e,1} & 1 & 0 & 0 \\ \vdots & & \ddots & 0 \\ C_{e,e} & C_{e-1,e-1} & \cdots & 1 \end{pmatrix}}_{=:\underline{C}_e} = \begin{pmatrix} d_{2e} & \cdot & \cdot & \cdot \\ 0 & \ddots & \cdot & \cdot \\ 0 & 0 & d_2 & \cdot \\ 0 & 0 & 0 & d_0 \end{pmatrix}. \tag{34}$$

$$\Longrightarrow \underline{C}_e^T \cdot \underline{S}_e \cdot \underline{C}_e = \begin{pmatrix} d_{2e} & 0 & 0 & 0 \\ 0 & \ddots & 0 & 0 \\ 0 & 0 & d_2 & 0 \\ 0 & 0 & 0 & d_0 \end{pmatrix} \tag{35}$$

The promised factorization follows:

$$\boxed{\underline{S}_e^{-1} = \underline{C}_e \underline{D}_e^{-1} \underline{C}_e^T \quad \text{or} \quad \underline{S}_e = (\underline{C}_e^T)^{-1} \underline{D}_e \underline{C}_e^{-1}} \tag{36}$$

$$\underline{C}_e = \begin{pmatrix} 1 & 0 & 0 & 0 \\ C_{e,1} & 1 & 0 & 0 \\ \vdots & & \ddots & 0 \\ C_{e,e} & C_{e-1,e-1} & \cdots & 1 \end{pmatrix} \quad \underline{C}_e^T = \begin{pmatrix} 1 & C_{e,1} & \cdots & C_{e,e} \\ 0 & 1 & & \vdots \\ 0 & 0 & \ddots & \vdots \\ 0 & 0 & 0 & 1 \end{pmatrix} \tag{37}$$

$$\underline{D}_e = \begin{pmatrix} d_{2e} & 0 & 0 & 0 \\ 0 & \ddots & 0 & 0 \\ 0 & 0 & d_2 & 0 \\ 0 & 0 & 0 & d_0 \end{pmatrix} \quad \underline{D}_e^{-1} = \begin{pmatrix} 1/d_{2e} & 0 & 0 & 0 \\ 0 & \ddots & 0 & 0 \\ 0 & 0 & 1/d_2 & 0 \\ 0 & 0 & 0 & 1/d_0 \end{pmatrix}, \tag{38}$$

where the d_{2l} are the discrepancies before the last length change and the columns of $\underline{C}_e$ are the corresponding coefficients of the error locator polynomial.

An interesting point is that the Cholesky-method achieves a similar triangular 'square-root' factorization.

By using relation (36), bounds for $\| \underline{S}_e^{-1} \|_\infty$, $\| \underline{S}_e \|_\infty$ and $\kappa\{\underline{S}_e\}$ are given by:

$$\| \underline{S}_e^{-1} \|_\infty \geq \frac{1}{|d_{2e}|}(1 + |C_{e,1}| + \cdots + |C_{e,e}|) \tag{39}$$

$$\| \underline{S}_e \|_\infty \geq |d_0| \tag{40}$$

$$\kappa\{\underline{S}_e\} \geq \frac{|d_0|}{|d_{2e}|}(1 + |C_{e,1}| + \cdots + |C_{e,e}|) \tag{41}$$

and likewise

$$\kappa\{\underline{S}_e\} \leq \kappa\{\underline{C}_e\}\kappa\{\underline{C}_e^T\}\frac{Max\,|d_{2i}|}{Min\,|d_{2i}|} \tag{42}$$

For this derivation of bounds only parameters are used, that are directly available from BMA.

6 Conclusions

After pointing out the expected advantages of complex coding compared to usual RS- or BCH-codes over finite fields, it has been shown that *"Analog Codes"* are able to correct multiple errors (bursts) also if additional background noise is superimposed. Simulations made obvious that the amplitude of the noise has to be of considerably lower amplitude than the 'bursts' to be corrected. Furthermore, it has been stated that intermediate solutions during execution of the recursive Berlekamp Massey-Algorithm are not meaningless but represent a measure for the conditioning of the corresponding sub-Toeplitz system.

References

[1] Wolf, J.K., "Redundancy, the Discrete Fourier Transform, and Impulse Noise Cancellation", *IEEE Trans. on Comm.*, vol. COM-31, No. 3, pp. 458-461, March 1983.

[2] Wolf, J.K., "Analog Codes", *IEEE Int. Conf. on Comm. (ICC '83)*, Boston, MA, USA, 19-22 June 1983, pp. 310-12 vol. 1.

[3] Hildebrandt, F.B., *Introduction to numercal analysis*, McGraw-Hill, New York, Toronto, London, 1956.

[4] Golub, G.H., van Loan, C.F., *Matrix Computations*, North Oxford Academic Publishing, Oxford, 1983

[5] Massey, J.L., "Shift-Register Synthesis and BCH Decoding", *IEEE Trans. on Inf. Theory*, vol. IT-15, No. 1, pp. 122-127, January 1969.

[6] Marshall, T.G., "Real number transform and convolutional codes", in *Proc. 24th Midwest Symp. Circuits Syst.*, S. Karne, Ed., Albuquerque, NM, June 29-30, 1981.

[7] Maekawa, Y., Sakaniwa, K., "An Extension of DFT Code and the Evaluation of its Performance", *Int. Symp. on Information Theory*, Brighton, England June 24-28, 1985.

[8] Cybenko,G, "The numerical stability of the Levinson-Durbin Algorithm for Toeplitz systems of equations", *SIAM J. Sci. Stat. Comput.*, vol. 1, No. 3, pp. 303-319, September 1980.

On the Complexity of Satisfiability Problems for Algebraic Structures (Preliminary Report)

H.B. Hunt III[1]
R.E. Stearns[2]

Computer Science Department
State University of New York at Albany
Albany, New York 12222
U.S.A.

Abstract

For several different algebraic structures S, we study the computational complexity of such problems as determining, for a system of equations on S,

1. if the system has a solution,
2. if the system has a unique solution, and
3. the number of solutions of the system.

Three types of results are obtained. For rings S, we show that these problems are either **NP**- or **coNP**-hard, when S is unitary, and in addition, are **SAT**-hard(npoylogn,n), when S is commutative and unitary. We also show that determining the number of solutions of a system of equations on S is **#P**-complete, when S is finite and unitary, and in addition, is **#SAT**-complete(npolylogn,n), when S is finite, commutative, and unitary. Second for all ordered unitary rings S, we show that determining if a system of equations on S has a unique solution is intuitively "as hard as" determining if a system of equations on S has a solution. Third, we characterize the complexity of these problems for additional algebraic structures including each finite or finite-depth lattice.

1. Introduction

For several algebraic structures S, we study the computational complexities of the problems **SAT**(S), **NONZERO_SAT**(S), {0,1}_**SAT**(S), **UNIQUE_SAT**(S), and **NUMBER_SAT**(S) defined as follows.

1. **SAT**(S) is the problem of determining if a system of equations on S has a solution on S.
2. **NONZERO_SAT**(S) is the problem of determining if a system of equations on S has a solution on S, where not all values taken on by the variables are zero.
3. {0,1}_**SAT**(S) is the problem of determining if a system of equations on S has a solution on S, where all variables take on the values 0 or 1.
4. **UNIQUE_SAT**(S) is the problem of determining if a system of equations on S has a unique solution on S.

---

[1] This research was supported in part by NSF Grant No. DCR 86-03184.

[2] This research was supported in part by NSF Grant No. DCR 83-03932.

5. **NUMBER_SAT(S)** is the problem of determining the number of solutions of a system of equations on S.

In the case that S is the field of real numbers, no efficient (i.e. deterministic polynomial time) algorithm is known to solve the problem **UNIQUE_SAT(S)**. In the case that S is the ring of integers, the problem **SAT(S)** is known to be undecidable by an easy reduction of Hilbert's Tenth Problem [Da,Ma] In the case that S is a field of characteristic two, the problem **SAT(S)** is known to be **NP**-hard [FY]. Finally, we showed in [HS1] that the problem **SAT(S)** is **NP**-hard, whenever S is a unitary ring.

Here, we extend our earlier results for **SAT(S)** in [HS1] in the following three directions. First, we show that each of the problems 1-4 above is either **NP**- or **coNP**-hard for S, whenever S is a unitary ring, and moreover, is **SAT**-hard(npolylogn,n) (defined below) when S is a commutative unitary ring. For finite rings S, we also show that problem 5 above is **#P**-complete, whenever S is unitary, and moreover, is **#SAT**-complete(npolylogn,n) (again, defined below), whenever S is commutative and unitary. Second, we show that the problem **SAT(S)** is deterministic linear time reducible to the problem **UNIQUE_SAT(S)**, whenever S is an ordered unitary ring; and we show that both problems are deterministic linear time interreducible, whenever S is an ordered field. This shows intuitively, for all ordered unitary rings S, that the problem **UNIQUE_SAT(S)** is "as hard as" the problem **SAT(S)**, whenever S is an ordered unitary ring. Thus in particular, the problem **UNIQUE_SAT(S)** is undecidable, when S is the ring of integers. Third, we present analogous complexity results for the above five problems for a number of additional algebraic structures including all (not necessarily distributive) finite or finite-depth lattices and many different semirings including all positive and all idempotent semirings [Zi]. Here due to size limitations we stress our new results for rings and for lattices. A fuller development of these results will appear in [HS3]. Selected proof sketches of some of our results appear in the appendix. Throughout we restrict our attention to structures with domains of cardinality ≥ 2.

2. An Observation on Intractability

It is often assumed that a problem is computationally "tractable" only if it has a deterministic polynomial time algorithm [Co, Ka, AHU, GJ]. It is also often assumed that no **NP**- or **coNP**-hard problem is "tractable". These assumptions are actually false, even if

$*$ **P** $\neq$ **NP**, and moreover, the satisfiability problem for 3CNF formulas (henceforth, abbreviated by **SAT**) requires deterministic time $2^{\Omega(n/\text{polylog}n)}$.

To see how easy particular **NP**-complete problems can be, even assuming $*$, one need only make two easy observations. The first observation is that NP-complete problems of deterministic time complexity $2^{O(n^{\epsilon})}$ exist, for all $\epsilon > 0$. Actually for all integers $k > 1$, there is an **NP**-complete problem that is provably solvable in deterministic time $O(n\log n) + O(2^{n^{1/k}} \cdot n^{1/k})$. The second observation is that many of these problems can be solved on today's computers for a significant ranges of inputs. In fact, slow-growing exponentials can be bounded above by slow-growing polynomially-bounded functions, for all inputs of subastronomical size. For example, $2^{n^{1/10}} \cdot n^{1/10} < 18n\log n$, for all $n < 2^{60}$. Problems whose deterministic time complexities are bounded above by such slow-growing exponential functions must be viewed as being effectively in **P**.

One might be tempted to conjecture, again assuming $*$, that the "natural" **NP**-complete problems in the literature require deterministic time similar to that assumed for **SAT** in $*$. Again, this assumption is false. In [LT,SH,RH] it is shown that a number of **NP**- or **coNP**-complete problems in the literature are solvable in deterministic time $2^{O(n^{1/2})}$, not just in deterministic time $2^{O(n)}$. These "easier" **NP**- or **coNP**-complete problems include the CLIQUE, SET PARTITION, KNAPSACK, RULER-FOLDING, PLANAR SAT, PLANAR 3DM, PLANAR HAMILTONIAN, and PLANAR TRAVELING SALESMAN Problems. These results suggest that even assuming $*$, a deeper analysis of the relative deterministic time complexities of individual **NP**- and **coNP**-complete problems is necessary in order to infer their practical computational intractability.

The usual method of proving that a problem B is **NP**-complete is to find a deterministic polynomial time reduction of some known **NP**-complete problem A to it. The emphasis in the literature has been placed on the deterministic time complexity of the reduction. Actually, the *size* of the reduction, discussed below, is more important in evaluating the value of the reduction as "evidence of practical computational intractability".

Definition 2.1. Let R be a reduction from set A to set B, which takes inputs w and produces outputs R(w) such that $w \in A$ iff $R(w) \in B$ [iff $R(w) \notin B$]. We say that R has *size complexity* $O(L(n))$, $\Omega(L(n))$, and $\theta(L(n))$ iff $||R(w)||$ is $O(L(||w||))$, $\Omega(L(||w||))$, or $\theta(L(||w||))$, respectively, where $||w||$ and $||R(w)||$ are the numbers of occurrences of symbols in w and R(w), respectively. In this paper, we are primarily concerned with the *sizes* of a formula and of a system of equations on an algebraic structure **S**. The *size* of a formula F on **S**, denoted $||F||$, is the number of occurrences of variables, constants, operators, or parentheses in F. The *size* of a system **E** of equations $f_i = g_i$ $(1 \le i \le m)$ on **S**, denoted $||\mathbf{E}||$, equals $\sum_{i=1}^{m} (||f_i|| + ||g_i|| + 1)$.

Let $k > 1$ and R be a deterministic polynomial time reduction of **SAT** to a language L such that R is of size complexity $\theta(n^k)$. Then even assuming $*$, the *most* that can be inferred about B is that B requires deterministic time $2^{\Omega(n^{1/k})}$. This observation leads us to introduce the concept of "t(n) time and s(n) size reducibility" and of **SAT**-hardness and **SAT**-completeness "modulo t(n) time and s(n) size reducibility".

Let Σ and Δ be finite alphabets, and let $L \subset \Sigma^*$ and $M \subset \Delta^*$ be languages.

Definition 2.2. We say that L is *t(n) time and s(n) size reducible to* M, if there is a function F computable by an O(t(n)) time-bounded deterministic Turing machine such that

1. either, for all $x \in \Sigma^*$, $x \in L$ iff $F(x) \in M$ or, for all $x \in \Sigma^*$, $x \in L$ iff $F(x) \in \Delta^* - M$, and
2. there exists c>0 such that, for all $x \in \Sigma^*$, $||F(x)|| \le s(||x||)$.

Note that condition 1 of Definition 2.2 is nonstandard. Our motivation for this is that the deterministic complexities of a set and of its complement are equal.

Definition 2.3. We say that L is **SAT**-*hard modulo* t(n) *time and* s(n) *size reducibility*, written "L is **SAT**-hard (t(n),s(n))", if **SAT** is t(n) time and s(n) size reducible to L. We say that L is **SAT**-*complete modulo* t(n) *time and* s(n) *size reducibility*, written "L is **SAT**-complete (t(n),s(n))", if L is both **SAT**-hard (t(n),s(n)), and t(n) time and s(n) size reducible to **SAT**.

Note that one immediate corollary of Definition 2.3 is that each **SAT**-hard(p(n),q(n)) language L, where p(n) and q(n) are polynomials, is either **NP**- or **coNP**-hard, but is not necessarily **NP**-hard unless **NP** = **coNP**.

Definition 2.4. Let **S** be an algebraic structure. We say that the problem **NUMBER_SAT(S)** is **#SAT**-*hard modulo* t(n) *time and* s(n) *size reducibility*, written "L is **#SAT**-hard(t(n),s(n))", if **SAT** is t(n) time and s(n) size parsimoniously reducible to **SAT(S)**. We say that the problem **NUMBER_SAT(S)** is **#SAT**-*complete modulo* t(n) *time and* s(n) *size reducibility*, written "L is

#SAT-complete(t(n),s(n))", if **NUMBER_SAT(S)** is both **#SAT-hard(t(n),s(n))**, and is **SAT(S)(t(n)** time and s(n) size parsimoniously reducible to **SAT**.

Note that one immediate corollary of Definition 2.4 is that each **#SAT-hard(p(n),q(n))** language L, where p(n) and q(n) are polynomials, is **#P-hard** [Val, GJ].

Here, we are primarily concerned with the cases when t(n) and s(n) are O(n) or are $O(n(\log n)^k)$, for some $k \geq 1$. Henceforth, we abbreviate "$O(n(\log n)^k)$, for some $k \geq 1$" by "npolylogn". Some easily verified properties of SAT-hardness and SAT-completeness(npolylogn,n) are the following. Let L be **SAT-hard(npolylogn,n)**. Then,

i. for all reals r > 1, **SAT** $\notin$ Dtime(n^r) => L $\notin$ Dtime(n^r), and

ii. for all reals r > 0,**SAT** $\notin$ Dtime($2^{O(n^r)}$) => L $\notin$ Dtime($2^{O(n^r)}$).

Let L be **SAT-complete(npolylogn,n)**. Then.

iii. for all reals r > 1, **SAT** $\in$ Dtime(n^r) $\Leftrightarrow$ L $\in$ Dtime(n^r), and

iv. for all reals r > 0,**SAT** $\in$ Dtime($2^{O(n^r)}$) $\Leftrightarrow$ L $\in$ Dtime($2^{O(n^r)}$).

Exactly analogous results hold between the problems **#SAT** and **NUMBER_SAT(S)**, whenever the problem **NUMBER_SAT(S)** is **#SAT-hard** or **#SAT-complete(npolylogn,n)**. Thus regardless of whether **P=NP** or **P≠NP**, each SAT-hard(npolylogn,n) problem intuitively "requires as much deterministic time as **SAT**", and each SAT-complete(npolylogn,n) problem intuitively "requires the same deterministic time as **SAT**". Also, each **#SAT-hard(npolylogn,n)** number problem intuitively "requires as much deterministic time as **#SAT**", and each **#SAT-complete(npolylogn,n)** number problem intuitively "requires the same deterministic time as **#SAT**".

3. Satisfiability Problems for Unitary Rings

Our first theorem, together with its proof, shows how to efficiently cryptomorphically embed [Bi] a nondegenerate Boolean algebra into any ring with a nonzero element idempotent under multiplication, and hence a fortiori, into any unitary ring. (Unlike [MB] but following [van] we do not assume that all rings are unitary, i.e. have a multiplicative identity.)

Theorem 3.1. Let $\mathbf{R} = (S,+,-,\cdot,0)$ be a ring with a nonzero element idempotent under the operation $\cdot$. Let D be any subset of S of cardinality ≥ 2 containing 0 such that, for all $x,y \in D$, $x = x\cdot x$ and $x\cdot y = y\cdot x$. Let $a \in D - \{0\}$. Let $D_a = \{b\cdot a \mid b \in D\}$. Let g,h: S×S $\rightarrow$ S be defined by $g(x,y) = x+y+-(x\cdot y)$ and $h(x,y) = x\cdot y$. Let i: S $\rightarrow$ S be defined by $i(x) = a+-x$. Let $\hat{D}_a$ be the closure of D_a under the operations g,h, and i. Let $\hat{g}$ and $\hat{h}$ be the restrictions of g and h, respectively, to $D_a \times D_a$; and let $\hat{i}$ be the restriction of i to D_a. Then, the algebraic structure $(\hat{D}_a,\hat{g},\hat{h},\hat{i},0,a)$ is a nondegenerate Boolean algebra.

In [HS2] we used Theorem 3.1 and its proof to characterize the complexities of the equivalence problems for formulas and for straight-line programs for many different commutative rings. Here, we show that each of the problems 1-4 above is either **NP-** or **coNP-hard**, for each unitary ring, and in addition, is **SAT-hard(npolylogn,n)**, for each commutative unitary ring. We also show that problem 5 above is **#P-complete**, for each finite unitary ring, and in addition, is **#SAT-complete(npolylogn,n)**, for each finite commutative unitary ring.

Theorem 3.2. Let $S = (S,+,-,\cdot,0,1)$ be a unitary ring. Then, each of the problems 1-4 above is either **NP-** or **coNP-hard** for S. Moreover if in addition S is commutative, then each of these four problem is also **SAT-hard(npolylogn,n)**. If S is a finite unitary ring, then the problem **NUMBER_SAT(S)** is **#P-complete**. If in addition S is commutative, the problem

NUMBER_SAT(S) is **#SAT**-complete(npolylogn,n).

We note that the conclusions of Theorem 3.2 for **NONZERO_SAT** and **UNIQUE_SAT** hold, even for systems of *constant-free* quadratic equations on S. For ordered unitary rings S, we can say significantly more about the relative complexities of SAT(S) and UNIQUE_SAT(S).

Theorem 3.3. Let S = (S,+,−,·,0,1) be an ordered unitary ring. Then, the **SAT(S)** problem is O(n) time and O(n) size reducible to the **UNIQUE_SAT(S)** problem.

We note that our proof of Theorem 3.3 actually shows that the problem **SAT(S)** for a system of m equations in n unknowns is deterministic linear time reducible to the problem **UNIQUE_SAT(S)** for a system of m+1 equations in n+1 unknowns. When S is an ordered field, e.g. the field of reals, a partial converse of Theorem 3.3 also holds.

Theorem 3.4. Let S = (S,+,−,·,0,1) be an ordered field. Then, the **UNIQUE_SAT(S)** problem for satisfiable systems of equations on S is O(n) time and O(n) size reducible to the **SAT(S)** problem for S.

Theorem 3.3 and its proof show that the problem **UNIQUE_SAT(S)** always requires "essentially at least as much deterministic time as" the problem **SAT(S)**, whenever S is an ordered unitary ring. Together, Theorems 3.3 and 3.4 show that the problems **SAT(S)** and **UNIQUE_SAT(S)** require "essentially the same deterministic time", whenever S is an ordered field. We have shown that several other satisfiability problems are deterministic linear time interreducible to the problem **SAT(S)**, whenever S is a field.

4. Satisfiability Problems for Lattices and Semirings

We present several complexity results for satisfiability problems that hold for each finite (not necessarily distributive) lattice. For lattices S, 0 and 1 denote the minimal and the maximal element of S, respectively. We consider the problems 1-5 above as well as the problem **NOT_ALL_EQUAL_SAT(S)** defined as follows.

6. **NOT_ALL_EQUAL_SAT(S)** is the problem of determining if a system of equations on S has a solution, where not all variables take on the same value.

Theorem 4.1. Let S = (S,∨,∧) be a finite lattice. The problems **SAT(S)**, **NON_ZERO_SAT(S)**, **{0,1}-SAT**, and **NOT_ALL_EQUAL_SAT(S)** are **SAT**-complete(npolylogn,n). The problem **UNIQUE_SAT(S)** is both **coNP**-hard and **SAT**-hard(npolylogn,n). The problem **NUMBER_SAT(S)** is both **#P**-complete and **#SAT**-complete(npolylogn,n.

Theorem 4.1 shows that, for all finite lattices S, the problems 1-6 for S require "essentially the same deterministic time as" the satisfiability problem for 3CNF formulas or as the problem of determining the number of satisfying assignments for 3CNF formulas. Thus, intuitively, algebra on finite lattices is "essentially as hard as, but no harder than" Boolean algebra. We conclude by noting that the proof of Theorem 4.1 can be generalized easily so as to apply to many additional lattices, including all finite depth lattices.

Analogues of Theorem 4.1 hold for many of the semirings studied in the literature [BC, Ei, Ta, Zi]. For example, we have the following theorem.

Theorem 4.2. Let S = (S,+,·) be any algebraic structure with binary operations + and · such that + is commutative and associative, · is associative and distributes on both sides over +, and

there exist distinct elements a,b, ∈ S such that $a = a + a = a \cdot b = b \cdot a$ and $b = b + b = b \cdot b = a + b$.

Then, the problem **SAT(S)** is both **NP**-hard and **SAT**-hard(npolylogn,nlogn).

Structures satisfying the conditions of Theorem 4.2 include $(2^{\{0,1\}^*},\cup,\bullet)$, for all nonempty sets A,$(2^{A\times A},\cup,\circ)$, $(R^+\cup\{\infty\},MIN,\oplus)$, various algebraic structures for path-finding problems in [BC, Ta, Zi], various Kleene algebras in [Con], and each finite unitary semiring, that is not a ring.

References

[AHU] A.V. Aho, J.E. Hopcroft, and J.D. Ullman, *The Design and Analysis of Computer Algorithms,* Addison-Wesley, Reading, Mass., 1974.

[BC] R.C. Backhouse and B.A. Carre, "Regular algebra applied to path finding prolems," *J. Inst. Maths. Applics.*, vol. 15, pp. 161-186, 1975.

[Bi] G. Birkhoff, *Lattice Theory* (3rd Edition), Am. Math. Soc., Providence, R.I., 1967.

[Con] J.H. Conway, *Regular Algebra and Finite Machines,* Chapman and Hill, Ltd., London, 1971.

[Co] S.A. Cook, "The complexity of theorem-proving procedures," *Proc. Third Annual ACM Symp. on Theory of Computing*, pp. 151-158, 1971.

[Da] M. Davis, *Computability and Unsolvability,* Dover Publications, Inc., New York, 1982.

[Ei] S. Eilenberg, *Automata, Languages, and Machines, vol. A,* Academic Press, New York, 1974.

[FY] A.S. Fraenkel and Y. Yesha, "Complexity of problems in games, graphs, and algebraic equations," *unpublished manuscript*, 1977.

[GJ] M.R. Garey and D.S. Johnson, *Computers and Intractability: A Guide to the Theory of NP-Completeness,* W.H. Freeman, San Francisco, Ca., 1979.

[HS1] H.B. Hunt III and R.E. Stearns, "Nonlinear algebra and optimization on rings are "hard"," *SICOMP*, vol. 16, pp. 910-929, 1987.

[HS3] H.B. Hunt III and R.E. Stearns, "The complexities of satisfiability problems for algebraic structures," *in preparation*, 1988.

[HS2] H.B. Hunt III and R.E. Stearns, "On the complexities of equivalence for commutative rings," Technical Report 87-22, SUNY Albany, 1987 (submutted for publication). Also see *Abstracts of Communications at 5-th International Conference on: Applied Algebra, Error-Correcting Codes and Cryptography, Applied Algorithms and Combinatorics, Computer Algebra and Complexity Theory*, pp. 43-44, Menorca, Spain.

[HSR] H.B. Hunt III, R.E. Stearns, and S.S. Ravi, "**SAT**, Finite Algebra, and the structure of **NP**," *in preparation*, 1988.

[Ka] R.M. Karp, "Reducibility among combinational problems," in *Complexity of Computer Computations*, ed. R.E. Miller and J.W. Thatcher, pp. 85-103, Plenum Press, New York, 1972.

[LT] R.L. Lipton and R.E. Tarjan, "Applications of a planar separator theorem," *SICOMP*, vol. 9, pp. 615-627, 1980.

[MB] S. MacLane and G. Birkhoff, *Algebra,* MacMillan, New York, 1967.

[Ma] Yu.V. Matisjasevic, "Enumerable sets are diophantine," *Soviet Math. Dokl.*, vol. 77, pp. 354-357, 1970. (English translation).

[RH] S.S. Ravi and H.B. Hunt III, "Application of planar separator theorem to counting problems," *IPL*, vol. 25, pp. 317-321, 1987.

[SH] R.E. Stearns and H.B. Hunt III, "On the complexity of the satisfiability problem and the structure of NP," Technical Report 86-21, Department of Computer Science, SUNY at Albany, Albany, New York, 1986.

[Ta] R.E. Tarjan, "A unified approach to path problems," *J. ACM*, vol. 28, pp. 577-593, 1980.

[Val] L.G. Valiant, "The complexity of enumeration and reliability problems," *SICOMP*, vol. 8, pp. 410-421, 1979.

[van] B.L. van der Waerden, *Modern Algebra* volumes 1 and 2, Frederick Ungar Publishing Co., New York, 1953.

[Zi] U. Zimmermann, *Linear and Combinatorial Optimization in Ordered Algebraic Structures,* North Holland, Amsterdam, 1981.

Appendix: Selected Proof Sketches

I. Proof sketch of Theorem 3.1. Since $0,a \in D_a$, $|D_a| \geq 2$. For all nonnegative integers i, let the sets D_i $(i \geq 0)$ be defined inductively by

i. $D_0 = D_a$, and

ii. $D_i = D_{i-1} \cup \{z \in S \mid \exists\, x,y \in D_{i-1}$ with $z{=}g(x,y)$, $z{=}h(x,y)$, or $z{=}i(x)\}$, for $i \geq 1$.

Then $\hat{D}_a$ equals $\bigcup_{i \geq 0} D_i$. It can be shown by mathematical induction that, for all $i \geq 0$ and for all $x,y \in D_i$, $x{=}x{\cdot}x$, $x{\cdot}y{=}y{\cdot}x$, and $x{\cdot}a{=}y{\cdot}a{=}x$. Given this, it is straight-forward to verify that the operations $\hat{g}$ and $\hat{h}$ are associative, commutative, idempotent, and absorptive, and that, for all $x,y,z \in \hat{D}_a$, $\hat{h}(x,\hat{g}(y,z)) = \hat{g}(\hat{h}(x,y),\hat{h}(x,z))$ and $\hat{g}(x,\hat{h}(y,z)) = \hat{h}(\hat{g}(x,y),\hat{g}(x,z))$. Thus, the structure $(\hat{D}_a,\hat{g},\hat{h})$ is a nondegenerate distributive lattice. Moreover for all $x \in \hat{D}_a$, $\hat{g}(0,x) = x$, $\hat{g}(a,x) = a$, $\hat{g}(x,\hat{i}(x)) = a$, and $\hat{h}(x,\hat{i}(x)) = 0$. Thus, the structure $(\hat{D}_a,\hat{g},\hat{h},\hat{i},0,a)$ is a complemented distributive lattice with universal bounds 0 and a, and complementation operator $\hat{i}$. Hence, this structure is a Boolean algebra [Bi].

II. Proof sketch of Theorem 3.2, when S is commutative.

A. Proof for **SAT(S)** and **{0,1}-SAT(S)**. Let $T = \{b \in S \mid b = b^2\}$, and let g, h, and i be defined as in the proof of Theorem 3.1 above with a=1. Let $\hat{T}$ be the closure of T under g, h, and i. Let $\hat{g},\hat{h}$, and $\hat{i}$ be the restrictions of g, h , and i, respectively, to $\hat{T}$. By the proof of Theorem 3.1, the structure $(\hat{T},\hat{g},\hat{h},\hat{i},0,1)$ is a nondegenerate Boolean algebra.

Let $f = c_1 \wedge ... \wedge c_m$ be a 3CNF formula. Let $x_1, \ldots,$ and x_p be the variables occurring in f; and let V_j, for $1 \leq j \leq m$, be the set of literals of f occurring in c_j. For $1 \leq i \leq p$, let $F[x_i]$ be x_i and $F[\bar{x}_i]$ be $1 +\!\!- x_i$. For $1 \leq j \leq m$, let $F[c_j]$ be $F[l_{j1}] + F[l_{j2}] +\!\!- F[l_{j1}] \cdot F[l_{j2}]$, if $V_j = \{l_{j1},l_{j2}\}$, and let $F[c_j]$ be

$$F[l_{j1}] + F[l_{j2}] + F[l_{j3}] +\!\!- F[l_{j1}] \cdot F[l_{j2}] +\!\!- F[l_{j1}] \cdot F[l_{j3}] +\!\!- F[l_{j2}] \cdot F[l_{j3}] + F[l_{j1}] \cdot F[l_{j2}] \cdot F[l_{j3}],$$

if $V_j = \{l_{j1},l_{j2},l_{j3}\}$. Let S[f] be the system of equations on S given by: $x_i = x_i^2$, $(1 \leq i \leq p)$ and $F[c_j] = 1$ $(1 \leq j \leq m)$. Clearly the system S[f] is constructible from f in deterministic linear time. Moreover, f is satisfiable iff S[f] has a {0,1}-valued solution iff S[f] has a solution on S.

B. Proof for **NONZERO_SAT(S)** and **UNIQUE_SAT(S)**. The proof is similar to that above, except that S[f] is given by: $u = u^2$, $x_i = x_i^2$, $1 \leq i \leq p$, $x_i = x_i \cdot u$, $1 \leq i \leq p$, and $u \cdot F[c_j] = u$ $(1 \leq j \leq m)$. Again, clearly, S[f] is constructible from f in deterministic linear time. Moreover, f is satisfiable iff S[f] has a nonzero {0,1}-valued solution iff S[f] has a nonzero solution on S. Note that S[f] is constant-free and it is satisfied, when all its variables equal 0. We only sketch why vi → iv. Suppose S[f] has a nonzero solution v on S. Then, $v[u] \neq 0$. Let $T = \{v[x] \mid x$ is a variable of $S[f]\} \cup \{0\}$. Let g,h, and i be defined as in the proof of Theorem 3.1 above with $a = v[u]$. Let $\hat{T}$

be the closure of T under g, h, and i. Let $\hat{g}, \hat{h}$, and $\hat{i}$ be the restrictions of g, h, and i, respectively, to $\hat{T}$. For all $w \in \hat{T}$, $w \cdot a = w$. Thus as above, the structure $(\hat{T}, \hat{g}, \hat{h}, \hat{i}, 0, a)$ is a nondegenerate Boolean algebra on which the system of equations $F[c_j] = 1$, $(1 \leq j \leq m)$ is satisfiable. Hence, f is satisfiable. (Recall that a formula involving only parentheses, variables, $\wedge, \vee, \neg, 0$, and 1 is equivalent to 0 on a Boolean algebra if and only if it is equivalent to 0 on the two element Boolean algebra.)

C. Proof for **NUMBER_SAT(S)**. By assumption the ring S is commutative, unitary, and finite. Let the set T and the Boolean algebra $\mathbf{T} = (\hat{T}, \hat{g}, \hat{h}, \hat{i}, 0, 1)$ be defined from **S** as in part A. Since **S** is finite, **T** is finite, and hence, **T** is atomic. Let a be an atom of **T**. Let $f = c_1 \wedge \ldots \wedge c_m$ be a 3CNF formula. Let the variables of f be x_i $(1 \leq i \leq p)$; and for $1 \leq j \leq m$, let V_j be the set of literals of f occurring in c_j. Let **a** be a constant symbol denoting the element a of $\hat{T}$. For $1 \leq i \leq p$ and $1 \leq j \leq m$, Let $F[x_i]$, $F[\overline{x}_i]$, and $F[c_j]$ be as in part A.

Let S[f] be the system of equations on S given by:

$$\mathbf{a} \cdot x_i = x_i \ (1 \leq i \leq p), \ x_i = x_i^2 \ (1 \leq i \leq p), \text{ and } F[c_j] = 1 \ (1 \leq j \leq m).$$

Clearly, the system S[f] is constructible from f in deterministic linear time. Moreover, the following holds.

Let v be any assignment of values from {0,1} to the variables $x_1, \ldots, x_p$ such that $v[f] = 1$. Then the assignment w of values from {0,a} to the variables $x_1, \ldots, x_p$ given by: $w[x_i] = 0$, if $v[x_i] = 0$, and $w[x_i] = a$, if $v[x_i] = 1$, satisfies S[f], and conversely.

This is easily seen by observing that any solution of S[f] is {0,a}-valued and that the restriction of $\hat{T}$ to {0,a} is isomorphic to the two element Boolean lattice.

III. Proof sketch of Theorem 3.3. Let **E** be a system of m equations in n unknowns on S. Thus, **E** is of the form: $f_1(x_1,\ldots,x_n) = b_1, \ldots, f_m(x_1,\ldots,x_n) = b_m$. Let **E'** be the system of m+1 equations in n+1 unknowns on S given by: $w = w^2$, $w \cdot (\sum_{i=1}^{n} x_i^2) = 0$, $(1-w) \cdot f_1(x_1,\ldots,x_n) = (1-w) \cdot b_1, \ldots, (1-w) \cdot f_m(x_1,\ldots,x_n) = (1-w) \cdot b_m$. Then, **E** has a solution on S if and only if **E'** has a unique solution on **S**. This follows by noting that:

a. $w=1$ and $x_1 = \ldots = x_n = 0$ is a solution of E', and

b. for any other solution of E', $w=0$, and the values taken on by $x_1, \ldots,$ and x_n are a solution of E and conversely.

IV. Proof sketch of Theorem 4.1. We present the proofs of SAT-hardness(npolylogn,n) and #SAT-hardness(npolylogn,n). We have shown that the set of all pairs of monotone Boolean formulas F and G such that $F \leq G$ is SAT-complete(npolylogn,n). Let S be any finite (nondegenerate) lattice. Let a be an atom of S; and let **a** be a constant symbol denoting a. Let F' and G' be the formulas on S, that result from F and G, respectively, by replacing each occurrence of *and* by $\wedge$, each occurrence of *or* by $\vee$, and each occurrence of a variable, say x, by $(x \wedge \mathbf{a})$. Then, the following are equivalent.

i. $\sim(F \leq G)$.

ii. The system of two equations $F' = \mathbf{a}$ and $G' = 0$ on S has a solution, has a nonzero solution, has a {0,1}-valued solution, or has a solution such that not all variables take on the same value.

iii. Let $x_1, \ldots,$ and x_n be the variables occurring in F or G. Let $y_1, \ldots,$ and y_n be n additional variables. The system of equations

$$x_i \wedge \mathbf{a} = x_i \ (1 \leq i \leq n),\ y_i \wedge \mathbf{a} = y_i \ (1 \leq i \leq n),$$

$$(x_1 \wedge y_1) \vee \ldots \vee (x_n \wedge y_n) \vee G = 0, \text{ and } F \vee (y_1 \wedge \ldots \wedge y_n) = \mathbf{a}$$

has a unique solution on S.

Finally, let $x_i (1 \leq i \leq n)$ be the variables occurring in F or G. The number of solutions of the system of equations $F' = \mathbf{a}$, $G' = 0$, $x_i \wedge \mathbf{a} = x_i$ $(1 \leq i \leq n)$ equals the number of assignments **v** of values from {0,1} to the variables $x_i (1 \leq i \leq n)$ such that **v**[F] = 1 and **v**[G] = 0. But determining this latter number is a #SAT-complete(npolylogn,n) problem. This follows since we can show that there are npolylogn time and linear size parsimonious reductions between the sets **SAT** and **UNATE_SAT** (the set of satisfiable 3CNF formulas in which no clause contains both complemented and uncomplemented literals), and the sets **UNATE_SAT** and the set of pairs of monotone Boolean functions (F,G) such that $\sim(F \leq G)$.

A COMPUTATIONAL PROOF OF THE NOETHER NORMALIZATION LEMMA

Alessandro Logar

1. Introduction

The Noether normalization lemma (see for instance [A-M] or [Z-S]) states that an ideal I in a polynomial ring $k[X_1, X_2, ..., X_n]$ can be transformed by a suitable change of coordinates

$$L : k[X_1, X_2, ..., X_n] \longrightarrow k[X_1, X_2, ..., X_n], \quad L(X_i) := \sum_{j=1}^{n} a_{ij}X_j, \quad a_{ij} \in k$$

in such a way that (if we put $X'_i := L(X_i)$):

- $X'_1, X'_2, \dots, X'_d$ are algebraically independent mod I,

and

- $X'_{d+1}, X'_{d+2}, \dots, X'_n$ are integral over $k[X'_1, X'_2, \dots, X'_d]$ mod I.

(We say that L puts I into *Noether normal position*).
The classical proofs of the Noether normalization lemma are constructive. For instance Seidenberg (see [Se1] or [Se2]) shows that almost all linear changes of coordinates $X'_i := \sum_{j=1}^{n} a_{ij}X_j$ put I in Noether normal position. From an algorithmic point of view, such a generic change of coordinates in which more entries than necessary are non-zero has the obvious disadvantage of making the polynomials in the bases totally dense. An inductive approach has been proposed by Vasconcelos ([Vas]), who suggested an algorithm which is (in a rough way) the following:

Step 1 Call $\mathcal{X}$ the set of variables $\{X_1, X_2, ..., X_n\}$ and define $J := I$;
Step 2 Perform (if necessary) a linear change of variables on-line and reorder $\mathcal{X}$ so that there exists an element of J monic in the first variable of $\mathcal{X}$;
Step 3 Set $\mathcal{X} := \mathcal{X}$ - 1st variable, and define $J := J \cap k[\mathcal{X}]$;
Step 4 if $J \neq (0)$ **then** go to step 1.
end.

Here the advantage is that if a variable is already integral over the previous one, no change of coordinates in step 2 is needed. The disadvantage is that step 3 (which may be done height(I) times) requires the computation of a Gröbner basis of the ideal (see [Bu1] and [Bu2]) w.r.t. the new variables.

In this paper we propose a Las Vegas probabilistic algorithm which puts a prime ideal $\mathfrak{P}$ into Noether normal position with the computation, in the average case, of two Gröbner bases, the second one being done after a change of coordinates given by a matrix with many zero entries (in some cases as many zero entries as possible).
From a computational point of view, the advantage of having a prime ideal $\mathfrak{P}$ in Noether normal position is that, as an immediate consequence, $k[X_1,X_2,\dots,X_n]/\mathfrak{P}$ is then a finitely generated $k[X_1,X_2,\dots,X_d]$ - module. Assume you want to do ideal theoretic computations in $k[X_1,X_2,\dots,X_n]/\mathfrak{P}$. One "default" approach is to associate to $J \subseteq k[X_1,X_2,\dots,X_n]/\mathfrak{P}$ its pre-image J^* in $k[X_1,X_2,\dots,X_n]$, and to perform computations with such pre-images in $k[X_1,X_2,\dots,X_n]$, (an example of this approach is given by the construction of the normalization of an algebraic variety, see [Se1] and [Tra]). If $\mathfrak{P}$ is in Noether normal position, another approach is to consider J as a submodule of the module representation of $k[X_1,X_2,\dots,X_n]/\mathfrak{P}$ and to perform computations with such representations. Since G-bases computations, on average, are exponential in the number of variables and the few computational experiences seem to suggest that computing G-bases in a free module is not much more expensive than in the polynomial ring, this approach should give a computational advantage.

2. Preliminary results

Let $\mathfrak{P} \subseteq k[X_1,X_2,\dots,X_n]$ be a prime ideal and let G be the Gröbner basis w.r.t. the lexicographical term ordering such that $X_1<X_2<\dots<X_n$.
Let $T_0:=\emptyset$ and, for $0 < j \le n$ let

$$T_j:=\begin{cases} T_{j-1} & \text{if } G\cap(k[X_1,X_2,\dots,X_j] \setminus k[X_1,X_2,\dots,X_{j-1}])\neq\emptyset \\ T_{j-1}\cup\{X_j\} & \text{otherwise.}\end{cases}$$

Then define $T_{\mathfrak{P}} := T_n$.

We recall that a set S of variables is termed algebraically independent mod $\mathfrak{P}$ if $\mathfrak{P} \cap k[S] = \varnothing$.

Proposition 2.1 If $T_{\mathfrak{P}}$ is given as above, we have:

1) $T_{\mathfrak{P}}$ is algebraically independent mod $\mathfrak{P}$;
2) $\dim \mathfrak{P} = \mathrm{card}(T_{\mathfrak{P}})$.

Before proving prop.2.1, we recall the following:

Lemma 2.2 Let g be an irreducible polynomial in k[X] and let $f \in k[Y_1, Y_2, ..., Y_d, X]$ be such that $f \notin (g)$. Then $(f,g) \cap k[Y_1, Y_2, ..., Y_d] \neq 0$.

Proof. Suppose on the contrary that $(f,g) \cap k[Y_1, Y_2, ..., Y_d] = 0$. Then consider the principal ideal $I := (f,g)k(Y_1, Y_2, ..., Y_d)[X]$ $(I \neq (1))$. Choose a generator h of I in $k[Y_1, Y_2, ..., Y_d, X]$ such that h is not divisible by a polynomial in $k[Y_1, Y_2, ..., Y_d]$. Then $g = \frac{a}{b} h$ with $a \in k[Y_1, Y_2, ..., Y_d, X]$, $b \in k[Y_1, Y_2, ..., Y_d]$, gcd(a,b)=1. Therefore $b \mid h$ and hence b=1, but g is irreducible and so g=h and this contradicts the fact that $f \notin (g)$. □

Proof of proposition. By induction on n. If n=1 then the proposition is clear. Suppose n>1.

Case 1. X_1 is algebraically independent mod $\mathfrak{P}$. Then let $K := k(X_1)$ and $\mathfrak{Q} := \mathfrak{P} K[X_2, ..., X_n]$. It is easy to verify that G is a Gröbner basis for $\mathfrak{Q}$ w.r.t. the lexicographical term ordering $X_2 < ... < X_n$, and moreover the variables $\{X_1, X_{i_1}, X_{i_2}, ..., X_{i_\delta}\}$ $(i_j \neq 1)$ are algebraically independent mod $\mathfrak{P}$ in $k[X_1, X_2, ..., X_n]$ if and only if $\{X_{i_1}, X_{i_2}, ..., X_{i_\delta}\}$ are algebraically independent mod $\mathfrak{Q}$ in $K[X_2, ..., X_n]$, hence we have that $T_{\mathfrak{P}} = T_{\mathfrak{Q}} \cup \{X_1\}$ and $\dim \mathfrak{P} = \dim \mathfrak{Q} + 1 = \mathrm{card}(T_{\mathfrak{Q}}) + 1$ and in this case the proposition is proved.

Case 2. Let X_1 be algebraic mod $\mathfrak{P}$. Then there exists $g \in k[X_1] \cap G$ irreducible. Let $K := \frac{k[X_1]}{(g)}$ and $\mathfrak{Q} := \phi(\mathfrak{P})$ where $\phi : k[X_1, X_2, ..., X_n] \longrightarrow K[X_2, ..., X_n]$ is the quotient map. $\mathfrak{Q}$ is prime and hence $T_{\mathfrak{Q}}$ is algebaically independent mod $\mathfrak{Q}$ and $\mathrm{card}(T_{\mathfrak{Q}}) = \dim \mathfrak{Q}$.

Claim: $\{X_{i_1}, X_{i_2}, ..., X_{i_\delta}\}$ $(i_j \neq 1)$ are algebraically independent mod $\mathfrak{P}$ in $k[X_1, X_2, ..., X_n]$ if and only if $\{X_{i_1}, X_{i_2}, ..., X_{i_\delta}\}$ are algebraically independent mod $\mathfrak{Q}$ in $K[X_2, ..., X_n]$.

In fact if there exists $F \in \mathcal{Q} \cap K[X_{i_1}, X_{i_2}, \dots, X_{i_\delta}] \setminus 0$ then let $f \in k[X_{i_1}, X_{i_2}, \dots, X_{i_\delta}, X_1] \cap \mathcal{P}$ be such that $\phi f = F$ and $f \notin (g)$, then we can apply lemma 2.1 and we get that $\{X_{i_1}, X_{i_2}, \dots, X_{i_\delta}\}$ are algebraically dependent mod $\mathcal{P}$ in $k[X_1, X_2, \dots, X_n]$. The converse is obvious.
Hence $\dim \mathcal{P} = \dim \mathcal{Q}$; moreover it is possible to verify that G is a Gröbner basis for $\mathcal{Q}$ w.r.t. the lexicographic term ordering $X_2 < \dots < X_n$. So we get that $T_{\mathcal{P}} = T_{\mathcal{Q}}$ and this proves the proposition. □

Remark. If I is a non-prime ideal, then it is false that $\dim I = \mathrm{card}(T_I)$. Take for instance $I := (X^2, XY) \subseteq k[X,Y]$, $X < Y$; for a correct definition of T_I in these cases see [K-W].

3. Gröbner bases and the Noether lemma

In this paragraph we will suppose that k is an infinite field.

We recall that a variable X_s is termed integral over $k[X_1, X_2, \dots, X_{s-1}]$ mod $\mathcal{P}$ if there exists a polynomial $F \in \mathcal{P} \cap k[X_1, X_2, \dots, X_s] \setminus (0)$ monic in X_s.

Proposition 3.1. X_s is integral over $k[X_1, X_2, \dots, X_{s-1}]$ mod $\mathcal{P}$ if and only if there exists a polynomial in G whose maximal monomial is a power of X_s.

Proof. Let X_s be integral over $k[X_1, X_2, \dots, X_{s-1}]$ mod $\mathcal{P}$. Then there exists a polynomial
$F := g_0(X_1, X_2, \dots, X_{s-1}) + g_1(X_1, X_2, \dots, X_{s-1})X_s + \dots + g_{m-1}(X_1, X_2, \dots, X_{s-1})X_s^{m-1}$
$+ X_s^m \in \mathcal{P}$ with $g_0, g_1, \dots, g_{m-1} \in k[X_1, X_2, \dots, X_{s-1}]$. But F is reducible mod G, and $M(F) = X_s^m$, hence there exists an element in G whose maximal monomial is a power of X_s. The converse is trivial. □

Hence, with the use of the Gröbner basis G we can establish which indeterminates are integral over the previous ones.

Proposition 3.2. If X_s is integral over $k[X_1, X_2, \dots, X_{s-1}, X_{i_j}, X_{i_{j+1}}, \dots, X_{i_\delta}]$ mod $\mathcal{P}$, (where $X_{i_j}, X_{i_{j+1}}, \dots, X_{i_\delta}$ are the indeterminates in $T_{\mathcal{P}}$ greater than X_s), then X_s is integral over $k[X_1, X_2, \dots, X_{s-1}]$ mod $\mathcal{P}$.

Proof. Let $F:=X_s^m+H(X_1,X_2,...,X_s,X_{i_j},...,X_{i_\delta})\in \mathcal{P}$, with the degree of H in X_s less then m. If $M(F)\in k[X_1,X_2,...,X_s]$, then $F\in k[X_1,X_2,...,X_s]$, $M(F)=X_s^m$ and X_s is integral over $k[X_1,X_2,...,X_{s-1}]$ mod $\mathcal{P}$.

Let us suppose then that $M(F)=\mu\nu$ where μ is a monomial in $k[X_1,X_2,...,X_s]$, $\nu\neq 1$ is a monomial in $k[X_{i_j},...,X_{i_\delta}]$. Note that X_s appears in μ with degree less than m. Observe that, by construction of $T_{\mathcal{P}}$, if a polynomial $L\in k[X_1,X_2,...,X_s,X_{i_j},...,X_{i_\delta}]\cap G$, then $L\in k[X_1,X_2,...,X_s]$. In particular $F\notin G$ and so there exists $P\in G$ such that M(F) is a multiple of M(P). Clearly $P\in k[X_1,X_2,...,X_s]$. If we reduce F we obtain $F_1:=F-\alpha\xi\nu P$ with a suitable $\alpha\in k$, and with ξ a monomial in $k[X_1,X_2,...,X_s]$. Hence:

$F_1=X_s^m+H_1(X_1,X_2,...,X_{s-1},X_{i_j},...,X_{i_\delta})\in \mathcal{P}$, with the degree of H_1 less then m in X_s. If $M(F_1)=X_s^m$, the proposition is proved. Otherwise $M(F_1)=\mu_1\nu_1$ where μ_1 is a monomial in $k[X_1,X_2,...,X_s]$, $\nu_1\neq 1$ is a monomial in $k[X_{i_j},...,X_{i_\delta}]$ and so we can repeat the above considerations. After a finite number of times, we get a polynomial $F'\in \mathcal{P}\cap k[X_1,X_2,...,X_s]$ s.t. $M(F')=X_s^m$. Hence F' gives an integral relation for X_s. □

Suppose that $T_{\mathcal{P}}=\{X_{i_1},X_{i_2},...,X_{i_d}\}$, let $X_{i_{d+1}},X_{i_{d+2}},...,X_{i_n}$ be the variables not in $T_{\mathcal{P}}$ ordered in such a way that $i_1 < i_2 < ... < i_d$ and $i_{d+1} < i_{d+2} < ... < i_n$; then define the following permutation of the variables: $Y_j := X_{i_j}$. With this permutation we have now that the first d variables $Y_1,Y_2,...,Y_d$ are algebraically independent mod $\mathcal{P}$. It is clear that after this permutation G is (possibly) not a Gröbner basis of $\mathcal{P}$ w.r.t. the lexicographical order $Y_1<Y_2<...<Y_n$ but, as an immediate consequence of prop.3.1 and prop.3.2, we can compute the set $J := \{ s \mid Y_s$ is integral over $k[Y_1,Y_2,...,Y_{s-1}]$ mod $\mathcal{P}\}$ since $Y_s = X_j$ is integral over $k[Y_1,Y_2,...,Y_{s-1}]$ iff X_j is integral over $k[X_{i_1},X_{i_2},...,X_{i_{j-1}}]$.

Let now A be the following matrix:

$$A := \left(\begin{array}{c|c} I(d\times d) & \begin{matrix} 0 \ldots\lambda_1\ldots 0 \\ \cdots \\ 0\ldots\lambda_d\ldots 0\end{matrix} \\ \hline 0 & I(n\text{-}d\times n\text{-}d)\end{array}\right)$$, the λ's are in the rth column with $r>d$

A gives the transformation of coordinates:

$$X'_i := X_i \text{ if } i > d, \quad X'_i := X_i + \lambda_i X_r \text{ if } i \le d.$$

Proposition 3.3. For almost every λ_i (that is, for all λ_i which are not a zero of a suitable polynomial), one has:

1) $X'_1, X'_2, \ldots, X'_d$ are algebraically independent mod $\mathfrak{P}$;
2) $X'_r (= X_r)$ is integral over $k[X'_1, X'_2, \ldots, X'_d]$ mod $\mathfrak{P}$;
3) if $X'_s (= X_s)$ ($s>d$, $s\neq r$) is integral over $k[X_1,\ldots,X_d,X_{d+1},\ldots,X_{s-1}]$ mod $\mathfrak{P}$, then it is integral over $k[X'_1, X'_2, \ldots, X'_d, X_{d+1},\ldots,X_{s-1}]$ mod $\mathfrak{P}$;
4) if $r < s$ and X_s is algebraic but not integral over $k[X_1,\ldots,X_d,X_{d+1},\ldots,X_{s-1}]$ mod $\mathfrak{P}$, then $X'_s (= X_s)$ is algebraic but not integral over $k[X'_1, X'_2, \ldots, X'_d, X_{d+1},\ldots,X_{s-1}]$ mod $\mathfrak{P}$.

Proof. It is clear that $k[X'_1, X'_2, \ldots, X'_d, X'_r] = k[X_1, X_2, \ldots, X_d, X_r]$. Let us call $\mathfrak{Q}$ the ideal $\mathfrak{P} \cap k[X_1,X_2,\ldots,X_d,X_r]$; $\mathfrak{Q} \neq (0)$, $\dim \mathfrak{Q} = d$ and so it is a principal ideal; let $\mathfrak{Q} = (F)$. We can write $F = F_0 + F_1 + \ldots + F_m$, where F_j is a homogeneous form of degree j. In particular

$$F_m(X_1, X_2,\ldots,X_d,X_r) = \sum_{i_1+\ldots+i_d+i_r=m} a_{i_1,\ldots,i_d,i_r} X_1^{i_1}\cdot X_2^{i_2}\cdot\ldots\cdot X_d^{i_d}\cdot X_r^{i_r}$$ with

$a_{i_1,\ldots,i_d,i_r} \in k$.

Let us express F w.r.t. $X'_1, X'_2, \ldots, X'_d, X_r$. If we use the relations $X_i = X'_i - \lambda_i X_r$ for $i=1,\ldots,d$, and if we regard F as a polynomial in X_r with coefficients in $k[X'_1, X'_2, \ldots, X'_d]$, we obtain that $F = F_m(-\lambda_1,-\lambda_2,\ldots,-\lambda_d,1)X_r^m + F'$ with the degree of F' in X_r less than m. Hence, if λ_i ($i=1,\ldots,d$), are such that $F_m(-\lambda_1,-\lambda_2,\ldots,-\lambda_d, 1)\neq 0$, then X_r is integral over $k[X'_1, X'_2, \ldots, X'_d]$ mod $\mathfrak{Q}$ and therefore mod $\mathfrak{P}$, and this proves 2); moreover in this case $X'_1,X'_2,\ldots,X'_d$ are algebraically independent mod $\mathfrak{Q}$, and hence they are algebraically independent mod $\mathfrak{P}$, and so 1) is proved.

3) is as follows: Let X_s be integral over $k[X_1,\dots,X_d,X_{d+1},\dots,X_{s-1}]$ mod $\mathfrak{P}$; if $r < s$, then $k[X'_1,X'_2,\dots,X'_d,X_{d+1},\dots,X_{s-1}] = k[X_1,\dots,X_d,X_{d+1},\dots,X_{s-1}]$ and there is nothing to prove. If $r > s$, then there exist $g_0(X_1,\dots,X_{s-1}), g_1(X_1,\dots,X_{s-1}),\dots, g_{p-1}(X_1,\dots,X_{s-1}) \in k[X_1,X_2,\dots,X_{s-1}]$ s.t.:

$$g_0 + g_1X_s + \dots + g_{p-1}X_s^{p-1} + X_s^p \in \mathfrak{P}.$$

Hence:

$$g_0(X'_1-\lambda_1X_r,\dots,X'_d-\lambda_dX_r, X_{d+1},\dots,X_{s-1}) + \dots + $$
$$+ g_{p-1}(X'_1-\lambda_1X_r,\dots,X'_d-\lambda_dX_r, X_{d+1},\dots,X_{s-1})X_s^{p-1} + X_s^p \in \mathfrak{P}.$$

This shows that X_s is integral over $k[X'_1,\dots,X'_d,X_{d+1},\dots,X_{s-1},X_r]$ mod $\mathfrak{P}$; but X_r is integral over $k[X'_1,X'_2,\dots,X'_d]$ mod $\mathfrak{P}$, hence the transitivity of the integral dependence implies that X_s is integral over $k[X'_1,X'_2,\dots,X'_d,X_{d+1},\dots,X_{s-1}]$ mod $\mathfrak{P}$.

4) is immediate because, if $r < s$, then $k[X'_1,X'_2,\dots,X'_d,X_{d+1},\dots,X_{s-1}] = k[X_1,\dots,X_d,X_{d+1},\dots,X_{s-1}]$. □

Let us consider the following matrix:

$$A := \left(\begin{array}{c|ccc} & a_{1d+1} & \cdots & a_{1n} \\ I(d\times d) & & \cdots & \\ & a_{dd+1} & \cdots & a_{dn} \\ \hline 0 & & I(n-d\times n-d) & \end{array}\right) \quad \text{with } a_{1s}=a_{2s}= \dots =a_{ds}=0 \text{ if } s\in J.$$

Then A gives the following transformation of coordinates:

$$X'_i := X_i + \sum_{s\notin J, s>d} a_{is}X_s \qquad \text{for } i=1,\dots,d.$$

Proposition 3.4. For almost every choice of the entries a_{ij}, $(i=1,\dots,d,$ $j\notin J,\ j>d)$ of A the following holds:

1) $X'_1,\dots,X'_d$ are algebraically independent mod $\mathfrak{P}$;

2) X_{d+i} is integral over $k[X'_1,\dots,X'_d]$ mod $\mathfrak{P}$ for $i=1,\dots,n-d$.

Proof. Let $r := \min\{ s \in \mathbb{N} \mid s \notin J \}$. Let us consider the transformation of coordinates given by $U_i := X_i + a_{ir}X_r$ $i=1,\dots,d$. After this transformation the situation is as follows:

- if $s\notin J$ and $s\neq r$, then X_s is algebraic but not integral over $k[U_1,\dots,U_d,X_{d+1},\dots,X_{s-1}]$ mod $\mathfrak{P}$ (prop.3.3, 4));

- if $s \in J$, then X_s is integral over $k[U_1,\ldots,U_d,X_{d+1},\ldots,X_{s-1}]$ mod $\mathcal{P}$ (prop.3.3, 3));
- if $s=r$, X_r is integral over $k[U_1,\ldots,U_d]$ mod $\mathcal{P}$ (prop.3.3, 2)).

Hence $\{ s > d \mid X_s$ is integral over $k[U_1,\ldots,U_d, X_{d+1},\ldots,X_{s-1}]$ mod $\mathcal{P} \} = J \cup \{r\}$; moreover $U_1,\ldots,U_d$ are algebraically independent mod $\mathcal{P}$ (prop.3.3, 1)).
If we repeat this procedure a sufficient number of times, then we obtain a coordinates transformation matrix as requred, and X_s is integral over $k[U_1,\ldots,U_d, X_{d+1},\ldots,X_{s-1}]$ mod $\mathcal{P}$ for every $s \geq d+1$, hence, using the transitivity of the integral dependence, X_s is integral over $k[U_1,\ldots,U_d]$ mod $\mathcal{P}$ for every $s \geq d+1$ and $U_1,\ldots,U_d$ are algebraically independent mod $\mathcal{P}$. □

As an immediate consequence of prop.3.4 and of the previous considerations, we get the following probabilistic algorithm which puts the ideal $\mathcal{P}$ in Noether normal position:

Algorithm 3.5

(<u>Input</u>: a prime ideal $\mathcal{P} \subseteq k[X_1,X_2,\ldots,X_n]$;
<u>Output</u>: an $n \times n$ matrix A s.t. $\mathcal{P}$ is in Noether normal position w.r.t. the variables $X := AX$).

Step 1 Compute a Gröbner basis G of $\mathcal{P}$ w.r.t. the lex. ord. $X_1<X_2<\ldots<X_n$ and $T_{\mathcal{P}}$; let $A:=I(n \times n)$;

Step 2 Call $X_1,X_2,\ldots,X_d$ the elements of $T_{\mathcal{P}}$ and $X_{d+1},X_{d+2},\ldots,X_n$ the others; set on $k[X_1,X_2,\ldots,X_n]$ the lex. ord. such that $X_1<X_2<\ldots<X_n$ call A_1 the matrix which gives this transformation; $A:=A_1A$;

Step 3 Compute $J := \{ s \mid X_s$ is integral over $k[X_1,X_2,\ldots,X_{s-1}]$ mod $\mathcal{P}\}$ (use prop.3.1 and prop.3.2);

Step 4 **For** $r:=1$ **to** d
    let $a_{rr}:=1$; $a_{ir}:=0$ if $i \neq r$;
  for $r:=d+1$ **to** n
    if $r \in J$ **then**
      let $a_{rr}:=1$; $a_{ir}:=0$ if $i \neq r$
    else
      choose $a_{1r}, a_{2r}, \ldots, a_{dr} \in k$ randomly;
      $X_i:= X_i+ a_{ir}X_r \quad i=1,\ldots,d$;
      let $a_{rr}:=1$; $a_{ir}:=0$ if $i \neq r$, $i > r$;
  $A_2 :=(a_{ij})$;
  $A:=A_2A$;

Step 5 compute a Gröbner basis G of $\mathfrak{P}$ w.r.t. the lex. ord. $X_1<X_2<\ldots<X_n$;
Step 6 **if** $(G\cap k[X_1,X_2,\ldots,X_d]\neq 0)$ **or** (there exists s: $d+1\leq s\leq n$ s.t. a power of X_s does not appear as a maximal monomial of an element of G (prop.3.1)) **then go to step 2;**
End.

At the end of this algorithm one has $\mathfrak{P}$ in Noether normal position w.r.t. the new variables $X:=AX$.

<u>Example</u> Consider the prime ideal $\mathfrak{P}:=(X_2^2+X_1X_2+1, X_1X_2X_3X_4+1) \subseteq \mathbb{Q}[X_1,X_2,X_3,X_4]$. The reduced Gröbner basis of $\mathfrak{P}$ w.r.t. the lex. ord. $X_1<X_2<X_3<X_4$ is $G:=\{X_2^2+X_1X_2+1, X_1X_3X_4-X_2-X_1\}$, hence $T_{\mathfrak{P}}=\{X_1,X_3\}$. As in step 2 we permute the coordinates putting $X_1:=X_1$, $X_2:=X_3$, $X_3:=X_2$, $X_4:=X_4$ hence with this new notation $J=\{X_3\}$ and $\{X_1,X_2\}$ are algebraically independent mod $\mathfrak{P}$. The matrix A_2 is:

$$\begin{pmatrix} 1 & 0 & 0 & a_{14} \\ 0 & 1 & 0 & a_{24} \\ 0 & 0 & 1 & 0 \\ 0 & 0 & 0 & 1 \end{pmatrix}, \text{ and hence } A=\begin{pmatrix} 1 & 0 & 0 & a_{14} \\ 0 & 0 & 1 & a_{24} \\ 0 & 1 & 0 & 0 \\ 0 & 0 & 0 & 1 \end{pmatrix}.$$

We take for instance $a_{14}=a_{24}=1$, and so we consider the new change of coordinates given by $X_1:=X_1+X_4$; $X_2:=X_2+X_4$; $X_3:=X_3$, $X_4:=X_4$, then we get the following reduced Gröbner basis for $\mathfrak{P}$ w.r.t. the lex. ord. $X_1<X_2<X_3<X_4$:

$$G:=\{X_4 - X_3^5 + X_3^4X_2 - 2X_3^4X_1 + X_3^3X_2X_1 - X_3^3X_2 - 3X_3^3 + 2X_3^2X_2 - 4X_3^2X_1 + X_1X_2X_3 - X_3X_1^2 - X_3 + X_2 - X_1;\quad X_3^6 + X_3^5X_2 + 2X_3^5X_1 - X_3^4X_2X_1 + X_3^4X_1^2 + 3X_3^4 - 2X_3^3X_2 + 4X_3^3X_1 - X_3^2X_2X_1 + X_3^2X_1^2 + 2X_3^2 - X_2X_3 + 2X_1X_3 + 1\}.$$

From prop.3.1 we have that w.r.t. these new coordinates the prime ideal $\mathfrak{P}$ is in Noether normal position.

<u>Remarks.</u> 1) It is easy to see that if we take in the previous example $a_{14}=0$ or $a_{24}=0$, then $\mathfrak{P}$ is not in Noether normal position, hence in this case the matrix A of proposition 3.4 is minimal.

2) The assumption that $\mathfrak{P}$ is prime is needed just in the proof of prop.2.2; also if $\mathfrak{P}$ is not prime the knowledge of a Gröbner basis of $\mathfrak{P}$ allows to commute both $\dim(\mathfrak{P})$ and a maximal set of algebraically independent variables $S_{\mathfrak{P}}$ [K-W]. So if $\mathfrak{P}$ is not prime

and $\dim(\mathcal{P})=\mathrm{card}(T_{\mathcal{P}})$, algorithm 3.5 can still be applied. If $\mathcal{P}$ is not prime and $\dim(\mathcal{P})>\mathrm{card}(T_{\mathcal{P}})$ the following modifications are required: i) after step 1 one has to compute a second Gröbner basis of the ideal after a permutation of variables such that $S_{\mathcal{P}}$ consists of the first $\dim(\mathcal{P})$ variables; ii) Step 3 makes use of prop3.1 only; iii) if the algorithm fails at step 6, it must be restarted after step 3 discarding the transformation done in step 4.

Suppose now that char(k)=0 and that $X_1,X_2,\dots,X_d$ are algebraically independent mod $\mathcal{P}$ (d is, as usual, the dimension of $\mathcal{P}$) and let G be a Gröbner basis of $\mathcal{P}$ w.r.t. the lex. ord.. From the primitive element theorem we have that $k(x_1,x_2,\dots,x_n)$ (that is the field of fractions of $k[X_1,X_2,\dots,X_n]/\mathcal{P}$) is a simple extension of $k(X_1,X_2,\dots,X_d)$ $(=k(x_1,x_2,\dots,x_d))$. Moreover we have:

Proposition 3.6. The following conditions are equivalent:

1) x_{d+1} is a primitive element of $k(x_1,x_2,\dots,x_n)$ over $k(X_1,X_2,\dots,X_d)$;
2) $\forall\, i>1\ \exists\, F_i \in G\cap k[X_1,X_2,\dots,X_{d+i}]$ s.t. $\deg_{X_{d+i}}F_i=1$ ($\deg_{X_{d+i}}(\cdot)$ is the degree of the polynomial w.r.t. the variable X_{d+i}).

Proof. If x_{d+1} is a primitive element then $k(X_1,X_2,\dots,X_d,x_{d+1})=k(x_1,x_2,\dots,x_n)$, so in particular $x_{d+i}\in k(X_1,X_2,\dots,X_d,x_{d+1})$ $\forall\, i>1$ which gives:

$$x_{d+i}=\frac{f(X_1,X_2,\dots,X_d,x_{d+1})}{h(X_1,X_2,\dots,X_d,x_{d+1})} \quad \text{and therefore:}$$

$Q := H(X_1,X_2,\dots,X_d,X_{d+1})X_{d+i}-F(X_1,X_2,\dots,X_d,X_{d+1})\in\mathcal{P}$, (where F and H are such that f=[F] and h=[H] mod $\mathcal{P}$). But $Q\in\mathcal{P}\cap k[X_1,X_2,\dots,X_{d+i}]$ and $M(Q)=\mu X_{d+i}$ where $\mu\in k[X_1,X_2,\dots,X_{d+i-1}]$ is a suitable monomial, hence there must exist a polynomial $F_i\in G\cap k[X_1,X_2,\dots,X_{d+i}]$ whose maximal monomial is linear in X_{d+i} , and therefore F_i is the desired polynomial.
Conversely, by 2) we have: $k(X_1,X_2,\dots,X_d,x_{d+1})=k(X_1,X_2,\dots,X_d,x_{d+1},x_{d+2})=\dots=k(x_1,x_2,\dots,x_n)$. (See also [GTZ]). □

As a straightforward modification of the proof of the primitive element theorem (see for instance [Nag]) we obtain that for almost all $\lambda_1,\dots,\lambda_{n-d}\in k$ (that is, for all $\lambda_1,\dots,\lambda_{n-d}$ that are not a zero of a suitable polynomial), the polynomial $\lambda_1X_{d+1}+\dots+\lambda_{n-d}X_n$ is a primitive element. This shows that a slight modification of the algorithm 3.5 allows us to obtain the ideal $\mathcal{P}$ in Noether normal position and with X_{d+1} primitive element.

Aknowledgments. The author is grateful to Teo Mora for many illuminating conversations on the subject, to the referees for having pointed out some errors in an earlier version of the paper and to Joos Heintz and Leandro Caniglia for having suggested the complexity arguments given below.

REFERENCES

[A-M] M.F. Atiyah, I.G. MacDonald, *Introduction to Commutative Algebra*, Addison-Wesley, Reading, Mass. (1969), IX+128 pp.

[Bu1] B. Buchberger, *Ein Algorithmus zum Auffinden der Basiselemente des Restklassenringes nach einem nulldimensionalen Ideal,* Ph.D. Thesis, Innsbruck, (1965).

[Bu2] B. Buchberger, *Gröbner bases: an algorithmic method in polynomial ideal theory*, Recent trends in multidimensional systems theory Bose N.K. Reidel (1985).

[G-M] P. Gianni, T. Mora, *Algebraic solution of systems of polynomial equations using Gröbner bases.* Communication at A.A.E.C.C. 5, Menorca 1987.

[GTZ] P. Gianni, B. Trager, G. Zacharias, *Gröbner bases and primary decomposition of polynomial ideals.* Preprint (1986).

[K-W] H. Kredel, V. Weispfenning, *Computing dimension and independent sets for polynomial ideals.* Preprint (1986).

[Nag] J. Nagata, *Field theory* , Marcel Dekker, INC, New York and Basel (1977).

[Se1] A. Seidenberg, *Construction of the integral closure of a finite integral domain*, Rend. Sem. Mat. Fis. Milano 40, 101-120, (1970).

[Se2] A. Seidenberg, *Constructions in algebra,* Trans. A.M.S.197, 273-313 (1974).

[Tra] C. Traverso, *A study of algebraic algorithms: the normalization,* Report Pisa, (1986)

[Vas] V. Vasconcelos, *What is a prime ideal?* Preprint.

[Z-S] O. Zariski, P. Samuel, *Commutative algebra,* I, II, Springer-Verlag (1975).

Note added in proof.

The recent results [Bro, CGH1, CGH2, Kol] about the effective Nullstellensatz allow both to improve the algorithm defined here and to give a good bound for its complexity.
In most applications of the effective Nullstellensatz, given an ideal $I:=(F_1,\dots,F_s)\subseteq k[X_1,\dots,X_n]$, it is shown that the existence of an element $G\in I$ satisfying some property P is equivalent to the existence of $G'\in I$, $a_i\in k[X_1,\dots,X_n]$ such that G' satisfies P, $G'=\sum a_iF_i$, $\deg(a_iF_i)$ is bounded by some explicit "good" function in n and $\delta:=\max\deg(F_i)$.
Therefore, given such a bound D, such an element G' can be computed by means of linear algebra.
From the point of view of practical efficiency, it is important to remark that in many instances truncated homogeneous Gröbner basis computations can be used instead of linear algebra, with a comparable theoretical complexity.

Let $^h-: k[X_1,X_2,\dots,X_n]\longrightarrow k[X_1,X_2,\dots,X_n]$ be the homogeneization function and let $*I:=(^hF_1,\dots,{}^hF_s)$. Suppose that $g=\sum a_iF_i$ and $\deg(a_iF_i)\le D$ then clearly there exist t, $t_i\in\mathbb{N}$ such that $X_0^t{}^hg=\sum X_0^{t_i}{}^ha_i{}^hF_i$ and $\deg(X_0^t{}^hg)=D=\deg(X_0^{t_i}{}^ha_i{}^hF_i)$. Fix a term-ordering < on $k[X_1,X_2,\dots,X_n]$ and let $<_h$ be the associated degree compatible term-ordering on $k[X_0,X_1,\dots,X_n]$ (i.e. $\phi<\psi$ iff $\deg(\phi)<\deg(\psi)$ or $\deg(\phi)=\deg(\psi)$ and $^a\phi<{}^a\psi$, where ^{a} is the affinization (see [Laz], [M-M]). M(g) (resp. $M_h(g)$) denotes, as usual, the maximal monomial of g w.r.t. < (resp. $<_h$).

Lemma A.1 There exists $g\in I$ s.t. $M(g)\in k[X_{i_1},X_{i_2},\dots,X_{i_j}]$ and $g=\sum a_iF_i$ with $\deg(a_iF_i)\le D$ if and only if there exists a homogeneous polynomial g' in the Gröbner basis of *I w.r.t. $<_h$ such that $\deg(g')\le D$, and $M_h(g')\in k[X_0,X_{i_1},\dots,X_{i_j}]$.

Proof. The "if" part is obvious: g is the affinization of g'. As for the "only if" part, as we have seen above, for some t, $X_0^t{}^hg\in *I$, $\deg(X_0^t{}^hg)=D$, $M_h(X_0^t{}^hg)\in k[X_0,X_{i_1},X_{i_2},\dots,X_{i_j}]$. So there exists g', homogeneous in the Gröbner basis of *I w.r.t. $<_h$ s.t. $M_h(g')$ divides $M_h(X_0^{t_h} g)$. Such a g' verifies the thesis. □

If Buchberger algorithm is applied to an ideal given through an homogeneous basis, it is easy to verify that the reduction of the S-polynomial of F, G (if $\neq 0$) does not change its degree, which is the degree of l.c.m.$(M_h(F),M_h(G))$. So clearly, if one applies Buchberger algorithm on a homogeneous basis without performing the S-polynomials S(F,G) s.t. l.c.m.$(M_h(F),M_h(G))>D$, then all elements in a Gröbner basis whose degree is not greater than D are still obtained. Therefore:

Corollary A.2. Assume that $g \in I$ is such that $M(g) \in k[X_{i_1},\dots,X_{i_j}]$, $g=\sum a_i F_i$ with $\deg(a_i F_i) \leq D$. Then the existence of such a g can be tested in polynomial time over D, if field operations require constant time.

Proof. Use the lemma above to translate the problem in the homogeneous case, and then compute the truncated Gröbner basis of *I w.r.t. $<_h$ as described above. □

To apply effective-Nullstellensatz techniques and truncated Gröbner basis computations to the Noether normalization problem, we need the following results from [F-G]:

Let $I:=(F_1,\dots,F_s) \subseteq k[X_1,\dots,X_n]$ and let $V:=\{\ \xi \in \bar{k} \mid F_i(\xi)=0,\ i=1,\dots,s\}$, where $\bar{k}$ denotes the algebraic closure of k. Set $d:=\dim(V)$, $\delta:=\max\deg(F_i)$ and $D:=3\cdot\max(3^n,\delta^n)$, and remark that $\deg(V)\leq\delta^{n-d}$ by Bezout's inequality ([Hei], [F-G]).

Proposition A.3. It holds:

1) If $I \cap k[X_{i_1},X_{i_2},\dots,X_{i_j}] \neq (0)$ then there is $g \in I \cap k[X_{i_1},X_{i_2},\dots,X_{i_j}]$ s.t. $g=\sum a_i F_i$ with $\deg(a_i F_i) \leq D\big(\deg(V)+1\big)$;

Assume now $I \cap k[X_1,X_2,\dots,X_d]=(0)$, where $d=\dim I$. Then:

2) if X_i is integral over $k[X_1,X_2,\dots,X_{i-1}]$ mod I, then there exists $g \in k[X_1,X_2,\dots,X_i] \cap I$ monic in X_i s.t. $g=\sum a_i F_i$ with $\deg(a_i F_i) \leq D\big(\deg(V)+1\big)$;

3) let k_0 be a finite subset of k s.t. $card(k_0)>D\big(deg(V)+1\big)$; let $J:=\{j \mid X_j$ is integral over $k[X_1,X_2,...,X_{j-1}]$ mod $I\}$, $AV:=\{s \mid d<s\leq n, s\notin J\}$. Then there are $a_{ij}\in k_0$, i=1,...,d, $j\in AV$ s.t. setting $X'_i:=X_i+\sum_{j\in AV} a_{ij}X_j$, i=1,...,d, X_i is integral over $k[X'_1,X'_2,...,X'_d]$ mod I for every i>d.

Proof. See [F-G]. □

Corollary A.4. If $\{X_{i_1},X_{i_2},...,X_{i_j}\}$ is a maximal set of variables s.t.:

(*) there is no $g\in I$: $g=\sum a_iF_i$, $deg(a_iF_i)\leq D\big(deg(V)+1\big)$ and $M(g)\in k[X_{i_1},..., X_{i_j}]$

then $\{X_{i_1},X_{i_2},...,X_{i_j}\}$ is a maximal set of algebraically independent variables mod I (and hence j=d).

Proof. Let $\{X_{i_1},X_{i_2},...,X_{i_j}\}$ be such that $I\cap k[X_{i_1},X_{i_2},...,X_{i_j}]\neq(0)$, then there is $g\in I$ with $g=\sum a_iF_i$, $deg(a_iF_i)\leq D\big(deg(V)+1\big)$ and $M(g)\in k[X_{i_1},..., X_{i_j}]$ by prop.A3, 1). Therefore, if $\{X_{i_1},X_{i_2},...,X_{i_j}\}$ is a maximal set of variables satisfying (*), then it is a set of algebraically independent variables mod I, hence dim I≥j. Since dim I=dim M(I) [G-T] and dim M(I)≤j (any set of algebraically independent variables for M(I) satisfies (*)), $\{X_{i_1},X_{i_2},...,X_{i_j}\}$ is a maximal set algebraically independent variables mod I. □

From prop A.3, cor. A.4 and the previous considerations, we realize that we can substitute Gröbner bases computations with truncated homogeneous Gröbner bases computations in the algorithm 3.5. We obtain the following algorithm, which, by the discussion above, is polynomial over δ^n (if field operations require constant time):

Algorithm A.4

Step 1 Compute by a homogeneous Buchberger algorithm truncated at degree $D(\delta^n+1)$ a maximal set of algebraically independent variables mod I; let $A:=I(n\times n)$;

Step 2 Perform a permutation of the variables so that $X_1,X_2,...,X_d$ is the maximal set of step 1. Call A_1 the matrix which gives this transformation; $A:=A_1A$;

Step 3 Compute, by a homogeneous Buchberger algorithm w.r.t. the lexicographic order truncated at degree $D(\delta^n+1)$ the set $J := \{ s \mid X_s$ is integral over $k[X_1,X_2,\dots,X_{s-1}]$ mod $I\}$ and $AV:=\{s \mid d<s\leq n, s\notin J\}$;

Step 4 Let $\mathcal{A}:=\{ A\in M_{n\times n}(k) \mid a_{ii}=1\ \forall i, a_{ij}=0$ if $i>d$ or $j\notin AV$, $a_{ij}\in k_0$ if $i\leq d$ and $j\in AV\}$;

while $X_1,X_2,\dots,X_d$ are not algebraically independent or $\exists s$: X_s is not integral over $k[X_1,X_2,\dots,X_{s-1}]$ **do**

choose $A_2\in\mathcal{A}$;

$A':=A_2A$;

$\mathcal{A}:=\mathcal{A}\setminus\{A_2\}$;

$A:=A'$

End.

[Bro] D. Brownawell, *Bouds for the degree in the Nullstellensatz,* Annals of Math. 2nd ser. 126, n.3,1987, p. 577-591.

[CGH1] L. Caniglia, A. Galligo, J. Heintz, *Borne simple exponentielle pour les degrés dans le théorème des zéros sur un corps de caractéristique quelconque.* C.R.Acad.Sci. Paris, t.307 Ser.I (1988) 255-258.

[CGH2] L. Caniglia, A. Galligo, J. Heintz, *Some new effectivity bounds in computationa geometry.* These Proceedings.

[F-G] N. Fitchas, M. Giusti, *The membership problem in the unmixed case is in NC.* Manuscript (1988).

[G-T] A. Galligo, C. Traverso, *Practical determination of the dimendion of an algebraic variety.* Preprint (1988).

[Hei] J. Heintz, *Definability and fast quantifier elimination in algebraically closed fields.* Theoret.Comput.Sci. 24 (1983) 239-277.

[Kol] J. Kollar, *Sharp effective Nullstellensatz.* Manuscript (1988).

[Laz] D. Lazard, *Gröbner bases, Gaussian elimination and resolution of systems of algebraic equations.* Proc. EUROCAL 1983, Springer L.N.Comp.Sci. 162 (1983).

[M-M] M. Möller, F. Mora, *Upper and lower bounds for the degree of Gröbner bases,* Proc. EUROSAM 84 Springer L.N.Comp.Sci. 174 (1984).

Alessandro Logar
Dipartimento di Matematica
p.le Europa, 1
34100 Trieste
ITALY

A GRAPHICS INTERFACE TO REDUCE

Jed Marti

The RAND Corporation

Abstract Graphical representation of symbolic equations is important for visualization, presentation, and education. This paper describes the design of a portable graphics subsystem for REDUCE that supports a wide range of graphics output devices and standards. Its sufficiency is demonstrated by a number of graphics display formats and demonstration programs. These include: two dimensional graphs, contour plots, wire frame images, and three dimensional solids with hidden surfaces removed. The graphics package supports simultaneous output to devices ranging from laser printers to high performance surface rendering color displays.

INTRODUCTION

For lack of accessible and adequate hardware, REDUCE [Hearn 85] has suffered from the lack of graphical output. The availability of inexpensive personal work-stations, lap top personal computers and laser printers has changed all this. The computer algebra user now has low resolution graphics on his lap-top or personal computer and high resolution 3-D color display and hardcopy for a modest portion of his research budget. It now seems fitting to investigate a portable graphics interface to the REDUCE system.

The users of computer algebra systems will benefit greatly from the presence of graphics display. Graphics output is an important visualization tool. Like estimation of numerical results, the display serves as confirmation of characteristics of the results of symbolic manipulation. Likewise, hard-copy output has an important role in the presentation of results. An increasing number of younger computer algebra users will greatly benefit from the graphical display of equations and the animation of algorithms.

With these users in mind, we designed and implemented both a graphics interface and low level driver routines for a number of different applications and hardware. The resulting system enables both users and implementors to migrate to new hardware with a short transition period and minimal implementation. It provides users with a stable, learnable set of functions that can be relied upon across machines. As a portable system in both code and data structures, it permits precalculated graphics data structures to be transferred between machines with different architectures across network and international boundaries. Our design relies on no particular graphics system or set of hardware. Around a set of primitive graphics functions we have wrapped a number of standard display algorithms. Display drivers for several different levels of hardware have been written to test the completeness and suitability of graphics interface functions.

This paper provides a short overview of existing interfaces. The second section examines available hardware and its influence on system design. The third section presents a range of applications supported by the system. The fourth section describes the features the interface supports. The concluding section discusses 4 existing implementations (Hewlett Packard Starbase, X windows, Postscript, and GKS). An appendix formalizes the display driver interface.

PERSPECTIVE

Most computer algebra systems have graphics display as a support package. These have been restricted by the hardware displays available during the implementation of the algebra system and hence are generally line drawings of various sorts: contour plots, wire frame images, and graphs. Lack of suitable hardware has assigned secondary roles to 3 dimensional display of solids and animation. However, much can be learned from the interfaces developed for existing systems. I divide these into two types: those supporting symbolic algebra systems and those existing as graphing packages in their own right.

MACSYMA [Macsyma 87] appears have the most complete symbolic algebra graphics support. The current manual lists a number of supported output devices, and several styles of display. The program supports standard 2D graphics of equations and parametric equations, 3D wire frame images, and contour plots. Its design features of interest to any computer algebra system implementor are: accepting more than one equation to be plotted, saving intermediate results to avoid repeated computation, a mechanism for storing computations externally that can be passed between machines, and mechanisms for modifying the graph display. The systems few draw backs are lack of portability, lack of support for non-standard input devices, and insufficient support for animation and color. Though many different output devices are mentioned in an older manual [MACSYMA 83], these appear to have been dropped from the supported package. Likewise the list of functions defined as primitives are not sufficient to support color or 3D surface rendering machines nor do they allow the applications programmer to automatically adjust his output to the characteristics of the display device.

The SMP manual [Cole 81] lists a similar set of capabilities. Like MACSYMA, there is no support for color or 3D surface rendering. The interface appears to have less control over the resulting display than MACSYMA. Like MACSYMA's interface, there is no provision for the programmer constructing his own non-standard display from graphics primitives.

Several other attempts are underway to add graphics output to computer algebra systems. The Bittencourt system [Bittencourt 87] relies on GKS for portability. Other packages have been implemented specifically for solving particular problems. For example, Seymour's CONFORM [Seymour 86] system for displaying and manipulating conformal mappings, is based on the CORE graphics standard. The use of CORE as a standard directly in the CONFORM code means that converting the program to another environment requires significant source code modification. Use of an extra layer in this and other similar systems increases portability.

There are a legion of stand-alone plotting packages many with very powerful features. Some examples (not exhaustive): GRAFL-II [McNamara 85], MATHGRAF [Chato 86], MATLAB [MGA Inc]. Most graphics hardcopy devices come with reasonably powerful subroutine libraries and packages. The most important aspect of these systems for symbolic algebra systems is the close attention paid to the human interface. Most are designed for unsophisticated users that want power without particular attention to the reference manual. These features are categorized as follows:

1. Explanatory user interface.
2. Intelligent defaults.
3. "Analog" control of presentation (use of the mouse, knob and button boxes).
4. Immediate feedback.
5. Color

We separate graphics display from window systems and other similar interfaces. There is considerable work already underway in this area, in particular, MathScribe [Smith 86], and Iris [Leon 86]. We make a distinction between the window system and and the methods for displaying non-textual information in them. It has been our experience that most graphics systems are not reasonable mechanisms for implementing window systems. Likewise, few window systems have the ability to support high performance, 3D surface rendering display devices.

THE COMPUTER ALGEBRA ENVIRONMENT

Hardware supporting computer algebra programs has progressed beyond the time-sharing main-frames used in their original construction. The design of the REDUCE graphics interface is influenced by the following four hardware classifications.

The Personal Computer

The personal computer, be it Lap-top, portable, or transportable, is characterized by low cost, a modest operating system and the following hardware: Low to medium resolution graphics varying from 320x200 pixels in a few colors (usually 4 to 8) or monochrome, to 640x400 with 8 to 16 colors. Normal keyboard input may be supplemented with a mouse. For main memory, it will have from 512k to a few megabytes. Disk storage ranges from low density floppies (360k per drive) to small hard disks (40 megabytes or less). Peripherals include a dot-matrix printer, and possibly a modem. The operating system does not generally support multiple tasks or virtual memory. Multi-tasking systems do not offer memory protection. The operating system and display hardware are generally insufficient to support a full-featured window system.

Several versions of REDUCE have been implemented on these small systems. Though the entire REDUCE system is implemented, it cannot generally all be loaded at one time because of insufficient storage. To accommodate these systems, the display algorithms recompute values to avoid wasting storage. Since few of these systems have anything in the way of graphics support, the REDUCE graphics driver routines include code for drawing lines with single points, filling areas, and dithering written in terms of very low level functions for setting and resetting pixels.

The Personal Work-Station

The personal work-station is an advanced version of the personal computer. In contrast, it is not easily portable as it has a much larger display and either a large local disk or network connections to a file server. Its capabilities are generally those of a time sharing minicomputer.

The work-station monitor features medium to high resolution, from 1024x780 to 1600x1280 or more individually addressable pixels. It will be either monochrome or have limited color ability (8 to 16 colors). The hardware will sometimes support fast block transfers, but any graphics support will be in software. To make the best use of the large display area, the operating system (usually UNIX or a clone) will support a window management function (X windows, NEWS, etc.). Operation of the window system is supported by a pointing device, usually a mouse or similar device. The personal work-station will support larger programs than the personal computer with more memory (4 to 128 megabytes of fast RAM) and will have either 100 megabytes or more of local disk or access to a large centralized file server. Many such systems have hardware floating point assist.

The personal work-station will support very large programs through the use of virtual memory and large local fast memory. Most of these systems already support REDUCE with sufficient storage for loading and executing the graphics package with the most complex equations. However, the lack of 3D graphics support software limits output to 2D representations of the surfaces (wire frame images and contour plots).

The personal graphics work-station is simply a work-station with extended graphics support. This can be as simple as a graphics package (PHIGS, and GKS for example), or a larger more expensive graphics accelerator with multiple display heads. The number of display planes will be 8 or more and the software (and hardware) will support 3 dimensional surface rendering. Since of a number of these systems are readily available, we have included primitives for initializing the surface rendering hardware and algorithms.

Time-sharing System

The user of the centrally located time-sharing system will most likely have indirect access to many different I/O devices: plotters, laser-printers, low speed display devices. In many cases these devices will be shared among a large user community and cannot be used effectively in an interactive fashion. In some extreme cases, large scale computing will be performed in a batch mode, with plotting done off line at irregular intervals.

True batch computing is done at relatively few installations and is not considered in our design strategy. Algebraic computation is by its very nature an interactive process. However, for the user of the time sharing system we have opted to support only terminals capable of displaying bit map or vector based graphics. We feel that graphs made of asterisks and dashes discourage potential users when the price of rudimentary display devices is very low. With a medium or low resolution display terminal, the user can interact with the graphics system until satisfied with a graph and then transfer the results to a shared higher resolution device with reasonable confidence in the output. Our design makes intelligent decisions about the resolution and color capabilities of the different output devices.

THE DESIGN OF THE PORTABLE GRAPHICS INTERFACE

Rather than concentrating on the algebraic interface, we summarized the facilities from the GKS [GKS 85], PHIGS [PHIGS], X [Gettys], and Postscript [Adobe 85] graphics systems. We then enumerated the applications we felt most useful to traditional computer algebra users.

1. Display of univariate, simple and parametric equations in two dimensions.

2. Wire frame images of multivariate, simple and parametric equations.

3. Contour plots of multivariate functions.

4. Surface rendered multivariate, parametric equations with hidden surface removal.

5. Interactive formation of graphical images.

To this list we added a number of facilities useful to a growing class of computer algebra users.

1. Animation of mathematical algorithms and processes.

2. Animated display of trivariate and parametric functions.

3. The use of color to increase the visual impact.

4. Graphical devices to monitor the stages of algebraic computation.

We then went through a number of iterations, implementing graphical applications and augmenting the standard to support the application. The resulting portable interface works with a number of graphics output devices and supports the above applications.

Design Hierarchy

A design hierarchy of 4 or more levels permits simultaneous use of more than one output device driver. As seen in figure 1, the top level is comprised of application programs such as wire frame images, contour plots, specific animations, and the like. The graphics driver program directs data and control to a selected output device driver program. Only one of these may be selected at a time, however more than one can be active at a time. The graphics interface program connects the LISP system to the graphics output primitives. These are often implemented in C or Fortran.

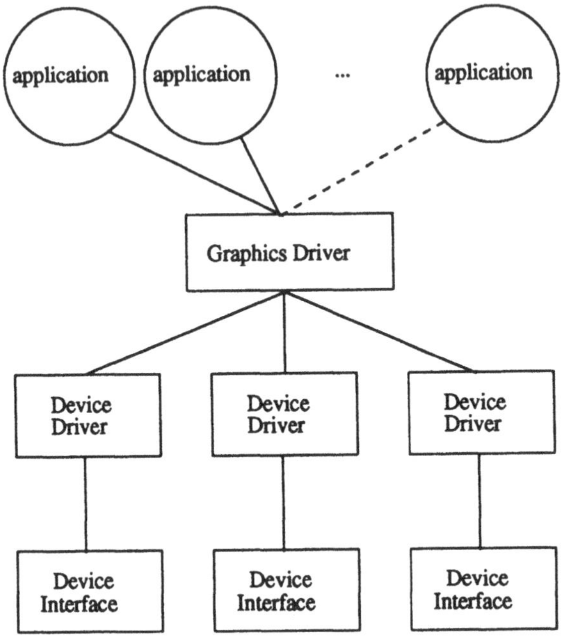

Figure 1. Design Hierarchy.

ALGEBRAIC INTERFACE

Like algebraic computation, converting equations into graphical displays is a time consuming process requiring much trial an error. The graphics application will generally be a series of refinements from an initial attempt. This paper examines the requirements of these applications, not their implementation. We first enumerate the types of display and their requirements.

Line/Scatter Plots

In figures 2 and 3 are typical examples of line and scatter plots generated using only the graphics primitives in the Appendix.

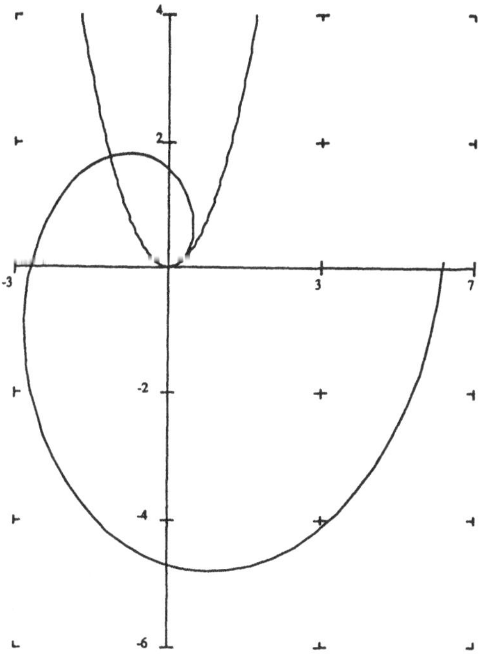

Figure 2. Line plot of function and parametric equation.

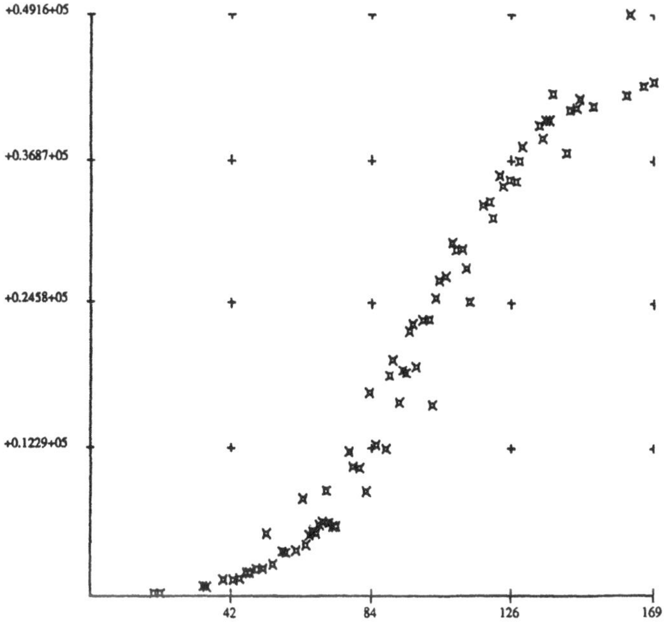

Figure 3. Scatter plot of data from experiment.

The important features required of the graphics primitives relate to scaling the display to the output device. We need to know the size of characters for proper axes labelling. Likewise, the display resolution affects the size of the markers in both plots. The ability to draw different line styles is significant for line plots. These plots were debugged with an X windows display and then plotted with the Postscript driver.

Contour Plots

If reasonably continuous, the behavior of a function of two variables can be investigated by display contours. In figure 4, is the Postscript display of a simulated wave tank with a barrier down the center with two holes.

Figure 4. Contour plot of wave tank.

Contour plots do not have significant hardware requirements and can be displayed in reasonable amounts of time on simple display devices. The most significant feature required (though not shown here) is the ability to display text along an irregular contour line. The interface routines allow text to be oriented at any angle and displayed at any position.

Wire Frame Plots

Simple 3 dimensional plots can be created with lines by drawing wire frame images and hiding lines where appropriate [Wright 73, Williamson 72]. In figure 5, we see a screen dump from an X windows display showing several hidden line displays as well as the REDUCE interaction window.

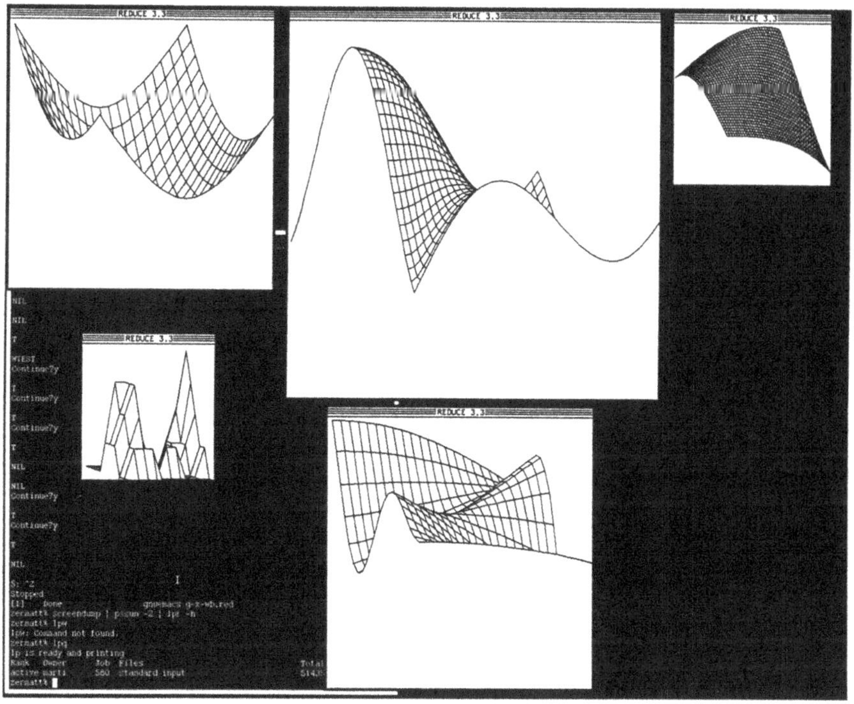

Figure 5. Wire frame displays.

This style of plot can be be written on virtually any output device. The important graphics requirements are knowing the display device resolution so that hidden lines can be clipped at a pixel boundary.

Non-Standard Displays

A quick glance through any computer algebra conference proceedings reveals a large number of pictures that don't fit the above criteria. Graphics are used to illustrate and animate algorithms. For example, figure 6 is a Postscript representation of an animated program demonstrating a crude form of numerical integration.

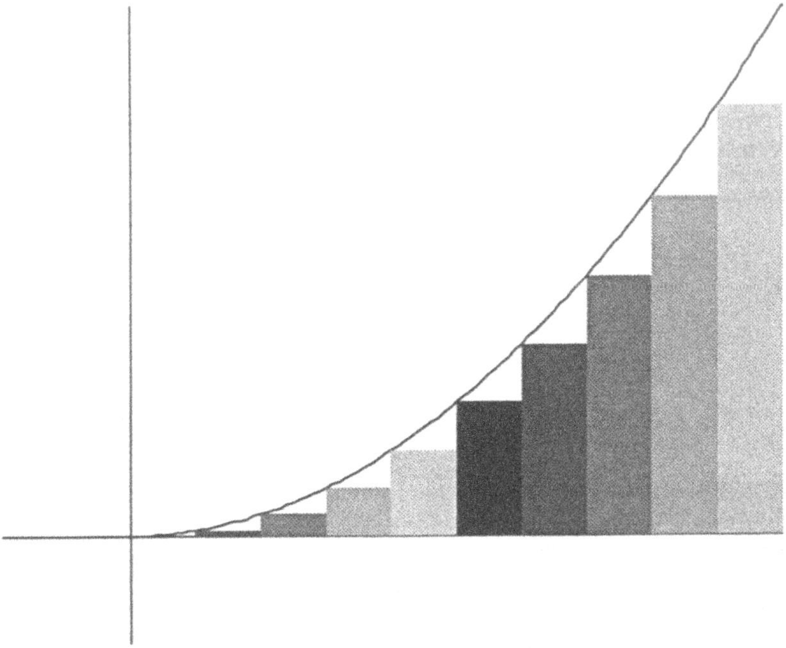

Figure 6. Numerical integration.

The features required for this type of display greatly exceed those required for the simple functional forms. This includes: locator, valuator, and button input, the ability to replicate and transform instances of graphical data objects, and the intelligent use of color.

Surface Rendering

The algorithms and hardware for displaying surfaces and solids has improved greatly over the years. Even inexpensive hardware an be used to create images such as those from the Hewlett Packard Renaissance display device. In figures 7-10 are black and white photographs taken directly from the color display. Figures 7 and 8 are part of an animation sequence showing properties of parabolas and cones. Figure 9 is a simple 3 dimensional surface and Figure 10 shows a simulated wave tank.

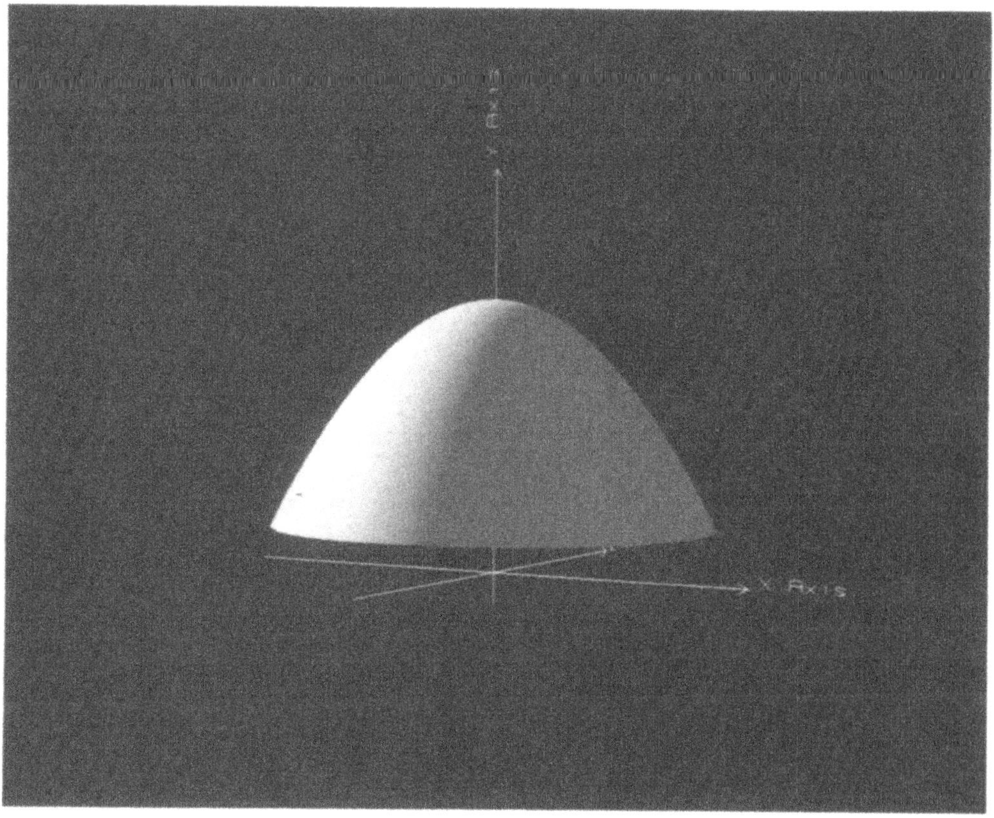

Figure 7. Parabola with light source and axes.

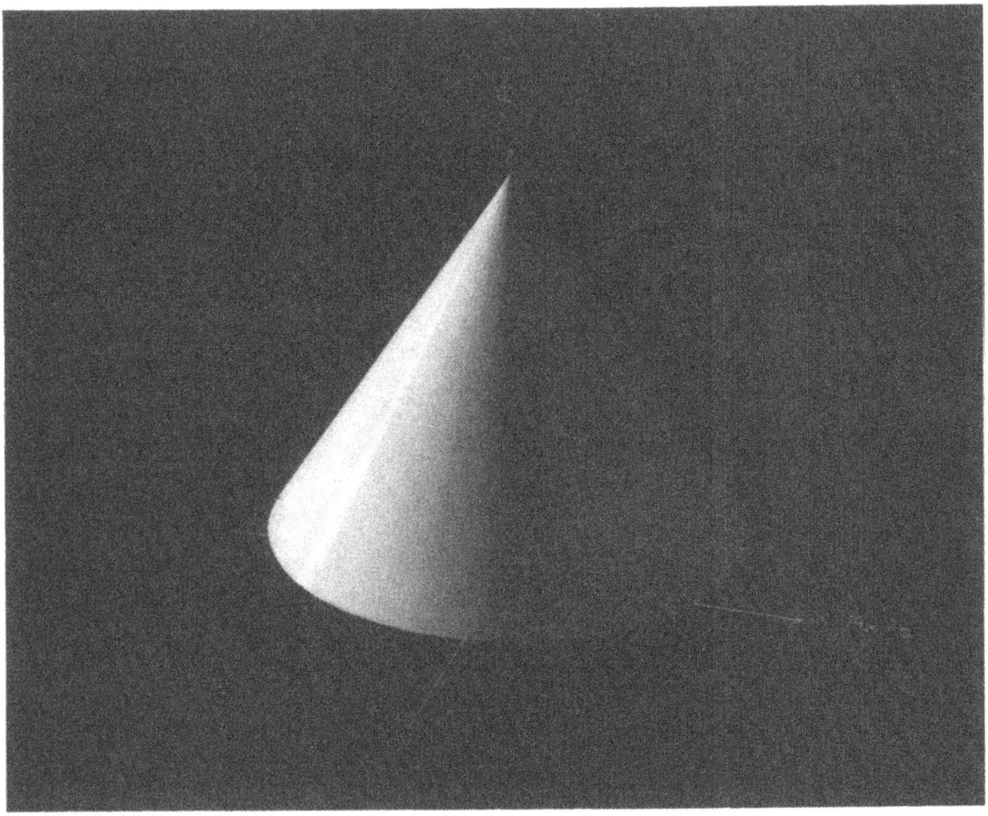

Figure 8. Cone with light source, axes, specular highlights.

Figure 9. Super-quadratic surface with altered viewpoint.

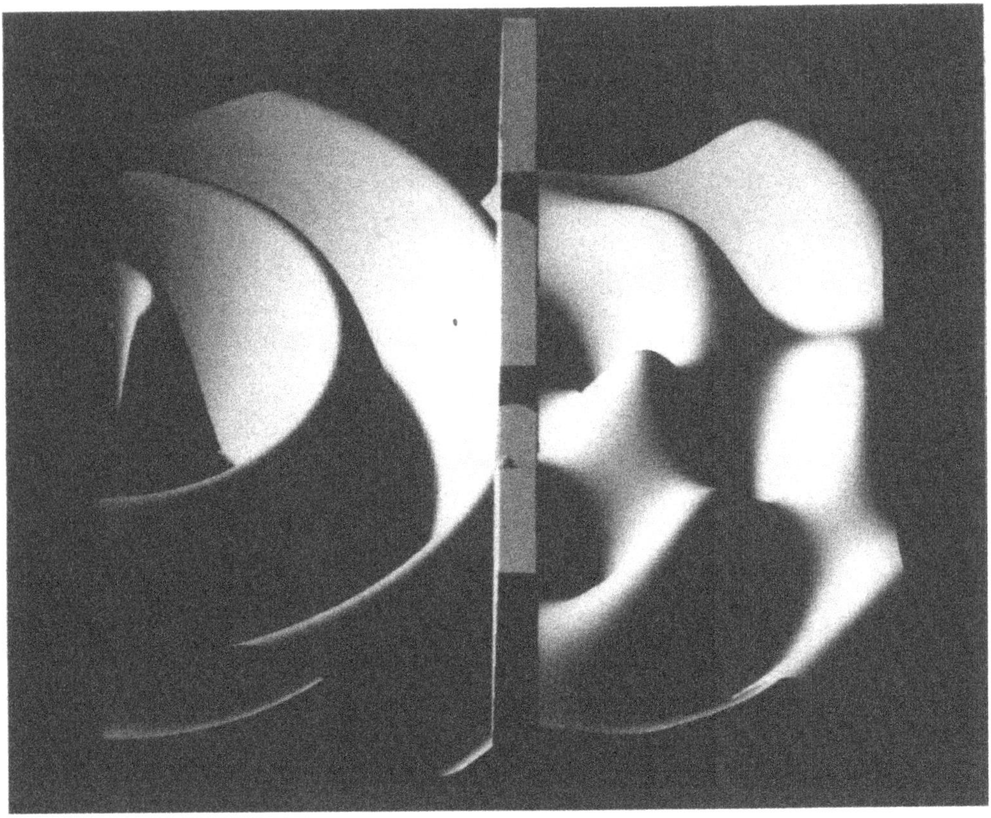

Figure 10. Simulated wave tank.

Though graphics standards exist for display in 3 dimensions, the commonality between them is rather limited. The features most usable by the computer algebra user are: Setting the view point and clipping volume, hidden surface removal, orientation of text in 3 dimensions, animation, and interactive mechanisms for setting the viewpoint. We have avoided the issues of light sources and shadows since these are relatively non-standard. When a particular graphics driver supports these activities (as in the case of HP Starbase), they can be modified with non-standard software escape sequences.

GRAPHICS ACTIVITIES

The graphics support routines are designed to support the activities depicted in the previous section. We examine each of the the features of the graphics system and discuss its implementation and portability.

Text

Across the spectrum of display devices, there are two character display styles which we characterize as *bitmap* and *stroked*. The bitmap font is characterized by a library of fixed size fonts stored as bits in a matrix. Switching font sizes is accomplished by switching to a new font library (X windows, most personal computers). Stroked fonts are drawn either as line segments (GKS) or as filled polygons of arbitrary size (Postscript and Starbase for example). The stroked fonts permit arbitrary registration, fill color, and style.

To allow a program to take advantage of at least some of these features, a standard query function distinguishes between the two types. On devices supporting bitmap fonts, the programmer is allowed to change fonts at any time. For simplicity, we assume fixed width fonts.

Color

We have divided support for color into four classifications: **monochrome**, **multi-chrome**, **grey-scale**, and **continuous** colors. Monochrome systems support only a foreground and background color. Multi-chrome systems support a restricted number of colors that can be easily maintained in a list. Grey-scale systems support a number of shades of a single color. Continuous color systems support a larger number of colors than can be conveniently maintained in a list.

The inquiry function returns the type of system as above. This information can be used to tailor program output appropriately. The following conventions are usually acceptable for application programs:

1. For monochrome devices, line style is normally used to differentiate between different lines.

2. Lines drawn on grey-scale devices are not easily distinguishable by intensity. Line style is normally used to differentiate. Grey scales are more useful for differentiating polygons.

3. Continuous colors are only useful in shading polygons. Lines should be differentiated by sticking to primary colors.

Lines

Though most graphics packages support line drawing, there are a number of facilities required beyond simply dropping a pen and drawing a straight line to some point. In particular, animating algorithms requires that lines be erased and some 3 dimensional displays require perspective transformations. We have separated line drawing ability into two styles: fixed and segments. In a fixed scheme, bits are transferred to the display device to form a line, or a pen is moved across a surface leaving a line. Examples of this type are X windows, Starbase, Core Graphics [CORE] and most inexpensive CRT display devices. More complex graphics devices and packages support graphical segments. When a line is drawn, its structure is maintained in a display list. A pointer or index of this segment completely identifies the bits. The segment can be translated and rotated by the graphics hardware given the pointer. Examples of this style are devices run by GKS and PHIGS.

Line characteristics that can be changed are the line style (dotted, dashed, dot-dash, etc.) and color. We have opted not to require support for lines wider than a single pixel. The standard defines several functions for display of lines both 2 and 3 dimensional.

Input

We examined the various common input devices that might be useful for computer algebra systems. Though many personal computers have no input device other than the keyboard, most workstations will have at lease a mouse with one or more buttons. High performance graphics workstations have additional input capabilities in the form of knobs and buttons. These are classified as *buttons*, *locators*, and *valuators*.

Buttons are either up or down. The programmer can inquire what buttons are available. Waiting for a button to be depressed is accomplished by having the program busy wait. Locators are characterized by having both an X and Y location when queried. These are given in world coordinates in relation to the currently selected view surface. Valuators have a single floating point value from 0.0 to 1.0. Since most systems with valuators have more than one, an inquiry function returns a list of selectable valuators.

Input devices are only useful for interactive control of the display. Thus programs must check for availability before attempting such actions as defaults from the devices have little meaning.

Polygons

Most reasonable graphics packages support display of filled polygons. Those systems that cannot reasonably be made to do so (pen plotters and the like) can at least draw wire frame images. Like line drawing, polygons are either fixed or segments. Systems that support segments for polygons use hardware to translate and rotate the polygon.

We have opted to to support interior fill styles for polygons. The number of vertices is limited to 256 by many implementations. Systems that support 3 dimensional polygons may also support hidden surface removal. Pseudo hidden surface removal can be done in 2 dimensions by drawing planes farthest away from the viewer first.

Viewpoints and Windows

To take advantage of graphics accelerators and remove the burden from the applications programmer, the interface provides an automatic method for mapping between **world coordinates** and **device coordinates**. The programmer works in world coordinates appropriate to the application. The graphics interface maps these coordinates into pixels or pen movements in the display area. When 3 dimensional display areas are created, the mapping is from 3 space coordinates to 2 space through the perspective transformations.

When the graphics output medium is interfaced to a window system, the interface supports multiple simultaneous output areas (see figure 5 for example) with differing transformations in each window. System dependent features such as different color maps in each window are not specified in the standard. To reuse a display area, the system allows the programmer to redefine the world coordinate to device coordinate transformation.

Segments

Supporting animation requires reasonable sophistication on the part of the graphics interface. Rather than require the user to manage a transformation stack and manipulate graphical objects himself, we have opted to include this facility in the graphics interface itself. Systems like GKS and PHIGS support this concept in both hardware and software providing *segments* as a handle on an internal representation of a graphical object. Systems that do not support segments will either not support animation or require construction of a display list and segment support.

IMPLEMENTATIONS

To test the applicability of the interface, we implemented a number of test display algorithms and an interface to the REDUCE algebraic mode. Examples of these appear in various places in the paper. To avoid the high cost of color reproduction, most of these are monochrome output from the postscript driver.

We tested a number of widely available standard graphics packages as well as those specific to a manufacturer. These include: GKS, Postscript, Starbase, and X Windows.

GKS

Though GKS is the basis of other work on integrating graphics and computer algebra, it is by no means the most appropriate choice [Bittencourt]. In particular, the standard [GKS], supports only two dimensional graphics. GKS does support segments: a useful artifact for animation. Of the three implementations available to us, one supports a hardware display list for fast transformation of segments, the others have a software display list with less performance. Applications have generally proved easy to move amongst the implementations.

As the GKS libraries are written in C (or Fortran) our implementation consists of a number of layers. The lowest levels translate Lisp data structures to and from GKS data structures. The driver supports only the following features: a single window (usually special hardware), locators, buttons, and valuators as available, 2D filled polygons, different line styles, segments, and color (as available).

Postscript

Postscript is the exclusive output from *Technique* [Wolfram]. The output is acceptable by both laser printers and window systems. Its only drawbacks are poor support for color and lack of acceptance on high-performance graphics workstations.

The Postscript implementation presupposes paper output, consequently it is of little worth for animation (not Postscript's fault). It supports several grey levels (rather than color), a single window per page, no locators or valuators, 2D filled polygons, different line styles, and no segments. Text processing is well supported with a large number of fonts and options. The implementation creates a symbolic file of Postscript commands (these can be quite large for even simple figures). The file is sent to a laser printer for display when the program executes a **Make-Current** function.

Starbase

Starbase is Hewlett Packard's proprietary interface to the Renaissance display device [HP]. This device is a medium speed surface rendering graphics engine with a high resolution display. The hardware supports Goraud shading, Z buffer hidden surface removal, up to 7 light sources, spline surfaces, hardware dithering, and animation. It does not have a display list (no segments).

Our implementation supports the full set of functions in the appendix. However, erasing is difficult in a 3 dimensional graphics, so this is accomplished by redrawing the entire scene. A library of C functions interfaces Lisp data structures to those required by Starbase. The large number of colors available in our 32 plane system prompted the model of the color functions used in the interface. This system was used to film 3 dimensional solid modelling animation sequences.

X Windows

The X Windows system is a standard accepted on a large number of workstations and time sharing machines. The system is particularly well suited for 2 dimensional graphics on medium performance work stations. We have tested our implementation on SUN and Hewlett Packard machines. Hardware and implementation differences are nearly transparent.

We implemented both a monochrome and color driver. The implementation uses the C library of routines rather than generating the packets required by the X server. It allows multiple windows to be active at a time (see Figure n) as well as: 2 dimensional polygon fill, several line styles, erasure rather than segment display, locators and buttons, and a large font library. The portability of this system is a standard against which all others can be judged.

FUTURE WORK

Several drivers need to be implemented to test the standard. Of particular importance are that for the IBM PC, the Apple MacIntosh, and PHIGS. These have the greatest potential for the largest number of users as well as being supported by a large number of hardware vendors.

Beyond the rudimentary programs demonstrated, additional application packages are contemplated. These include: user interaction with 3 dimensional surfaces, scatter plots and curve fitting, and creating graphical data bases. The existing application remain to be tested on smaller machines. In all cases, the application programs need to be tested in many environments.

Animation is of particular importance to educational users. The interface addresses the low level software issues of animation, but not how to create animated sequences. Promising research in this area is being conducted in the simulation and graphics communities.

CONCLUSIONS

Our initial experiences with the interface assure us that application program portability is possible at the same level as REDUCE algebraic applications. The sole cost is recognizing the differences in display characteristics and dynamically modifying program behavior for appropriate output.

ACKNOWLEDGEMENTS

The author would like to thank Tony Hearn, Arthur Norman, and Barbara Gates for many helpful discussions. This work was supported in part by an equipment grant from the Hewlett Packard Corporation.

APPENDIX

Terminology

Area: A view area. This can be a window, piece of paper, or a physical output device.

Device Coordinates: Coordinates presented to the hardware device. These are usually integers representing a particular pixel on a screen.

Segment Descriptor: On systems that support graphical objects, this is a pointer, index or what ever of the graphical representation of an object (GKS and PHIGS implementations).

World Coordinates: Coordinates that are transformed into Device Coordinates. The World Coordinate system is initialized by INITIALIZE2D and INITIALIZE3D.

Functions

(BUTTON DEV:any):id

DEV is a button descriptor returned by the INQUIRE-BUTTONS function. If the named button is depressed, DOWN is returned otherwise UP is returned. By convention DEV is a dotted-pair of (<description> . <value>) where <description> is a string that can be used in explanatory text and <value> is system dependent. If no such button device exists, UP is returned.

(CLEAR-AREA):area

If the output device supports this operation, the viewing area currently selected is cleared to its background color. For hardcopy devices, this should have the effect of moving to a new piece of paper or some physical action. This function is required. The value returned should be the the area descriptor (as returned by INITIALIZE2D or INITIALIZE3D).

(CLOSE-AREA):NIL

If the output device is a CRT, the graphics connection to the currently selected portion of the CRT (perhaps all) is severed. The graphics area can not subsequently be drawn to without a new INITIALIZEnD function call. If the area is on a hardcopy device, final printing is initiated and the graphics connection is closed. The value returned should be NIL, not the area descriptor since it is now invalid.

(DELETE-SEGMENT N:any):NIL

On devices (or in implementations that support segments), deletes the segment from the display list (see INQUIRE-SEGMENTSP). The actual deletion from the device occurs by calling the MAKE-CURRENT function. If segments are not supported, this function does nothing or generates an error.

(DRAW-LINE2D PLIST2D:number-list):any
(DRAW-LINE3D PLIST3D:number-list):any

Draw a line in 2 or 3 dimensional space. PLIST2D is a list of ((x y) (x y) ...) points and PLIST3D is a list of ((x y z) (x y z) ...) points. If segments are supported, the graphics segment is returned, otherwise NIL is returned.

(DRAW2D X:number Y:number):any
(DRAW3D X:number Y:number Z:number):any

Move the "pen" from its current position to World Coordinates X,Y and draw a line. Additional characteristics of the line may be set if the output device supports them (for example color, dotted-lines, etc.).

This function is required, but line type (color, style, width) is dependent upon the hardware being used. In systems that support segments, these functions return a segment descriptor.

INITIALIZE2D DEVICE:any
BITX:integer BITY:integer BITSX:integer BITSY:integer
LLX:number LLY:number URX:number URY:number OPTIONS:any):any
DEFAULT2D DEVICE:any
LLX:number LLY:number URX:number URY:number OPTIONS:any):any

Initializes a 2 dimensional viewing surface at BITX,BITY on the display DEVICE. The viewing area will be BITSX by BITSY in size. The lower left hand corner will be mapped into LLX, LLY and the upper right hand corner into URX,URY. The functions INQUIRE-WIDTH-BITS and INQUIRE-HEIGHT-BITS return the size of this area and INQUIRE-DISPLAY returns a data structure describing the display device.

OPTIONS is a list of parameters for a particular device. DEFAULT2D is an equivalent function that gives a default display area with the specified world coordinate system.

(INITIALIZE3D DEVICE:any
BITX:integer BITY:integer BITSX:integer BITSY:integer
LLX:number LLY:number LLZ:number
URX:number URY:number URZ:number OPTIONS:any):any
(DEFAULT3D DEVICE:any
LLX:number LLY:number LLZ:number
URX:number URY:number URZ:number OPTIONS:any):any

Initialize a 3 dimensional viewing area at BITX,BITY on the display DEVICE. The viewing area on the display will be BITSX by BITSY in size (device coordinate bits). The area being viewed will be clipped into the box with a lower left hand corner in the front at LLX, LLY, LLZ and a far upper right hand corner at URX, URY, URZ. The viewpoint of this area must be selected by SET-VIEW before display can take place. OPTIONS is a list of parameter names and parameters specific to the display device. This can be things like setting fonts, shading algorithms, back face clipping, double buffering and the like. The functions INQUIRE-WIDTH-BITS and INQUIRE-HEIGHT-BITS return the bit sizes of the display area. Other INQUIRE-functions indicate amounts of support for 3D. Required 3D support includes lines and text. Polygons, hidden surface, and shading are optional. DEFAULT3D builds a default display area.

(INQUIRE-BUTTONS):any-list

Returns a list of possible "button" input device identifiers. By convention these device identifiers are dotted-pairs (<description> . <value>) where <description> is a string that can be used in explanatory text. These buttons can be queried by the BUTTON function and return a state of DOWN or UP. If no such button devices exist, NIL is returned.

(INQUIRE-COLOR):boolean

INQUIRE-COLOR returns the type of color support provided by the currently selected output device. INQUIRE-COLOR returns MONOCHROME, MULTICHROME, GREYSCALE, or CONTINUOUS as defined above.

(INQUIRE-COLORS):id-list

INQUIRE-COLORS returns a list of colors accepted by the currently selected output device. By convention, the first element is the default background color, the second the foreground color. Subsequent elements, if any, are identifiers naming accepted colors. For CONTINUOUS systems, this list contains at least the standard defaults: (BLACK WHITE RED YELLOW GREEN CYAN BLUE MAGENTA).

(INQUIRE-DISPLAY):list

Returns a list describing the current display area. Separate INQUIRE functions are defined for each item.

1. (INQUIRE-DISPLAY-TYPE) Display type, one of the following:

   a) CRT - full screen CRT, only one display area.
   b) WINDOW - multiple window system CRT.
   c) DOTMATRIX - Dot matrix printer.
   d) POSTSCRIPT - Postscript display device.
   e) PLOTTER - hardcopy device, only one window, stroke graphics.

2. (INQUIRE-IMPLEMENTATION-NAME) Implementation name (author etc.)

3. (INQUIRE-POLYGON2D-FILLP) 2D polygon fill supported (T/NIL).

4. (INQUIRE-POLYGON3D-FILLP) 3D polygon fill supported (T/NIL).

5. (INQUIRE-COLOR) Color support (MONCHROME/MULTICHROME/CONTINUOUS).

6. (INQUIRE-WINDOWSP) Multiple simultaneous view areas (windows) supported (T/NIL).

7. (INQUIRE-SEGMENTSP) Segments supported (T/NIL).

8. (INQUIRE-TEXT-STYLE) Returns either PIXMAP or STROKE.

(INQUIRE-FONT-HEIGHT):number
Returns the height, in world coordinates, of the output font.

(INQUIRE-FONT-WIDTH):number Returns the width, in world coordinates, of the output font.

(INQUIRE-HEIGHT-BITS):number
Returns the number of vertical bits (or increments for a plotter) in the currently selected display area.

(INQUIRE-LINE-STYLES):any-list
Returns a list of possible line styles for use by the SETLINESTYLE function.

(INQUIRE-LOCATOR):any
Returns the a system specific identification for a "locator" input device. By convention this is a dotted-pair (<description> . <value>) where <description> is a string that can be used in explanatory text and <value> is system specific. Locators are characterized by returning an (X . Y) value from the LOCATE function. NIL is returned if no locator device exists. Only one locator device can be accessed by the system.

(INQUIRE-VALUATORS):any-list
Returns a list of identifiers of "valuator" input devices. These input devices are characterized by returning a single value from the VALUATOR function. If no valuator devices exist, NIL is returned.

(INQUIRE-WIDTH-BITS):number
Returns the number of horizontal bits (in increments for a plotter) in the currently selected display area.

(LOCATE DEV:any):dotted-pair
Returns a dotted-pair (X . Y) in the world coordinate system of the currently selected display area. DEV is the value returned by INQUIRE-LOCATOR function. If no locator device exists, (0.0 . 0.0) is returned. If no display area has been opened, the function should return some reasonable value. If a display area has been opened on some other driver, the value returned is in relation to the last area opened on the device the locator is associated with.

(MAKE-CURRENT):any

If commands are being buffered, this function completes the display process. In systems supporting graphics segments, this routine swaps buffers (for animation effects). Other systems might cause a display of buffered commands. For hard copy systems, the current output is sent to the display device.

(MOVE2D X:number Y:number):NIL
(MOVE3D X:number Y:number Z:number):NIL

Move the "pen" to the World Coordinates X,Y without drawing a line. This function must be supported.

(MOVE-SEGMENT2D S:segment X:number Y:number R:rotation):segment
(MOVE-SEGMENT3D S:segment X:number Y:number PITCH:number ROLL:number YAW:number):segment

For systems that support graphics segments, these cause translation and rotation of graphical objects. In systems that do not support segments these cause an error.

(POLYGONFILL2D PLIST2D:number-list):any
(POLYGONFILL3D PLIST3D:number-list):any

Draw an area and fill it with the current color. PLIST2D is a list of X and Y coordinates ((x y) (x y) ...). The last coordinates need not be the same as the first. PLIST3D is a list of ((x y z) (x y z) ...) points, the last need not be the same as the first. Shading options are set as part of the machine specific options in INITIALIZE3D. POLYGONFILL3D need not be supported.

(RESCALE2D LLX:number LLY:number URX:number URY:number):any
(RESCALE3D LLX:number LLY:number LLZ:number URX:number URY:number URZ:number):any

Reset the scale on the current display area.

(SELECT-AREA A:any):any

Selects a view area for display. The argument is that returned by INITIALIZE2D or INITIALIZE3D.

(SET-COLOR COLOR:{id, float-list}):any

SET-COLOR sets the output color for all subsequent output operations. If COLOR is an identifier, it must be the name of a color acceptable by the system. If the system cannot support the named color, an appropriate substitution is made. When COLOR is a list of three floating point numbers, the selected color is that closest to the corresponding (red green blue) values. Here 1.0 means full intensity color, 0.0 means black.

(SET-FONT FNT:{string, any}):fontid

Sets the output font to a font style of the argument. This is normally a system dependent name, a string naming a specific font. SET-FONT returns a font identifier that can be used in subsequent SET-FONT calls. If the font cannot be changed, the call is ignored. If the requested font is not available, an error or warning message is given.

(SET-FONT-ORIENTATION A1:float A2:float):any

For implementations that support stroke based character sets, this function sets the 2 and 3 dimensional orientation of subsequent displayed text. A1 is the angle (in degrees) in the X - Y plane (the vertical plane facing the viewer in 3D). A2 is the angle in degrees in the X-Z plane. This argument is ignored for 2 dimensional text.

(SET-FONT-SCALE WIDTH:float HEIGHT:float SLANT:float THICKNESS:float):any

For implementations that support stroke based character sets, this function alters the size and style of the output characters. WIDTH, HEIGHT, and THICKNESS are given as world coordinate values. SLANT in degrees causes letters to lean out of perpendicular. If stroke characters are not supported, this function call is ignored.

(SET-LINE-STYLE STYLE:any):any

Sets the output line style. STYLE is selected from any of those returned by INQUIRE!-STYLES function.

(SET-VIEW EYEX:number EYEY:number EYEZ:number
REFX:number REFY:number REFZ:number ANG:number):NIL

Set the viewpoint for a 3 dimensional display with the view point at location EYEX, EYEY, EYEZ centering the point REFX, REFY, REFZ in the middle of the viewing area. The viewing plane is perpendicular to the line between these two points. ANG is the angular field of view in degrees.

(TEXT2D X:number Y:number MSG:string):string
(TEXT3D X:number Y:number Z:number MSG:string):string

Display the text of MSG at world coordinates X,Y. The color is set by SetColor, the font is selected by INITIALZE2D (possibly modified by a parameter in OPTIONS) or INITIALIZE3D. 3D text is displayed parallel to the viewing plane with location subject to the perspective transformation but characters not scaled. The size of the font can be determined with the INQUIRE-FONT-HEIGHT and INQUIRE-FONT-WIDTH functions.

(VALUE DEV:any):float

Returns the value of the DEV valuator device. DEV is one of the list returned by the INQUIRE-VALUATORS function and is by convention (<description> . <value>) where <value> is system dependent and <description> is a string that can be used in explanations. If no such valuator device exists, 0.0 is returned.

List of References

[Adobe 85] Adobe Systems, *Postscript Language Tutorial and Cookbook*, Addison-Wesley Publishing Company, Inc., Menlo Park, California, 1985.

[GKS] *Computer Graphics - Graphical Kernel System (GKS) Functional Description*, American National Standard, ANSI X3.124-1985.

[BittenCourt] Guilherme BittenCourt, *Integration of Graphical Tools in a Computer Algebra System*, private communication.

[Chato 86] Donna M. Chato, *MATHGRAF Makes Math FUN!*, Caren Company, 12137 Midway Drive, Tracy, CA, 95376, 1986.

[Cole 81] Chris A. Cole, S. Wolfram, et al, *SMP A Symbolic Manipulation Program*, California Institute of Technology, July 1981, Chapter 10.

[Gettys] Jim Gettys, R. Newman, T. D. Fera, *Xlib - C Language X Interface Protocol Version 10*, MIT Project Athena, 1986.

[Hearn 87] A. C. Hearn, *REDUCE User's Manual, Version 3.3*, RAND Publication CP78 (Rev 10/87).

[HP] Hewlett Packard Corporation, *Starbase Reference*, HP Part Number 98592-90061, 1987.

[Leong 86] B. L. Leong, *Iris: Design of a User Interface Program for Symbolic Algebra*, Proceedings of the 1986 Symposium on Symbolic and Algebraic Computation, Bruce Char Ed., ACM 1986, pp. 1-6.

[McNamara 85] Brendan McNamara, P. A. Willman, *GRAFL-II: User Oriented Science Graphics*, Lawrence Livermore National Laboratory, UCID-20467, 1985.

[MGA Inc.] MGA Inc., *Pro-MATLAB*, 73 Junction Square Drive, Concord, MA, 01742, 1987.

[MIT 83] Mathlab Group, *MACSYMA Reference Manual*, The Mathlab Group, Laboratory for Computer Science, January 1983, Volume II, Sections 1,2.

[Seymour 86] Harlan R. Seymour, *Conform: A Conformal Mapping System*, Proceedings of SYMSAC '86, July, 1986, pp. 163-168.

[Smith 86] Carolyn J. Smith, N. M. Soiffer, *MathScribe: A User Interface for Computer Algebra Systems*, Proceedings of the 1986 Symposium on Symbolic and Algebraic Computation, Bruce Char Ed., ACM 1986, pp. 7-12.

[Symbolics 87] Symbolics, Inc., *MACSYMA Reference Manual*, Computer-Aided Mathematics Group, Symbolics Inc., Eleven Cambridge Center, Cambridge, MA 02142, Chapter 14.

[Williamson] Hugh Williamson,*Algorithm 420 Hidden-Line Plotting Program [J6]*, CACM **15**, No. 2, February 1972, pp. 100-103.

[Wolfram] Stephen Wolfram, *Some Preliminary Technique Examples*, private communication.

[Wright] Thomas J. Wright, *A Two-Space Solution to the Hidden Line Problem for Plotting Functions of Two Variables*, IEEE Transactions on Computers, Vol. C-22, No. 1, January 1973, pp. 28-33.

[PHIGS] *Programmers Hierarchical Interface.*

VLSI DESIGNS FOR MULTIPLICATION OVER FINITE FIELDS $GF(2^m)$

Edoardo D. Mastrovito
Department of EE, Linköping University
S-581 83 Linköping, Sweden

ABSTRACT

The finite fields $GF(2^m)$ play a central role in the implementation of BCH/Reed-Solomon coders and decoders. Also, these fields are attractive in some data encryption systems. In this paper we describe a method for designing a parallel multiplier for $GF(2^m)$ that is both speed and area efficient. The multiplier proposed is based on the conventional (or polynomial) base representation. From our multiplier we can derive the one introduced by Bartee and Schneider [9]. Their multiplier has been considered unsuitable for VLSI because of lack of modularity. Our approach shows that this multiplier is indeed modular and can also exhibit a high degree of regularity. It is thus well suited for VLSI. Compared to the best parallel design available today, our design requires, roughly, only half the number of gates and still achieves a high operational speed. The speed, size and regularity of our design depends on the irreducible polynomial used to generate the field. In the paper we derive two simple selection criteria for choosing the irreducible polynomial in order to obtain a good design. Also, we present a list of best polynomials for $m \leq 16$.

1. INTRODUCTION

The finite fields $GF(2^m)$ play a central role in the implementation of BCH/Reed-Solomon coders and decoders [1], [2]. The most efficient decoding algorithms for BCH/RS codes involve a considerable amount of computations in finite fields [3], [4]. There is, thus, a need for compact and fast multipliers which are easily implemented in VLSI. Addition in $GF(2^m)$ is readily realized by bitwise exclusive-OR (XOR) addition while multiplication is a difficult task that requires much more complex devices. We are primarily concerned with fast parallel multipliers. The multipliers described in this paper multiply two field elements in a single clock cycle. Another property which is of great importance to us is the size of the multipliers since multiplication is frequent in most known decoding algorithms. In the Berlekamp-Massey algorithm, for example, $2t$ multipliers are required where t is the number of errors that can be corrected by the code [5].

A number of VLSI designs have been proposed [6], [7], [8]. These designs are regular and require in the order of $4m^2$ logical gates or more. The multipliers described in this paper are modular and almost as regular but require, often, only in the order of $2m^2$ logical gates. From our multiplier we can derive the one introduced by Bartee and Schneider in 1963 [9]. References [6] [7] [8] consider the multiplier in [9] as unsuitable for VLSI because of lack of modularity. Our approach shows that this multiplier is indeed modular and can also exhibit a high degree of regularity. It is thus well suited for VLSI. In section 2 we present the algorithm, in section 3 we analyse it and and in section 4 we describe the realization of the multiplier.

2. MULTIPLICATION IN $GF(2^m)$

Each element in the field $GF(2^m)$ will be represented by a polynomial of degree m-1 or less, whose coefficients are the binary digits 1 and 0. Each element is a residue mod $P(x)$ where $P(x)$ is an irreducible polynomial over $GF(2)$ of degree m, and all arithmetic operations are performed modulo 2. Alternatively, we will make use of the vector notation $A = (a_{m-1}, \ldots, a_0)$, where $a_{m-1}, \ldots, a_0$ are the coefficients of the polynomial associated with the field element A.

Let $A(x)$ and $B(x)$ be two elements of $GF(2^m)$. The product $A(x)B(x)$ mod $P(x)$ is denoted by $C(x)$ and can be expressed in the following way:

$$C(x) = A(x)B(x) \bmod P(x) = \sum_{i=0}^{m-1} c_i x^i$$

$$\text{where} \quad c_i = a_{m-1} f_{m-1}^i(B) + a_{m-2} f_{m-2}^i(B) + \ldots + a_1 f_1^i(B) + a_0 f_0^i(B) \tag{1}$$

The functions $\{f_j^i\}$ are linear in B, i.e. they are of the form

$$f_j^i(B) = \sum_{\substack{\text{for some } k \\ k \in [0,1,\ldots,m-1]}} b_k \tag{2}$$

Multiplication can now be described through the functions f_j^i which are dependent on the choice of irreducible polynomial $P(x)$. In matrix notation we can write the product as follows

$$C = \begin{pmatrix} f_{m-1}^{m-1} & f_{m-2}^{m-1} & \cdots & f_0^{m-1} \\ f_{m-1}^{m-2} & f_{m-2}^{m-2} & \cdots & f_0^{m-2} \\ \vdots & \vdots & & \vdots \\ f_{m-1}^0 & f_{m-2}^0 & \cdots & f_0^0 \end{pmatrix} \begin{pmatrix} a_{m-1} \\ a_{m-2} \\ \vdots \\ a_0 \end{pmatrix} = M A^t \tag{3}$$

We will call the m by m matrix M the *product matrix*. In general the determination of M is straightforward but the calculations quickly become rather tedious with growing m. Explicit formulas for computation of the functions f_j^i will be derived in the next section.

3. ANALYSIS OF THE DEPENDENCY OF M ON $P(x)$

We will see, in the next section, that the size, speed and regularity of the multiplier will depend heavily on the functions f_j^i, that is on the polynomial $P(x)$. Simply observe that the worst f_j^i we could ever encounter is the function $b_{m-1}+b_{m-2}+\ldots+b_0$ which requires m-1 XOR-gates, while the best f_j^i is the function b_u which requires no gates at all. Also, it takes obviously much longer time to compute $b_{m-1}+b_{m-2}+\ldots+b_0$ than b_u. The *number* of coefficients b_v involved in a function f_j^i will be called the *width* of f_j^i. It is then clear that it is desirable to

a) minimize the maximum width, i.e. maximize the speed
b) minimize the number of distinct f_j^i of width > 1

The aim of this section is to find simple criteria to determine wether a polynomial $P(x)$ is good or bad in the sense of properties a) and b). Also, we will prove the goodness of some particular classes of polynomials. We will start by formalizing the algorithm of the previous section and, in order to simplify the notation, we define the following symbol

$$s_k = \sum_{u+v=k} a_u b_v \qquad u,v \in [0, 1, 2, \dots, m\text{-}1] \tag{5}$$

The product $C(x) = A(x)B(x) \bmod P(x)$ can now be expressed in compact form by

$$C(x) = \sum_{k=0}^{2m-2} x^k s_k \bmod P(x) = \sum_{k=0}^{m-1} x^k s_k + \sum_{k=m}^{2m-2} [x^k \bmod P(x)] s_k \tag{6}$$

Let q_n^{k-m} be defined by

$$\sum_{n=0}^{m-1} q_n^{k-m} x^n = x^k \bmod P(x) \qquad k = m, m+1, \dots, 2m-2 \tag{7}$$

where the coefficients $\{q_v^u\}$ are elements of $GF(2)$. In matrix notation (7) becomes

$$\begin{pmatrix} x^m \\ x^{m+1} \\ \vdots \\ x^{2m-2} \end{pmatrix} = \begin{pmatrix} q_{m-1}^0 & q_{m-2}^0 & \cdots & q_0^0 \\ q_{m-1}^1 & q_{m-2}^1 & \cdots & q_0^1 \\ \vdots & \vdots & & \vdots \\ q_{m-1}^{m-2} & q_{m-2}^{m-2} & \cdots & q_0^{m-2} \end{pmatrix} \begin{pmatrix} x^{m-1} \\ x^{m-2} \\ \vdots \\ 1 \end{pmatrix} = Q \begin{pmatrix} x^{m-1} \\ x^{m-2} \\ \vdots \\ 1 \end{pmatrix} \qquad \bmod P(x) \tag{8}$$

where Q is an m-1 by m binary matrix which we will call the *reduction matrix*. We can now insert (7) into (6) and obtain

$$C(x) = \sum_{k=0}^{m-1} x^k s_k + \sum_{k=m}^{2m-2} \Big(\sum_{n=0}^{m-1} q_n^{k-m} x^n \Big) s_k = \sum_{k=0}^{m-1} x^k s_k + \sum_{k=0}^{m-2} \Big(\sum_{n=0}^{m-1} q_n^k x^n \Big) s_{k+m} =$$

$$= \sum_{k=0}^{m-1} x^k s_k + \sum_{k=0}^{m-2} q_{m-1}^k x^{m-1} s_{k+m} + \sum_{k=0}^{m-2} q_{m-2}^k x^{m-2} s_{k+m} + \dots + \sum_{k=0}^{m-2} q_0^k s_{k+m} \tag{9}$$

Let us look at the product coefficient c_i in (9)

$$c_i = s_i + \sum_{k=0}^{m-2} q_i^k s_{k+m} = s_i + q_i^0 s_m + q_i^1 s_{m+1} + \dots + q_i^k s_{m+k} + \dots + q_i^{m-2} s_{2m-2} \tag{10}$$

It is now time to utilize the definition of s_k given in (5) and organize the terms in the right-hand side in (10) in a more convenient form which allows us to see all the products $a_u b_v$ involved in (10). This convenient form is shown in Table 1. The first row shows the possible range of the index u. The second row shows the range of the index v for the term s_i, the third row shows the range of the index v for the term $q_i^0 s_m$ and so on for the other rows.

With the help of Table 1 we can write (10) in the following way

$$c_i = a_0 b_i + a_1 (b_{i-1} + q_i^0 b_{m-1}) + a_2 (b_{i-2} + \sum_{t=0}^{1} q_i^{1-t} b_{m-1-t}) + \dots +$$

$$+ a_i(b_0 + \sum_{t=0}^{i-1} q_i^{i-1-t} b_{m-1-t}) + a_{i+1}(\sum_{t=0}^{i} q_i^{i-t} b_{m-1-t}) + \dots + a_{m-1}(\sum_{t=0}^{m-2} q_i^{m-2-t} b_{m-1-t}) \tag{11}$$

		u:	0	1	2	3	...	i	i+1	...	m-2	m-1
s_i	=>	v:	i	i-1	i-2	i-3	...	0				
$q_i^0 s_m$	=>	v:		m-1	m-2	m-3	...	m-i	m-i-1	...	2	1
$q_i^1 s_{m+1}$	=>	v:			m-1	m-2	...	m-i+1	m-i	...	3	2
	:	:				:	:	:	:	:	:	:
$q_i^i s_{m+i}$	=>	v:							m-1	...	i+2	i+1
	:	:								:	:	:
$q_i^{m-3} s_{2m-3}$	=>	v:									m-1	m-2
$q_i^{m-2} s_{2m-2}$	=>	v:										m-1

Table 1

Comparison of (11) with the expression for c_i given in (1) yields the functions $f_j^i(B)$

$$\begin{cases} f_0^i(B) = b_i & i=0,\ 1,\ \dots,\ m-1 \quad (12) \\ f_j^i(B) = \sigma(i-j)\ b_{i-j} + \sum_{t=0}^{j-1} q_i^{j-1-t} b_{m-1-t} & i=0,\ 1,\ \dots,\ m-1\ ,\ j=1,\ \dots,\ m-1 \quad (13) \end{cases}$$

where $\sigma(k)$ is a step function defined by

$$\sigma(k) = \begin{cases} 1 & ,\ k \geq 0 \\ 0 & ,\ k < 0 \end{cases}$$

In vector notation (13) becomes

$$f_j^i(B) = \sigma(i-j) b_{i-j} + (q_i^0, q_i^1, \dots, q_i^{j-1}) \begin{pmatrix} b_{m-j} \\ b_{m-j+1} \\ \vdots \\ \vdots \\ b_{m-1} \end{pmatrix} \qquad \begin{matrix} i=0,\ 1,\ \dots,\ m-1 \\ j=1,\ 2,\ \dots,\ m-1 \end{matrix} \tag{14}$$

With the help of (14) we can make some considerations about the reduction matrix Q.

I. The vector $\bar{q} = (q_i^0, q_i^1, \ldots\ldots, q_i^{j-1})$ is a subset of column number i in the reduction matrix Q. Since we are interested in having functions f_j^i of small width, the vector $\bar{q}$ should contain a low number of ones, i.e. it should have a low (Hamming) weight. For $j = m$-1 this vector coincides with column number i in Q. Hence, a low maximum column weight of the matrix Q is desirable.

II. Let the diagonal of Q be the entries $q_0^0, q_1^1, \ldots, q_{m-2}^{m-2}$. The term $\sigma(i\text{-}j)b_{i-j}$ is non-zero only when $i \geq j$, that is f_j^i includes b_{i-j} only if $i \geq j$. Let w_1 be the maximum column weight of the matrix Q and w_2 the maximum column weight of the matrix Q with all the entries *on* and *below* the diagonal set to zero. Also, recall that the last component in the vector $\bar{q}$ is q_i^{j-1}. Then the maximum width is given by

$$\mathrm{MAX}\,\{w_1, w_2+1\} \tag{15}$$

III. For each i the term $q_i^0 b_{m-j}$ is the only term *always* being involved in the expression of f_j^i for $j = 1, \ldots, m$-1. Since we are interested in having few functions f_j^i of width >1, the vector $(q_{m-1}^0, q_{m-2}^0, \ldots, q_0^0)$ should contain as few ones as possible. That is, the first row of Q should have as low weight as possible. The first row of Q is nothing but the m least significant coefficients of $P(x)$. Hence, we should choose a polynomial $P(x)$ of lowest weight. Binary irreducible polynomials of degree >1 must have (odd) weight at least 3.

IV. It is easily verified that, no matter how we choose $P(x)$, the maximum width in (15) will never be less than 2. Hence, whenever we find a polynomial whose matrix Q has maximum width 2 we know we have found a good polynomial.

We are now ready to summarize our analysis and express some selection criteria. We suppose we are given a set I of irreducible polynomials of degree m and we would like to know which polynomial(s) should be selected in order to obtain a good multiplier design. Obviously, one way to find the best polynomial(s) is to compute the product matrix M for each polynomial in I and select the best one. However, the number of irreducible polynomials grows quickly with m. For example there are 4080 irreducible polynomials of degree 16. Therefore, we would like to reduce the set I by applying some simple selection criteria. The criteria we suggest are the following:

C1. Out of the set I select the subset I_1 of polynomials of minimum weight. If the subset I_1 is small, simply compute the product matrix M for each polynomial in I_1 with the help of (12) and (13) and select the best one. If I_1 is large then apply C2 on it.

C2. For each polynomial in the given set compute the maximum width according to (15) and select the subset I_2 of polynomials with smallest maximum width. If the subset I_2 is small, simply compute the product matrix M for each polynomial in I_2 with the help of (12) and (13) and select the best one. If I_2 is large select the subset I_3 (out of I_2) of polynomials of minimum weight, compute the associated product matrices and select the best one.

These two criteria reduce to one much simpler criterion whenever an irreducible polynomial of weight 3 (trinomial) exists. In these cases the selection criterion is simply:

C3. If $m = 2{\cdot}3^l$, $l = 0, 1, 2, \ldots$ then choose the polynomial $x^m+x^{m/2}+1$, otherwise the polynomial x^m+x^k+1 of lowest k.

In [14] we see that at least one irreducible trinomial x^m+x^k+1 exists for about half of the values in the range $2 \leq m \leq 1000$. As a partial motivation for C3 we introduce the following propositions:

<u>*Proposition 1*</u>: Suppose x^m+x+1 is irreducible over $GF(2)$. Then the associated product matrix M has the following form

$$M = \begin{pmatrix} b_0 + b_{m-1} & b_1 & b_2 & \dots & b_{m-2} & b_{m-1} \\ b_{m-1} + b_{m-2} & b_0 + b_{m-1} & b_1 & \dots & b_{m-3} & b_{m-2} \\ b_{m-2} + b_{m-3} & b_{m-1} + b_{m-2} & b_0 + b_{m-1} & \dots & b_{m-4} & b_{m-3} \\ \vdots & \vdots & \vdots & \vdots & \vdots & \vdots \\ b_2 + b_1 & b_3 + b_2 & b_4 + b_3 & \dots & b_0 + b_{m-1} & b_1 \\ b_1 & b_2 & b_3 & \dots & b_{m-1} & b_0 \end{pmatrix}$$

The maximum width is 2 and there are m-1 distinct functions f_j^i of width > 1. These m-1 functions appear in both the left-most column and the next-to-last row. Hence, only m-1 XOR gates are required to generate the m^2 entries of M.

Proof: The reduction matrix is easily found to be

$$Q = \begin{pmatrix} 0 & 0 & 0 & 0 & \dots & 0 & 0 & 1 & 1 \\ 0 & 0 & 0 & 0 & \dots & 0 & 1 & 1 & 0 \\ 0 & 0 & 0 & 0 & \dots & 1 & 1 & 0 & 0 \\ \vdots & \vdots & \vdots & \vdots & \vdots & \vdots & \vdots & \vdots & \vdots \\ 0 & 1 & 1 & 0 & \dots & 0 & 0 & 0 & 0 \\ 1 & 1 & 0 & 0 & \dots & 0 & 0 & 0 & 0 \end{pmatrix}$$

which contains ones only on the diagonal (as defined in II) and the entries soon above it. That is

$$q_i^k = 1 \qquad \text{if } i = k \text{ or } i = k+1 \text{ , } k = 0, 1, \dots, m\text{-}2$$

The maximum width according to (15) is then 2, i.e. the best possible. We look at the coefficient c_i as we did in (10) and obtain

$$c_i = s_i + \sum_{k=0}^{m-2} q_i^k s_{k+m} = \begin{cases} s_0 + s_m & i = 0 \\ s_i + s_{m+i-1} + s_{m+i} & i = 1, 2, \dots, m-2 \\ s_{m-1} + s_{2m-2} & i = m-1 \end{cases} \tag{16}$$

With the help of (16) and (5) we can create a table similar to Table 1 for each c_i and look again at all products $a_u b_v$. This is done in Table 2. It is now easy to see the actual functions f_j^i involved in each product coefficient c_i. The first row of M corresponds to c_{m-1} which is seen to be

$$c_{m-1} = a_{m-1}(b_0 + b_{m-1}) + a_{m-2} b_1 + \dots + a_1 b_{m-2} + a_0 b_{m-1}$$

the second row is given by c_{m-2}

$$c_{m-2} = a_{m-1}(b_{m-1} + b_{m-2}) + a_{m-2}(b_0 + b_{m-1}) + \dots + a_1 b_{m-3} + a_0 b_{m-2}$$

and so on for the other rows of M. A closer look at Table 2 reveals that there are m-1 distinct functions f_j^i and that they can be read out from c_1. These functions are

$$b_2 + b_1 \text{ , } b_3 + b_2 \text{ , } b_4 + b_3 \text{ , } \dots \text{ , } b_{m-1} + b_{m-2} \text{ , } b_0 + b_{m-1}$$

This completes the proof. □

			u:	0	1	2	3	4	.....	m-3	m-2	m-1
c_0	s_0	=>	v:	0								
	s_m	=>	v:		m-1	m-2	m-3	m-4	.....	3	2	1
c_1	s_1	=>	v:	1	0							
	s_m	=>	v:		m-1	m-2	m-3	m-4	.....	3	2	1
	s_{m+1}	=>	v:			m-1	m-2	m-3	.....	4	3	2
c_2	s_2	=>	v:	2	1	0						
	s_{m+1}	=>	v:			m-1	m-2	m-3	.....	4	3	2
	s_{m+2}	=>	v:				m-1	m-2	.....	5	4	3
	:		:				:	:	:	:	:	:
c_{m-2}	s_{m-2}	=>	v:	m-2	m-3	m-4	m-5	m-6	.....	1	0	
	s_{2m-3}	=>	v:								m-1	m-2
	s_{2m-2}	=>	v:									m-1
c_{m-1}	s_{m-1}	=>	v:	m-1	m-2	m-3	m-4	m-5	.....	2	1	0
	s_{2m-2}	=>	v:									m-1

Table 2

Even if most trinomials x^m+x+1 are not irreducible there are several ones of practical interest. The following is a list of all irreducible trinomials x^m+x+1 of degree 2 through 127, [14]:

m = 2, 3, 4, 6, 7, 9, 15, 22, 28, 30, 46, 60, 63, 127

Proposition 2: Suppose $m = 2{\cdot}3^l$, $l = 0, 1, 2, \ldots$. Then $x^m+x^{m/2}+1$ is irreducible and the associated product matrix M has maximum width 2. Also there are $m/2$ distinct functions f_j^i of width > 1. Hence, only $m/2$ XOR gates are required to generate the m^2 entries of M.

Proof: The irreducibility of the trinomial for $m = 2{\cdot}3^l$, $l = 0, 1, 2, \ldots$ is proved in [13, p.12, Th.1.1.28]. The rest is proved applying the same technique as in Prop. 1. □

Whenever the trinomial $x^m+x^{m/2}+1$ is irreducible it is in fact the best polynomial one can choose. Unfortunately it is rarely irreducible. The first five irreducible ones are

m = 2, 6, 18, 54, 162

$x^m+x^{m/2}+1$ and x^m+x+1 are the only trinomials achieving the lowest maximum width, i.e. 2. It is a rather simple matter to check that the maximum width can not be less than 3 for all trinomials x^m+x^k+1 where $k \geq 2$, $k \neq m/2$. In fact the maximum width is exactly 3 for $2 \leq k < m/2$, increases for $k > m/2$ and reaches the highest value (m) for $k = m$-1. By experimenting we have found that the lowest k for which x^m+x^k+1 is irreducible will minimize the number of distinct functions f_j^i of width ≥ 3 and consequently the number of gates required. In fact, we have observed that the number of gates is $2m^2$-2+k for $1 \leq k < m/2$.

We conclude this section by presenting another class of polynomials which achieves the lowest maximum width. It is in fact the class of maximum (!) weight polynomials.

Proposition 3: Suppose the polynomial $x^m+x^{m-1}+x^{m-2}+ \ldots +x+1$ is irreducible over $GF(2)$. Then the associated product matrix M has maximum width 2 and the number of distinct f_j^i of width > 1 is $(m^2-m)/2$. Hence, $(m^2-m)/2$ XOR gates are required to generate the m^2 entries of M.

Proof: Apply the same technique as in Prop.1. □

The high number of distinct f_j^i exhibited by this class is due to the high weight of the polynomials, see consideration III. By [12, p.21, Th.33] we find that the first eleven irreducible all-ones polynomials are

m = 2, 4, 10, 12, 18, 28, 36, 52, 58, 60, 66

In many of these cases there will be other polynomials requiring fewer gates (but of higher maximum width) than the ones above. However, if speed is of main concern these polynomials should be considered. Finally, we can point out that the three classes presented above are the only ones we have found to achieve the lowest maximum width, i.e. the highest speed. We believe firmly there are no other ones.

4. REALIZATION OF THE MULTIPLIER

We will now describe the implementation of the multiplication algorithm of the previous sections. The hardware presented here will make use only of 2-inputs AND-gates and 2-inputs XOR-gates. A multiple-input gate will be realized as a tree of 2-inputs gates since this is, in general, the fastest approach [10, ch.5.3.3].

A parallel multiplier can be realized utilizing expression (1). We look at the coefficient c_i, that is the coefficient of x^i in (1)

$$c_i = a_{m-1}f_{m-1}^i(B) + a_{m-2}f_{m-2}^i(B) + \ldots + a_1 f_1^i(B) + a_0 f_0^i(B) \tag{17}$$

Since modulo 2 multiplication is the same as logical AND and modulo 2 addition the same as XOR, c_i can be computed using the combinational network of figure 1. Since the expression of each c_i in (17) has the same structure, only the f_j^i are changing, we can use the block r_i m times and simply change its inputs. This is shown in figure 2. The block r_i consists of m AND-gates and m-1 XOR-gates, totally $2m$-1 gates. This block is repeated m times which means that the space complexity of the multiplier, so far, is $2m^2$-m gates. To complete the complexity analysis we need to determine the functions f_j^i in order to compute the number of gates needed to generate them. However, before we proceed with the analysis we need to introduce a few parameters:

$N_r = 2m$-1 is the number of gates in *one* of the m identical blocks r_i. It does *not* depend on $P(x)$
N_f is the number of gates required to generate all functions f_j^i. It depends on $P(x)$

$N = N_f + mN_r$ is the total number of gates, i.e. N is a measure of the space complexity of the multiplier
$R = mN_r/N$ is the *regularity* of the multiplier
D is the number of gates along the worst path through the multiplier. This is a measure of the speed of the multiplier and will be called the *depth*

In the previous section we presented three classes of polynomials which are particularly good for our purposes:

$\underline{x^m+x^{m/2}+1}$ This is the best class with $N_f = m/2$ and maximum width 2. Then

$$N = 2m^2 - m/2 \qquad R = \frac{4m-2}{4m-1}$$

and since the depth of an n-inputs tree of gates (see fig.3) is $\lceil \log_2 n \rceil$ we obtain

$$D = 2 + \lceil \log_2 m \rceil$$

$\underline{x^m+x+1}$ This class is next best with $N_f = m-1$ and maximum width=2. Then

$$N = 2m^2-1 \qquad R = \frac{2m^2-m}{2m^2-1} \qquad D = 2 + \lceil \log_2 m \rceil$$

$\underline{x^m+x^{m-1}+\dots+x+1}$ This is the worst of these three classes. Here, we have $N_f = (m^2-m)/2$ and the maximum width 2. Then

$$N = \frac{5m^2-3m}{2} \qquad R = \frac{4m-2}{5m-3} \qquad D = 2 + \lceil \log_2 m \rceil$$

These results should be compared with the normal basis multiplier in [6] which is the fastest VLSI design we have found published and whose space complexity, for certain values of m, is $4m^2-3m$. Table 3 shows a list of best polynomials for the fields $GF(2^m)$, $m \leq 16$, with the relevant parameters of the associated multipliers. The percentual save in gates relative to the best normal basis multipliers is given under *Save* (%). The polynomials generating the best normal basis multipliers are listed in the Appendix . It should be pointed out that some of the figures of Table 3 could, in some cases, be improved. This improvement is obtained by doing some simple optimization work (reutilization of partial sums) in the design of the block generating the functions f_j^i . For some values of m this improvement can be sensible. These improved figures have been determined for m = 8,13,16 and are shown between parenthesis in Table 3. For those fields where the trinomial x^m+x^k+1, $2 \leq k < m/2$, is irreducible, experiments have shown that the number of gates can be reduced from $2m^2-2+k$ to $2m^2-1$ by doing the optimization work mentioned above.

Inspection of Table 3 shows that

i) our multipliers require between 30 (m=2) and 80 (m=16) percent fewer gates than those in [6]. The average save is 50.6 percent. Also, we see that, for these fields, the number of gates N is less than $2.2m^2$.

ii) the regularity of our multipliers is high, 92.2 percent in average. The designs in [6] are, according to our definition of regularity, always 100 percent regular.

m	$P(x)$	D (CB)	D (NB)	N	*Save* (%) (in gates)	R (%)
2	2,1,0	3	3	7	30.0	85.7
3	3,1,0	4	4	17	37.0	88.2
4	4,1,0	4	4	31	40.4	90.3
5	5,2,0	6	5	50	41.2	90.0
6	6,3,0	5	5	69	45.2	95.7
7	7,1,0	5	6	97	62.5	93.8
8	8,7,5,1,0	6	6	154 (140)	53.0 (57.3)	77.9 (85.7)
9	9,1,0	6	6	161	45.8	95.0
10	10,3,0	7	6	201	45.7	94.5
11	11,2,0	7	6	242	46.3	95.5
12	12,3,0	7	6	289	46.5	95.5
13	13,7,6,1,0	7	7	393 (371)	66.0 (67.9)	82.7 (87.6)
14	14,5,0	7	6	395	46.8	95.7
15	15,1,0	6	7	449	66.4	96.9
16	16,11,6,5,0	7	8	563 (537)	79.2 (80.1)	88.1 (92.4)

Table 3. The best polynomials of degree $m \leq 16$. For each polynomial only the actual *powers* of x are given. CB indicates Conventional Basis, NB indicates Normal Basis.

iii) we can expect a very high operational speed (depth ≤ 7 for $m \leq 16$!). Also, the achievable speed is often the same as in [6], sometimes slightly higher, sometimes slightly lower.

iv) For those fields where the polynomial x^m+x+1 is not irreducible we obtain still significant gains in complexity while keeping a high degree of regularity and a high operational speed. In particular this is true even for those fields where no irreducible trinomials exist, i.e. for m=8,13,16 [14].

The best polynomials listed in Table 3 have been found by checking all possible polynomials of degree $m \leq 16$. This gives us an opportunity to test the selection criteria C1 and C2 of Sec.3. We tested the criteria for $m = 8,13,16$. The results are given in Table 4. The first column shows the total number of irreducible polynomials, the second column the number of polynomials selected by C1 and the third column the final set of polynomials after C1 and C2.

m	Total # of polynomials	After C1	After C1 & C2 (Final set)	Is best polynomial included in final set ?
8	30	17	3	Yes
13	630	67	1	Yes
16	4080	94	1	Yes

Table 4

We see that the final set is very small. Also, in all cases the best polynomial is found in the final set (in two cases *only* the best is left !). We conclude this section by showing the design of a complete multiplier over $GF(16)$. It is depicted in Fig. 4.

5. CONCLUSIONS

In this paper, we have given a new view of the multiplication algorithm in $GF(2^m)$ which was first introduced by Bartee and Schneider [9]. This algorithm utilizes the conventional (or polynomial) basis representation of the elements in $GF(2^m)$. We have used this algorithm to implement a fast bit-parallel multiplier. In contrast to what has been believed ,[6] [7] [8], our view shows that this multiplier is both modular and regular and thus well suited for VLSI implementation. The number of gates required by this multiplier depends on the irreducible polynomial used to generate the field. In this paper, we have derived some simple criteria which can spot the good polynomials out of a given set of irreducible polynomials. Applying these criteria, we found that the number of gates required by this multiplier is less than $2.2m^2$ for $2 \leq m \leq 16$. In particular, for those fields where the trinomial x^m+x^k+1, $1 \leq k < m/2$, is irreducible, the number of gates is $2m^2$-1 and for $k = m/2$ this number is $2m^2$-$m/2$. Hence, for these fields and for $m \leq 16$ this multiplier requires less chip area than those in [6] [7] [8].

APPENDIX

The design of a normal basis multiplier is centrated around a certain product function f [6] which is a sum of terms of the type $a_u b_v$. The number of such terms is denoted here by T. The amount of gates required is computed in [6] as $m(2T$-1). Since nothing is said in [6] about how to find the irreducible polynomials $P(x)$ giving the minimum T value, we had to check all possible polynomials of degree ≤ 16. The following is a list of best polynomials and the corresponding T values. Again, only the actual *powers* of x are indicated.

m	$P(x)$	T
2	2,1,0	3
3	3,2,0	5
4	4,3,2,1,0	7
5	5,4,2,1,0	9
6	6,5,4,1,0	11
7	7,6,5,2,0	19
8	8,7,5,3,0	21
9	9,8,6,5,4,1,0	17
10	10,9,8,7,6,5,4,3,2,1,0	19
11	11,10,8,4,3,2,0	21
12	12,11,10,9,8,7,6,5,4,3,2,1,0	23
13	13,12,10,7,4,3,0	45
14	14,13,12,9,8,1,0	27
15	15,14,12,9,7,5,4,2,0	45
16	16,15,13,12,11,10,8,7,5,3,2,1,0	85

REFERENCES

[1] F.J. MacWilliams and N.J. Sloane, *The Theory of Error-Correcting Codes*, Amsterdam: North-Holland 1986.

[2] W.W. Peterson and E.J. Weldon, *Error-Correcting Codes*, Cambridge, MA: MIT Press, 1972.

[3] R.E. Blahut, *Theory and Practice of Error Control Codes*, Reading, MA: Addison-Wesley, 1984.

[4] R.E. Blahut, *A Universal Reed-Solomon Decoder*, IBM J. Res. Develop., vol.28 no.2, pp.150-158, 1984.

[5] K.Y. Liu, *Architecture for VLSI Design of Reed-Solomon Decoders*, IEEE Trans. Comput., vol. C-33, pp.178-189, 1984.

[6] C.C. Wang, T.K. Truong, H.M. Shao, L.J. Deutsch, J.K. Omura and I.S. Reed, *VLSI Architectures for Computing Multiplications and Inverses in $GF(2^m)$*, IEEE Trans. Comput., vol. C-34, pp. 709-717, Aug. 1985.

[7] C.S. Yeh, I.S. Reed and T.K. Truong, *Systolic Multipliers for Finite Fields $GF(2^m)$*, IEEE Trans. Comput., vol. C-33, pp. 357-360, Apr. 1984.

[8] B.A. Laws and C.K. Rushforth, *A Cellular-Array Multiplier for $GF(2^m)$*, IEEE Trans. Comput., vol. C-20, pp. 1573-1578, Dec. 1971.

[9] T.C. Bartee and D.I. Schneider, *Computations with Finite Fields*, Inform. Contr., vol. 6, pp. 79-98, Mar. 1963.

[10] N. Weste and K. Eshraghian, *Principles of CMOS VLSI Design*, Reading, MA: Addison-Wesley, 1985.

[11] A. Gill, *Linear Sequential Circuits*, New York: MacGraw-Hill, 1967.

[12] L.E. Dickson, *Linear Groups with an Exposition of the Galois Field Theory*, New York: Dover, 1958.

[13] J.H. v. Lint, *Introduction to Coding Theory*, New York: Springer, 1982.

[14] N. Zierler, J. Brilluart, *On Primitive Trinomials (Mod 2)*, Inform. Contr. vol. 13, pp. 541-554, 1968.

[15] H.O. Burton, *Inversionless Decoding of Binary BCH Codes*, IEEE Trans. Inform. Theory, vol. IT-17, no.4, pp. 464-466, July 1971.

[16] R. Lidl, H. Niederreiter, *Finite Fields*, Reading, MA: Addison-Wesley, 1983.

[17] E.D. Mastrovito, *VLSI Designs for Computations over Finite Fields* $GF(2^m)$, Internal Report LiTH-ISY-I--- (to be printed), Linköping University, Sweden, 1988.

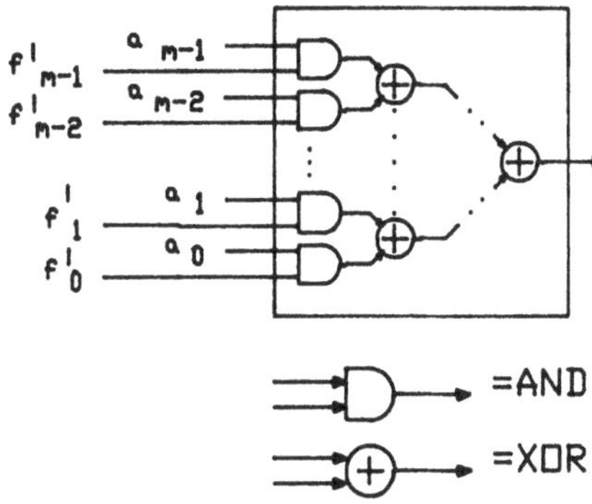

Fig. 1 The block r_i

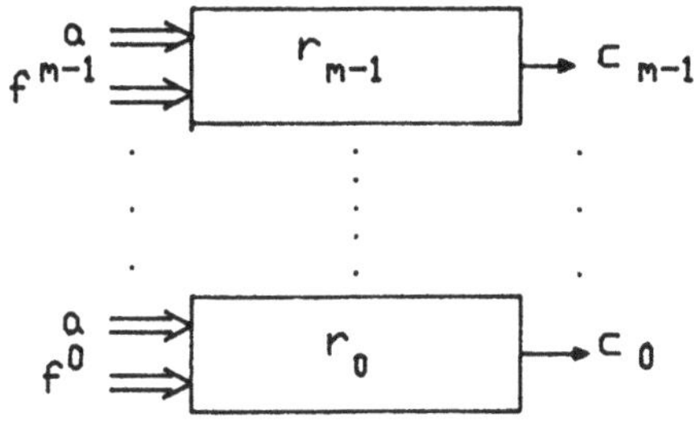

Fig. 2

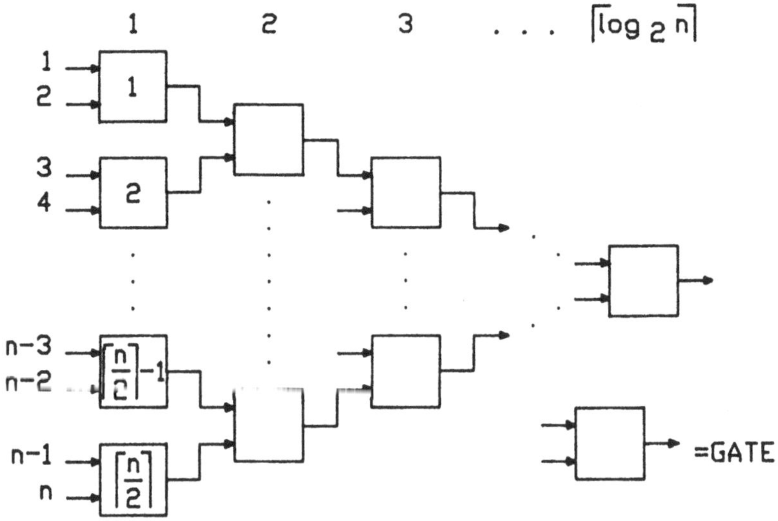

Fig. 3

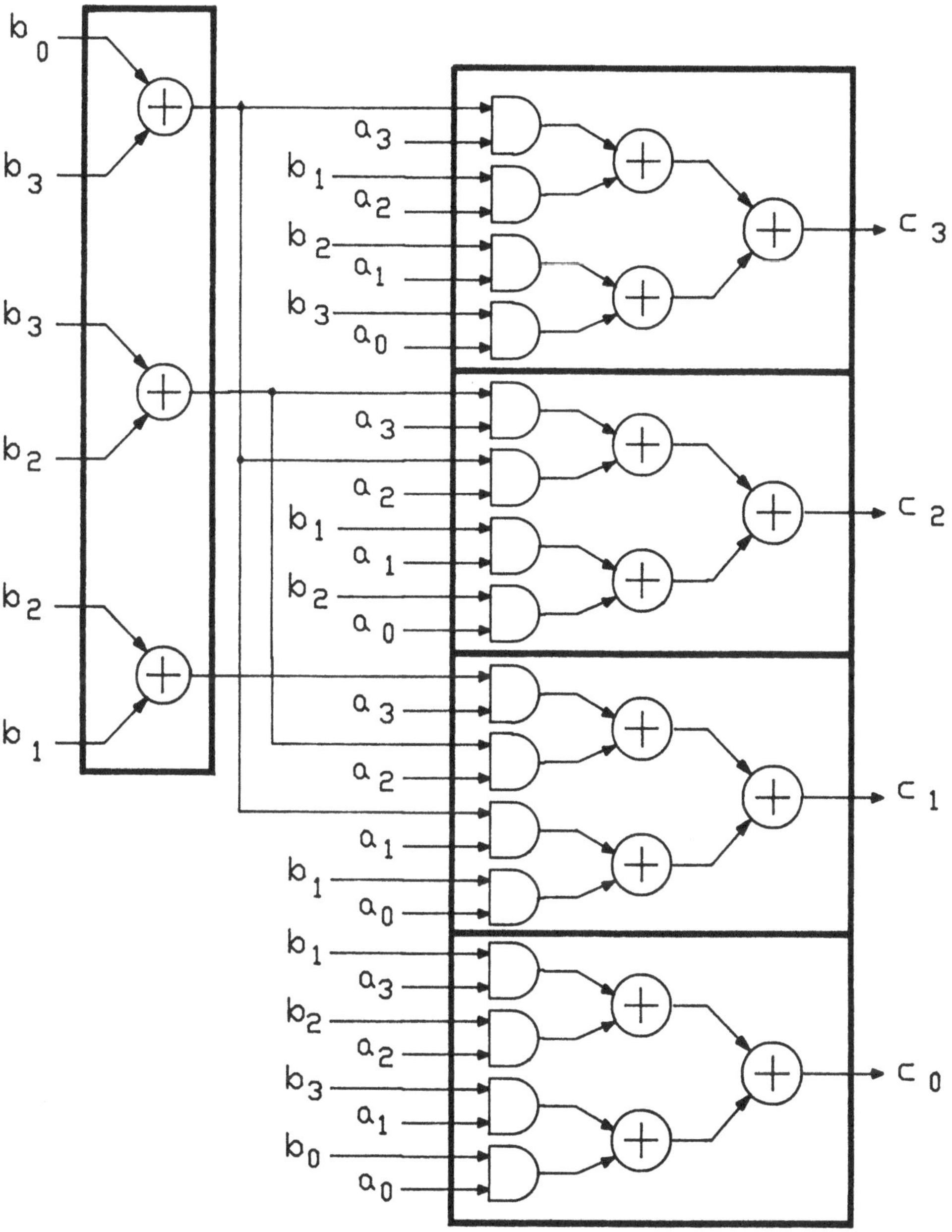

Fig. 4. A multiplier over GF(16) with $P(x) = x^4+x+1$.

A Primality Test using Cyclotomic Extensions

Preda Mihailescu - Seminar für angewandte Mathematik
Swiss Federal Institute of Technology (ETH)
CH 8092 Zürich

ABSTRACT : The cyclotomic polynomial $\Phi_s(x)$ (where s is an integer >1) is the irreducible polynomial over $\mathbf{Q}$, having the primitive s-th roots of unity as zeroes. If $\mathbf{K}$ is the field $\mathbf{Q}$ or $\mathbf{F}_p$, with p a prime, an s-th cyclotomic extension of $\mathbf{K}$ is the splitting field of $\Phi_s(x)$ over $\mathbf{K}$. Every cyclic (actually: every abelian) extension of $\mathbf{K}$ is included in some cyclotomic one. We present in §2 a procedure for constructing cyclic and cyclotomic extensions of fields. Cyclic fields are used in [BS] in the context of a factoring algorithm. For $\mathbf{K} = \mathbf{F}_p$, this procedure can be used to produce irreducible polynomials of given degree over $\mathbf{F}_p$.
H.W.Lenstra, Jr. has extended in [Le2] the concept of cyclotomic extensions to rings $\mathbf{Z}/(n\mathbf{Z})$, with n>1 an integer, and showed that existence of such extensions of degree $s>\sqrt{n}$ implies a drastical constraint upon possible prime factors of n. He proposes a primality test based upon factoring $\Phi_s(x)$ over $\mathbf{Z}/(n\mathbf{Z})$, using the Berlekamp algorithm. Using our algorithm for constructing cyclotomic extensions over a finite field and some pseudoprime test involving Jacobi sums, we improve in §5,6 Lentra's approach to proving existence of an s-th cyclotomic extension of $\mathbf{Z}/(n\mathbf{Z})$ in polynomial time, when $s>\sqrt{n}$ and $\mathrm{ord}_s(n) = O(\log(n)^{c.\log\log\log(n)})$. This leads to a new primality test, which is an algorithmical realisation of the sketches in [Le2]. The intimate connection with the Adleman test ([APR],[Le1],[CoLe]) becomes evident by the very similar algebraic techniques used in that test and in ours. The new algorithm is comparable to [CoLe] in assymptotic runtime and capacity to prove primality of a test number; it is slightly superior by the fact that, (a) in the most time consuming steps of both algorithms, the set of operations required by CE is a subset of the set of operations required by Jacobi-sum test as described in [CoLe] and (b) the new algorithm provides a proof of primality in all cases the test [CoLe] does so, and also in some further cases.

§1 Introduction

Primality testing - subject which deeply fascinated mathematical mind along the centuries - found in the last decade a new dramatical development. We refer to articles in [LT], to [Ri] and [Le3] for a survey of the history and recent developments in this field.

For practical purposes very efficient Monte-Carlo algorithms ([SSt],[Ra]) are known, which can in most cases prove compositeness of a natural number n. Numbers which are not found to be composite upon repeated testing, are prime with probability very close to 1.

Primality can be proved under special hypothesis concerning factors of $(n^k - 1)$ (k=1,2,3,4,6) using tests from the "Lucas-Lehmer" family; a wide bibliography and a thorough overview of these methods can be found in [Wi]. Up to now, only one efficient deterministic primality test - i.e. a test which proves n to be composite or prime, without conditions on n - is known. This is the Jacobi-sum test proposed by Adleman [APR], [Le1]; its running time is $O(\log(n)^{c.\log\log\log(n)})$, for a positive real constant c: slightly slower than the one of Monte - Carlo and Lucas-Lehmer tests, which are polynomial. A faster Las-Vegas version of this test is described in detail in [CoLe].

One should also mention the latest development in the domain of primality testing: tests using elliptic curves and abelian varieties. The first one in order of chronology, [GoKi], is of theoretic interest but was never expected to be implemented, since it uses the Schoof algorithm for computing the order of an elliptic group; this algorithm is polynomial, but practically very slow $O(\log(n)^{7+\varepsilon}$, raising the runtime of the test of [GoKi] to $O(\log(n)^{10+\varepsilon})$. The abelain variety test [AHu] was designed for avoiding the use of an unproved conjecture in the run time analysis of [GoKi] and is more ineffective than [GoKi]. A practically performant test using elliptic curves was proposed and implemented by Atkins [Co], who used a different approach from [GoKi], thus avoiding the use of the Schoof algorithm. We refer to [Le3] and [LL] for a survey of these algorithms.

The Jacobi-sum test brought powerful algebraic methods - such as Gauss- and Jacobi-sums - into the field of computational number theory. In [Le1] and [Le2], Lenstra suggested using these methods together with a theory of cyclotomic extensions of finite rings -developed in [Le2] - for primality tests that have both the qualities of Lucas - Lehmer and the Adleman tests.

Independently, we have developed a test that fits in this program. The new tool of this test is an algorithm for constructing cyclic and cyclotomic extensions of fields. This gives an efficient solution to the problem of factoring cyclotomic polynomials over $\mathbb{Z}/(n\mathbb{Z})$ ($n > 1$, a natural number), problem which was solved in [Le2] with the Berlekamp or Rabin algorithm. These algorithms are however inefficient in this context, not taking advantage of specific properties of cyclotomic polynomials. We remark that our algorithm splits cyclotomic polynomials over finite fields. This algorithm can be used over $\mathbb{Z}/(n\mathbb{Z})$ in the context of primality testing; generally it does however not split a cyclotomic polynomial over $\mathbb{Z}/(n\mathbb{Z})$, unless the factorization of n is known or a cyclotomic extension of $\mathbb{Z}/(n\mathbb{Z})$ in the sense of [Le2] exists.

Tests with elliptic curves use a recursive proof. A sequence of probable primes (primality being tested with some fast probabilistic test) of decreasing sizes is built. If the smalest number in this sequence can be proved with some deterministic test to be prime, than the primality of all elements of the sequence follows. In [GoKi], the Jacobi-Sum test described in [CoLe] is proposed for the test of primality of this smalest probable prime. The test we propose - called CE for simplicity - is a Las Vegas test which has similar properties with [CoLe] and can be used with more success than this test. This is a consequence of the fact that for testing a given integer n for primality, the computations needed by the CE are essentially a strict subset of the computations needed in [CoLe]; in all the cases when this last test succeeds to prove primality, CE does so too (see §7). Furthermore, from a theoretical point of view, CE has the interesting characteristic of generalizing Lucas-Lehmer tests, so that no restricitve conditions on the test number n are required. Generalizing Lucas-Lehmer tests by using the concept of cyclotomic extensions of rings was the reason for which this concept was developed in [Le2].

In §2 we discuss cyclotomic extensions of fields and rings $\mathbb{Z}/(n\mathbb{Z})$ and state a primality criterion of Lenstra. In §3 we define Gauss and Jacobi-sums and give some essential properties. The algorithm for splitting cyclotomic polynomials over a field is presented in §4 together with some applications independent of primality testing. We use this algorithm in §5 to state in terms of Jacobi-sums some conditions on n, which imply the criterion of Lenstra. This leads to the primality test presented in §6. In §7 we make conclusive remarks regarding some aspects which could not have been treated by this brief paper.

§2 Cyclotomic extensions

We recall some basic properties of cyclotomic extensions of $\mathbb{Q}$ and Galois fileds $\mathbb{F}_p$ ([La,ChVIII,§3],[Cn,pp.34-35]) and give a generalization of this concept to rings $\mathbb{Z}/(n\mathbb{Z})$ (n an integer $n>1$) according to [Le2]. This generalization allows a useful primality criterion for n.

Let $s>1$ be a natural number. The cyclotomic polynomial $\Phi_s(x)$ is defined as the minimal polynomial over $\mathbb{Q}$ of a primitive s-th root of unity $\zeta_s \in \mathbb{C}$. The zeroes of $\Phi_s(x)$ are the primitive s-th roots of unity in $\mathbb{C}$ and $\deg(\Phi_s(x)) = \varphi(s)$ (φ the Euler totient function).

The field $\mathbb{Q}[\zeta_s] / (\Phi_s(x))$ is an abelian extension of degree $\varphi(s)$, called the s-th cyclotomic extension of $\mathbb{Q}$. If p^k ($k>1$) is a primepower, then $\Phi_{p\wedge k}(x) = \Phi_p(x^{p\wedge(k-1)})$ (we shall often use the notation $a\wedge b = a^b$ for double exponents or exponents in the index). Furthermore, if $(p,s)=1$, then $\Phi_{p.s}(x) = \Phi_s(x^p)/ \Phi_s(x)$.

We consider more closely the cyclotomic field $\mathbb{L}_q = \mathbb{Q}[\xi_q]$, where q is a prime. The Galois group $\mathrm{Gal}(\mathbb{L}_q, \mathbb{Q})$ is isomorphic to $\mathbb{F}_q^*$: $\mathrm{Gal}(\mathbb{L}_q, \mathbb{Q}) = \{\sigma_v \mid (v,q) = 1, \sigma_v(\xi_q) = \xi_q^{\,v}\}$. Let g be a generator of $\mathbb{F}_q^*$ and f be any divisor of $(q-1)$. Then there is a unique (cyclic) extension $\mathbb{K}$ of $\mathbb{Q}$ with $\mathbb{L}_q \supset \mathbb{K} \supset \mathbb{Q}$ and $[\mathbb{K}:\mathbb{Q}] = f$. If $t =(q-1)/f$, $H= \mathbb{F}_q^*/<g^f \bmod q>$, where $<g^f \bmod q>$ is the cycle generated by g^f in $\mathbb{F}_q^*$, then $\mathbb{K}$ is fixed by the automorphisms $\tau \in \{ \sigma_v \mid v \in <g^f \bmod q> \}$ and $\mathrm{Gal}(\mathbb{K}, \mathbb{Q})$ is isomorphic to H. Let $n=g^f$ and the *Gauss period of degree* f be defined by:

(2.1) $\eta_j = \sum_{1\le i\le t} \zeta_q^{g^{\wedge}j.n^{\wedge}i}$ where ζ_q is a primitive q-th root of unity and j=1, 2, ... ,f

(Note that η_0 is defined with (2.1), then $\eta_0 = \eta_f$. We shall use this observation to reduce cyclically the indices of the η).

The 'length' t of the Gauss cycles is equal to $[\mathbb{L}_q : \mathbb{K}]$, while $f = [\mathbb{K}: \mathbb{Q}]$. Furthermore, $\eta_j = \mathrm{Tr}\,[\mathbb{L}_q : \mathbb{K}]\,(\zeta_q^{g^{\wedge}j})$ for j=1, 2, ... , f. According to Dirichlet's theorem on repartition of primes in arithmetic progressions, for any positive integer f >1, there is a prime q=1 mod f and consequently for any integer f>1, there is a cyclic extension of degree f of $\mathbb{Q}$ which is included in a cyclotomic one of prime order. This fact is used in [BS].

While $\Phi_s(x)$ is irreducible over $\mathbb{Q}$, it splits over a Galois field $\mathbb{F}_p$ (with p a prime and (s,p)=1) in f irreducible factors of degree t, where $t=\mathrm{ord}_s(p)$ and $f.t = \varphi(s)$. We consider the case where s=q is a prime ($\neq$p) and let the degree of Gauss periods in (2.1) be $f=(q-1)/\mathrm{ord}_q(p)$. Let **R** be the ring $\mathbf{R} = \mathbb{Z}[\zeta_q]/(p.\mathbb{Z}[\zeta_q])$, $\mathbb{K}$ the splitting field of $\Phi_q(x)$ over $\mathbb{F}_p$, ω a q-th root of unity in $\mathbb{K}$ and τ_j, the Frobenius automorphisms of $\mathbb{K}$ sending ω to $\omega^{p^{\wedge}j}$ (j=1, 2, ... , t). The coefficients of the polynomials $\Psi_i(x) = \prod_{1\le j\le t}(x-\tau_j(\omega^{g^{\wedge}i}))$ are invariant under the τ_j and hence $\Psi_i \in \mathbb{F}_p[x]$ for i=1, 2 , ...,f. Furthermore, Ψ_i are irreducible and have the product Φ_q, showing that:

(2.2) $\Phi_q(x) = \prod_{1\le i\le f} \Psi_i(x)$ with $\Psi_i(x)) = \prod_{1\le j\le t}(x - \tau_j(\zeta'^{g^{\wedge}i}))$

(2.3) $\mathbb{F}_p[x]/(\Psi_i(x)) \cong \mathbb{K}$ (i=1,2,...,f)

From (2.2) and (2.3) follows: $\mathbf{R} \cong \prod_{1\le i\le f} \mathbb{F}_p[x]/(\Psi_i(x)) = \prod_{1\le i\le f} \mathbb{K}$. Let λ_i be the projection from **R** to $\mathbb{F}_p[x]/(\Psi_i(x)) = \mathbb{F}_p[\omega^{g^{\wedge}i}]$ (i=1, 2, ... , f) . Note that $\lambda_i(\eta_j)$ are invariant under the Galois group <q mod p> of $\mathbb{K}$ over $\mathbb{F}_p$ (for i=1, 2, ... ,f; j=1, 2, ... ,t) and are consequently elements of $\mathbb{F}_p$. The automorphism σ_g of $\mathbb{Z}[\zeta_q]$ sending ζ_q to ζ_q^g induces a cyclic permutation of the set $\{\lambda_i(\eta_j)\}$ of projections of η_j (i fixed). An essential observation for our method of constructing cyclotomic extensions is the fact that the coefficients of the irreducible polynomials $\Psi_i(x)$ (i=1, 2, ... ,f) are lineary dependent of the set $\{\lambda_i(\eta_j)\}$(j=1, 2, ... ,t), for a fixed i , say i=f, to fix the notions; as we shall show, the coefficients can actually be determined as solutions of some linear systems, provided q<p. Note that a change of the index j is equivalent to letting a power of σ_g act upon $\{\lambda_i(\eta_j)\}$. According to (2.2) the conjugates of ω are $\tau_j(\omega)= \omega^{p^{\wedge}j}$ (j=1,2,...,t). Let k be a positive integer, k<t and $s'_k = \sum_{1\le i\le t} \tau_j(\omega)^k$ be the sum of the k-th powers of the conjugates of ω; let $h\in\{1,2,...,f\}$ be such that $(k/g^h) \in$ <p mod q>. Then one verifies that $s'_k=\lambda_f(\eta_h)$, showing that in $\mathbb{F}_p$ the sums of k-th powers of the zeroes of $\Psi_f(x)$ are equal to the projections of some Gauss period η_h. The Newton formulæ

(2.4) $s'_k - s'_{k-1}.\varepsilon_1 + ... + (-1)^{k-1} s'_1.\varepsilon_{k-1} + (-1)^k.k.\varepsilon_k = 0 \quad k\le t$

relate the sums of powers to the elementary symmetric polynomials ε_j in the variables $\tau_i(\omega)$ (i,j=1,2,...,t), which are up to a sign, the coefficients of $\Psi_f(x)$. Suppose that q<p, such that k=1,2,...,t are all units in $\mathbb{F}_p$. Then (2.4) is a regular triangular linear system of equations in $\mathbb{F}_p$

allowing to compute the coefficients of $\Psi_f(x)$, given $\lambda_f(\eta_h)$, h=1,2,...,f. Of course, knowing $\lambda_f(\eta_h)$ implies knowledge of $\Psi_f(x)$, but we shall see in §4 how to bypass this circular argument.

Consider now a finite local ring $\mathbf{R} = \mathbb{Z}/(p^k.\mathbb{Z})$, k>1, p prime, instead of a field $\mathbb{Q}$ or $\mathbb{F}_p$. Let s>1 be some positive integer, (s,p)=1 and $\mathbb{K} = \mathbb{F}_{p^\wedge t}$ with $t=ord_s(p)$ be a cyclotomic extension of order s of $\mathbb{F}_p$, defined by some irreducible factor $\Psi(x)$ of $\Phi_s(x)$ over $\mathbb{F}_p$. Then there exists a Galois extension **A** of **R**, unique up to isomorphism, such that Gal(**A**, **R**) = Gal($\mathbb{K}$, $\mathbb{F}_p$) and $\mathbf{A}/p.\mathbf{A} \cong \mathbb{K}$ [Le2,2.10]. We shall call **A** an s-th cyclotomic extension of **R** if (k,t)=1.

The natural generalization of the notion of s-th cyclotomic extension to finite rings $\mathcal{N} = \mathbb{Z}/(n.\mathbb{Z})$ with n an integer >1 was given by Lenstra in [Le2,§4]. In this general setting, the existence of an s-th cyclotomic of $\mathcal{N}$ is a property of n, which makes the concept suitable for primality testing.

We introduce some **notations** which we shall use throughout this paper. Script capitals will denote the ring of integers modulo some natural number >1; thus $\mathcal{N} = \mathbb{Z}/(n.\mathbb{Z})$, $\mathcal{M} = \mathbb{Z}/(m.\mathbb{Z})$, $\mathcal{S} = \mathbb{Z}/(s.\mathbb{Z})$, etc. We may also write $\mathcal{P} = \mathbb{Z}/(p.\mathbb{Z})$,etc. when p is a prime, but the multiplicative rather then the field structure is of interest.

Suppose (s, n) = 1; then following two sets of conditions are equivalent:

(I) There is an $\mathcal{N}$ algebra Ω, an automorphism σ and an element $\omega \in \Omega$ such that:

(i) Ω is Galois over $\mathcal{N}$ with group generated by σ

(ii) $\Phi_s(\omega) = 0$ and $\sigma\omega = \omega^n$.

(iii) $\sigma^t = id_\Omega$ with $t=ord_s(n)$

(II) There is a polynomial $\Psi(x) \in \mathcal{N}[x]$ with degree $t = ord_s(n)$ and dividing $\Phi_s(x)$ mod n. With $\Omega = \mathcal{N}[x]/(\Psi(x))$ and $\omega = x + (\Psi(x))$ following conditions hold:

(i) Let $H(s) = \mathcal{S}^*/\langle n \bmod s\rangle$. The polynomials $\Psi'_h(x) = \prod_{1 \le k \le t}(x - \omega^{h.n^\wedge k})$, regarded as polynomials in $\Omega[x]$ have all the coefficients in $\mathcal{N}$, for $\forall h \in H(s)$.

(ii) Let $\Psi_h(x) \in \mathcal{N}[x]$ be the polynomials maped upon $\Psi'_h(x)$ by the canonical imbedding of $\mathcal{N}[x]$ in $\Omega[x]$. Then $\Psi_1(x) = \Psi(x)$.

(iii) $\Phi_s(x) = \prod_{h \in H(s)} \Psi_h(x) \mod n$.

For a proof of the equivalence, remark that if (I) holds, then the polynomial $\Psi'(x) = \prod_{1 \le k \le t}(x - \sigma^k(\omega)) \in \Omega[x]$ has coefficients in $\mathcal{N}$ and it can be used to show that (II) holds. Conversely, if (II) holds, it can be shown there is an automorphism σ of Ω with $\sigma\omega = \omega^n$ and Ω is Galois over $\mathcal{N}$ with group generated by σ.

Definition: If (I) or (II) hold , (Ω,σ,ω) is called an s-th cyclotomic extension of $\mathcal{N}$.

The restriction on n imposed by the existence of an s-th cyclotomic extension of $\mathcal{N}$ is given by following:

Theorem1 [Le2,(4.2)]

An s-th cyclotomic extension of $\mathcal{N}$ exists if and only if

(2.5) $r \in \langle n \bmod s\rangle \quad \forall r \mid n$

If $s>\sqrt{n}$, one has following primality criterion:

Corollary 1

If an s-th cyclotomic extension of $\mathfrak{N}$ exists and $r_i \in \mathbb{Z}$, $i=1,2,\dots,t$ are such that

(i) $0 \le r_i < s$ and $r_i = n^i \bmod s$

(ii) $n/r_i \notin \mathbb{N}$ (r_i does not divide n) for $i=1,2,\dots,t$

then n is prime.

Proof: If n were composite, there is a prime $r \mid n$, with $r<\sqrt{n}<s$ and thus $r = r_i$ for some i, which contradicts (ii). •

This criterion is interesting in connection with a theorem of Prachar, Odlyzko and Pomerance ([APR],[CoLe,§3,4] which implies that $\forall n > e^e$, there exists a positive constant c and a natural number $s>\sqrt{n}$ such that $\lambda(s)=O(\log(n)^{c.\log\log\log(n)})$, where λ is the Carmichael function. For such s, a fortiori, $\mathrm{ord}_s(n) = O(\log(n)^{c.\log\log\log(n)})$.

This completes the motivation for proving existence of cyclotomic extension for primality testing.

§3 Gauss- and Jacobi-sums

Let m be an odd positive integer, $\lambda(m)$ the Carmichael function, $f \mid \lambda(m)$, $f>1$ and $\zeta_f \in \mathbb{C}$ an primitive f-th root of unity. A *multiplicative character* χ of *conductor* m and *order* f is a multiplicative group homomorphism $\chi: \mathfrak{M} \to <\zeta_f>$ with $\chi(x) = 0$ for $(x,m) \neq 1$ and $\chi(\mathfrak{M}) = <\zeta_f>$. χ is primitive if there is no $d \mid m$ such that $\chi(x)=1$ $\forall x \in \{y \in \mathfrak{M}^* \mid y = 1 \bmod m/d\}$. For simplicity of notation, we let $\zeta = \zeta_f$ and ξ be a complex primitive root of unity of order m throughout this paragraph. We shall write $\mathfrak{M}^\wedge$ for the set of all characters $\chi: \mathfrak{M} \to <\zeta_{\lambda(m)}>$ and $\#\mathfrak{M}^\wedge = \varphi(m)$, which is a consequence of $\mathfrak{M}^\wedge \cong \mathfrak{M}^*$. We denote the automorphism of $\mathbb{Q}[\xi]$ sending ξ to ξ^a ($(a,m)=1$) by σ_a. The Gauss sum $\tau(\chi)$ is defined by:

(3.1) $\tau(\chi) = \sum_{x \in \mathfrak{M}} \chi(x).\xi^x$

and the action of σ_a upon $\tau(\chi)$ is given by:

(3.2) $\sigma_a(\tau(\chi)) = \chi^{-1}(a).\tau(\chi)$

The following identities give the absolute value of $\tau(\chi)$ in $\mathbb{C}$ and relate $\tau(\chi)$ to $\tau(\chi^{-1})$:

(3.3) $\tau(\chi).\overline{\tau(\chi)} = m$ and $\tau(\chi).\tau(\chi^{-1}) = \chi(-1).m$

If $\chi, \chi' \in \mathfrak{M}^\wedge$ are primitive , the Jacobi-sum is defined by:

(3.4) $j(\chi,\chi') = \sum_{x \in \mathfrak{M}} \chi(x).\chi'(1-x)$

and verifies:

(3.5) $j(\chi,\chi').\ \tau(\chi.\chi') = \tau(\chi).\ \tau(\chi')$ for χ, χ', $\chi.\chi'$ primitive.

Let χ be a fixed primitive character of order f and define $J_k(\chi)$ (k=1,2,...,f-1) inductively by:

(3.6) $J_1(\chi)=1;\ J_{k+1}(\chi) = J_k(\chi).\ j(\chi,\chi^k)$ (k=1,2,...,f-2)

It follows from (3.5) that

(3.7) $J_k(\chi)= \tau(\chi)^k / \tau(\chi^k)$

Furthermore, putting k=f-1 in (3.7) and using (3.3) together with the fact that $\chi^{f-1}=\chi^{-1}$ for a primitive character χ of order f, one gets:

(3.8) $(\tau(\chi))^f = \chi(-1).m.J_{f-1}(\chi)$

Note that (3.2) and (3.7) imply that $J_k(\chi)$ are invariant under the action of σ_a, $\forall a \in \mathfrak{M}^*$.
If m is squarefree, properties of Gauss and Jacobi-sums can be reduced to those of sums of prime conductor and primepower order by some identities we shall call reduction formulæ:

Reduction of conductor

Let $m = m_1.m_2$ with coprime divisors m_i (i=1,2), both different from 1, and $\chi \in \mathfrak{M}$^. If $x \in \mathfrak{M}$, we use the Chinese remainder theorem to identify x with the pair of its remainders mod m_i (i=1,2): $x=(x_1,x_2)$, with $x=x_i$ mod m_i (i=1,2). Herewith, we define two characters χ_i with conductors m_i (i=1,2) by $\chi_1(x) = \chi((x,1))$ and $\chi_2(y) = \chi((1,y))$ with $x \in \mathfrak{M}_1$ and $y \in \mathfrak{M}_2$, $\mathfrak{M}_i = \mathbf{Z}/(m_i.\mathbf{Z})$ (i=1,2). Let $e_1=(1,0) \in \mathfrak{M}$, $e_2=(0,1) \in \mathfrak{M}$ and ξ_i be the primitive m_i-th roots of unity given by $\xi_i = \xi^{e_i}$. Then

(3.9) $\tau(\chi)= \tau(\chi_1).\ \tau(\chi_2)$

the Gauss-sums on the right hand side being taken with respect to ξ_i respectively (i=1,2). Furthermore, if $\chi' \in \mathfrak{M}$^ and χ'_i are defined like χ_i (i=1,2) then:

(3.10) $j(\chi,\chi') = j(\chi_1,\chi'_1) \cdot j(\chi_2,\chi'_2)$

Reduction of order

Let $f = f_1.f_2$ with coprime divisors f_i (i=1,2), both different from 1. Suppose $u.f_1 + v.f_2 = 1$ and let $\zeta_1= \zeta^{f2}$ and $\zeta_2= \zeta^{f1}$ be two primitive roots of unity of orders f_i (i=1,2). Let $\chi_1(x) = \chi^{f2}(x)$ and $\chi_2(x) = \chi^{f1}(x)$ or equivalently $\chi_i(x) = \zeta_i^{\,a}$ (i=1,2; a an integer) if and only if $\chi(x) = \zeta^a$, showing that χ_i have orders f_i (i=1,2). Then (3.5) implies:

(3.11) $\tau(\chi) = j(\chi_1^{\,v},\chi_2^{\,u}).\ \tau(\chi_1^{\,v}).\ \tau(\chi_2^{\,u})$

the characters on the right hand side having orders f_1 and f_2.

We now let the conductor of Gauss- and Jacobi-sums be a prime q, $\mathfrak{Q}= \mathbf{Z}/(q.\mathbf{Z})$ and f |(q-1); ζ and ξ are complex primitive roots of unity of respective orders f and q. We recall the definition

(2.1): $\eta_j = \sum_{1\le i\le t} \zeta_q^{g\wedge j.n\wedge i}$, j=1,2,...,f ; t=(q-1)/f and n = g^f mod q, where g generates $\mathbb{Q}^*$. We let χ be the multiplicative character $\chi:\mathbb{Q}\to<\zeta>$ given by $\chi(g)=\zeta$ and $\chi(0)=0$ and the Gauss-sum $\tau(\chi)$ be given by $\tau(\chi) = \sum_{x\in \mathbb{Q}} \chi(x).\xi^x$. Regrouping this sum according to powers of ζ (and using $\chi(g)=\zeta$), we get :

$$(3.12)\quad \tau(\chi) = \sum_{1\le i\le t} \zeta^i.\eta_i$$

which, together with the action (3.2) of σ_g on the Gauss-sum, shows that Gauss-sums are Lagrange resolvents for η_j .

We shall be interested in Gauss- and Jacobi-sums computed modulo a prime p. Let q, f, t, g, ζ and ξ be as before, p a prime $\neq$q and $\mathbf{R} = \mathbf{Z}[\zeta, \xi]/(p. \mathbf{Z}[\zeta, \xi])$. The elements of $\mathbf{R}$ can be written as $\alpha = \sum_{1\le i\le f,1\le j\le q} a_{ij}\, \zeta^i.\xi^j$ with a_{ij} elements of the field $\mathbb{F}_p$ and using the identity $\alpha^p = \sum_{1\le i\le f,1\le j\le q} a_{ij}\, \zeta^{i.p}.\xi^{j.p}$ one gets following identity in $\mathbf{R}$ ([CoLe,§5]):

$$(3.13)\quad \tau(\chi)^p = \chi^{-p}(p).\tau(\chi^p) \qquad \text{in } \mathbf{R}$$

If $t=\mathrm{ord}_q(p)$, iterating (3.13) t times yields:

$$(3.14)\quad \tau(\chi)^{p\wedge t-1} = \chi^{-1}(p) \qquad \text{in } \mathbf{R}$$

Naturally, (3.13) and (3.14) hold when replacing χ by a power $\chi'= \chi^u$ (u = 0 mod f is allowed under the definition $\tau(1)=-1$, 1 being the all-one character: $1(x)=1\ \ \forall x\in \mathbb{Q}^*$.)

§4 Constructing cyclotomic extensions

We shall only consider the case of cyclotomic extensions of prime orders q over fields $\mathbb{L}$.

If $\mathbb{L}=\mathbb{Q}$, $\mathbb{Q}[\zeta_q] = \mathbb{Q}[x]/(\Phi_q(x))$. We shall give here also a construction of cyclic extensions $\mathbb{K}$, $\mathbb{Q}[\zeta_q]\supset\mathbb{K}\supset\mathbb{Q}$. Let f, t, g, n, η_j (j=1,2,...,f), $\mathbb{K}$, H be as in §2 and substitute $\xi = \zeta_q$ (for agreement of notations with §3). The automorphism σ_g of $\mathbb{Q}[\xi]$ sending ξ to ξ^g acts upon the set $\{\eta_j | j=1,2,...,f\}$ transitively by $\sigma_g(\eta_j) = \eta_{j+1}$ (indexes j+1 taken cyclically modulo f). This shows that the η_j are zeroes of a cyclical polynomial and can be found by means of Lagrange resolvents, which are in this case Gauss-sums, according to (3.12). We let like in §3, ζ be an f-th primitive root of unity, χ be the multiplicative character $\mathbb{Q} \to <\zeta>$ given by $\chi(0)=0$ and $\chi(g)= \zeta$; put $\tau(\chi) = \sum_{x\in \mathbb{Q}} \chi(x).\xi^x$ and define $\tau(\chi^f)= \tau(1)=-1$. Then:

$$(4.1)\ f.\ \eta_f = \sum_{1\le j\le f} \tau(\chi^j)$$

and η_j , j=1,2,...,f-1 can be determined using (3.2) and the action of σ_g on η_f. Let $\mathbb{L}_f = \mathbb{Q}[\zeta]$ and $\alpha = \chi(-1).q.J_{f-1}(\chi) = \tau(\chi)^f \in \mathbb{L}_f$. Let β be any f-th root of α in $\mathbb{C}$. Then $\beta= \zeta^{-j}.\tau(\chi) = \sigma_g^{\,j}(\tau(\chi))$ for some j, $1\le j\le f$. Define $\eta\in \mathbb{L}_f[\beta]$ by:

(4.2) $\eta = (\sum_{1\le k\le f} \beta^k / J_k(\chi))/f$

By (4.1) and construction of β, $\eta = \sigma_g^{\,j} (\eta_f) = \eta_j$. Since $\mathbb{K} = \mathbb{Q} [\eta_j]$ for any j, $1\le j\le f$, we conclude $\mathbb{K} = \mathbb{Q} [\eta]$. Note that this construction only involves $\mathbb{L}_f$ and not $\mathbb{Q} [\xi]$; the number j need not be known.

For $\mathbb{L} = \mathbb{F}_p$ (p an odd prime >q) let $t=\mathrm{ord}_q(p)$ and $f=(q-1)/t$ and perform the same steps for computing the Gauss periods of degree f as in the case $\mathbb{L} = \mathbb{Q}$; computations can be performed in the ring $\mathbf{R}_f = \mathbb{Z}[\zeta_f]/(p.\ \mathbb{Z}[\zeta_f])$. $\mathbf{R}_f$ being a direct product of isomorphic fields, the extraction of an f-th root of α can be dealt with by algorithms like in [Hu]. Note that the Gauss periods are elements of $\mathbb{F}_p$ and consequently, the equation $\beta^f = \alpha$ has a solution in $\mathbf{R}_f$. The extraction of f-th roots is naturally reducible to successive extraction of roots of prime power orders r^k ($k\ge 1$), with $r^k \| f$. This is always possible by mere exponentiation of α in $\mathbf{R}_f$, provided $p^{r-1} \ne 1 \bmod r^2$. Otherwise, if $v = v_r (p^{r-1} - 1) >1$ and $\omega = \zeta_f^{\,f/r}$, $\mathbf{R}_r$ the ring $\mathbf{R}_r = \mathbb{Z}[\omega]/(p.\ \mathbb{Z}[\omega])$, then an element $\varepsilon \in \mathbf{R}_r$ must be found, with $\varepsilon^{(P-1)/r} \ne 1$, where $P=p^{r-1}$ (e.g., by trial and error). This yields a r^v-th root of unity ρ in $\mathbf{R}_r$ and the r^k-th root of α can be consequently computed by exponentiation and subsequent division by powers of ρ. This kind of algorithm will be crucial for the case when the modulus n (rather then p) is not known to be prime. The coefficients of an irreducible factor $\Psi(x) | \Phi_q (x)$ over $\mathbb{F}_p$ can be found given the Gauss periods of degree f by using (2.4), which completes the construction of a q-th cyclotomic extension of $\mathbb{F}_p$. Constructing extensions of composite order s implies nontrivial operations, even when extensions of each prime power order $q^k \| s$ are given. We shall however not consider this topic before §5, and there in a different setting.

§5 Proving existence of cyclotomic extensions over $\mathfrak{N}$

Let $n, s \in \mathbb{Z}_{>1}$, s odd and squarefree. In this and the next section we shall use congruences modulo different natural numbers a and use following **notations** : $G(a) = \langle n \bmod a \rangle$, $t(a) = \#G(a)$ and $H(a)= (\mathbb{Z} /a.\mathbb{Z})^*/G(a)$; like in §3, $\mathfrak{A}$ are the integers modulo a and $\mathfrak{A}$^ its dual, the set of the multiplicative characters defined on $\mathfrak{A}$ (precisely, $\mathfrak{A}$^ is dual to $\mathfrak{A}^*$, as mentioned in §3). The dual of H(a) is H(a)^ = $\{\chi \in \mathfrak{A}\text{^} \mid \chi(n)=1\}$. Let $T=\mathrm{lcm}\ \{q-1 \mid q \text{ prime}, q|s \}$. We shall suppose in this paragraph that following conditions on n are fulfilled:

(5.1) $(n,Ts) = 1$ and $(n,p)=1, \forall p<T$, p prime.

Our purpose is to prove the existence of an s-th cyclotomic extension of $\mathfrak{N}$ - without explicitly constructing one.

If $f, m \in \mathbb{Z}_{>1}$ are such that $m|s$ and $f|\lambda(m)$, $t=\lambda(m)/f$ and ξ is a complex primitive root of unity of respective orders f and m, we shall consider more general Gauss periods of degree f with respect to $v \in \mathfrak{M}^*$, where $\mathrm{ord}_m(v)=t$; they are given by:

(5.2) $\eta_h (\xi, v) = \sum_{1\le i\le t} \xi^{h.v\text{^}i}$ $\qquad h \in (\mathbb{Z} /m.\mathbb{Z})^*/ \langle v \bmod m \rangle$

For m=q a prime and $v=g^f$, where g generates $\mathfrak{Q}^*$, this is (2.1); when m is composite, not all v of order t belong to the same cycle. If v = n,

(5.3) $(\varphi(m)/t(m)).\ \eta_h(\xi, n) = \sum_{\chi\in H(m)^\wedge} \chi^{-1}(h).\tau(\chi)$ $\quad\quad \forall\, h\in H(m)$

establishes the relation between Gauss periods and Gauss sums. This relation can be derived by changing the order of summation on the right hand side and using the fact that H^(m) is a group such that if $x\notin$ <n mod m>, $\exists\chi\in$ H^(m) with $\chi(x)\neq 1$.

We shall use (3.8) together with (5.3) in order to calculate $\eta_h(\xi, n)$. The reduction formulæ of §3 will allow us to consider only characters $\chi = \chi_{q,p\wedge k}$ of prime power order p^k and prime conductor q. In this case (3.8) provides a p^k-th power of $\tau(\chi_{q,p\wedge k})$ for a fixed q-th root of unity ξ; we have therefore to consider the problem of taking p^k-th roots in adequate extensions of $\mathfrak{R}$, where $1\le k\le w$ and $w=v_p(T)$ where v_p is the p-adic valuation: $p^w \| T$.

Let t=t(p) and $v=v_p(n^t-1)$. It is easy to verify that:

(5.4) $v_p(n^{t.p\wedge i} - 1) = i+v$ $\quad$ $i\ge 0$ if p is odd or p=2 and v>1

and

(5.4') $v_p(n^{2\wedge i} - 1) = i+2$ $\quad$ $i>0$ if $p=2$ and $v=1$ (or equivalently, n=3 mod 4)

Suppose first that v = 1 and p is odd. In this case we write: $\mathbf{P}_k = \mathbf{Z}[\zeta] / (n.\mathbf{Z}[\zeta])$ - with $\zeta = \zeta_{p\wedge k}$ a primitive p^k-th root of unity - and $\mathbf{Q}_k = \mathbf{Z}[\zeta,\eta_0(p^k, \xi)] / (n.\mathbf{Z}[\zeta, \eta_0(p^k, \xi)])$. The roots $\zeta_{p\wedge k}$ are taken such that $\zeta_{p\wedge(k-1)} = \zeta^p{}_{p\wedge k}$ and $\mathbf{P}_w \supset \dots \mathbf{P}_2 \supset \mathbf{P}_1$ and $\mathbf{Q}_w \supset \dots \mathbf{Q}_2 \supset \mathbf{Q}_1$. Let further λ_k be the homomorphisms $\mathbf{Z}[\zeta] \to \mathbf{P}_k$ and $\mu_k: \mathbf{Z}[\zeta, \eta_0(p^k, \xi)] \to \mathbf{Q}_k$.

We consider now the case v>1. Let $\mathbf{P}_1$ be like before and $\zeta = \zeta_p$ be a primitive p-th root of unity in $\mathbf{P}_1$; suppose

(5.5) $\exists x\in \mathbf{P}_1$ such that $\quad x^{n\wedge t-1} = 1$ and $x^{(n\wedge t-1)/p} \neq 1$

Then $\omega = x^{(n\wedge t-1)/p\wedge v}$ is a primitive p^v-root of unity in $\mathbf{P}_1$. We shall define λ_1 as before and put $\mathbf{P}_k = \mathbf{P}_1[x]/((x^{p\wedge(k-1)} - \omega^{p\wedge(v-1)}))$; we let λ_k be the natural homomorphism $\mathbf{Z}[\zeta] \to \mathbf{P}_k$ and $\mu_k: \mathbf{Z}[\zeta, \eta_0(p^k, \xi)] \to \mathbf{Q}_k = \mathbf{P}_k[\eta_0(p^k, \xi)]$.

For p=2 and v=1, $v_2(n^2-1) = 3$; we shall show how to construct a primitive 8-th root of unity in $\mathbf{P}_2 = \mathbf{Z}[i]/(n.\mathbf{Z}[i])$ with $i = \sqrt{-1}$. If n is prime,

(5.6) $2^{(n-1)/2} = \pm 1$ mod n,

the sign on the right hand side being + or - according to 2 being a quadratic residue or nonresidue mod n. If (5.6) holds for a given n (to be tested for primality), a primitive 8-th root of unity in $\mathbf{P}_2$ is given by $\omega = (1+\hat{\imath})/\sqrt{2}$, where $\hat{\imath}$ is the image of i in $\mathbf{P}_2$, $\sqrt{2} = 2^{(n+1)/4}$ mod n, if in (5.6) the sign + is taken and $\sqrt{2} = \hat{\imath}.(-2)^{(n+1)/4}$ otherwise. We define $\mathbf{P}_k$ for k>2 by $\mathbf{P}_k = \mathbf{P}_2[x] / (x^{k-1} - \omega^2)$.

Consider now $\alpha\in \mathbf{P}_k$ such that

(5.7) $\alpha^{(N-1)/p\wedge k} = 1$ in $\mathbf{P}_k$, $\quad$ with $N = n^{t'}$, t' = t(p) if $k\le v$ and $t' = t.p^{k-v}$ for k>v; for p=2 and v=1, $t' = t.p^{k-2}$ for all k>2.

If v=1, $((N-1)/p^k, p)=1$ and there is a $\gamma\in(\mathbb{Z}/p^k.\mathbb{Z})^*$ such that $\gamma.(N-1)/p^k = -1 \bmod p^k$. Let $\delta = (\gamma.((N-1)/p^k)+1)/p^k$ and put $\beta = \alpha^\delta$ in $\mathbf{P}_k$. By its definition and (5.7), β is a p^k-th root of α in $\mathbf{P}_k$. If $k\geq v$, reasoning like in the case v=1, $((N-1)/p^k, p)=1$, there is $\gamma\in(\mathbb{Z}/p^k.\mathbb{Z})^*$ with $\gamma.((N-1)/p^k)=-1 \bmod p^k$ and $\delta=(\gamma.((N-1)/p^k)+1)/p^k$. Put $\beta=\alpha^\delta$ in $\mathbf{P}_k$. By definition of β and (5.7), $\beta^{p\wedge k} = \alpha$ and β is a p^k-th root of α in $\mathbf{P}_k$. If $1<k<v$, notice that, by construction, there is a primitive p^v-th root of unity $\omega'\in\mathbf{P}_k$ and $\alpha^{(N-1)/p\wedge v} = \omega'^{\nu.p\wedge k}$ for some ν coprime to p. With v instead of k, one finds γ like above, $\delta=(\gamma.((N-1)/p^v)+1)/p^k$ and $\beta=\omega'^{-\nu.\gamma}\alpha^\delta$ is a p^k-th root of α.
We have thus proved:

Lemma1

Let $p^k|T$, $v = v_p(n^{t(p)}-1)$ and $\mathbf{P}_1 = \mathbb{Z}[\zeta] / (n.\mathbb{Z}[\zeta])$ - with $\zeta = \zeta_p$ a complex primitive p-th root of unity. Suppose if $v>1$, that there is an $x\in\mathbf{P}_1$ such that (5.5) holds and if p=2 and v=1, that (5.6) holds. Than the rings $\mathbf{P}_k$ can be constructed as before; furthermore, if $\alpha\in\mathbf{P}_k$ is such that (5.7) holds, than there is a β in $\mathbf{P}_k$ with $\beta^{p\wedge k} = \alpha$ and β is a p^k-th root of α in $\mathbf{P}_k$.

We shall suppose from now on in this paragraph, that the conditions (5.5) or (5.6) are fulfilled for each p|T and that the towers of extensions $\mathbf{P}_w \supset ... \mathbf{P}_2 \supset \mathbf{P}_1$ and $\mathbf{Q}_w \supset ... \mathbf{Q}_2 \supset \mathbf{Q}_1$ can be constructed. We let $\mathbf{P}$ and $\mathbf{Q}$ be the push-outs of the sets $\{\mathbf{P}_w \mid p^w \| T\}$ and $\{\mathbf{Q}_w \mid p^w \| (q-1)\}$ and λ and μ be the homomomorphisms λ: $\mathbb{Z}[\zeta_T] \to \mathbf{P}$ and μ: $\mathbb{Z}[\zeta_T, \eta_0(n, \xi_s)] \to \mathbf{Q}$ such that the restrictions to $\mathbf{P}_k$ and $\mathbf{Q}_k$ are λ_k and μ_k, $\forall p^k \| T$.

Let q|s be a fixed prime and f=(q-1)/t(q). For all prime powers $p^k \| f$, consider the characters $\chi=\chi_p$: $\mathfrak{Q} \to <\zeta>$ (with $\zeta\in\mathbb{C}$ a p^k-th root of unity) given by $\chi(g) = \zeta$ (g generating $\mathfrak{Q}^*$) and put $\alpha_p = \lambda(\chi(-1).q.J_{p\wedge k-1}(\chi))$. We can use α_p to state a condition for existence of a q-th cyclotomic extension of $\mathfrak{N}$:

Theorem 2

If (5.7) is satisfied with $\alpha = \alpha_p$ for all $p^k \| f$, a q-th cyclotomic extension of $\mathfrak{N}$ exists.

Proof (sketch). Let r |n, $\mathfrak{r}$ a maximal ideal of $\mathbf{P}_k$ containing r and $\mathfrak{R}$ a maximal ideal of $\mathbf{Q}_k$ containing $\mathfrak{r}$, $\mathbb{L}\supset\mathbb{K}\supset\mathbb{F}_r$ the corresponding fields and ν a homomorphism $\mathbf{P}_k \to \mathbb{K}$ $\nu' = \lambda \circ \nu$. Let $\tau(\chi) = \sum_{x\in\mathfrak{Q}} \chi(x).\xi^x \in \mathbb{Z}[\zeta, \eta(p^k, \xi)]$; by Lemma 1, there exists $\beta\in\mathbf{P}_k$ with $\beta^{p\wedge k} = \alpha$ and $\beta' = \nu(\beta)$ is a p^k-th root of $\alpha'_p = \nu(\alpha_p)$ in $\mathbb{K}$. But $\nu'(\tau(\chi))$ is a p^k-th root of α'_p in $\mathbb{K}$ and since $\mathbb{K}$ is a field, $\nu'(\tau(\chi))\in\mathbb{K}$. According to §2 and §4, $\mathbb{K} = \mathbb{L}$ and $\eta(p^k, \xi)\in\mathbb{F}_r$ follows. Finally, by (3.11) and (5.3), $\eta(f, \xi)\in\mathbb{F}_r$ and using (2.4) (this linear system is regular due to (5.1)) $\Phi_q(x)$ splits over $\mathbb{F}_r$ in f factors (not necesserily irreducible) of degree t. This shows that $r\in<n \bmod q>$ and since this holds for all primes r |n, a q-th cyclotomic extension of $\mathfrak{N}$ exists by theorem 1.
•

Of course, the existence of q-th cyclotomic extensions of $\mathfrak{N}$ for all prime q|s does by far not imply existence of an s-th extension; this can be seen already by comparing the degree $\prod_{q|s} t(q)$ of the product of the q-th extensions with the degree $t(s) = \mathrm{lcm}_{q|s}(t(q))$ of an s-th extension of $\mathfrak{N}$.

We shall inductively prove the existence of extensions of degrees m with products of 2, 4, ... primes. In order to describe the induction process let $\#\{q|\ q$ prime and $q|s\} = u$ and h be such that $2^{h-1} < u \leq 2^h$; if $u<2^h$ one can expand the list of primes dividing s with 2^h-u factors 1. A binary tree of divisors of s is defined as follows: write on the lowest level (level 0) all primes q|s, completing the level with 1 up to 2^h vertices. We define a binary tree in which each vertex is identified (labeled) by some divisor of s and the terminal vertices are the level 0 just described. Suppose levels 0, 1, ..., k-1 are labeled for some k≤h. The vertices on level k are then labeled by the product of the labels of the two connected vertices on level k-1. A vertex is called trivial if it is 1 or one of the connected vertices on level k-1 is 1.

This completes the description of a binary tree with following properties:

(i) the root (level h) is s
(ii) the terminal vertices are primes q|s or trivial
(iii) all vertices on level k (0≤k<h) are pairwise coprime and products of 2^k primes or 1s.

Let m|s, $m=m_1.m_2$ with coprime divisors m_i (i=1,2), both different from 1. Suppose m_i-th cyclotomic extensions of $\mathfrak{N}$ exist. Under this premise, we shall give sufficient conditions for an m-th cyclotomic extension to exist.

Let $\xi \in \mathbb{C}$ be an m-th primitive root of unity, $\xi_i = \xi^{m/m_i}$ primitive roots of order m_i (i=1,2) and let $\eta_h(f, \xi, n)$ be given by (5.2), h∈H(m), f=λ(m)/t(m) and $\eta_h(\xi_i, n)$, h∈H(m_i) defined by analogy. Let r|n be a prime and $\mathfrak{r}$ a maximal ideal of $\mathbb{Z}[\xi]/(n.\mathbb{Z}[\xi])$ containing r, $\mathbb{K}$ the factor field and κ the natural homomorphism $\mathbb{Z}[\xi] \to \mathbb{K}$. The induction hypothesis implies $\kappa(\eta_h(\xi_i,n)\in \mathbb{F}_r$ for all h∈H(m_i). If $\mathfrak{R}$ is a maximal ideal of $\mathbb{Z}[\zeta, \xi]/(n.\mathbb{Z}[\zeta, \xi])$ containing $\mathfrak{r}$, $\mathbb{L}\supset\mathbb{K}$ the corresponding factor field and κ' the extension of κ to $\mathbb{Z}[\zeta, \xi]$, then:

(5.8) $\kappa'(\chi^{-1}(h).\tau(\chi)) \in \mathbb{F}_r[\zeta']$ $\quad \forall\ \chi\in H(m_1)^\wedge \times H(m_2)^\wedge$

where $\zeta'= \kappa'(\zeta)$ and the direct product $H(m_1)^\wedge \times H(m_2)^\wedge$ is induced by (3.9). The relation (5.8) is a consequence of the induction hypothesis. Note that the factor (φ(m)/t(m)) occuring in (5.3) is a unit in $\mathfrak{N}$ by (5.1), since all its prime factors divide T.

We want to improve (5.8) and show it holds for all χ∈H(m)^. This would imply $\kappa(\eta_h(\xi,n))\in \mathbb{F}_r$ and $\Phi_m(x)$ splits in f factors of degree t(m). Consequently r∈<n mod m>, showing that an m-th cyclotomic extension of $\mathfrak{N}$ exists.

Let $\chi\in H(m)^\wedge - H(m_1)^\wedge \times H(m_2)^\wedge$ and $\chi=(\chi_1,\chi_2)$ be the splitting of χ according to the conductors m_1, m_2. Since $\chi(n)=\chi_1(n).\chi_2(n)=1$ and $\chi_i(n)\neq 1$ (i=1,2), we have $\chi_1(n)=\chi_2^{-1}(n)=\rho$, where ρ is a root of unity of order $e = (t(m_1), t(m_2))$ (this follows from $\chi_i(n)^{t(m_i)}=1$). We can now follow the path of theorem 2 in order to give an existence criterion for the m-th cyclotomic extension of $\mathfrak{N}$.

Theorem 3

For all primepowers $p^k \| e$, find a character $\chi_i\in \mathfrak{M}_i^\wedge$ with $\chi_1(n)=\chi_2^{-1}(n)=\zeta$, where ζ is a root of unity of order p^k. Put $\alpha_p = \lambda\,(\chi_1(-1).q.J_{f_1-1}(\chi_1)\,\chi_2(-1).q.J_{f_2-1}(\chi_2))$, where χ_i are primitive of order f_i respectively. (i=1, 2)

If (5.7) holds for α_p, for each p|e, then an m-th cyclotomic extension of $\mathfrak{N}$ exists.

Let $p^k \| e$ in theorem 3. Then there are primes $q_i|m_i$ with $p^k|(q_i -1)$ (i=1,2), by definition of e.

The characters χ_i in theorem 3 can be chosen with conductors q_i. If χ_i is such a character, χ'_i is the (reducible) character of conductor m, which is $= \chi_i \mod q_i$ and the trivial character modulo all other primes dividing m_i and if $\kappa_i \in H(m_i)^\wedge$ then $\chi^*_i = \chi'_i \cdot \kappa_i$ is primitive and verifies the hypothesis of theorem3. Furthermore, if $\upsilon_i \in \mathfrak{M}_i^\wedge$ is any character with $\upsilon_i(n)=\zeta$ (or $=\zeta^{-1}$ according to i=1 or i=2), then $\kappa_i . \upsilon_i \in H(m_i)^\wedge$, which shows that it is indeed sufficient to test (5.5') once for every $p^k \| e$.

Theorems 2 and 3 allow to prove the existence of an s-th cyclotomic extension of $\mathfrak{R}$ by induction along a binary tree of divisors of s. The computational work involved in the proof takes checking (5.7) in the rings $\mathbf{P}_k$ (note that the index k varyes for each test and is bounded by $v_p(T)$) for elements α_p which are products of Jacobi-sums.

§6 Primality test

Let n, s $\in \mathbf{Z}_{>1}$, s odd and squarefree, $s>\sqrt{n}$ and T=lcm $\{q-1 \mid q \text{ prime}, q|s\}$. We describe an algorithm for testing the primality of n, which is based on the theorems 1-3.

(A) check $(n,Ts) = 1$.

(B) for all primes $p|T$ with $n^{p-1} \equiv 1 \mod p^2$ let $t = \text{ord}_p(n)$, $v = v_p(n^t-1$

- (i) if $\exists q|s$, q prime and $n^{(q-1)/p} \neq 1 \mod p$ let χ be a character of conductor q and order p with $\chi(n)\neq 1$ and $\alpha = \lambda_p\,(\chi(-1).q.J_{p-1}(\chi)) \in \mathbf{P}$. Check that (5.5) is verified with $x = \alpha$. In this case, go to (C), otherwise declare n composite and stop.
- (ii) find by trial and error $x \in \mathbf{P}_1$ such that (5.5) is verified. If this is possible within a chosen number of trials, go to (C).
- (iii) the test stops with no decision on n. (s and T may be changed, so that p does not divide T).

(B')If $n \equiv 3 \mod 4$, check (5.6) and go to (C) if it holds or else declare n composite and stop.

(C) For all $q|s$ put $f = (q-1)/t(q)$. For all $p^k \| f$ do following: let χ be a primitive multiplicative character of conductor q and order p^k. Put $\alpha \in \mathbf{P}_k$, $\alpha = \lambda\,(\chi(-1).q.J_{p\wedge k-1}(\chi))$ and check (5.7) holds. If this is not the case declare n composite and stop.

(D) Consider a binary tree of divisors of s. For every non-terminal, non-trivial node m connected to m_i (i=1, 2) on the next lower level, let $e = (t(m_1), t(m_2))$. For every $p^k \| e$ find q_i with $p^k|(q_i-1)$ (i=1,2) and two characters χ_i of conductors q_i such that $\chi_1(n) = \chi_2^{-1}(n) = \zeta$ ($\zeta \in \mathbb{C}$ primitive root of unity of order p^k). Let $\alpha = \lambda\,(\chi_1(-1).q.J_{f_1-1}(\chi_1)\,\chi_2(-1).q.J_{f_2-1}(\chi_2))$, where χ_i are primitive of order f_i respectively (i=1, 2). Check that (5.7) holds. If this is not the case, declare n composite and stop.

(E) For i=1, 2, ... ,t(s) let r_i be such that $1<r_i<s$ and $r_i \equiv n^i \mod s$ and check that r_i does not divide n. If i is prime, also check that i does not divide n.
If n passes all these tests, declare n prime and stop.

Remarks (A) and (B) check the premises of §5 and the ones of lemma 1. If (C) and (D) are passed, an s-th cyclotomic extension of $\mathfrak{N}$ exists by theorems 2 and 3. If (E) is passed, n is prime by theorem 1. In (A) - (D) the existence of an s-th cyclotomic extension of $\mathfrak{N}$ is proved in polynomial time $(\log(n))^{4+\varepsilon}$, whereas the methods proposed in [Le2] take overpolynomial time.

§7 Conclusions

We aimed to give a comprehensive insight in primality testing with cyclotomic extensions. Detailed proofs and refinement yielding more effective variations of the algorithm could not fit in this paper. We shall only enumerate here such possible variations:

(7.1) keeping the same structure of the algorithm, the rings P_k could be chosen using roots of unity over $\mathfrak{N}$ (rather than projecting complex roots by homomorphism). A discussion of such alternatives can be found in [CoLe, §10].

(7.2) the exponent in (5.7) can be easily reduced to n by using (3.13) rather than (3.14), with the necessary adaptation for n not being known to be prime. This is done in [CoLe].

(7.3) if t(s) is smaller than some constant k (e.g. k=50) some generalized Lucas-Lehmer tests can be given, which use theorems 1-3. In such a case proving existence of a cyclotomic extension can be done more efficiently by finding roots of unity than by considering the splitting of the cyclotomic polynomial.

(7.4) an interesting idea might consist in considering in stage (C) of the algorithm polynomials $\Gamma(x)=\prod_{1\leq i\leq p^k} (x - \eta_i) \in \mathfrak{N}[x]$ and show they have at least one root in $\mathfrak{N}$. Similar polynomials can be defined also for stage (D), so that (B) and the tests of (5.7) become superfluous; one would not have to perform any computations in extensions of $\mathfrak{N}$ in this case. The disadvantage of this approach is that the proof of existence of roots of $\Gamma(x)$ requires probabilistic algorithms, thus increasing the amount of probabilistic steps of the global test.

(7.5) the only non-polynomial stage in this algorithm is (aswell as for the Adleman algorithm) stage (E). This fact still holds, if s is chosen to be much larger - e.g. $s=O(n^{\log(n)})$. It is not clear if the information gained by proving existence of such a large cyclotomic extension could not be used to reduce the number of tests in stage (E).

The use of Jacobi-sums for proving existence of cyclotomic extensions shows the intimate similarity of this testing approach to the test of Adleman and this more dramatically than just by the fact that both procedures share the same final stages. The proposal of Lenstra [Le2] to implement a combination of both algorithms becomes still more appealing. For a given n, the same s and T can be used for both algorithms. Although a thorough comparison of the two algorithms is difficult - in spite of the deep similarity - following observation makes cyclotomic extensions interesting: the computation intensive test of (5.7) (or an equivalent form thereof - (7.2)) is performed in [CoLe] for all pairs (p,q) with $p^k \| (q-1) | T$ and $q|s$, whereas using cyclotomic extensions, only some of these pairs need to be considered. Since it is known ([CoLe]) that these stages of the algorithm are the most time consuming, one can expect the cyclotomic extension test to run slightly faster than the Jacobi-sum one.

The *prerpocessing stage* (B) is the only nondeterministic stage of the algorithm; the condition $\exists q|s$, q prime and $n^{(q-1)/p} \neq 1 \bmod p$ of (B) (i) is necessary for running the Jacobi sum test [CoLe] (otherwise p is discarded, like in the option in (B) (iii)). Consequently, whenever [CoLe] can give a decision about a number n, the step (B)(i) is run successfully and our test ends with a decision too. It also ends with a decision when (B)(ii) is run successfully and (B)(i) not. This shows that the new test ends with a decision at least whenever [CoLe] does so. Note that the test in [CoLe] is deterministic under assumption of the generelized Rieman hypothesis. Therefore, the cases when it might not end with a decision in predicted time should not be a cause of worry: such a case is either a serious hint to falsity of the generelized Rieman hypothesis or to a flaw in the program. In both cases an important information would come out of such an exception.

Acknowledgements The list of people who encouraged this work is very long and could hardly be exhausted here. The help of all of them has been important and I express my deep gratitude to all those who, at one stage or another helped me push the load further. I am most specially grateful to prof. H.W.Lenstra, Jr. who, by his attentive reading and criticisms helped my intuitions become mathematical thinking.

References

[APR] L.M.Adleman, C.Pomerance, R.S.Rumely: "On Distinguishing Prime Numbers from Composite Numbers" *Ann. of. Math.* v. **117**, 1983, pp. 173-206

[AHu] L.M.Adleman, M.A.Huang :"Recognizing Primes in Random Polynomial Time", **STOC** 1987, pp. 462-469

[BS] E.Bach, J.Shallit: "Factoring with Cyclotomic Polynomials", **FOCS**, 1985, pp. 443-450

[Cn] H. Cohn: "A Classical Invitation to Algebraic Numbers and Class Fields", Springer *Universitext* , 1978

[Co] H. Cohen: "Atkin's Primality Test", lecture in *Bonn Workshop of Foundation of Computing* , Bonn, June 28-July 3, 1987

[CoLe] H.Cohen, H.W.Lenstra,Jr.: "Primality Testing and Jacobi Sums", *Math of Comp.*, vol **42**, 1984, pp. 297-330

[GoKi] S.Goldwasser, J.Killian : "Almost all Primes Can Be Quickly Certified", **STOC** 1986, pp. 316-329

[Hu] M.A.Huang: " Riemann Hypothesis and Finding Roots over Finite Fields" , **STOC** 1984, pp. 121-130

[La] S.Lang: "Algebra" , Addison & Wesley

[Le1] H.W.Lenstra,Jr.: "Primality Testing Algorithms", in *Lecture Notes in Mathematics* vol **901** , pp. 243-258

[Le2] H.W.Lenstra,Jr.: "Galois Theory and Primality Testing", in *Lecture Notes in Mathematics* vol **1142** , pp. 243-258

[Le3] H.W.Lenstra,Jr.: "Elliptic Curves and Number Theoretic Algorithms", *Report* nr. **86-19**, University of Amsterdam, Dept. of mathematics

[LL] A.K.Lenstra, H.W.Lenstra,Jr.: "Algorithms in Number Theory", University of Chicago, Technical Report **87-008**, May 1987

[LT] H.W.Lenstra,Jr. and R.Tijdeman (eds), " Computaional Methods in Number Theory " , Mathematical Centre Tracts, **154/155** , Mathematisch Centrum, Amsterdam 1986

[Ra] M.O.Rabin : "Probabilistic Algorithms for Testing Primality", *J. of Number Theory* vol. **12**, 1980, pp. 128-138

[Ri] H.Riesel: " Prime Numbers and Computer Methods for Factorization", *Birkhäuser Progr. Math.* vol. **57/85**

[SSt] R.Solovay, V.Strassen: "A fast Monte-Carlo Test for Primality" *SIAM J. of Comput.* vol. **6**, 1977, pp. 84-85. erratum, ibid., vol. **7** ,1978, p. 118

[Wi] H.C.Williams: "Primality Testing on a Computer", *Ars Combin.* **5** (1978) pp.127-185

THE METHOD OF SYNDROME CODING AND ITS APPLICATION FOR DATA COMPRESSION AND PROCESSING IN HIGH ENERGY PHYSICS EXPERIMENTS

N.M. Nikityuk
Joint Institute for Nuclear Research
Laboratory of High Energies, Head
Post Office, P.O. Box 79, Moscow
USSR

1. PROBLEM

Agreat number of multichannel position-sensitive detectors is used in high energy physics experiments. There is a small number (10-20%) of signals for processing with the aid of electronics and special-purpose processors. The processing of physics information is hierarchical in nature. If it is considered in time, the time of the first level is 100 ns when background information is filtered according to the selection criteria of useful events. The particle multiplicity is the main selection criterion for useful events registered in a coordinate detector. The second criterion is to determine the interaction coordinates of particle with sources in the usual binary code for a minimum time.

Figure 1 presents a simple scheme containing a target T and a detector. The detector consists of 31 sources (for example, scintillators). Two particles interact with the scintillators which generate two signals simultaneously. After amplifying and shaping, the signals are supplied to the inputs of a majority coincidence circuit and an encoder (Fig.2). Sometimes several neighbouring scintillators (cluster events) generate from one particle. In this case there arises the problem of cluster registration and identification. The multiplicity of particles is determined with the aid of the majority coincidence circuit. The interaction coordinates X_1 and X_2 are measured by means of the parallel encoder.

If we know wittingly that $t=1$, the construction of an encoder based on combination circuits presents no difficulties. To solve the problem at $t > 1$, shift registers and priority encodes are commonly used. Synchronization pulses are required for the operation of these circuits. This means that very much time and a large number of circuits are needed for a great variety of registration channels. Modern PROMs have a limited number of address inputs, and so they cannot solve this problem. The use of correcting code theory and particle helps to answer the following question: "How is a parallel encoder constructed for $t > 1$"[1,2].

2. SYSTEM OF ANALOGIES

To make the best use of coding theory and practice, the author has suggested a system of analogies for coding theory and the theory of multichannel hodoscopic systems[3] (see Table)

Table

Coding Theory 1	Theory of Hodoscopic Systems 2
1. Code vector of the block codes having n symbols, $K = n-tm$ information symbols.	Word read out from n position-sensitive sources
2. Mistake vector e	Physical events
3. Cluster of mistakes	The number of position-sensitive sources t operated in the detector simultaneously
4. The number of parity check bits $N=mt$	The number of digits, N, at the outputs of the parallel encoder
5. Code efficiency $V = k/n$	Compression coefficient, $K_c=n/N$
6. Coding device	Encoder
7. Parity check matrix H^T	Connection matrix H
8. Coding distance d	Coding distance d
9. Weight of the code word, W	Row weight of the connection matrix, W
10. Column weight parity check matrix, b	Branching coefficient for a signal from a sensitive source, b
11. Two-dimensional iterative code	Data coding in a two-dimensional hodoscope
12. Completely asymmetric channel	Multichannel data transmission system

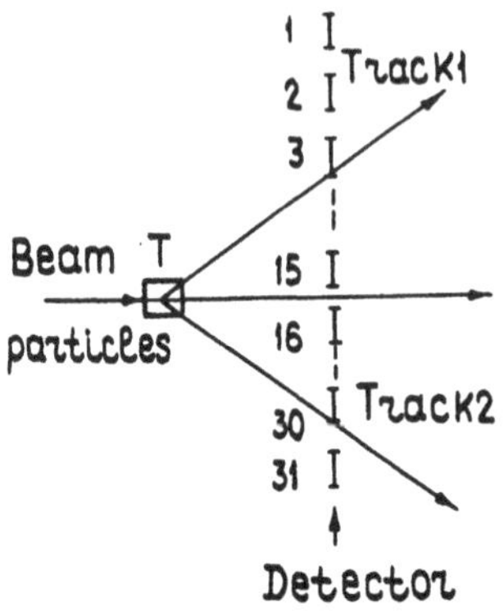

Fig.1. Simple diagram for the registration of two particles by the scintillation hodoscope. T - target, 1-31 - scintillators.

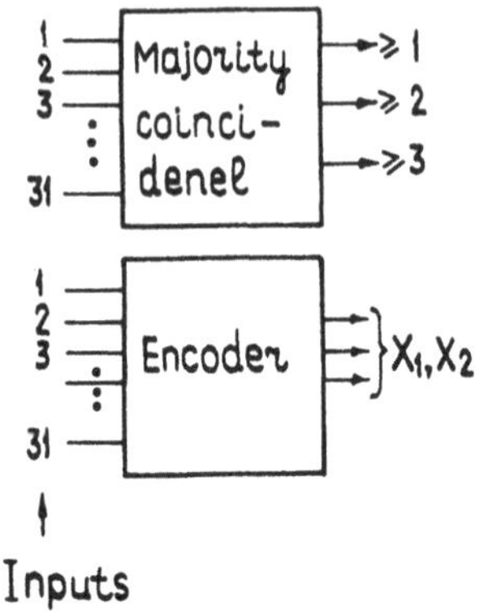

Fig.2. Block-diagram for fast event selection.

3. SYMDROME METHOD

The essence of the method is the following[1,2,4,6]. As assumed in coding technique, let us consider a typical multichannel data transmission system (Fig.3a). There is a 1-bit register on the transmitting side. A K-bit syndrome is added to the 1-bit register according to a given rule. The 1 + K-bit word is transmitted to a receiver. Encoding is carried out by the encoder. If there are mistakes during the transmission, the decoder finds and corrects them using the syndrome.

Now let us consider a simpler transmission scheme (Fig.3b). Assume that the transmitted n-bit word is always zero. Then ohes occuring when the sources operate are considered as an error vector of the code word. For example, the 31-bit zero word 000......000 is transmitted, and on the receiver side we get the following word:

0000000001000000000001000000000

This means that there are mistakes in positions 10 and 22 if the count of the positions is carried out from left to right. Further, if the BCH code correcting two mistakes is chosen, the number of bits at the encoder outputs is

$$N = 2\log_2 31 \simeq 10.$$

In other words, we have the effect of compression from a 31-bit unitary position code to a 10-bit cyclic code. In this case the compression effect grows with increasing the number n. As a result, it is possible to use PROMs for solving the problem by the arithmetic method.

It should be stressed that one circumstance is important: according to theory, the syndrome carries information of the multiplicity and coordinates.

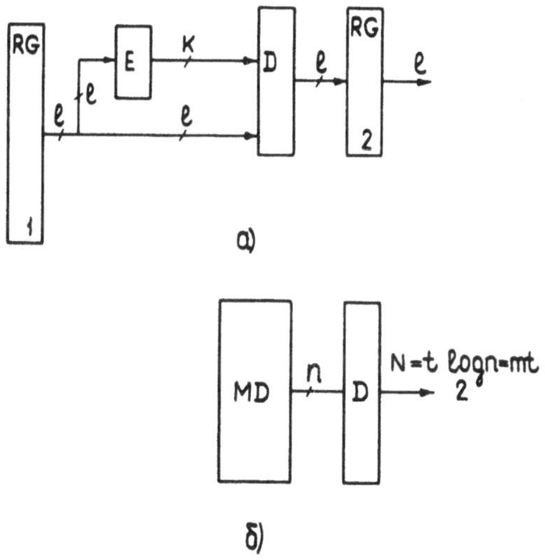

Fig.3. Explanation of the Syndrome Coding Method. a) Conventional multichannel system where the correcting code is used. E - encoder, D - encoder. b) System of data registration where the method of syndrome coding is used. MD - detector.

4. APPLICATION OF THE SYNDROME METHOD

This method is used to construct fast electronic units for the selection of events and special-purpose processors in nuclear physics experiments. Let us consider several examples.

1. <u>Use of the algebraic theory of correcting codes.</u> Assume that there is a position-sensitive detector containing $n = 31$ sources, and it is necessary to determine the multiplicity $(t \leqslant 2)$ and coordinates of particle interaction. To draw a principal diagram of the parallel encoder, the matrix H^T (connection matrix) should be constructed. The syndrome code is calculated with the aid of this matrix.

The elements of the Galois field $GF(2^5)$ generated by an irreducible polynomial $x^5 + x^2 + 1$ are presented on the right of Fig.4. Figure 5 shows a principal scheme used to calculate the syndrome S_1. The numbers of the channels (position-sensitive sources), which logic pulses are sent from, are denoted by numerals. For $n = 31$ and $t = 2$ there is a 10-bit cyclic code at the outputs of the encoder, This code carries information on the multiplicity and interaction coordinates of two particles. As it follows from the known theorem[6,8], for finding t it is necessary to analyse the following determinations:

$\det L_1 = S_1$ and $\det L_2 = S_1^3 + S_3$.

$H^T =$

Channels			S_{10}	S_{11}	S_{12}	S_{13}	S_{14}	S_{30}	S_{31}	S_{32}	S_{33}	S_{34}
1	α^0	α^0	1	0	0	0	0	1	0	0	0	0
2	α^1	α^3	0	1	0	0	0	0	0	0	1	0
3	α^2	α^6	0	0	1	0	0	0	1	0	1	0
4	α^3	α^9	0	0	0	1	0	0	1	0	1	1
5	α^4	α^{12}	0	0	0	0	1	0	1	1	1	0
6	α^5	α^{15}	1	0	1	0	0	1	1	1	1	1
7	α^6	α^{18}	0	1	0	1	0	1	1	0	0	0
8	α^7	α^{21}	0	0	1	0	1	0	0	0	1	1
9	α^8	α^{24}	1	0	1	1	0	0	1	1	1	1
*10	α^9	α^{27}	0	1	0	1	1	1	1	0	1	0
11	α^{10}	α^{30}	1	0	0	0	1	0	1	0	0	1
12	α^{11}	α^2	1	1	1	0	0	0	0	1	0	0
13	α^{12}	α^5	0	1	1	1	0	1	0	1	0	0
14	α^{13}	α^8	0	0	1	1	1	1	0	1	1	0
15	α^{14}	α^{11}	1	0	1	1	1	1	1	1	0	0
16	α^{15}	α^{14}	1	1	1	1	1	1	0	1	1	1
17	α^{16}	α^{17}	1	1	0	1	1	1	1	0	0	1
18	α^{17}	α^{20}	1	1	0	0	1	0	0	1	1	0
19	α^{18}	α^{23}	1	1	0	0	0	1	1	1	1	0
20	α^{19}	α^{26}	0	1	1	0	0	1	1	1	0	1
21	α^{20}	α^{29}	0	0	1	1	0	1	0	0	1	0
*22	α^{21}	α^1	0	0	0	1	1	0	1	0	0	0
23	α^{22}	α^4	1	0	1	0	1	0	0	0	0	1
24	α^{23}	α^7	1	1	1	1	0	0	0	1	0	1
25	α^{24}	α^{10}	0	1	1	1	1	1	0	0	0	1
26	α^{25}	α^{13}	1	0	0	1	1	0	0	1	1	1
27	α^{26}	α^{16}	1	1	1	0	1	1	1	0	1	1
28	α^{27}	α^{19}	1	1	0	1	0	0	1	1	0	0
29	α^{28}	α^{22}	0	1	1	0	1	1	0	1	0	1
30	α^{29}	α^{25}	1	0	0	1	0	1	0	0	1	1
31	α^{30}	α^{28}	0	1	0	0	1	0	1	1	0	1

(S_{10} … S_{14} = S_1; S_{30} … S_{34} = S_3)

Fig.4. BCH-code parity check matrix for the correction of two mistakes (or for the registration of two events).

In addition, we introduce some other signs EVEN and ODD. These signs are obtained very simply if all outputs of the detector after amplifying and shaping are connected to the inputs of the parallel checker. Then we have the following algorithm for event selection. If $S_1 \neq 0$, there is at least one signal at the outputs of the detector. If det $L_1 \neq 0$, det $L_2 = S_1^3 + S_3 = 0$ and there is a sign ODD, then $t = 1$. For det$L_1 \neq 0$, det $L_2 \neq 0$ and a sign EVEN, $t = 2$. If det $L_1 \neq 0$, det $L_2 \neq 0$ and there is a sign ODD, then $t > 3$. So, in our example $S_1 = a^9$ and $S_3 = a^{27}$. Then $S_1 + S_3^3 = a^{(9)3} + a^{27} = 0$ (mod. 2). But for $t = 2$, $S_1 = a'$, $S_3 = a^{29}$ and det $L_2 \neq 0$. Using the parallel methods of calculation in the Galois field GF(2^4) and fast ECL-microcircuits, the following parameters for special-purpose processors have been obtained /7/: at $n = 15$ the solution time for $t = 1$, $t = 2$, $t = 3$ and also for the determination of three interaction coordinates X_1, X_2, and X_3 does not exceed 40 ns.

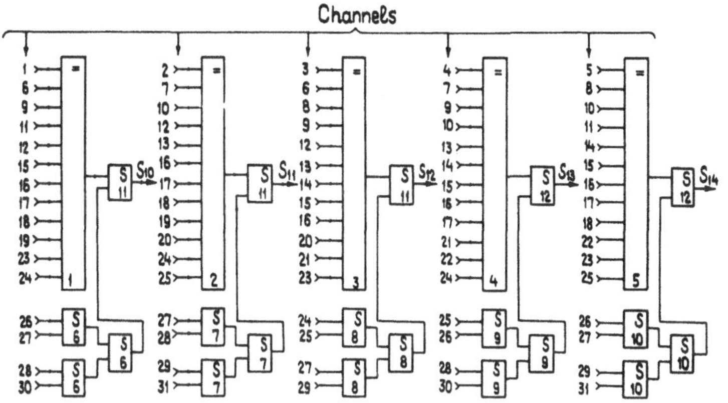

Fig. 5. Principal scheme used to calculate the syndrome S_1.

Figure 6 presents a scheme of the processor which can execute three functions: a majority coincidence circuit, a parallel counter and a processor for calculation of three event coordinates X_1, X_2, and X_3. Since $t \leqslant 3$, according to the theory of BCH-code decoding, which corrects three mistakes, the processor analyses the following determinants:

det $L_1 = S_1$, det $L_2 = S_1^3 + S_3$

and

det $L_3 = S_1^6 + S_1^3 S_3 + S_3^2 + S_1 S_5$.

For a fast solution of such determinants, parallel algorithms are used for calculation in the Galois field GF(2^m) /14/. The coordinates are calculated with the help of PROMs /7/.

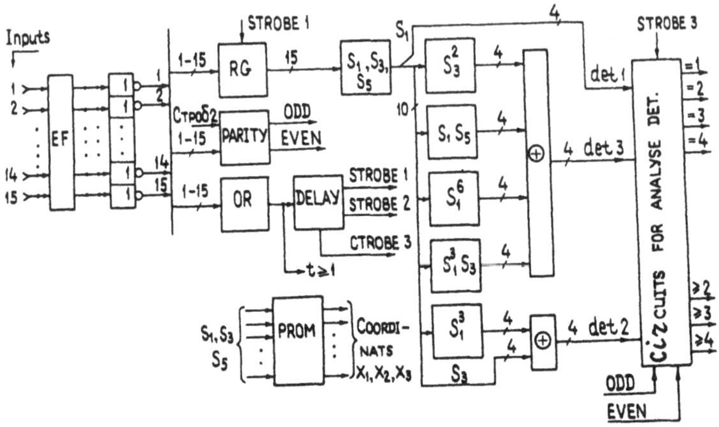

Fig. 6. Scheme of the majority coincidence circuit and of the special-purpose processor with algebraic structure. EF - emitter followers, T - delay, RG - register.

2. Superimposed codes[3,9,10-12]. It is convenient to use these in light coding systems and when signals have a small amplitude (analog signals). It is important that light and electronic amplifiers-mixers can be used to calculate the syndrome. However, K_c of such codes is smaller than for BCH-codes since coding words are added not by modulo-2 but by the Boolean sum rules. Superimposed codes having a constant weight are used for data compression in scintillation hodoscopes and for the creterion of majority coincidence circuits and parallel counters. Two coding schemes are presented in Fig.7. The first scheme is used at CERN, and the other is suggested by the author. It is obvious that the number of registration channels in the second scheme is smaller at other equal parameters. And its economical efficiency rises as C_N^2 /2N. Optical fibers can be also used for coding (Fig.8). The selection of events is performed as follows[11,12]. If the weight of the syndrome code at the outputs is 2, t is always equal to 1. If the weight is equal to 3 or 4 and there is a sign EVEN, then t=2 or 4 and so on. The Hemming codes and the Gray code are shown to be superimposed ones[3,10,15].

3. Use of iteration codes. This class of codes is very wide. Iteration codes are extensively used for a large number of registration channels (t > 100) and in studies of complicated topologies of events with clusters.

Let us consider some examples. Figure 9 shows a complicated event registered in the detector. The detector is composed of 1296 position

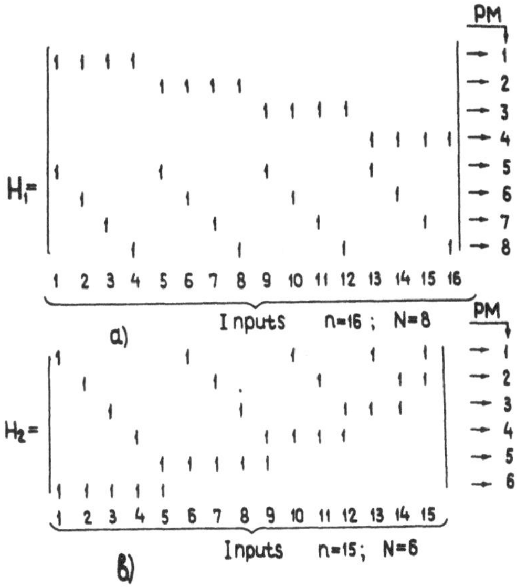

Fig.7. Parity check matrix for
a) the known coding scheme ($n = 2N = 16$) and
b) the coding scheme ($n = C_N^2 = C_6^2 - 15$)
suggested by the author, PM - photomultipliers.

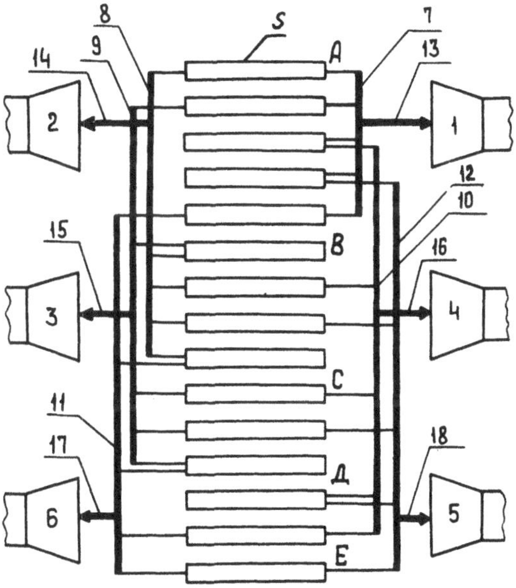

Fig.8. Example of the economical coding scheme for the scintillation hodoscope where optical fibers and optical mixers are used according to the matrix H_2. 1-6 - photomultipliers; 7-12 - optical fibers; 13-18 - optical mixers; S - scintillators.

sensitive sources presented as a matrix comprising 36 rows and 36 columns. The syndrome of this code has 128 bits, and the code efficiency grows with increasing the number n. So, $K_c = 1296/128 = 10$ for $n = 1296$ and $K_c = 4096/256 = 16$ for $n = 4096$. Short algorithm for event recognition given in the figure consists in the following. The coincidence of signals EVEN and OR in rows 1 and 2 in column 6 indicates unambiguously that pulses are supplied from sources 6, 7, 42 and 43. The signals coming from sources 752-755, etc., are also registered unambiguously. It should be emphasized that any codes, e.g. superimposed codes, can be taken for iteration.

4. <u>Application of the abundant Gray code for increasing the space resolution of a scintillation hodoscope</u> /15/. In accordance with the table of analogous, we can present a coding scheme of the scintillation hodoscope containing $n = 15$ scintillators and $N = 4$ photomultipliers in the form of a Hemming code parity check matrix:

$$H_3 = \left|\begin{array}{ccccccccccccccc} 1&0&1&0&1&0&1&0&1&0&1&0&1&0&1 \\ 0&1&1&0&0&1&1&0&0&1&1&0&0&1&1 \\ 0&0&0&1&1&1&1&0&0&0&0&1&1&1&1 \\ 0&0&0&0&0&0&0&1&1&1&1&1&1&1&1 \\ n_1&n_2&n_3&n_4&n_5&n_6&n_7&n_8&n_9&n_{10}&n_{11}&n_{12}&n_{13}&n_{14}&n_{15} \end{array}\right| \begin{array}{c} N_1 \\ N_2 \\ N_3 \\ N_4 \\ \text{Outputs} \end{array}$$

Inputs

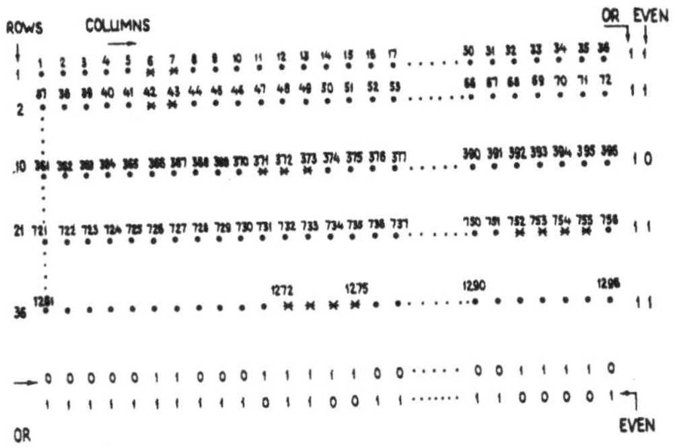

Fig.9. Example of using the iteration code OR-PARITY for complicated event registration. The events are marked by *.

If we add columns 7 and 8 or 13 and 14 by the Boolean sum rules in the matrix H_3, we get a similar result 1111. In practice, this means that using such a coding scheme, it is impossible to register the coordinates of double clusters because the Hemming code can correct only one mistake. Rearranging the columns of the matrix H_3, we obtain 15 words of the Gray code presented as a matrix H_4.

$$H_4 = \left|\begin{array}{ccccccccccccccc} 1&1&0&0&1&1&0&0&1&1&0&0&1&1&0 \\ 0&1&1&1&1&0&0&0&0&1&1&1&1&0&0 \\ 0&0&0&1&1&1&1&1&1&1&1&0&0&0&0 \\ 0&0&0&0&0&0&0&1&1&1&1&1&1&1&1 \\ n_1&n_2&n_3&n_4&n_5&n_6&n_7&n_8&n_9&n_{10}&n_{11}&n_{12}&n_{13}&n_{14}&n_{15} \end{array}\right| \begin{array}{l} N_1 \\ N_2 \\ N_3 \\ N_4 \\ \text{Outputs} \end{array}$$

Inputs

The scintillator is put according to each column of the matrix H_4 and the photomultiplier according to each row. It is not difficult to check that such Boolean sums as $n_1 \vee n_2$ and $n_2 \vee n_3, n_4\ n_5$ and $n_5\ n_6, n_9 \vee n_{10}$ and $n_{10} \vee n_{11}$ are coincident in the matrix H_4. This factor leads to decreasing the space resolution of the double cluster coordinates in these positions. To solve such uncertainties, additional bits to the classical Gray code have been suggested by the author. These bits are arranged in definite positions:

$$1\ 0\ 0\ 1\ 0\ 0\ 0\ 0\ 1\ 0\ 0\ 1\ 0\ 0\ 0$$

If one needs for register the coordinates of triple clusters, it is necessary to add abundant bits

$$\left|\begin{array}{ccccccccccccccc} 1&1&0&0&1&1&0&0&1&1&0&0&1&1&0 \\ 0&1&1&1&1&0&0&0&0&1&1&1&1&0&0 \\ 0&0&0&1&1&1&1&1&1&1&1&0&0&0&0 \\ 0&0&0&0&0&0&0&1&1&1&1&1&1&1&1 \\ 1&0&0&1&0&0&0&0&1&0&0&1&0&0&0 \\ 0&0&1&0&0&0&1&0&1&0&0&0&0&1&0 \\ 0&1&0&0&0&0&0&1&0&0&0&0&1&0&0 \\ n_1&n_2&n_3&n_4&n_5&n_6&n_7&n_8&n_9&n_{10}&n_{11}&n_{12}&n_{13}&n_{14}&n_{15} \end{array}\right| \begin{array}{l} N_1 \\ N_2 \\ N_3 \\ N_4 \\ N_5 \\ N_6 \\ N_7 \\ \text{Outputs} \end{array}$$

Inputs

The number of abundant bits does not depend on the one of bits in the Gray code. In other words, the efficiency of the abundant Gray code grows with increasing the number n. So, $K_c = 15/7$ for $n = 15$ and $K_c = 127/110$ for $n = 127$.

Different coding schemes created as a mask of transparent glass and scintillators are given in Fig.10. One can see that the space resolution for cluster registration improves with increasing the number of additional bits. Such a light coding scheme can be created if optical fibers are used.

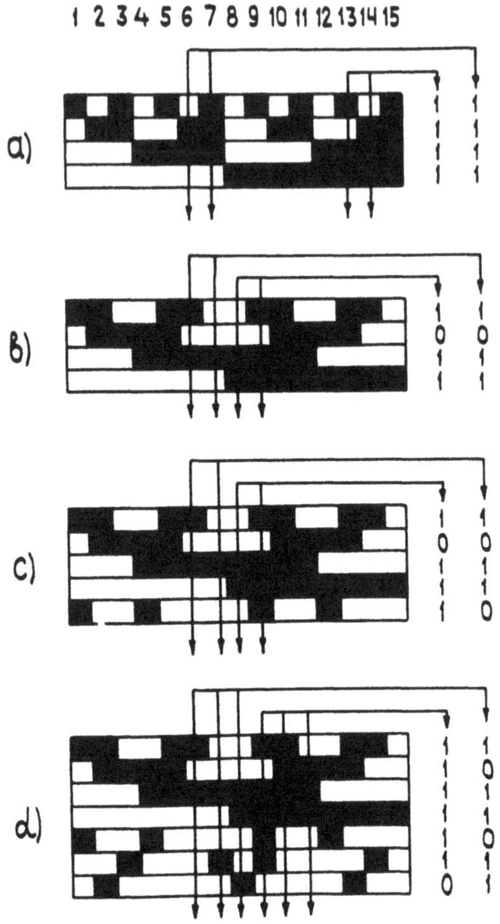

Fig.10. Coding scheme of the scintillation hodoscope: a) usual binary code; b) classical Gray code; c,d) Gray code with additional bits.

CONCLUSION

Increasing information from multichannel detectors of nuclear particles has generated a need for studying the questions of optimal coding data readout and processing methods. The consideration of specific examples shows that there is a close connection between digital signal processing and correcting code theory. The use of the syndrome coding methods allows one to construct devices for data compression and processing registered in a great number of position-sensitive sources.

The property of the coding word syndrome to carry information on the number and coordinates of errors arising in the process of data transmission is used. It is clear that the method of syndrome coding can be effectively used in other fields of science and engineering technology where information processing from a large number of position-sensitive sources is needed.

Thus, we use both theoretical and practical aspects of the problem to connect such fields as digital data processing and the theory of correcting codes[16,17].

The method of syndrome coding can be commonly used in devices for signature analysis and syndrome testing of large scale-integration microcircuits and microprocessors[18].

REFERENCES

1. Nikityuk N.M., Radzhabov R.S., Shafranov M.D. Nucl.Instr. and Meth., 1978, v.155, N.1, p.485.
2. Nikityuk N.M. JINR, P11-80-484, Dubna, 1980.
3. Nikityuk N.M. JINR, P11-81-784, Dubna, 1981, p.13.
4. Nikityuk N.M. JINR, E11-87-10, Dubna, 1987.
5. Nikityuk N.M., Radzhabov R.S., Shafranov M.D. Pribori i Tekhnika Eksperimenta, 1978, N.4, p.95.
6. Gaidamaka R.I., Kalinnikov V.A., Nikityuk N.M. JINR, P13-82-628,Dubna,1982, p.11.
7. Nikityuk N.M. JINR, P10-87-254, Dubna, 1987, p.14.
8. Massey J.L. IEEE Trans. on Inform. Theory, 1965, v.IT-11, N.4, p.580-585.
9. Kauts W.H. IEEE Trans. on Inform.Theory, 1964, v.IT-10, N.5, p.363.
10. Nikityuk N.M. Pribori i Teknika Eksperimenta, 1986, N.6, p.78-81.
11. Nikityuk N.M., Selikov A.B. JINR, P10-86-481, Dubna, 1987.
12. Nikityuk N.M. Pribori i Tekhnika Eksperimenta, 1987, N.3, p.59-65.
13. Nikityuk N.M. JINR, P10-87-266, Dubna, 1987.
14. Nikityuk N.M. JINR, P11-87-54, Dubna, 1987.
15. Komolov L.N. et al. Pribori i Tekhnika Eksperimenta, 1987, N.4.
16. Ancheta T.C. IEEE Trans. on Inform. Theory, 1976, v.IT-22, N.4, p.432.
17. Blauhut R.E. Proc IEEE, 1985, v.73, N., p.30-53.
18. Frohwerk R.A. Hewlett Packard Journal, 1977, v.28, N.9, p.2-8.

NONASSOCIATIVE DEGREE FIVE IDENTITIES NOT IMPLIED BY COMMUTATIVITY: A COMPUTER APPROACH

G.M.Piacentini Cattaneo*

Department of Mathematics, Second University of Rome, Italy

Introduction

In this paper a complete classification is given of all (reducible or irreducible) nonassociative degree five identities not implied by commutativity, thus generalizing [3]. To do this one needs only classify all homogeneous multilinear identities of that kind, and, since any multilinear identity is equivalent to a collection of minimal ones, it suffices to find a way of writing out all minimal multilinear degree five identities not implied by commutativity. We can do this in a straightforward way with the Young tableaux approach described in [4], which enables a computer to generate all minimal multilinear identities of any degree. We also compare our work with Osborn's classification [3] of the degree five irreducible identities, and determine the corresponding values of the coefficients.

Young tableaux and polynomials

Let $K\langle X\rangle$ be the free nonassociative algebra on a countable set of generators, over a field K of characteristic zero. By "nonassociative" we will mean "not necessarily associative". An element f of $K\langle X\rangle$ is a nonassociative polynomial identity of a K-algebra R if every K-morphism of $K\langle X\rangle$ in R sends f to zero. An algebra satisfying a non zero nonassociative polynomial identity is called a nonassociative P.I. algebra. Since in characteristic zero the ideal of all identities of an algebra is homogeneous and generated by the multilinear polynomials it contains, to classify identities of a given degree n amounts to classify all homogeneous multilinear identities of that degree. Now, if V_n is the set of all homogeneous multilinear polynomials in n nonassociative variables, one has $V_n = \bigoplus_{A\in \mathcal{A}(n)} V_n^A$, $\mathcal{A}(n)$ being the set of all association forms of n factors, and V_n^A the span of all degree n monomials in the association form A.

We assume familiarity with Young tableaux and the connection with representations of S_n. Following [2], [4] or [5], each V_n^A may be identified with KS_n, the group algebra of the symmetric group S_n over K. For each V_n^A we may use the decomposition of KS_n into a direct sum of two-sided ideals I_λ, λ running over all partitions of n, and each I_λ being a direct

* Partly supported by GNSAGA of CNR and MPI.

sum of minimal left ideals $(KS_n)e_j$, $j = 1$ to d_λ, where $e_j = e_{T_{\lambda_j}}$ is the semi-idempotent associated to the standard tableau $T_{\lambda,}$, $\dim I_\lambda = d_\lambda^2$, $\dim(KS_n)e_j = d_\lambda$, d_λ being the number of standard tableaux of shape D_λ. An identity of an algebra is called minimal if it does not imply any other strictly weaker identity of the same degree. We obtain all minimal identities of a given degree n by taking as a basis over K the polynomials associated to the standard tableaux; the algorithm, which has been implemented by Pittaluga (see [4]) on a Micro VAX of our Department using Common Lisp, is the following:

1) Fix an integer n and a partition λ of n.

2) Write down the "canonical" standard tableau T_λ

$$\begin{array}{lllll} 1 & d_1+1 & d_1+d_2+1 & \ldots & \ldots \\ 2 & d_1+2 & d_1+d_2+2 & \ldots & \\ \vdots & \vdots & \ldots & & \\ d_2 & d_1+d_2 & \ldots & & \\ \vdots & & & & \\ d_1 & & & & \end{array}$$

3) Construct the "canonical" polynomial

$$G(x_1, x_2, \ldots, x_{d_1}) = s_{d_1}(x_1, x_2, \ldots, x_{d_1}) s_{d_2}(x_1, x_2, \ldots, x_{d_2}) \ldots$$

where $s_d(x_1, x_2, \ldots, x_d) = \sum_{\sigma \in S_d} \operatorname{sgn}(\sigma) x_{\sigma(1)} x_{\sigma(2)} \ldots x_{\sigma(d)}$.

4) Write all standard tableaux T'_λ of the given shape.

5) Determine the permutation σ such that $T'_\lambda = \sigma T_\lambda$.

6) For each possible standard tableau $T'_\lambda = \sigma T_\lambda$ construct $G'_\sigma(x_1, \ldots, x_{d_1}) = G(x_1, x_2, \ldots, x_{d_1})\sigma^{-1}$, where G is the canonical polynomial.

7) For any set S of forms of associations, every minimal identity corresponding to the given shape is a sum (extended to S) of linear combinations of the G'_σ's corresponding to the different standard tableaux of that shape.

The classification

We will now study degree five identities. In this case characteristic zero may be weakened to characteristic not 2, 3 or 5. Our aim is the classification of all degree five identities not implied by commutativity. We will therefore work mod commutativity: the fourteen forms of association of a word of length five reduce to only three: $((** \cdot *)*)*$, $(** \cdot **)*$ and $(** \cdot *) \cdot **$.

In Table 1 we give the complete classification by listing all possible minimal nonassociative degree five identities not implied by commutativity. Had we used the technique that has been used in [1] to determine all degree four identities not implied by commutativity, we would not have succeeded, or, certainly, not with the small amount of computing time; the listing of each minimal degree five identity, for every diagram, takes, by our method, only a few seconds.

In [3] Osborn had determined all irreducible degree five identities not implied by commutativity: an identity is irreducibile if, in the presence of unity element, it does not imply any identity of lower degree type. Now, a minimal identity in our sense, in the presence of a unity element, may very well imply an identity of lower degree, hence it may be reducible. For example, the identity $2x_1^2x_3 - x_1x_3 \cdot x_1 - x_3x_1 \cdot x_1$ is minimal but reducible, since in the presence of unity it implies commutativity. To limit one's discussion to irreducible identities seems too restrictive, since often the class of algebras satisfying some reducible identity may contain significant algebras: for example, in [1] we proved that the class of algebras that satisfy the degree four identity $(yx \cdot x)x - y(x^2x)$ is a true generalization of right alternativity and it includes simple algebras. In the presence of a unity element this identity is equivalent to right alternativity. For this reason it seems important to find a way to classify all nonassociative identities, irreducible or not. Furthermore, knowing that an identity is minimal is important because it gives information on how strong or how weak the identity is.

Examination of Table I yields the following theorem.

Theorem. Let R be a commutative nonassociative algebra over a field of characteristic not 2, 3 or 5 and let R satisfy an identity of degree 5 not implied by the commutative law. Then R satisfies at least one from the families of identities a), b), c), d), e) or f) of Table 1.

In the presence of a unity element 1, the following relations hold among the coefficients:

$$\alpha + \alpha' + \alpha'' = 0 \qquad (a)$$

$$\alpha + 3\beta + 6\gamma + 7\alpha' + \alpha'' = 0 \qquad \beta'' = \alpha' \qquad (b)$$

$$\alpha + \beta + 2\gamma + 2\alpha' + \alpha'' + 6\beta'' = 0 \qquad (c)$$

TABLE 1

		$((**\cdot *)*)*$		$(**\cdot **)*$		$(**\cdot *)\cdot **)$
a)	(5)	$\alpha(x^2x\cdot x)x$	+	$\alpha' x^2x^2\cdot x$	+	$\alpha''(x^2x)x^2$
b)	(4 1)	$\alpha[(x^2y\cdot x)x-((yx\cdot x)x)x]+$ $\beta[(x^2x\cdot y)x-((yx\cdot x)x)x]+$ $\gamma[(x^2x\cdot x)y-((yx\cdot x)x)x]$	+	$\alpha'[(x^2x^2)y-(yx\cdot x^2)x]$	+	$\alpha''[x^2y\cdot x^2-(yx\cdot x)x^2]+$ $+\beta''[x^2x\cdot yx-(yx\cdot x)x^2]$
c)	(3 2)	$\alpha[(x^2y\cdot y)x-((xy\cdot y)x)x-((yx\cdot x)y)x+(y^2x\cdot x)x]+$ $\beta[(x^2y\cdot x)y-((xy\cdot y)x)x-((yx\cdot x)x)y+(y^2x\cdot x)x]+$ $\gamma[(x^2x\cdot y)y-((xy\cdot x)y)x-((yx\cdot x)x)y+(y^2x\cdot x)x]$	+	$\alpha'[(x^2y^2)x-(xy\cdot xy)x]$	+	$\alpha''[x^2y\cdot yx-(xy\cdot y)x^2-(yx\cdot x)\cdot xy+(y^2x)x^2]$ $\beta''[x^2x\cdot y^2-2(xy\cdot x)\cdot xy+(y^2x)x^2]$
d)	(3 1 1)	$\alpha[(x^2y\cdot z)x-(x^2z\cdot y)x-((yx\cdot x)z)x+((yx\cdot z)x)x+((zx\cdot x)y)x-((zx\cdot y)x)x]+$ $\beta[(x^2y\cdot x)z-(x^2z\cdot x)y-((yx\cdot x)x)z+((yx\cdot z)x)x+((zx\cdot x)x)y-((zx\cdot y)x)x]+$ $\gamma[(x^2x\cdot y)z-(x^2x\cdot z)y-((yx\cdot x)x)z+((yx\cdot x)z)x+((zx\cdot x)x)y-((zx\cdot x)y)x]$	+	0	+	$\alpha''[x^2y\cdot zx-x^2z\cdot xy-(yx\cdot x)\cdot xz+(yx\cdot z)x^2+$ $+(zx\cdot x)\cdot xy-(zx\cdot y)x^2]$
e)	(2 2 1)	$\alpha[(x^2y\cdot z)y-((xy\cdot y)z)x-(x^2z\cdot y)y+((xy\cdot z)y)x-((yx\cdot x)z)y+(y^2x\cdot z)x+$ $((yx\cdot z)x)y-(y^2z)x)x+((zx\cdot x)y)y-((zy\cdot x)y)x-((zx\cdot y)x)y+((zy\cdot y)x)x]+$ $\beta[(x^2y\cdot y)z-((xy\cdot y)x)z-(x^2z\cdot y)y+((xy\cdot z)x)y-((yx\cdot x)y)z+(y^2x)x)z+$ $((yx\cdot z)y)x-(y^2z\cdot x)x+((zx\cdot x)y)y-((zy\cdot x)x)y-((zx\cdot y)y)x+((zy\cdot y)x)x$	+	$\alpha'[x^2y^2\cdot z-(xy\cdot xy)z-(x^2\cdot yz)y+$ $+(xy\cdot xz)y+(xy\cdot yz)x-(xz\cdot y^2)x]$	+	$\alpha''[x^2y\cdot yz-(xy\cdot y)\cdot xz-x^2z\cdot y^2+2(xy\cdot z)\cdot xy+$ $-(xy\cdot x)\cdot yz+y^2x\cdot xz-y^2z\cdot x^2+(xz\cdot x)y^2+$ $-(yz\cdot x)\cdot xy-(xz\cdot y)\cdot xy+(yz\cdot y)x^2]$
f)	(2 1 1 1)	$x^2s_3(y,z,t)+xy\cdot s_3(x,t,z)+xz\cdot s_3(x,y,t)+xt\cdot s_3(x,z,y)$	+	0	+	0

$$\alpha + \beta + 2\gamma + \alpha \ = 0 \qquad (d)$$

$$\alpha + \beta + \alpha' + \alpha'' = 0 \qquad (e)$$

We next compare our equations with Osborn's classification in [3], by listing the values of the coefficients of Table 1 which correspond to the irreducible identities (12), (13), (14), (15) and (16) of Theorem 5 in [3].

$$\alpha = 2 \quad \alpha' = -3 \quad \alpha'' = 1 \qquad (Eq.12)$$

$$\begin{cases} \alpha' = \beta'' = -\beta_2 \quad \alpha = 6\beta_1 - 9\beta_2 + \beta_3 \\ \beta = -4\beta_1 + 5\beta_2 \quad \gamma = \beta_1 \quad \alpha'' = \beta_2 - \beta_3 \end{cases} \qquad (Eq.13)$$

$$\begin{cases} \alpha = -\gamma_1 + 2\gamma_3 \quad \beta = \gamma_1 - \gamma_2 + 4\gamma_3 \\ \gamma = -2\gamma_3 \quad \alpha' = -4\gamma_3 \quad \alpha'' = \gamma_2 \quad \beta'' = \gamma_3 \end{cases} \qquad (Eq.14)$$

$$\alpha = -\delta_1 \quad \beta = \delta_1 + \delta_2 \quad \gamma = 0 \quad \alpha'' = -\delta_2 \qquad (Eq.15)$$

$$\alpha = -\epsilon_2 \quad \beta = \epsilon_1 + \epsilon_2 \quad \alpha' = 0 \quad \alpha'' = -\epsilon_1. \qquad (Eq.16)$$

Acknowledgements

I am very grateful to M. Pittaluga for having written ([4]) the program in Lisp on which these applications are based.

References

[1] L. CARINI, I.R.HENTZEL and G.M.PIACENTINI CATTANEO, *Degree four identities not implied by commutativity*, Comm. Algebra **16** (2) (1988), 339–356.

[2] V.S.DRENSKI, *Representations of the symmetric group and varieties of linear algebras*, Math. USSR Sb. **43** (1982), 85–101.

[3] J.M.OSBORN, *Identities in nonassociative algebras*, Can. J. Math. **17** (1965), 78–92.

[4] G.M.PIACENTINI CATTANEO and M.PITTALUGA, *Nonassociative polynomial identities, Young tableaux and computers*, Sigsam Bulletin **22** (2) (1988), 17–26.

[5] A. REGEV, *The representations of S_n and explicit identities for P.I. algebras*, J.Algebra **51** (1978), 25–40.

ON COMPLETELY REGULAR PROPELINEAR CODES

by
J.Rifà, J.M.Basart, L.Huguet
Dep.d'Informàtica. Fac.Ciències.
UNIVERSITAT AUTÒNOMA DE BARCELONA
CATALUNYA. (Spain)

ABSTRACT:

In a previous paper (see [7]) we found that given a distance regular e-latticed graph Γ we can associate with it a completely regular code C. We used this in order to solve a conjecture given by Bannai in [1].

In the present paper we introduce the **propelinear code** structure with the aim of studying the algebraic structure of completely regular codes (not necessarily linear) associated with distance-regular e-latticed graphs.

We give the basic properties of this structure. We construct, from a propelinear code C, an associate graph $\Omega(C)$ and we prove that C is a completely regular code if and only if $\Omega(C)$ is a distance-regular graph.

INTRODUCTION:

Sections 1 and 2 have not any new result but they contain some necessary definitions.

Section 3 is devoted to introduce the concept of e-covering and to obtain some results used later.

In section 4 we define the propelinear code structure and we obtain some of its basics properties.

In section 5 we study the algebraic structure of codes constructed from distance regular e-latticed graphs.

Finally in section 6 we present the main result of this paper: From a propelinear code C is possible to construct an associated graph $\Omega(C)$; C is a completely regular code if and only if $\Omega(C)$ is a distance regular graph. This result is a generalization of an analogous result that we obtained in [6] restricted to linear codes.

1. COMPLETELY REGULAR CODES

Let F^n be a n-dimensional vector space over the Galois field GF(q) of $q = p^r$ elements, p being a prime number.

The **weight**, $W_H(V)$, of a vector $V \in F^n$ is the number of its nonzero coordinates. The **Hamming distance** between two vectors $V, S \in F^n$ is $d(V,S) = W_H(V-S)$.

A q-ary **block-code** of length n, is a subset C of F^n. We call the elements of C **code words**.

The **minimum distance d** of a code C is the minimum value of d(a,b), where a, b $\in$ C, and $a \neq b$.

The **error correcting capability e** of a code C is the integer part of the quotient (d-1)/2. In this case C is an e-error correcting code.

We define $B(V,p) = \#\{ c \in C \mid d(V,c) = p \}$, the number of code words at distance p from $V \in F^n$, (see[2]).

Let $\rho(V)$ be the distance from $V \in F^n$ to the code C, that is to say: $\rho(V) = \min. \{ p \mid 0 \le p \le n , B(V,p) \neq 0 \}$.

The **covering radius** ρ of a code C is the maximum value of $\rho(V)$, for all $V \in F^n$

An e-error-correcting code C in F^n is called **completely regular** if for all $V \in F^n$ and for all $0 \le p \le n$, the value of $B(V,p)$ depends only on $\rho(V)$, that is to say, $\forall V \in F^n$ such that $\rho(V) = u$, its value is a constant number $B(V,p) = b_{up}$.

We refer to the numbers b_{up} as **the outer distribution numbers** of a completely regular code C, (see[2]).

2. DISTANCE-REGULAR GRAPHS

A **graph** $\Gamma(X,A)$ consist of a vertex set X and **edge** set A, where every edge is a subset of cardinality two of X.

Let Γ be a connected undirected graph on X.

A **path** of length r from x to y, $(x,y \in X)$, is a sequence of vertices $x_0 = x, x_1, x_2, \ldots, x_{r-1}, x_r = y$ such that each (x_i, x_{i+1}) is an edge of Γ.

The **distance** from x to y is the minimum length of paths from x to y, and is denoted by $d(x,y)$.

Let Γ, Γ' be two graphs. A map $\theta: \Gamma \longrightarrow \Gamma'$ is a **graph-isomorphism**, if θ is a bijective, distance-preserving map.

The **diameter** of Γ is the maximum distance between any two vertices and denoted by ρ.

Let R_i be the relation on X defined by $(x,y) \in R_i$ if and only if $d(x,y) = i$. Let D_i be the matrix indexed by the graph elements, whose entries (x,y) are "1" or "0" according to whether the distance is or is not $d(x,y) = i$, (see [1], [2]).

A graph is called a **distance-regular graph** if for every integers $i,j,k \in \{0,1,\ldots,\rho\}$, the number of $z \in X$ such that $(x,z) \in R_i$ and $(z,y) \in R_j$ is a constant number, denoted by p_{kij}, whenever $(x,y) \in R_k$, and $p_{kij} = p_{kji}$.

An equivalent definition would be to say that a graph is a distance-regular graph if the configuration $(X, \{R_i\})$ becomes an association scheme (commutative and symmetric), that is to say, if $\langle D_i \rangle$ is a $(\rho+1)$-dimensional algebra over $\mathbb{C}$, (see [1]).

The constant number p_{011} is called the **valency** of the graph.

If Γ is a distance-regular graph and $B_1 = (p_{k1j})$ is the intersection matrix with respect to R_1, then (see [1]) B_1 is a tridiagonal matrix with non-zero off-diagonal entries:

$$B_1 = \begin{bmatrix} a_o & c_1 & \cdots & 0 & \\ b_o & a_1 & \cdots & \cdots & \\ & b_1 & \cdots & c_{\rho-1} & \\ & \cdot\cdot & \cdots & a_{\rho-1} & c_\rho \\ 0 & & \cdots & b_{\rho-1} & a_\rho \end{bmatrix} \quad \begin{array}{l} a_i = p_{i1i} \ ,i = 0,1,\ldots,\rho \\ b_i = p_{i1(i+1)},\ i = 0,1,\ldots,\rho-1 \\ c_{i+1} = p_{(i+1)1i},\ i = 0,\ldots,\rho-1 \end{array}$$

2.1. <u>DISTANCE-REGULAR GRAPH e-LATTICED, e ≥ 3</u>

Definition: A distance-regular graph is **e-latticed**, of valency n, if the parameters p_{i1j} of its intersection matrix B_1 coincide with those of n-cube (see the definition of n-cube in 2.2), for $i+j \leq 2\cdot e - 1$.

That is to say:

A distance-regular graph will be e-latticed if its parameters satisfy:

$$\left[\begin{array}{l} p_{i1j} = n - i \ , \text{ for } \ 0 \leq i = j - 1 \leq e - 1 \\ p_{i1i} = 0 \\ p_{i1j} = i, \text{ for } 0 \leq i = j+1 \leq e \end{array}\right.$$

<u>**A distance-regular e-latticed graph, of valency n, is called an H(e,0,n)-graph**</u>

2.2. <u>**Partitions of the graph F^n**</u>, (see [3],[4]).

Let F be the field Z/2. We can consider a fixed basis e_1, e_2, ..., e_n, in F^n, and we can represent each element of F^n through their coordinates in the basis { e_i }. We will use the graph structure F^n, where the vertices are all the elements in F^n, and two vertices are adjacent if and only if they are at unit distance (distance means Hamming's distance). This graph structure is called <u>n-cube</u>.

Let R be a partition of the n-cube F^n. In the set of non-empty equivalence classes F^n/R, we can define an adjacency relation:

$\forall$ A,A' $\in F^n/R$, A is adjacent to A' if and only if there exist both representatives, in the classes A and A', which are adjacent in F^n.

2.2.1. Regular Partitions

A partition R is called **regular** if for any two classes A, A', the number of elements $y \in A$, which are adjacent to $x \in A'$, is a constant number b(A,A') independent of $x \in A'$.

2.2.2. Uniformly Regular Partitions

A regular partition F^n/R is **uniformly regular** if there are constant numbers b_0, b_1, such that:

$$b(A,A') = \begin{cases} b_0 & \text{if } A = A' \\ b_1 & \text{if } A, A' \text{ are adjacent.} \end{cases}$$

2.2.3. Completely Regular Partitions

A partition $F^n/\approx$ is called a **completely regular** partition if it is regular and all equivalence classes are completely regular codes with the same outer distribution numbers.

3. HAMMING'S COVERINGS

3.1.-**Definition**: Let Γ be a connected undirected graph on X. A map $\theta: F^n \longrightarrow \Gamma$ is called an **e-local isomorphism** if θ induces a graph isomorphism ($i = 0,1,\ldots,e-1$):

$$\theta: B_i(u) \longrightarrow B_i(\theta(u)) \quad \text{for every vertex } u \in F^n.$$

where $B_i(u) = \Gamma_o(u) \cup \Gamma_1(u) \cup \ldots \cup \Gamma_i(u)$,

and $\Gamma_j(u) = \{ x \in F^n / d(x,u) = j \}$

3.2.-**Definition**: An **e-covering** is a surjective, e-local isomorphism.

Remark: If $\theta: F^n \longrightarrow \Gamma$ is an e-covering, then its restriction to every $\theta: \Gamma_i(u) \longrightarrow \Gamma_i(\theta(u))$ is a graph isomorphism.

3.3.- **Definition**: Let θ be an e-covering, $\theta: F^n: \longrightarrow \Gamma$, and let $\alpha \in \Gamma$ be the image of $0 \in F^n$, $\theta(0) = \alpha$.

Let $C = \theta^{-1}(\alpha)$. C will be cal.ed the associated code of e-covering θ.

Remark: In [7] we prove that starting from a distance regular e-latticed graph Γ, $(e \geq 3)$, of valency n, that is to say, an H(e,0,n)-graph with $e \geq 3$, we can construct an e-covering, $e \geq 3$, θ_α: F^n ------> Γ , for every $\alpha \in \Gamma$, such that the set C of inverse images of α becomes a completely regular code.
$C = \theta_\alpha^{-1}(\alpha)$ is the associated code to e-covering θ_α.

3.4.-**Proposition**: Let θ be an e-covering, $e \geq 3$, and let a,b be two elements such that $\theta(a) = \theta(b)$. Then:

a): There exists only one graph isomorphism,
$\tau_{ab}: B_2(a) \longrightarrow B_2(b)$, such that $\forall\, x \in B_2(a)$, $\theta(x) = \theta(\tau_{ab}(x))$

b): There exists only one isometry,
$\pi_{ab}: F^n \longrightarrow F^n$, such that $\forall\, W_H(v) \leq 2$: $\theta(a+v) = \theta(b+\pi_{ab}(v))$

c): There exists only one isometry,
$\pi_{ab}: F^n \longrightarrow F^n$, such that $\forall\ v \in F^n : \theta(a+v) = \theta(b+\pi_{ab}(v))$

Proof:

a): From the e-covering definition, the restriction of θ to both, $B_2(a)$ and $B_2(b)$, is a graph isomorphism:

$$\theta: B_2(a) \longrightarrow B_2(\theta(a)) = B_2(\theta(b))$$

$$\theta: B_2(b) \nearrow$$

So there exists only one graph isomorphism, $\tau_{ab}: B_2(a) \longrightarrow B_2(b)$ which makes the previous diagram commutative.

b): The elements of $B_2(a)$ are $a + v$, where $W_H(v) \le 2$, and the elements of $B_2(b)$ are $b + w$, where $W_H(w) \le 2$.
Starting from the previous graph isomorphism τ_{ab}, we can construct a map $\pi_{ab}: B_2(0) \longrightarrow B_2(0)$ in the following way:
If $v \in F^n$ and $W_H(v) \le 2$, then $\pi_{ab}(v)$ is the only element in F^n such that: $\tau_{ab}(a+v) = b + \pi_{ab}(v)$.
It is clear that $W_H(v) = W_H(\pi_{ab}(v))$.
We can extend π_{ab} to F^n in the following way:
If $s \in F^n$, $s = \Sigma\, e_i$, then $\pi'_{ab}(s) = \Sigma\, \pi_{ab}(e_i)$.
To prove that this extension is well-defined we must see that $\pi_{ab}(e_i+e_j) = \pi_{ab}(e_i) + \pi_{ab}(e_j)$:
In fact, $b + \pi_{ab}(e_i+e_j)$ is the only element in $B_2(b)$ which is at distance two of b, and distance one of both $b + \pi_{ab}(e_i)$ and $b + \pi_{ab}(e_j)$. On the other hand, $b + \pi_{ab}(e_i) + \pi_{ab}(e_j)$ is the only element which is at distance two of b, and distance one of both, $b + \pi_{ab}(e_i)$ and $b + \pi_{ab}(e_j)$.

c): We take the previous isometry π_{ab} and we shall prove that the equality $\theta(a+v) = \theta(b+\pi_{ab}(v))$ is true for every $v \in F^n$:
The proof works by induction over the weight of v.
Case $W_H(v) = 1, 2$ is true by the previous paragraph.
If we suppose that the equality is true for $2 \le W_H(v) < r$, we must prove the equality for $W_H(v) = r$.
If $2 < W_H(v) = r$, then we can write $v = w + e + ç$, where

$W_H(w) = r-2$ and $W_H(e) = W_H(ç) = 1$.

By induction, both equalities $\theta(a+w) = \theta(b+\pi_{ab}(w))$ and $\theta(a+w+e) = \theta(b+\pi_{ab}(w+e))$, hold.

We must prove that $\theta(a+w+e+ç) = \theta(b+\pi_{ab}(w+e+e+ç))$.

We can apply the previous paragraph to elements $a' = a+w$ and $b' = b+\pi_{ab}(w)$. Thus we deduce that there exists only one isometry $\pi'_{a'b'}$ such that:

(*) $\qquad \theta(a'+v') = \theta(b'+\pi'_{a'b'}(v')),\ \forall\, W_H(v') \leq 2.$

We can take $v' = e+ç$, and then we see that the induction assertion holds if we prove that the isometries π_{ab} and $\pi'_{a'b'}$ are equal.

In (*), we could also do, for every e_i, $v' = e_i$, and then it would be:

(**) $\qquad \theta(a'+e_i) = \theta(b'+\pi'_{a'b'}(e_i)).$

But $(a'+e_i) = (a+w+e_i)$ and $W_H(w+e_i) \leq r-1$. Then by the induction hypothesis we will have:

$\theta(a'+e_i) = \theta(b+\pi_{ab}(w+e_i)) = \theta(b'+\pi_{ab}(e_i))$

If we compare this result with the equality (**), which holds for every e_i, we can conclude that:

$\pi'_{a'b'}(e_i) = \pi_{ab}(e_i)$. ∎

4. PROPELINEAR CODES

Let C be a subset (not necessarily linear) of F^n ($F = Z/2$), and **we will suppose that** $0 \in C$. We will also suppose that we have fixed an initial basis $\{ e_i \}$ in F^n.

Remark: If C is a linear set we will have $C + C = C$, or also: For every $v \in C$, $v + C = C$. In the case that C is not linear, we would like a permutation π_v for every fixed $v \in C$, on the n coordinates, such that: $v + \pi_v(C) = C$. In this case we will define the notion of **propelinear codes**.

4.1.- **Definition**: Let C be a subset of Fn, $0 \in C$. C is called a **propelinear code** if $\forall\, v \in C$ there exists an isometry:

$\pi_v: F^n \longrightarrow F^n$ such that:

i.- $\forall\, v \in C$: $v + \pi_v(s) \in C$ if and only if $s \in C$.

ii.- $\forall\, v,s \in C$: $\pi_v \cdot \pi_s = \pi_m$, where $m = v + \pi_v(s)$

From π_v we can construct the following mapping:

$\emptyset_v: F^n \longrightarrow F^n$ defined by $\emptyset_v(a) = v + \pi_v(a)$

We will call **PC(n,d)-code** any propelinear code C in F^n, where $d = \min.\{\, d(a,\emptyset_v(a)) > 0 \;/\; \forall\, a \in F^n,\ \forall\, v \in C \,\}$

4.2.- **Example**:

For example, in F^4, the set $C = \{0, v, w, s\}$, where

$0=(0000)$; $v = (1100)$; $w = (0110)$; $s = (0101)$

is a propelinear PC(4,2)-code, but not a linear code.

We can define $\pi_o = \text{Id}$.

$$\pi_v = \pi_{12} \cdot \pi_{34}$$
$$\pi_w = \pi_{23} \cdot \pi_{14}$$
$$\pi_s = \pi_{13} \cdot \pi_{24}$$

π_{ij} means the transposition that changes the coordinates i and j.

4.3.- **Properties**:

Let C be a propelinear code in F^n:

4.3.1.- **Every linear code is, also, a propelinear code.**

In fact, we can take $\forall\, v \in C$, $\pi_v = \text{Id}$.

4.3.2.- **$\emptyset_v$ is a bijective map and a distance-preserving map.**

In fact, if $\emptyset_v(a) = \emptyset_v(b)$ then $v + \pi_v(a) = v + \pi_v(b)$. But π_v is an isomorphism, so $a = b$.

Moreover: $d(\emptyset_v(a),\emptyset_v(b)) = d(v + \pi_v(a) , v + \pi_v(b)) = d(\pi_v(a),\pi_v(b)) = W_H(\pi_v(a-b)) = W_H(a-b) = d(a,b)$.

4.3.3.- **$\forall\, v \in C$, $\pi_v(0) = 0$.**

Clearly, because π_v is an isometry.

4.3.4.- $\pi_0 = Id.$ and $\emptyset_0 = Id.$

In fact $\pi_0 \cdot \pi_0 = \pi_m$, where $m = 0 + \pi_0(0) = 0$

Then $\pi_0 \cdot \pi_0 = \pi_0$, so $\pi_0 = Id.$

4.3.5.- **$\forall$ v $\in$ C there exists s $\in$ C such that** $\pi_s \cdot \pi_v = \pi_v \cdot \pi_s = Id.$ **and** $\emptyset_s \cdot \emptyset_v = \emptyset_v \cdot \emptyset_s = Id.$

In fact, π_v is an isomorphism and so there exists $s \in F^n$ such that $\pi_v(s) = v$. Then $\emptyset_v(s) = 0$ and by 4.1.i. $s \in C$.

By 4.1.ii. : $\pi_v \cdot \pi_s = \pi_m$, where $m = \emptyset_v(s) = 0$.

Then $\pi_v \cdot \pi_s = \pi_0 = Id.$ Now, from $s \in C$ we can find $t \in C$ such that $\pi_s \cdot \pi_t = Id.$ So $\pi_v = \pi_v \cdot \pi_s \cdot \pi_t = \pi_t$, and $\pi_s \cdot \pi_v = Id.$

Also, $\emptyset_v \cdot \emptyset_s(a) = \emptyset_v(\emptyset_s(a)) = \emptyset_v(s+\pi_s(a)) = v + \pi_v(s + \pi_s(a)) = v + \pi_v(s) + \pi_v \cdot \pi_s(a) = m + \pi_m(a) = 0 + \pi_0(a) = a$.

Then $\emptyset_v \cdot \emptyset_s = Id.$

4.3.6.- $\emptyset_v \cdot \emptyset_s = \emptyset_m$, **where** $m = \emptyset_v(s)$.

In fact,

$\emptyset_v \cdot \emptyset_s(a) = \emptyset_v(\emptyset_s(a)) = \emptyset_v(s + \pi_s(a)) = v + \pi_v(s + \pi_s(a)) = v + \pi_v(s) + \pi_v \cdot \pi_s(a) = m + \pi_m(a) = \emptyset_m(a)$, where $m = \emptyset_v(s)$.

4.3.7.- $G = \{ \pi_v / v \in C \}$ **and** $G' = \{ \emptyset_v / v \in C \}$ **are group structures. Group G' has the same cardinality as C.**

In fact, see 4.3.3, 4.3.4, 4.3.5, for first assertion.

Moreover, if we consider $\sigma: C \longrightarrow G'$ defined by

$$\sigma(v) = \emptyset_v$$

then we can see that σ is a bijective map.

If $\sigma(v) = \sigma(s)$ then $\emptyset_v = \emptyset_s$ and $\forall a \in F^n$:

$\emptyset_v(a) = \emptyset_s(a)$, that is to say: $v + \pi_v(a) = s + \pi_s(a)$

If we take $a = 0$, we can conclude $v = s$.

4.3.8.- **If code C is linear then G = { 0 } and G' is isomorphic to group C.**

In fact, we can consider the map $\sigma: C \longrightarrow G'$, where $\sigma(v) = \emptyset_v$.

In the previous paragraph we proved the bijectivity of σ.

Moreover if we consider, see 4.3.1, $\forall v \in C$, $\pi_v = Id.$; it is obvious that $G = \{0\}$.

And, on the other hand it is easy to see that σ is a morphism:

$\sigma(v+s) = Ø_{v+s}$, $\forall$ $a \in F^n$: $Ø_{v+s}(a) = v+s+\pi_{v+s}(a) = v+s+a$

$\sigma(v) \cdot \sigma(s) = Ø_v \cdot Ø_s$ $\forall$ $a \in F^n$: $Ø_v \cdot Ø_s(a) = Ø_v(s+a) = v+s+a$.

5. Code C associated to H(e,0,n)-graph, $e \geq 3$

5.1.- **Proposition**: Let $\Theta: F^n \longrightarrow \Gamma$ be an e-covering, $e \geq 3$. Let C be its associate code (see 3.3). Then C is a propelinear PC(n,d)-code, $d \geq 3$.

Proof:

For every $v \in C$ we have $\Theta(v) = \Theta(0) = \alpha \in \Gamma$. By 3.4 c), there exists just one isometry $\pi_v: F^n \longrightarrow F^n$, such that:

$\Theta(a) = \Theta(v+\pi_v(a)) = \Theta(Ø_v(a))$, for every $a \in F^n$.

i.- It is easy to see that $(v + \pi_v(a)) \in C$ if and only if:

$\Theta(a) = \Theta(v+\pi_v(a)) = \alpha \in \Gamma$, and this happens if and only if $a \in C$.

ii.- If $v,s \in C$, we shall show that $\pi_v \cdot \pi_s = \pi_m$, where $m \in C$ and $m = v + \pi_v(s)$:

Taking into account the definitions of π_v and π_s, we can write for every basis element $e_i \in F^n$:

$\Theta(e_i) = \Theta(\ s + \pi_s(e_i)\ ) = \Theta(\ v + \pi_v(\ s + \pi_s(e_i)\ )\ ) =$

$\Theta(m + \pi_v \cdot \pi_s(e_i))$, where $m = v + \pi_v(s) \in C$.

But from the definition of π_m, we know that it is the only isometry that $\forall\ e_i$, $\Theta(e_i) = \Theta(m+\pi_m(e_i))$.

Then $\pi_v \cdot \pi_s = \pi_m$.

iii.- If $d = \min.\{\ d(a,Ø_v(a)) > 0\ /\ a \in F^n,\ v \in C,\ \}$:

We want to prove that $d \geq 3$.

If we suppose that $0 < d \leq 2$ then there would exist $a \in F^n$, $v \in C$, such that $d(a,Ø_v(a)) \leq 2$, that is to say $Ø_v(a) \in B_2(a)$. But Θ is an e-covering, $e \geq 3$. Thus, the restriction of Θ to $B_2(a)$ is a graph isomorphism

$$\Theta: B_2(a) \longrightarrow B_2(\Theta(a)).$$

We know that $\Theta(a) = \Theta(Ø_v(a))$. Thus, we will have $a = Ø_v(a)$, which is contradictory with the fact that $d \neq 0$. ▪

5.2.- __Corollary__: The code C associated to an H(e,0,n)-graph, $e \geq 3$, as in [7] is a propelinear PC(n,d)-code, $d \geq 3$.

In fact, from an H(e,0,n)-graph, $e \geq 3$, in [7] we construct an e-covering, $e \geq 3$, which has an associate code C that satisfies the previous proposition.

6. The Ω(C)-graph associated to a propelinear code C

If C is a propelinear code, we can define in F^n an equivalence relation as:

a ≈ b iff there exists $v \in C$ such that $\emptyset_v(a) = b$.

6.1.- __Proposition__: The relation ≈ is an equivalence relation.

1.- **Reflexive**: By 4.3.3 $\emptyset_0(a) = a$, so $a \approx a$.

2.- **Symmetric**: If $a \approx b$, there exists $v \in C$ such that $\emptyset_v(a) = b$
By 4.3.4 there exists $s \in C$ such that $\emptyset_v \cdot \emptyset_s = \mathrm{Id}$.
Then $b = \emptyset_v(a) = \emptyset_v \cdot \emptyset_s(b)$. By 4.3.2 $\emptyset_v$ is a bijective map, so $a = \emptyset_s(b)$, and $b \approx a$.

3.- **Transitive**: If $a \approx b$ then there exists $v \in C$ and $\emptyset_v(a) = b$.
If $b \approx c$ then there exists $s \in C$ and $\emptyset_s(b) = c$.
Then $c = \emptyset_s(b) = \emptyset_s(\emptyset_v(a)) = \emptyset_m(a)$ (See 4.3.5)
So $a \approx c$. ∎

6.2.- __The Ω(C)-graph__:

We can consider the quotient set $\Omega(C) = F^n/\approx$ as a set of vertices, and we define the adjacency relation between vertices in the following way:

Two vertices $A, A' \in \Omega(C)$ are adjacent if and only if there exist representatives in the classes A and A', which are at unit distance (distance refers to Hamming's distance).

Remark: It is well-known that if C is a linear, binary, completely regular code, then Ω(C) is a distance-regular graph. (See [2], [6]). The same is true in the case that C be a binary, completely regular, propelinear code. We will prove this in 6.6.

6.3.- **Proposition**. The partition $F^n/\approx$ is a regular partition.

Proof: We have, for a fixed $a, b \in F^n$, $A = \{ \emptyset_v(a) \ / \ v \in C \}$ and $A' = \{ \emptyset_v(b) \ / \ v \in C \}$.

Starting from a fixed $s, t \in C$, let J_s, J_t be the sets:

$$J_s = \{ v \in C \ / \ d(\emptyset_v(b), \emptyset_s(a)) = 1 \}$$

$$J_t = \{ v \in C \ / \ d(\emptyset_v(b), \emptyset_t(a)) = 1 \}$$

We need to prove that $\#J_s = \#J_t$.

Let $\emptyset_k$ be the inverse map of $\emptyset_s$, and $\emptyset_m = \emptyset_t \cdot \emptyset_k$.

If $\delta: J_s \longrightarrow J_t$ is the restricted map $\emptyset_m$ on J_s, and since we know that $\emptyset_m$ is a bijective map, we only need to prove that $\delta(J_s) = J_t$:

If $v \in J_s$, then $d(\emptyset_v(b), \emptyset_s(a) = 1$ and by 4.3.2

$1 = d(\emptyset_m(\emptyset_v(b)), \emptyset_m(\emptyset_s(a)) =$(by 4.3.5)$= d(\emptyset_{\delta(v)}(b), \emptyset_t \cdot \emptyset_k \cdot \emptyset_s(a))$ $= d(\emptyset_{\delta(v)}(b), \emptyset_t(a))$. So $\delta(v) \in J_t$. ∎

6.4.- **Proposition**: Let C be a PC(n,d)-code, $d \geq 3$, then the partition $F^n/\approx$ is a uniformly regular partition.

In fact, in this case we have $b_o = 0$, $b_1 = 1$. ∎

6.5.- **Lemma**: Let R be a uniformly regular partition ($b_o = 0$, $b_1 = 1$) in F^n. Let $v, w \in F^n$. The number Q_{pvw} of elements $x \in F^n$, of weight p, such that $(v + x)$ and (w) are in the same equivalence class, is given by $[P_p(D)]_{\alpha\beta}$.

α and β represent, respectively, the classes where the elements v and w are.

$P_m(x)$ is defined by: $(m \geq 2)$

$P_o(x) = 1$; $P_1(x) = x$; $m \cdot P_m(x) = x \cdot P_{m-1}(x) - (n-m+2) \cdot P_{m-2}(x)$

Proof:

Case $p = 0$: Q_{ovw} is zero or one depending on whether $\alpha = \beta$.
We have $Q_{ovw} = [Id.]_{\alpha\beta}$

Case $p = 1$: Q_{1vw} is zero or one depending on whether α and β are adjacent. We have $Q_{1vw} = [D]_{\alpha\beta}$

We assume that lemma is true for $p < r$. We want to prove the assertion for r:

We have: $r \cdot \Omega_{rvw} = \Sigma_s \Omega_{(r-1)sw} \cdot [D]_{\tau\beta} - (n-r+2) \cdot \Omega_{(r-2)vw}$, where τ represents the class of $s \in F^n$.

By the induction hypothesis the lemma holds. ∎

6.6.- **Theorem**: Let C be a PC(n,d)-code, $d \geq 3$. Then $\Omega(C)$ is a distance-regular graph if and only if C is a completely regular code.

Proof:

Let β be a class in $F^n/\approx$. To prove that β is a completely regular code we need to see that for every $v \in F^n$, the value of $B(v,p)$ is a constant number b_{up}, where u is the distance from v to β. (see 1.).

Let $w \in \beta$, then:

$B(v,p, = \#\{ c \in \beta / W_H(v+c) = p \} =$

$= \#\{ s \in F^n / W_H(s) = p, (v+s) \in \beta \} =$

$= \Omega_{pvw}$ (see 6.5)

From 6.4 and 6.5, $\Omega_{pvw} = [P_p(D)]_{\alpha\beta}$, where D is the adjacency matrix of graph $F^n/\approx$, and α is the class where v is.

Then $B(v,p) = [P_p(D)]_{\alpha\beta}$

The condition that $B(v,p)$ only depends on u and p, where u is the distance from v to β, is equivalent to say that $B(v,p)$ only depends on p and the distance from α to β (see 6.3). Thus, the condition that $B(v,p)$ is a constant number which only depends on p and on the distance from α to β, is equivalent to say that for every p, $P_p(D)$ linearly depends on D, D_2, D_3, ..., D_ρ, where D_i is a matrix indexed by the elements of graph $\Omega(C)$, and their entries (α,β) are "1" or "0" according to whether the distance from α to β is or is not "i". $P_p(D) = \Sigma_{u=0}^{u=\rho} b_{up} \cdot D_u$. This last condition is the distance-regular graph condition (see 2.). ∎

REFERENCES:

[1] E.BANNAI & T.ITO, "Algebraic Combinatorics I". The Benjamin-Cummings Publishing Co.,Inc. California. (1984).

[2] P.DELSARTE, "An algebraic approach to the association schemes of coding theory". Philips Research Reps. Suppl. 10. (1973).

[3] G.ETIENNE, "Perfect Codes and Regular Partition in Graphs and Groups". Europ. J. of Comb, 8, pp. 139-144 (1987).

[4] D.G.HIGMAN, "Coherent Configuration I Ordinary Representation Theory". Geom. Dedic. 4, pp. 1-32 (1975).

[5] J.RIFÀ,"Equivalències entre estructures combinatòricament regulars: Codis, Esquemes i grafs". Tesi doctoral. Univ. Autònoma de Barcelona. (1987).

[6] J.RIFÀ & L.HUGUET, "Characterization of Completely Regular Codes through P-Polynomial Association Schemes". Springer-Verlag, LNCS, n.307, 157-167, (1988).

[7] J.RIFÀ & L.HUGUET, Classification of a Class of Distance-Regular Graphs via Completely Regular Codes. Proceed. CO87. Southampton 1987. To appear in Discrete Maths.

N-Dimensional Berlekamp-Massey Algorithm for Multiple Arrays and Construction of Multivariate Polynomials with Preassigned Zeros

Shojiro SAKATA
Toyohashi University of Technology
Dep. of Production Systems Engineering
Tempaku, 440 Toyohashi, JAPAN

Abstract: In this paper we propose an algorithm of finding a minimal set of linear recurring relations for a given finite set of n-dimensional arrays. This algorithm is an n-dimensional extension of the Berlekamp-Massey algorithm for multisequences as well as an extension of the n-dimensional Berlekamp-Massey algorithm for a single array. Our algorithm is used to obtain Groebner bases of ideals defined by preassigned zeros. The latter problem is an extension of that treated by Moeller and Buchberger in the sense that the zeros can be over any finite extension $\tilde{K}$ of the base field K. Our approach gives an efficient method of obtaining Groebner bases of ideals defined by zeros to construct n-dimensional cyclic codes (i.e. Abelian codes). In case that the dimension n is small, the computational complexity is of order $O((lLd)^2)$, where l, L and d are the degree of the extension of $\tilde{K}$ over K, the number of the zeros and the size of the independent point set for the Groebner basis, respectively.

1. Introduction

The Berlekamp-Massey algorithm [1], [6] is used in decoding BCH codes and also it yields a shortest linear feedback shift register (LFSR) capable of generating a given sequence. The problem of finding a shortest LFSR which will generate a given finite set of sequences has been treated by several authors [3], [4]. Among them, Feng and Tzeng [4] gave an Euclidean method and Dai [3] gave an iterative method in which one proceeds to examine component by component for the given set of sequences. Recently we [10], [12] gave an algorithm of finding a minimal set of linear recurring relations for a given n-dimensional (nD) array. That is an extension of the Berlekamp-Massey algorithm, i.e. a synthesis algorithm of nD LFSR's to generate nD cyclic codes (i.e. Abelian codes) [11]. In general an nD cyclic code is characterized by a finite number of (representative) codewords, i.e. nD arrays, and thus our previous algorithm for a single nD array does not always suffice.

In this paper we propose an nD Berlekamp-Massey algorithm for multiple nD arrays which has computational complexity of $O((ms)^2)$, i.e. an iterative algorithm of finding a minimal set of linear recurring relations for a given finite set of nD arrays, where m and s are the number and the size of the given nD arrays, respectively. This algorithm is an extension not only of the 1D Berlekamp-Massey algorithm for multisequences by Dai [3] w.r.t. the dimension n but also of the nD Berlekamp-Massey algorithm for a single nD array [10], [12] w.r.t. the number m of arrays. In both ways this is not a straightforward extension of them because several difficulties are contained in extending an algorithm pertinent to the univariate polynomial ring to that applicable to the multivariate polynomial ring as in extending the classical Euclidean algorithm to the Groebner basis algorithm [2].

The set-up realized in our algorithm is useful for obtaining Groebner bases where ideals are given by properties and not by bases. In particular, our algorithm can also be used to find a Groebner basis of the ideal of polynomials in the n-variate polynomial ring $K[z]:=K[z_1,\dots,z_n]$ which vanish at a preassigned set of points in $\tilde{K}^n$, where K is any field and $\tilde{K}$ is a finite extension of K. The problem is an extension of that treated by Moeller and Buchberger [7]. They gave an algorithm for constructing a Groebner basis of the ideal of K[z] defined by preassigned zeros in K^n, where any field extension is not considered, i.e. $\tilde{K}=K$.

Let l be the degree of the extension of $\tilde{K}$ over K. For a given set of points $V=\{\alpha^{(1)},\dots,\alpha^{(L)}\}$ in $\tilde{K}^n$, we inquire how to find a Groebner basis of the radical ideal of $K[z]$

$$I(V):=\{f\in K[z] \mid f(\alpha)=0,\ \alpha\in V\},$$

where any n-variate polynomial $f(z_1,\ldots,z_n)$ is simply written as $f(z)$ or $f=\Sigma_{a\in\Gamma_f} f_a z^a$, where $\Gamma_f:=\{a\in\Sigma_0 \mid f_a(\in K)\neq 0\}$ is a finite subset of the nD lattice which is identified with the set $\Sigma_0=Z_0{}^n$ of all n-tuples $a=(a_1,\ldots,a_n)$ of nonnegative integers $a_1,\ldots,a_n$, and z^a denotes the power product $z_1{}^{a_1}\ldots z_n{}^{a_n}$; $f(\alpha)=f(\alpha_1,\ldots,\alpha_n)$ and $\alpha^a=\alpha_1{}^{a_1}\ldots\alpha_n{}^{a_n}\in\tilde{K}$ for $\alpha=(\alpha_1,\ldots,\alpha_n)\in\tilde{K}^n$. It is easy to see that $\alpha\in\tilde{K}^n$ is a zero iff the linear recurring (LR) relation:

$$f[u]_p:=\Sigma_{a\in\Gamma_f} f_a u_{a+p}=0, \tag{1a}$$

is valid at every $p\in\Sigma_0$ for the nD array $u=(u_p)$ over $\tilde{K}$ which has the components $u_p:=\alpha^p\in\tilde{K}$. If we express each $u_p\in\tilde{K}$ by an l-tuple $(u_p{}^{(0)},\ldots,u_p{}^{(l-1)})$ of $u_p{}^{(k)}\in K$, $0\leqq k\leqq l-1$, every nD array $u^{(k)}=(u_p{}^{(k)})$ over K, $0\leqq k\leqq l-1$, must satisfy the same LR relation (1a). Consequently, we need the nD Berlekamp-Massey algorithm for multiple arrays, or more exactly, an algorithm of finding a Groebner basis of the ideal $I=I(U)$ which is composed of all $f\in K[z]$ satisfying the identity (1a) for every array $u=u^{(i)}$ over K in a prescribed set $U=\{u^{(i)} \mid i\in M:=\{0,1,\ldots,m-1\}\}$, where the fixed integer m depends on the number L of preassigned zeros and the degree l of the extension, i.e. in general $m\leqq Ll$.

The extended version of the Moeller-Buchberger problem has been treated also by Imai [5] and Poli [8] to construct semisimple nD cyclic codes defined by preassigned zeros. At any way, we are required to obtain certain linear dependencies among the powers of the zeros. Our approach gives an efficient method which has computational complexity of $O((lLd)^2)$, where l, L and d are the degree of the field extension, the number of the zeros and the size of the independent point set corresponding to the Groebner basis, respectively. Furthermore, our approach is different from their methods also in the sense that the problem is first converted to a problem for multiple nD arrays and then solved by applying our algorithm to the arrays. This situation is similar with the case of Fast Fourier Transform, where the original problem given in the time domain is first converted to the corresponding problem in the frequency domain, and then the solution is obtained in the latter domain. In this context, polynomials and arrays correspond to time functions and frequency functions, respectively.

The following is the structure of the present paper. In Section 2, we introduce several preliminary concepts briefly and present the formal statement of our problem. In Section 3, we give an algorithm for solving our problem and we verify it in Section 4. In Section 5, we show a simple example of application of our algorithm to the extended version of the Moeller-Buchberger problem. Concluding remarks appear in Section 6. In the following descriptions, we confine ourselves to the 2D case (n=2), which can be extended to any higher dimensions except some complications in more than two dimensions [12].

2. Preliminaries

Over the 2D integral lattice $\Sigma_0 := Z_0^2$, we have the ordinary partial ordering $\leqq$ defined as follows:

$$p=(p_1,p_2) \leqq q=(q_1,q_2) \text{ iff } p_1 \leqq q_1 \text{ and } p_2 \leqq q_2,$$

where $p<q$ iff $p \leqq q$ and $p \neq q$. On the other hand, to design an iterative algorithm, we fix a total ordering $<_T$ over Σ_0, e.g. the total degree ordering as follows:

$$(0,0)<_T(1,0)<_T(0,1)<_T(2,0)<_T(1,1)<_T(0,2)<_T(3,0)<_T\cdots$$

Besides, $p \leqq_T q$ iff $p <_T q$ or $p=q$. For a 'point' $p=(p_1,p_2) \in \Sigma_0$, we have the 'next point'

$$p+1 := (p_1-1, p_2+1) \text{ if } p_1>0;$$
$$:= (p_2+1, 0) \quad \text{ if } p_1=0,$$

according to the total order $<_T$. Furthermore, to treat a set $U=\{u^{(i)} \mid i \in M:=\{0,1,\ldots,m-1\}\}$ of m 2D arrays $u^{(i)}$, we also introduce a total ordering $<<_T$ over $\Sigma_0^M := \Sigma_0 \times M$ which is defined as follows:

$$((0,0),0) <<_T ((0,0),1) <<_T \cdots <<_T ((0,0),m-1) <<_T ((1,0),0) <<_T \cdots <<_T$$
$$((1,0),m-1) <<_T ((0,1),0) <<_T \cdots <<_T ((0,1),m-1) <<_T ((2,0),0) <<_T \cdots$$

Besides, $(q,l) \leqq_T (p,k)$ iff $(q,l) <<_T (p,k)$ or $(q,l)=(p,k)$. The order implies that, at each point p of Σ_0, we stop to examine the components $u_p^{(i)}$ of all given arrays $u^{(i)}$, $0 \leqq i \leqq m-1$, respectively, and then go to the next point p+1 of p in the order of $<_T$. Thus, for $(p,k) \in \Sigma_0^M$, we define the next point $(p,k)+1 \in \Sigma_0^M$ of (p,k) according to the total order $<<_T$:

$$(p,k)+1 := (p,k+1) \text{ if } k<m-1;$$
$$:= (p+1,0) \text{ if } k=m-1.$$

We assume that each element (2D array) u of U is defined on Σ_0 or a finite subset of Σ_0:

$$\Sigma_0{}^p := \{q \in \Sigma_0 \mid q <_T p\}$$

which is specified by a point $p \in \Sigma_0$. In the former case we call such an array u a perfect (infinite) array, and in the latter case we denote u as u^p and call the cardinality $|\Sigma_0{}^p|$ the 'size' of the finite array u. For an array u, let u^q denote the restriction (truncation) of u within $\Sigma_0{}^q$. We must inquire any linear recurring (LR) relations which are satisfied by every array u of U. For a bivariate polynomial f over K, i.e. an element $f=\Sigma_{a\in\Gamma_f} f_a z^a$ of $K[z]:=K[z_1,z_2]$ ($\Gamma_f \subset \Sigma_0$), we call the maximum element (w.r.t. $<_T$) of Γ_f the 'degree' of f and denote it as Deg(f) or $s=s_f$, which specifies the 'head term' $f_s x^s$ of f ($f_s \neq 0$). Instead of the form of the identity (1a), we consider a LR relation at a point $x \in \Sigma_s := \{x \in \Sigma_0 \mid s \leqq x\}$ (or $\in \Sigma_s{}^p := \{x \in \Sigma_0 \mid s \leqq x <_T p\}$) w.r.t. an array u (or u^p) in the following form:

$$f[u]_x := \sum_{a\in\Gamma_f} f_a u_{a+x-s} = 0, \tag{1b}$$

where s is specified implicitly by f in the left-hand expression. For a prescribed set $U=\{u^{(i)} \mid i \in M\}$ of m arrays $u^{(i)}$ which are defined at least on $\Sigma_0{}^p$, $q \in \Sigma_0{}^p$ and $l \in M$, let

$$\Sigma_0{}^{(q,l)} := \{(a,i) \in \Sigma_0{}^M \mid 0 \leqq i < l,\ a \in \Sigma_0{}^{q+1}\}$$
$$\cup \{(a,i) \in \Sigma_0{}^M \mid l \leqq i \leqq m-1,\ a \in \Sigma_0{}^q\}, \tag{2a}$$

$$\Sigma_s{}^{(q,l)} := \{(a,i) \in \Sigma_0{}^{(q,l)} \mid a \geqq s\}, \tag{2b}$$

$$U^{(q,l)} := \{u_a{}^{(i)} \mid (a,i) \in \Sigma_0{}^{(q,l)}\}, \tag{3}$$

$$\mathrm{VALPOL}(U^{(q,l)}) := \{f \in K[z] \mid f[u^{(i)}]_x = 0,\ (x,i) \in \Sigma_s{}^{(q,l)},\ s=\mathrm{Deg}(f)\}. \tag{4}$$

While an element f of $\mathrm{VALPOL}(U^{(q,l)})$ is said to be 'valid' up to $q \in \Sigma_0$ w.r.t. $u^{(l)}$, it is also said to be 'valid' up to $(q,l) \in \Sigma_0{}^M$ (w.r.t. U). We define the 'order' $\mathrm{Ord}(f)=(q,l) \in \Sigma_0{}^M$ of a polynomial f (w.r.t. U) by the condition:

$$f \in \mathrm{VALPOL}(U^{(q,l)}) \text{ and } f[u^{(l)}]_q \neq 0. \tag{5}$$

For any set U of perfect arrays, $\mathrm{VALPOL}(U)=\{f \in K[z] \mid f[u]_x=0,\ u \in U,\ x \in \Sigma_s,\ s=\mathrm{Deg}(f)\}$ is an ideal I(U) of K[z], but $\mathrm{VALPOL}(U^{(q,l)})$ is not necessarily an ideal of K[z]. Although our goal is to get a Groebner basis of I(U) for a set U of perfect arrays, we begin by exploring a subset F of

$VALPOL(U^{(q,l)})$ each element f of which has a minimal degree w.r.t. the partial order <. Now we have the main definition:

Definition 1: $F \subset K[z]$ is said to be a 'minimal polynomial set' for $U^{(q,l)}$ iff the following conditions are satisfied:

(1) $F \subset VALPOL(U^{(q,l)})$;

(2) For $S:=\{Deg(f) \mid f \in F\}$, there exists no couple of distinct elements $s,t \in S$ s.t. $s \leqq t$; S is said to be a 'non-degenerate' subset of Σ_0 and by it a subset of Σ_0 is defined as follows:

$$\Sigma_S:=\bigcup_{s \in S} \Sigma_s; \tag{6}$$

(3) For $\Delta(F):=\Sigma_0/\Sigma_S$, there exists no polynomial g s.t. $g \in VALPOL(U^{(q,l)})$ and $Deg(g) \in \Delta(F)$, where / denotes the set difference operator.

The class of all minimal polynomial sets F for $U^{(q,l)}$ is denoted as $FF(U^{(q,l)})$. While $F \in FF(U^{(q,l)})$ is not unique for $U^{(q,l)}$, but S and $\Delta(F)$ is unique for $U^{(q,l)}$. Thus, $\Delta(F)$ is written as $\Delta(U^{(q,l)})$, which is called the 'independent point set' of $U^{(q,l)}$. Trivially, if $(q,l) \leqq_T (p,k)$, then $\Delta(U^{(q,l)}) \subseteq \Delta(U^{(p,k)})$.

In general $\Delta(U^{(q,l)})$ is a finite subset of Σ_0. Let C be the set of all maximal points (w.r.t. <) in $\Delta(U^{(q,l)})$. Then, we can write

$$\Delta(U^{(q,l)})=\Gamma_C:=\bigcup_{c \in C} \Gamma_c, \tag{7a}$$

where $\Gamma_c:=\{a \in \Sigma_0 \mid a \leqq c\}$ for $c \in C$. For later convenience, we allow C to contain two infinite points $(\infty,-1)$, $(-1,\infty)$ outside of Σ_0, i.e.

$$C_\infty:=\{(\infty,-1), (-1,\infty)\} \subseteq C.$$

Thus, $\Gamma_C:=\bigcup_{c \in C}\{a \in Z^2 \mid a \leqq c\} \subset Z^2$ and

$$\Delta(U^{(q,l)})=\Gamma_C \cap \Sigma_0, \tag{7b}$$

where Z^2 is the set of all pairs of (non-negative and negative) integers. For each $s \in S$ and $i \in \{1,2\}$, there exists $c \in C$ s.t. $s_i=c_i+1$, $s_j \leqq c_j$, $j(\neq i) \in \{1,2\}$. Such relationship is denoted as $s \vdash c$ and s is said to be 'adjoined' to c.

In the previous paper [10], we have proven that, for an array $u=u^p$,

$$\Delta(u^p)=\bigcup_{t,q \in \Sigma_0^p} \Delta_{q-t}, \tag{8}$$

where

$\Delta_{q-t} := \Gamma_{q-t}$, if there exists a polynomial g having $Deg(g)=t$ and $Ord(g)=q$;

$:= \phi$, otherwise.

In the above, $Ord(g) \in \Sigma_0$ is defined w.r.t. the single array u. Thus, we have m distinct subsets $\Delta(u^{(i)p})$ for $u^{(i)} \in U$, which are defined independent of each other by the above identity. We must distinguish between $Ord^{(i)}(f) \in \Sigma_0$ and $Ord(f) \in \Sigma_0{}^M$, where $Ord^{(i)}(f)$ is defined w.r.t the array $u^{(i)}$ and $Ord(f)$ is defined w.r.t. U. Trivially, we have

$$\Big(\bigcup_{0 \leqq i < l} \Delta(u^{(i)q+1})\Big) \cup \Big(\bigcup_{l \leqq i \leqq m-1} \Delta(u^{(i)q})\Big) \subseteq \Delta(U^{(q,l)}), \qquad (9)$$

where the equality does not always hold.

Example 1: For the set $U=\{u^{(0)},u^{(1)}\}$ of arrays over K=GF(2) shown in Fig. 1 a, b and q=(2,1), we can verify that

$\Delta(U^{(q,0)}) = \Gamma_{(2,0)} \cup \Gamma_{(0,1)}$;

$\Delta(u^{(0)q}) = \Gamma_{(1,0)}$, $\Delta(u^{(1)q}) = \Gamma_{(1,0)} \cup \Gamma_{(0,1)}$,

and so we have the gap between $\Delta(U^{(q,0)})$ and $\Delta(u^{(0)q}) \cup \Delta(u^{(1)q}) = \Gamma_{(1,0)} \cup \Gamma_{(0,1)}$.

0	1	0	1	0
1	1	0	0	
0	1	0		
0	0	*		
0	0			
1				

a

1	1	1	1	0
0	1	0	1	
1	1	0		
0	1			
0	*			
0				

b

Fig. 1: 2D arrays $u^{(0)}$, $u^{(1)}$ over GF(2).

Thus, our problem is how to find $\Delta(U^{(q,l)})$ and $F \in FF(U^{(q,l)})$ at every $(q,l) \in \Sigma_0{}^{(p,k)}$ iteratively in the increasing order w.r.t. $<<_T$.

3. Finding a minimal polynomial set for $U^{(p,k)}$

Before attacking our problem, we have the following observations which follow from the lemmas in the previous paper [10], [12]:

(1) Let $F \in FF(U^{(q,l)})$. If every polynomial $f \in F$ satisfies $f[u^{(l)}]_q=0$, then $F \in FF(U^{(q',l')})$, where $(q',l')=(q,l)+1 \in \Sigma_0{}^M$.

(2) Let $F \in FF(U^{(q,l)})$ and, for some $f \in F$ with $Deg(f)=s$,

$$d_f := f[u^{(l)}]_q \neq 0, \qquad (10)$$

where d_f is called the 'discrepancy' of f. If there exists g $\in K[z]$ with $Deg(g)=t$ s.t. $Ord(g)=(a,l)$ for some $a \in \Sigma_0{}^q$ and $\hat{c}:=a-t \geqq q-s$, then the polynomial

$$h:=h(f,g):=f-(d_f/d_g)z^{s-q+\hat{c}}g \qquad (11a)$$

has $Deg(h)=s=Deg(f)$ and $h \in VALPOL(U^{(q',l')})$ for $(q',l')=(q,l)+1$, where d_g is the discrepancy of g, i.e. $d_g:=g[u^{(l)}]_a(\neq 0)$, and $\hat{c}$ is called the 'span' of g and denoted as Span(g).

These observations lead us to introduce the following iterative algorithm, which we will verify in the next section. During the process of the algorithm, we keep or update at the j-th iteration (j=0,1,2,...) the data:

$(q,l)=(q_j,l_j) \in \Sigma_0(p,k)$; $U_j:=U(q,l)$;

$F_j \in FF(U_j)$; $S_j:=\{Deg(f) \mid f \in F_j\}$;

C_j s.t. $\Delta(U_j)=\Gamma_{C_j} \cap \Sigma_0$, i.e. $\Sigma_{S_j}=\Sigma_0/\Gamma_{C_j}$;

$\bar{F}_j:=\{f \in F_j \mid Ord(f)=(q,l)$, i.e. $f[u^{(l)}]_q \neq 0$; $Deg(f) \leqq q-\hat{s}$ for some $\hat{s} \in \hat{S}_j\}$; $\bar{S}_j:=\{Deg(f) \mid f \in \bar{F}_j\}$;

$G_j:=\{g \in K[z] \mid g \in \bar{F}_i$, $Ord(g)=(q_i,l_i)$, $l_i=l_j$, $q_i \in \Sigma_0{}^q$ for some $i<j\} \cup \{0, 0\}$;

$\hat{C}_j:=\{Span(g) \mid g \in G_j\} \supseteq C_\infty$, where let $(\infty,-1)$, $(-1,\infty) \in C_\infty$ be Span(g) for two polynomials g=0, respectively; $\hat{S}_j$ s.t. $\Sigma_{\hat{S}_j}=\Sigma_0/\Gamma_{\hat{C}_j}$.

In the following, for $f \in \bar{F}_j$ with $Deg(f)=s$, $g \in G_j$ with $Span(g)=\hat{c}$ and $q=q_j$, let

$$r:=\max(s,q-\hat{c}),$$

where, for $a=(a_1,a_2)$, $b=(b_1,b_2) \in \Sigma_0$,

$$\max(a,b):=(\max(a_1,b_1),\max(a_2,b_2)) \in \Sigma_0.$$

Then, the polynomial

$$h:=h(f,g):=z^{r-s}f-(d_f/d_g)z^{r-q+\hat{c}}g \qquad (11b)$$

has $Deg(h)=r$, which is identical with the formula (11a) iff $q-\hat{c} \leqq s$.

Algorithm:

```
Step 1: j:=0; (q,l):=((0,0),0); F_0:={1}; S_0:={(0,0)}; C:=C_∞;
        G_0,...,G_{m-1}:={0, 0}; Ĉ_0,...,Ĉ_{m-1}:=C_∞; Ŝ_0,...,Ŝ_{m-1}:={(0,0)};
Step 2: F̄_j:={f∈F_j | Ord(f)=(q,l), Deg(f)≦q-ŝ for some ŝ∈Ŝ_j};
        S̄_j:={s=Deg(f) | f∈F̄_j}(⊆S_j);
        if F̄_j=ϕ, then
          begin
          F_{j+1}:=(F_j∪{h=h(f,g) | f∈F_j with Deg(f)=s s.t. Ord(f)=(q,l),
                         q-s≦ĉ for some g∈G_j with Span(g)=ĉ∈Ĉ_j})
                /{f∈F_j | Ord(f)=(q,l)};
          S_{j+1}:=S_j; C_{j+1}:=C_j;
          G_{j+m}:=G_j; Ĉ_{j+m}:=Ĉ_j; Ŝ_{j+m}:=Ŝ_j;
          end;
        else [F̄_j≠ϕ]
          begin
          F_{j+1}:=(F_j∪{h=h(f,g) | f∈F_j with Deg(f)=s s.t. Ord(f)=(q,l),
                         q-s≦ĉ for some g∈G_j with Span(g)=ĉ∈Ĉ_j}
                      ∪{h=h(f,g) | f∈F̄_j with Deg(f)=s, g∈G_j with
                         Span(g)=ĉ∈Ĉ_j s.t. s<q-ĉ}
                      ∪{h=h(f,g) | f∈F̄_j with Deg(f)=s, g∈G_j with
                         Span(g)=ĉ∈Ĉ_j s.t. q-ĉ≦c, q-ŝ≧s, s⊢c, ŝ⊢ĉ
                         for some c∈C_j and some ŝ∈Ŝ_j})
                /{f∈F_j | Ord(f)=(q,l)};
          S_{j+1}:=(S_j∪{q-ĉ | ĉ∈Ĉ_j s.t. s<q-ĉ for
                         some s∈S̄_j}
                      ∪{max(s,q-ĉ) | s∈S̄_j, ĉ∈Ĉ_j s.t. q-ĉ≦c, q-ŝ≧s,
                         s⊢c, ŝ⊢ĉ for some c∈C_j and some ŝ∈Ŝ_j}
                /{s=Deg(f) | f∈F̄_j s.t. Ord(f)=(q,l)};
          C_{j+1}:=(C_j∪{q-ŝ | ŝ∈Ŝ_j s.t. s≦q-ŝ for some s∈S̄_j})
                /{c∈C_j | c<q-ŝ for some ŝ∈Ŝ_j satisfying s≦q-ŝ
                        for some s∈S̄_j};
          G_{j+m}:=(G_j∪{f∈F̄_j})
                /{g∈G_j | Span(g)<q-Deg(f) for some f∈F̄_j};
          Ĉ_{j+m}:=(Ĉ_j∪{Span(f)=q-s | s=Deg(f), f∈F̄_j})
                /{ĉ∈Ĉ_j | ĉ<q-Deg(f) for some f∈F̄_j};
```

$$\hat{S}_{j+m} := (\hat{S}_j \cup \{q-c \mid c \in C_j \text{ s.t. } \hat{s} < q-c \text{ for some } \hat{s} \in \hat{S}_j \text{ satisfying}$$
$$s \leqq q-\hat{s} \text{ for some } s \in \bar{S}_j\}$$
$$\cup \{\max(q-c, \hat{s}) \mid c \in C_j,\ \hat{s} \in \hat{S}_j \text{ s.t. } q-c \leqq \hat{c},\ \hat{s} \leqq q-s,$$
$$s \vdash c,\ \hat{s} \vdash \hat{c} \text{ for some } s \in \bar{S}_j \text{ and some } \hat{c} \in \hat{C}_j\})$$
$$/\{\hat{s} \in \hat{S}_j \mid s+\hat{s} \leqq q \text{ for some } s \in \bar{S}_j\};$$

end;

Step 3: $j := j+1$; $(q,l) := (q,l)+1$;

if $(q,l)=(p,k)$ then stop; else go to Step 2.

In Step 2, the cases of $s' := q-\hat{c} \in S_{j+1}$ and $s' := \max(s, q-\hat{c}) \in S_{j+1}$ are illustrated in Fig. 2 a and b, respectively. Similarly, the case of $c' := q-\hat{s} \in C_{j+1}$ is illustrated in Fig. 3. Furthermore, the cases of $\hat{s}' := q-c \in \hat{S}_{j+m}$, $\hat{s}' := \max(q-c, \hat{s}) \in \hat{S}_{j+m}$ and $\hat{c}' := q-s \in \hat{C}_{j+m}$ are illustrated in Fig. 4 a, b and 5, respectively.

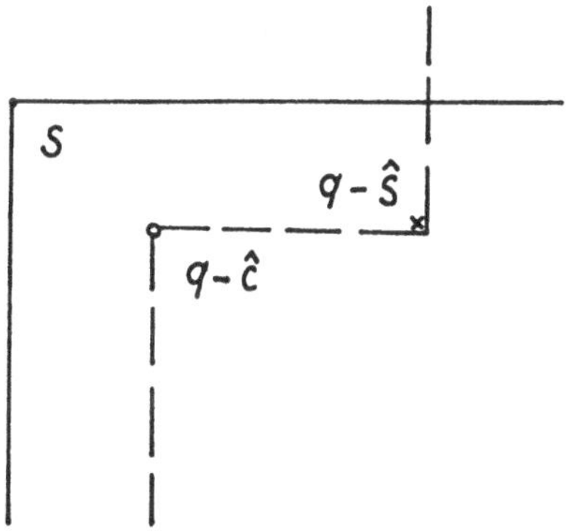

Fig. 2a: $s' := q-\hat{c} \in S_{j+1}$

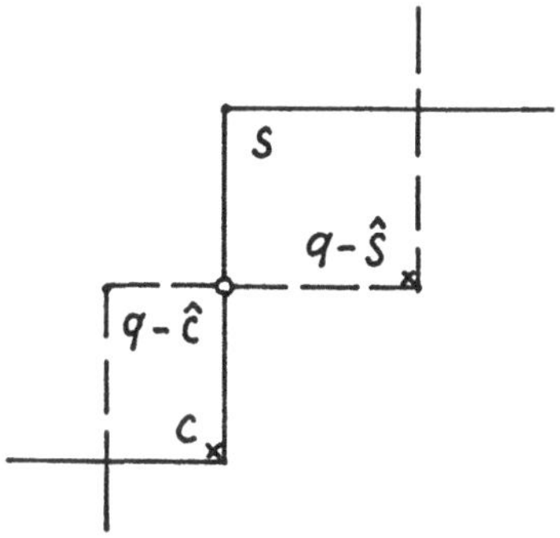

Fig. 2b: $s' := \max(s, q-\hat{c}) \in S_{j+1}$

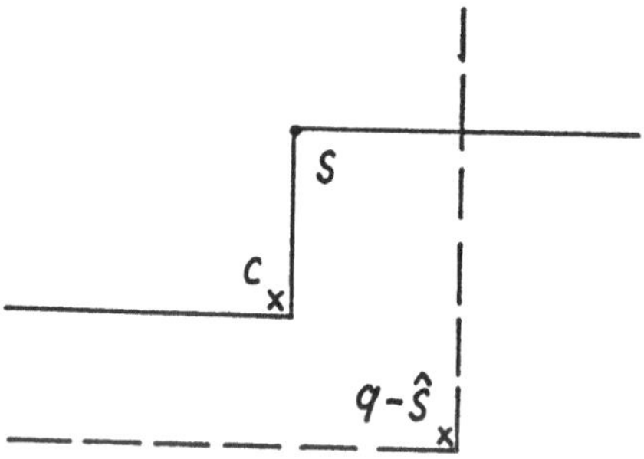

Fig. 3: $c' := q-\hat{s} \in C_{j+1}$.

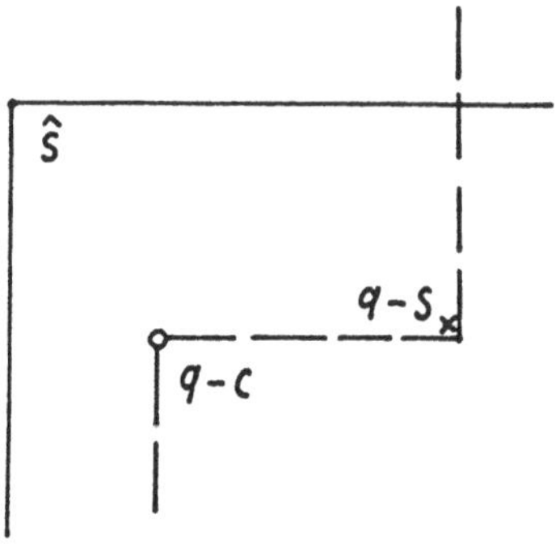

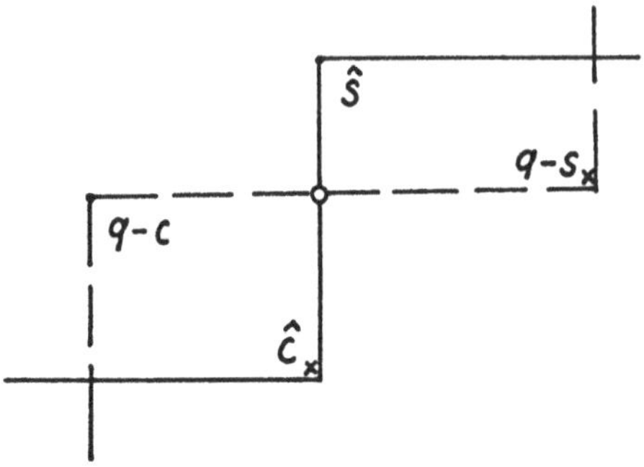

Fig. 4a: $\hat{s}' := q-c \in \hat{S}_{j+m}$

Fig. 4b: $\hat{s}' := \max(q-c, \hat{s}) \in \hat{S}_{j+m}$

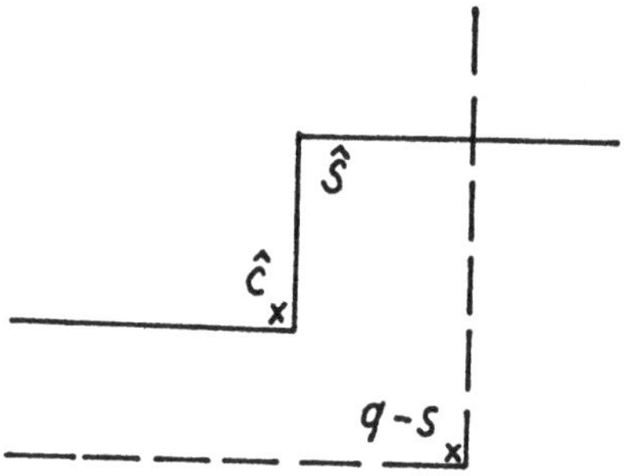

Fig. 5: $\hat{c}' := q-s \in \hat{C}_{j+m}$

Remark: For each $i \in M$, the sequences of subsets $\hat{C}_i$, $\hat{C}_{i+m}$, $\hat{C}_{i+2m}$, ... and $\hat{S}_i$, $\hat{S}_{i+m}$, $\hat{S}_{i+2m}$, ... are pertinent to the i-th array $u^{(i)}$. Thus, during the process, we have m+1 non-decreasing sequences of subsets:

$$\Delta(U_0) \subseteq \Delta(U_1) \subseteq \Delta(U_2) \subseteq \ldots;$$
$$\Gamma \hat{C}_0 \subseteq \Gamma \hat{C}_m \subseteq \Gamma \hat{C}_{2m} \subseteq \cdots,$$
$$\vdots$$
$$\Gamma \hat{C}_{m-1} \subseteq \Gamma \hat{C}_{2m-1} \subseteq \Gamma \hat{C}_{3m-1} \quad \cdots.$$

Every strict inclusion $\subset$ in the first sequence corresponds to a strict inclusion in one of the other sequences.

Example 2: For the same set $U=\{u^{(0)},u^{(1)}\}$ of arrays over $K=GF(2)$ as in Example 1, we have the computation by the algorithm shown in Table 1, where, for clarity, the columns of (q,l), G_j and C_j are divided into two parts which correspond to $u^{(0)}$ and $u^{(1)}$, respectively.

4. Verification of Algorithm

In this section, we verify the algorithm, i.e. we prove

Theorem 1: Let F_j, $j=0,1,2,\ldots$, be determined by the algorithm. Then,

$$F_j \in FF(U^j),\ j=0,1,2,\ldots$$

Since $F_0=\{1\}\in FF(U^0)$, we have only to prove that, if $F_j\in FF(U^j)$ for any $j\in Z_0$, then $F_{j+1}\in FF(U^{j+1})$, where F_{j+1} is determined by the procedure of Step 2 in the algorithm. In particular, if $\bar{F}_j=\phi$, then either we have $F_{j+1}=F_j$ or we can find each polynomial f in F_{j+1} by the formula (11a), where $S_{j+1}=S_j$ for either case. (For details, see Theorem 1 in the previous paper [12].) If $\bar{F}_j\neq\phi$, then there exists $\hat{s}\in\hat{S}_j$ s.t. $s\leqq q-\hat{s}$ for $f\in\bar{F}_j$ with $Deg(f)=s$, where $q=q_j$. Therefore, we have only to show that there exists no polynomial $h\in VALPOL(U^{j+1})$ s.t. $Deg(h)=r\leqq q-\hat{s}$, where we remark that $S_j\neq S_{j+1}$ (i.e. $\Delta(U^j)\subset\Delta(U^{j+1})$) iff $\hat{S}_j\neq\hat{S}_{j+m}$ (i.e. $\Gamma\hat{C}_j\subset\Gamma\hat{C}_{j+m}$).

Let $Ord(1)$ be $(q_{j(0)},l_{j(0)})$ for the polynomial $f=1\in F_0$. Then, it is easy to see that there exists no polynomial $h\in VALPOL(U^{j(0)+1})$ s.t. $Deg(h)\leqq q_{j(0)}-\hat{s}=q_{j(0)}$, where $\hat{s}=(0,0)\in\hat{S}_0=\hat{S}_{j(0)}$. Thus, we can assume that $j(0)<j$. Now, let

$$\begin{aligned} J&:=\{i\in Z_0 \mid 0\leqq i<j,\ S_i\neq S_{i+1}\}\\ &=\{i\in Z_0 \mid 0\leqq i<j,\ \hat{S}_i\neq\hat{S}_{i+m}\}\\ &=\{j(i)\in Z_0 \mid 0\leqq i\leqq\tau,\ j(0)<j(1)<\ldots<j(\tau)<j\},\end{aligned}$$

where $0\leqq\tau\ (\in Z_0)$. Any polynomial h with $Deg(h)=r$ s.t. $s\leqq r\leqq q-\hat{s}$ can be expressed in the following form:

$$h= h_{sr}z^{r-s}f + \sum_{(s',r')\in\Pi'} h_{s'r'}z^{r'-s'}f_{s'} + \sum_{0\leqq i\leqq\tau}\Big(\sum_{(\alpha(i),\beta(i))\in\Pi(i)} h_{\alpha(i)\beta(i)}z^{\beta(i)-\alpha(i)}f_{\alpha(i)}\Big), \quad (12)$$

Table 1: Computation for the arrays $u^{(0)}$, $u^{(1)}$ shown in Fig. 1a, b ($x:=z_1$, $y:=z_2$; The data $C_\infty=\{(\infty,-1), (-1,\infty)\}$ is omitted.)

j	q	l	F	G	$\hat{C}/C_\infty$
0	0,0	0	1	0	
1		1	1	0	
2	1.0	0	x y	0	
3		1	x^2 y	1	(0,0)
4	0,1	0	x^2 y	x	(0,0)
5		1	x^2 $y+x$	1	(0,0)
6	2,0	0	x^2 $y+x+1$	x	(0,0)
7		1	x^2 $y+x+1$	1	(0,0)
8	1,1	0	x^2+1 $y+x+1$	x	(0,0)
9		1	x^2+1 $y+x+1$	1	(0,0)
10	0,2	0	x^2+1 $y+x+1$	x	(0,0)
11		1	x^2+1 $y+x+1$	1	(0,0)
12	3,0	0	x^2+1 $xy+x+1$ y^2+y+x	x	(0,0)
13		1	x^3 $xy+x+1$ y^2+y+x	$y+x+1$	(0,1)
14	2,1	0	x^3 $xy+x+1$ y^2+y+x	x^2+1	(1,0)

15		1	x^3 $xy+x+1$ y^2+y+x	$y+x+1$	$(0,1)$
16	1,2	0	x^3 $xy+x+1$ y^2+y+x	x^2+1	$(1,0)$
17		1	x^3 $xy+x+1$ $y^2+x^2+y+x+1$	$y+x+1$	$(0,1)$
18	0,3	0	x^3 $xy+x+1$ $y^2+x^2+y+x+1$	x^2+1	$(1,0)$
19		1	x^3 $xy+x+1$ $y^2+x^2+y+x+1$	$y+x+1$	$(0,1)$
20	4,0	0	x^3 $xy+x+1$ y^2+x^2	x^2+1	$(1,0)$
21		1	x^3 $xy+x+1$ y^2+x^2	$y+x+1$	$(0,1)$
22	3,1	0	x^3 $xy+x+1$ y^2+x^2	x^2+1	$(1,0)$
23		1	x^3 $xy+x+1$ y^2+x^2	$y+x+1$	$(0,1)$
24	2,2	0	$x^3+y+x+1$ $xy+x+1$ y^2+x^2	x^2+1	$(1,0)$
. .					
32	4,1	0	$x^3+y+x+1$ $xy+x+1$ y^2+x^2	x^2+1	$(1,0)$
33	*				

where

$f \in \bar{F}_j$, Deg(f)=s, $s \leqq r \leqq q-\hat{s}$, $h_{sr}(\in K) \neq 0$;

$f_{s'} \in F_j$, Deg($f_{s'}$)=s', $s' \leqq r' <_T r$, $h_{s'r'} \in K$;

$\Pi' := \{(s'', r') \in \Sigma_0 \times \Sigma_0 \mid s' = \mathrm{Deg}(f'),\ f' \in F_j,\ s' \leqq r' <_T r\}$;

$f_{\alpha(i)} \in \bar{F}_{j(i)}$, $h_{\alpha(i)\beta(i)} \in K$,

$\Pi(i) := \{(\alpha, \beta) \in \Sigma_0 \times \Sigma_0 \mid \alpha = \mathrm{Deg}(g),\ g \in \bar{F}_{j(i)},$
$\alpha \leqq \beta \leqq q_{j(i)} - \hat{s}$ for some $\hat{s} \in \hat{S}_{j(i)}\}$, $0 \leqq i \leqq \tau$;

(For $\Pi(i)$, see Fig. 6.) We remark that

$\mathrm{Ord}(f) = (q, l) := (q_j, l_j)$;

$\mathrm{Ord}(f') \geqq >_T (q, l)$, $f' \in F_j$.

$\mathrm{Ord}(f_{\alpha(i)}) = (q_{j(i)}, l_{j(i)})$, $f_{\alpha(i)} \in \bar{F}_{j(i)}$, $0 \leqq i \leqq \tau$;

Now, we require $\mathrm{Ord}(h) >>_T (q, l)$ on the assumption that Deg(h)=r, i.e. $h_{sr} \neq 0$. To investigate the order of each term $h_{\alpha(i)\beta(i)} z^{\beta(i)-\alpha(i)} f_{\alpha(i)}$ in the expression (12), we define the order of a polynomial f as a polynomial with (virtual) degree $r \geqq \mathrm{Deg}(f)$ as follows:

$\mathrm{Ord}_r f = (p, k)$ iff $f[u^{(i)}]_x{}^r = 0$, $(x, i) \in \Sigma_r{}^{(p,k)}$ and $f[u^{(k)}]_p{}^r \neq 0$,

where, for an array u and $x \geqq r$,

$$f[u]_x{}^r := \sum_{a \in \Gamma_f} f_a u_{a+x-r}.$$

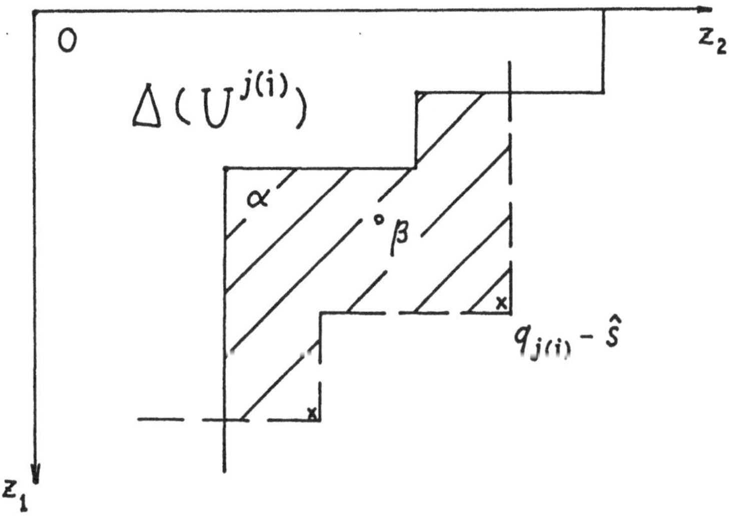

Fig. 6: $(\alpha, \beta) \in \Pi(i)$

In particular, $Ord_s(f)=Ord(f)$ for $s=Deg(f)$. Then, we have

Lemma 1: For any polynomial f with $Deg(f)\leqq r$,
$Ord_r(f)=(r,0)+Ord(f)-(Deg(f),0)$. (Its proof is in Appendix.)

By the way, we remark that $Ord(z^i f)=Ord(f)$, and that, for $r\geqq Deg(f)+i$,
$Ord_r(z^i f)=(r-i,0)+Ord(f)-(Deg(f),0)$.

Now we have the key

Lemma 2: For $(\alpha,\beta)\in\Pi(i)$, $(\alpha',\beta')\in\Pi(i')$, $0\leqq i,i'\leqq\tau$, s.t. $(\alpha,\beta)\neq(\alpha',\beta')$, we have

$$Ord_r(z^{\beta-\alpha}f_\alpha)\neq Ord_r(z^{\beta'-\alpha'}f_{\alpha'}).$$

(The proof is in Appendix.)

Furthermore, for $(\alpha,\beta)\in\Pi(i)$, $0\leqq i\leqq\tau$, $Ord_r(z^{\beta-\alpha}f_\alpha)\neq(q,l)$ and, for $(s',r')\in\Pi'$,

$$Ord_r(z^{r'-s'}f_{s'})=(r-r',0)+Ord(f_{s'})>>_T Ord(f_{s'})\geqq>_T Ord(f)=(q,l).$$

From the above considerations, we can rearrange the terms of the expression (12) in the increasing order of $Ord_r(\ )$. Among them, every term $h_{\alpha\beta}z^{\beta-\alpha}f_\alpha$ having $Ord_r(\ )<<_T(q,l)$ must vanish, i.e. $h_{\alpha\beta}=0$, since $h[u^{(i)}]_x{}^r=0$ for $(x,i)\in\Sigma_r{}^{(q,l)}$, and $Ord_r(z^{\beta-\alpha}f_\alpha)$ are distinct not only from each other but also from $Ord_r(z^{r-s}f)$. Therefore, the term $h_{sr}z^{r-s}f$ must also vanish, i.e. $h_{sr}=0$, which is a contradiction. Consequently, we have completed the proof of Theorem 1.

Example 3: For the same set of arrays as in Example 1, let the component $u_{(4,1)}{}^{(1)}=1$ be appended. Then, for j=32, $(q,l)=(q_j,l_j)=((4,1),1)$, we have $(x:=z_1,\ y:=z_2)$

$J=\{((0,0),1),\ ((1,0),0),\ ((0,2),1),\ ((3,0),0)\}$, $\tau=3$;
$\bar{F}_{j(0)}=\{1\}$, $\bar{F}_{j(1)}=\{x\}$, $\bar{F}_{j(2)}=\{y+x+1\}$; $\bar{F}_{j(3)}=\{x^2+1\}$;
$\bar{F}_j=\{x^3+y+x+1,\ xy+x+1\}$, $\hat{S}_j=\{(1,0),(0,2)\}$.

For $\hat{s}=(1,0)\in\hat{S}_j$, let $r=q-\hat{s}=(3,1)$. The values of $Ord_r(z^{\beta-\alpha}f_\alpha)$ for $f\in(\bigcup_{0\leqq i\leqq 3}\bar{F}_{j(i)})\cup\bar{F}_j$ which are arranged in the increasing order of $<<_T$ are shown in Table 2. From the table, it is easy to see that we cannot have $h \in VALPOL(U^{j+1})$ s.t. $Deg(h)=r=(3,1)$ for the appended data $u_{(4,1)}^{(1)}=1$.

Table 2: $Ord_r(z\beta - \alpha f_\alpha)$

$Ord_r(\ )$	$z\beta - \alpha f_\alpha$
((3,1),0)	x
((3,1),1)	1
((4.1),0)	x^2+1
((4,1),1)	$y(x^3+y+x+1)$ (or $x^2(xy+x+1)$)
((3,2),1)	$y+x+1$
((5,1),1)	$x(xy+x+1)$
((4,2),1)	$x^3+y+x+1$
((6,1),1)	$xy+x+1$

By the way, we have the uniqueness theorem (its proof is similar as in the case of a single array [11]):

Theorem 2: Let $q_j=q$ and $l_j=0$. If

$$\bigcup_{s\in S_j} \left(\bigcup_{i\in M} (s+\Gamma_{\hat{C}_{j+i}}) \right) \subset \Sigma_0{}^q,$$

then $F_j \in FF(U^j)$ is unique up to scalar factor in K, where $s+\Gamma_{\hat{C}_{j+i}} := \{s+a \mid a \in \Gamma_{\hat{C}_{j+i}}\}$.

The theorem assures that, after applying our algorithm in a finite number of iterations, we can get the desired unique solution, and so that we can solve the extended version of Moeller-Buchberger problem by our method.

5. Application of Algorithm

Now we apply our algorithm to the extended version of the Moeller-Buchberger problem.

Example 4: Let $K=GF(2)$ and α be a root of the irreducible polynomial over K: x^3+x+1, i.e. $\alpha \in \tilde{K}:=GF(2^3)$. For $V=\{(\alpha, \alpha^2)\}$, we have an array $u_{ij}=\alpha^{i+2j}$ over $\tilde{K}$, which can be represented by the set $U=\{u^{(0)}, u^{(1)}, u^{(2)}\}$

of arrays over K. While the exponents i of values α^i are shown in Fig. 7a, the components of arrays over K are shown in Fig. 7b-d. Applying the algorithm to U, we have at (q,l)=((3,0),0)

$$F=\{x^2+y,\ xy+x+1,\ y^2+y+x\} \in FF(U^{((3,0),0)}),$$

where $x:=z_1$, $y:=z_2$. Since the arrays of U are doubly periodic with the fundamental period parallelogram:

$$(7,0)\quad (5,1):=\{(i,0)\in\Sigma_0 \mid 0\leqq i<7\}\cup\{(i,1)\in\Sigma_0 \mid 5\leqq i<12\},$$

we can stop the computation at p=(4,0) or much earlier. In fact, the solution F obtained at ((3,0),0) just constitutes the reduced Groebner basis of I=I(V).

0	2	4	6	1	3	5	0
1	3	5	0	2	4	6	
2	4	6	1	3	5		
3	5	0	2	4			
4	6	1	3				
5	0	2					
6	1						
0							

Fig. 7a: 2D array of exponents i+2j

1	0	0	1	0	1	1	1
0	1	1	1	0	0	1	
0	0	1	0	1	1		
1	1	1	0	0			
0	1	0	1				
1	1	0					
1	0						
1							

Fig. 7b: component 2D array $u^{(0)}$

0	0	1	0	1	1	1	0
1	1	1	0	0	1	0	
0	1	0	1	1	1		
1	1	0	0	0			
1	0	1	1				
1	0	0					
0	1						
0							

Fig. 7c: component 2D array $u^{(1)}$

0	1	1	1	0	0	1	0
0	0	1	0	1	1	1	
1	1	1	0	0	1		
0	1	0	1	1			
1	1	0	0				
1	0	1					
1	0						
0							

Fig. 7d: component 2D array $u^{(2)}$

6. Concluding remarks

We have given the nD Berlekamp-Massey algorithm for multiple nD arrays which is a generalization not only of the nD Berlekamp-Massey algorithm for a single nD array [10], [12] but also of the (one-dimensional) Berlekamp-Massey algorithm for multiple sequences (one-dimensional arrays) [3]. Our algorithm has computational complexity of $O((ms)^2)$, where m and s are the number and the size of the given arrays, respectively. In our algorithm we fix the order among the arrays, i.e. at each point q of the nD lattice Σ_0, we stop to examine the components $u_q^{(0)}$, $u_q^{(1)}$, ..., $u_q^{(m-1)}$ of all the given arrays $u^{(i)}$, $0 \leqq i \leqq m-1$, in this order. It is necessary to fix the order of arrays as described above, for our algorithm does not work well in an indefinite order of arrays, e.g., in case of m=2,

..., (q,0), (q,1), (q+1,1), (q+1,0), ...

By our method we can also obtain multivariate polynomials with preassigned zeros. The latter problem is an extension of the Moeller-Buchberger problem [7] in the sense that the zeros can be elements of any extension field of the base field. In this case, our algorithm has computational complexity of order $O((lLd)^2)$, where l, L and d are the degree of field extension, the number of the given zeros and the size of the independent point set corresponding to the Groebner basis, respectively. Imai [5] and Poli [8] treated a similar problem to construct nD cyclic codes (Abelian codes). Our approach can be an alternative and more efficient method. Over a field K having characteristic 0, we cannot have any periodicity of arrays. Even in such cases, we can get the solution in a finite number of iterations in view of the uniqueness theorem.

REFERENCES

[1] Berlekamp, E. R., Binary BCH codes for correcting multiple errors: The Gorenstein-Zierler generalized nonbinary BCH codes for the Hamming metric, in "Algebraic Coding Theory," Chapters 7 and 10, pp.176-199, 218-240, McGraw-Hill Book Comp., New York, 1968.

[2] Buchberger, B.B, Groebner bases: An algorithmic method in polynomial

ideal theory, in "Multidimensional Systems Theory," (Ed.N.K.Bose), Chapter 6, pp.184-232, D. Reidel Publishing Comp., Dordrecht, 1985.

[3] Dai, Z. D, An iterative algorithm for synthesizing multi-sequences, preprint.

[4] Feng, G. L. and Tzeng, K. K., A genelarized Euclidean Algorithm for multiseqence shift-register synthesis, the Fourth Carribean Conference on Combinatorics and Computing, Puerto Rico, 1985.

[5] Imai, H., A theory of two-dimensional cyclic codes, Information and Control, vol. 34 (1977), pp.1-21.

[6] Massey, J. L., Shift-register synthesis and BCH decoding, IEEE Trans. Information Theory, vol.IT-15 (1969), no.1, pp.122-127.

[7] Moeller, H. M. and Buchberger, B., The construction of multivariate polynomials with preassigned zeros, in "Computer Algebra: EUROCAM '82 European Computer Algebra Conference, Marseille, France, April 1982," (Ed. G. Goos and J. Hartmanis), pp.24-31, Springer-Verlag, Berlin, 1982.

[8] Poli, A., Some algebraic tools for error-correcting codes, in "Algebraic Algorithms and Error-Correcting Codes: AEECC-3, Grenoble, France, July 1985," (Lecture Notes in Computer Science, No.229; Ed G. Goos and J. Hartmanis), pp.43-60, Springer-Verlag, Berlin, 1986.

[9] Sakata, S., On determining the independent point set for doubly periodic arrays and encoding two-dimensional cyclic codes and their duals, IEEE Trans. Information Theory, vol.IT-27 (1981), no.5, pp.556-565.

[10] Sakata, S., Finding a minimal set of linear recurring relations capable of generating a given finite two-dimensional array, J. of Symbolic Computation, vol. 5 (1988), no. 3, pp.321-338.

[11] Sakata, S., Synthesis of two-dimensional linear feedback shift registers, in "Proceedings of AAECC-5, Menorca, Spain, June, 1987," (to appear), Springer-Verlag.

[12] Sakata, S., Extension of the Berlekamp-Massey algorithm to n dimensions (submitted for publication in 'Information and Computation').

APPENDICES

Proof of Lemma 1:

$$f[u^{(i)}]_x{}^r=\sum_{a\in\Gamma_f} f_a u^{(i)}{}_{a+x-r}=\sum_{a\in\Gamma_f} f_a u^{(i)}{}_{a+(x-r+Deg(f))-Deg(f)}.$$

Therefore, $f[u^{(l)}]_x{}^r=0$, if $(x-r+Deg(f),i)<<_T Ord(f):=(p,k)$, and $f[u^{(k)}]_x{}^r\neq 0$, if $x-r+Deg(f)=p$. Thus, $Ord_r(f)=(r,0)+Ord(f)-(Deg(f),0)$. Q.E.D.

Proof of Lemma 2:

First, let $\beta\neq\beta'$, $(\alpha,\beta),(\alpha',\beta')\in\Pi(i)$, i.e. $Ord(f_\alpha)=Ord(f_{\alpha'})$. Then, $Ord_r(z^{\beta-\alpha}f_\alpha)=(r,0)+Ord(f_\alpha)-(\beta,0)$ is not equal to $Ord_r(z^{\beta'-\alpha'}f_{\alpha'})=(r,0)+Ord(f_{\alpha'})-(\beta',0)$.

Second, let $(\alpha,\beta)\in\Pi(i)$, $(\alpha',\beta')\in\Pi(i')$, $l_{j(i)}=l_{j(i')}$, $i<i'$. Then, $Ord(f_\alpha)-(\beta,0)$ and $Ord(f_{\alpha'})-(\beta',0)$ are distinct from each other, since, in view of the definition of $\hat{C}_{j+m}$, $\hat{S}_{j+m}$ from $\hat{C}_j$, $\hat{S}_j$ in Step 2, $Ord(f_\alpha)=(q_{j(i)},l_{j(i)})$, $q_{j(i)}-\beta\leqq q_{j(i)}-\alpha\leqq\hat{c}$ for some $\hat{c}\in\hat{C}_{j(i+1)}$, $i+1\leqq i'$, $Ord(f_{\alpha'})=(q_{j(i')},l_{j(i')})$, $q_{(i')}-\beta'\geqq q_{j(i')}-(q_{j(i')}-\hat{s})=\hat{s}$ for some $\hat{s}\in\hat{S}_{j(i')}$, i.e. $q_{j(i')}-\beta'\in\Sigma_{\hat{S}_{j(i')}}$, $q_{j(i)}-\beta\in\Gamma_{\hat{C}_{j(i+1)}}$, and $\Gamma_{\hat{C}_{j(i+1)}}\cap\Sigma_{\hat{S}_{j(i')}}=\phi$.

Third, let $(\alpha,\beta)\in\Pi(i)$, $(\alpha',\beta')\in\Pi(i')$, $l_{j(i)}\neq l_{j(i')}$. Then, obviously, $Ord(z^{\beta-\alpha}f_\alpha)=(*,l_{j(i)})$ and $Ord(z^{\beta'-\alpha'}f_{\alpha'})=(*,l_{j(i')})$ are distinct. Q.E.D.

COVERING RADIUS: IMPROVING ON THE SPHERE–COVERING BOUND

Juriaan Simonis

Delft University of Technology
Faculty of Mathematics and Informatics
P.O. Box 356, 2600 AJ DELFT, HOLLAND

Abstract: Currently, the best general lower bound for the covering radius of a code is the sphere covering bound. For binary linear codes, the paper presents a new method to detect cases in which this bound is not attained.

I. INTRODUCTION

Starting with a multiset[1] S of cardinality n in $\mathbb{F}_2^r$, we define a sequence of $r+1$ multisets S^ℓ of cardinality $\binom{n}{\ell}$:

$$S^\ell := \begin{cases} \{\underline{0}\} & \text{if } \ell = 0, \\ \{\sum_{\underline{x}\in U} \underline{x} \mid U \subset S \wedge |U| = \ell\} & \text{if } \ell > 0. \end{cases}$$

So S^ℓ consists of all linear combinations in which exactly ℓ vectors of S are involved.

Suppose that S **spans** the $\mathbb{F}_2$–vector space $\mathbb{F}_2^r$, i.e. that

$$\bigsqcup_{\ell=0}^{r} S^\ell \supset \mathbb{F}_2^r .$$

Then there is a well–defined smallest integer t such that

$$\bigsqcup_{\ell=0}^{t} S^\ell \supset \mathbb{F}_2^r \qquad (*).$$

This integer t is called the **covering radius** t(S) of S in $\mathbb{F}_2^r$.

---

[1] The notion of multiset will be explained in the appendix.

Remarks:

a) This notion of covering radius clearly is invariant under coordinate transformations of $\mathbb{F}_2^r$. So we could – and perhaps should – have replaced $\mathbb{F}_2^r$ by an arbitrary r–dimensional $\mathbb{F}_2$–vector space V.

b) The **usual** definition of covering radius can, for instance, be found in Cohen et al. [3]. Our number t(S) equals the covering radius (in their sense) of the **linear** code $C(S)^\perp$, to be defined in section II.

The major problems with respect to the covering radius are as easy to formulate as they are hard to solve:

1) to determine, for all r and t, the value of n[r,t], the cardinality of the smallest S in $\mathbb{F}_2^r$ for which t(S) equals t,

2) to describe these S, up to linear equivalence.

Remark:

Obviously, such a minimal S does not contain the zero vector and has no multiple elements, i.e. is a proper set.

Graham and Sloane [4] have drawn up a table that gives recent information on n[r,t] for small r and t. The **open** cases start with

$$16 \le n[7,2] \le 19$$

$$23 \le n[8,2] \le 28$$

$$15 \le n[9,3] \le 19.$$

One lower bound for n[r,t] is completely obvious: since the left hand side of (*) must contain at least as many vectors as the right hand side, we have

$$\sum_{\ell=0}^{t} \binom{n}{\ell} \geq 2^r$$

(the so–called **sphere–covering bound**).

The purpose of this paper is to present a new method to detect cases in which this bound is not attained.

II. THE CODES C(S) AND $C(S)^\perp$

To any multiset S of cardinality n in $\mathbb{F}_2^r$, we can associate two linear subspaces (**linear codes**) C(S), $C(S)^\perp$ in $\mathbb{F}_2^n$ as follows:

Order the elements of S arbitrarily:

$$S = \{\underline{x}_1, \underline{x}_2, \ldots\ldots, \underline{x}_n\},$$

and take the $\underline{x}_i$ as the columns of the $r \times n$–matrix

$$M(S) := [\underline{x}_1 \; \underline{x}_2 \; \ldots\ldots \; \underline{x}_n].$$

Then we **define** C(S) and $C(S)^\perp$ to be the null space and the row space of M(S) respectively.

Remarks:

a) C(S) and $C(S)^\perp$ are orthogonal complements with respect to the standard scalar product on $\mathbb{F}_2^n$.

b) If S spans an m–dimensional linear subspace of $\mathbb{F}_2^r$, then $\dim C(S)^\perp = m$ and $\dim C(S) = n-m$.

We shall avoid the – irrelevant – ordering of S by **identifying** $\mathbb{F}_2^n$ with the power set P(S): a vector

$$(c_1, c_2, \ldots, c_n) \in \mathbb{F}_2^n$$

corresponds to the submultiset

$$\{\underline{x}_i \mid c_i = 1\} \subset S.$$

Then the vectors of $C(S)$ correspond to the submultisets of S that add up to the zero vector.

As to $C(S)^\perp$, the i–th coordinate $\underline{y}^t\underline{x}_i$ of a vector

$$\underline{y}^t M(S) \in C(S)^\perp$$

is equal to 1 if and only $\underline{x}_i$ is contained in the affine hyperplane

$$\{\underline{\xi} \mid \underline{y}^t\underline{\xi} = 1\}$$

of $\mathbb{F}_2^r$.

Hence the non–zero vectors of $C(S)^\perp$ have to be identified with the intersections $H \cap S$ of S with the affine hyperplanes $H \subset \mathbb{F}_2^r$ that do not pass through $\underline{0}$.

The **weight** $|X|$ of an element $X \in P(S)$ simply is its cardinality as a submultiset of S.

Further we define

$$A_i(S) := |\{X \in C(S) \mid |X| = i\}|$$

and

$$B_i(S) := |\{X \in C(S)^\perp \mid |X| = i\}|.$$

These numbers are connected by the well–known **MacWilliams identities**:

$$\sum_{j=0}^{n} K_i(j;n)B_j(S) = 2^m A_i(S) \quad (i=0,\dots,n),$$

with

$$m := \dim C(S)^\perp$$

and

$$K_i(j;n) := \sum_{k=0}^{i} (-1)^k \binom{j}{k}\binom{n-j}{i-k}.$$

III. THE MAIN RESULT

Now suppose that S is a spanning multiset of cardinality n in $\mathbb{F}_2^r$ such that $t(S) = t$.

Then

$$T := \left(\bigsqcup_{\ell=0}^{t} S^{\ell} \right) \setminus \mathbb{F}_2^r$$

is a multiset of

$$N := \sum_{\ell=0}^{t} \binom{n}{\ell} - 2^r$$

elements.

Suppose that T spans a p–dimensional linear subspace of $\mathbb{F}_2^r$.

Claim: the $B_i(T)$ are completely determined by the $B_i(S)$!

This follows from the

Theorem

The linear mapping

$$C(S)^{\perp} \longrightarrow C(T)^{\perp}, \; H \sqcap S \longmapsto H \sqcap T,$$

maps vectors of weight j onto vectors of weight

$$\sigma(j) := \begin{cases} \sum_{\ell=0}^{t} \sum_{m \geq 0} \binom{j}{jm+1}\binom{n-j}{\ell-2m-1} - 2^{r-1} & \text{if } j \neq 0, \\ 0 & \text{if } j = 0. \end{cases}$$

Proof

First observe that

$$H \sqcap T = H \sqcap \{ (\bigsqcup_{\ell=0}^{t} S^{\ell}) \setminus \mathbb{F}_2^r \} = \{ \bigsqcup_{\ell=0}^{t} (H \sqcap S^{\ell}) \} \setminus H.$$

Now the only cosets of the linear hyperplane H^c are H and H^c. So

$$\sum_{\underline{x}\in U} \underline{x} \in H \text{ if and only if } |H \cap U| \equiv 1(2).$$

Hence, if $|H \cap S| = j$, then

$$|H \cap S^{\ell}| = |\{U \subset S \mid |U| = \ell \wedge |H \cap U| \equiv 1(2)\}| = \sum_{m\geq 0} \binom{j}{2m+1}\binom{n-j}{\ell-2m-1},$$

whence

$$|H \cap T| = \{\sum_{\ell=0}^{t} \sum_{m\geq 0} \binom{j}{2m+1}\binom{n-j}{\ell-2m-1}\} - 2^{r-1}.$$

□

Corollaries

a) $B_j(S) = 0$ if $\sigma(j) \notin [0,N]$.

b) $B_h(T) = 0$ If $h \notin \sigma([0,n])$.

c) The linear mapping

$$C(S)^{\perp} \longrightarrow C(T)^{\perp},\ H \cap S \longmapsto H \cap T,$$

is surjective, with an (r–p)–dimensional kernel. Hence

$$\sum_{\sigma(j)=h} B_j(S) = 2^{r-p} B_h(T).$$

d) Combining this result with the MacWilliams identities for the T–codes, we find

$$\sum_{j=0}^{n} K_g(\sigma(j);N)B_j(S) = 2^r A_g(T) \quad (g=0,1,\dots,n).$$

From these corollaries, and additional considerations,we may either infer the non–existence of an S with prescribed parameters or obtain knowledge that is instrumental in its construction.

For example, our method yields

$$n[7,2] \geq 17$$

$$n[8,2] \geq 24$$

$$n[9,3] \geq 16.$$

The last result has been treated in Simonis [5]. In the next section, we shall prove the inequality $n[7,2] \geq 17$ as an illustration of our approach.

This particular result has also been obtained by Brualdi, Pless and Wilson [1]. In fact, they use a lemma that is equivalent to our corollary a) in the case $t = 2$. More recently, Calderbank and Sloane [2] derived the improvement $n[7,2] \geq 18$.

IV. AN EXAMPLE

Suppose that a 16–multiset S in $\mathbb{F}_2^7$ existed with $t(S) = 2$.

Since the sphere covering bound implies that $n[7,2] > 15$, the multiset S does not contain $\underline{0}$ and does not have any multiple points. In other words:

$$A_1(S) = A_2(S) = 0.$$

The multiset

$$T := (\{\underline{0}\} \sqcup S \sqcup S^2) \setminus \mathbb{F}_2^7$$

has

$$1 + 16 + \binom{16}{2} - 2^7 = 9$$

elements.

The function

$$\sigma(j) := \begin{cases} j + j(16-j) - 2^6 & \text{if } j \neq 0, \\ 0 & \text{if } j = 0 \end{cases}$$

takes the following values:

$$\sigma(0) = 0$$

$$\sigma(1) = \sigma(16) = -48$$

$$\sigma(2) = \sigma(15) = -34$$

$$\sigma(3) = \sigma(14) = -22$$

$$\sigma(4) = \sigma(13) = -12$$

$$\sigma(5) = \sigma(12) = -4$$

$$\sigma(6) = \sigma(11) = 2$$
$$\sigma(7) = \sigma(10) = 6$$
$$\sigma(8) = \sigma(9) = 8.$$

Corollary a) gives:

$$B_1(S) = B_2(S) = B_3(S) = B_4(S) = B_5(S) =$$
$$B_{16}(S) = B_{15}(S) = B_{14}(S) = B_{13}(S) = B_{12}(S) = 0.$$

Corollary b gives:

$$B_1(T) = B_3(T) = B_4(T) = B_5(T) = B_7(T) = B_9(T) = 0.$$

Since $\sigma(j) \neq 0$ unless $j = 0$, we infer from corollary c) that $p = 0$, i.e. that $\dim C(T)^{\perp} = 7$. Moreover we find

$$B_6(S) + B_{11}(S) = B_2(T)$$
$$B_7(S) + B_{10}(S) = B_6(T)$$
$$B_8(S) + B_9(S) = B_8(T)$$

Finally, corollary d) yields

$$K_g(0;9)B_0(S) + K(2;9)\{B_6(S) + B_{11}(S)\} +$$
$$K_g(6;9)\{B_7(S) + B_{10}(S)\} + K_g(8;9)\{B_8(S) + B_9(S)\} = 2^7 A_g(T).$$

This information, combined with the MacWilliams identities, is more than sufficient to establish the non–existence of S. For example, the four MacWilliams equations

$$\begin{aligned} 1 + \quad B_2(T) + \quad B_6(T) + \quad B_8(T) &= 2^7 \\ 9 + \quad 5B_2(T) - \quad 3B_6(T) - \quad 7B_8(T) &= 2^7 A_1(T) \\ 84 + \qquad\qquad 8B_6(T) - \quad 28B_8(T) &= 2^7 A_3(T) \\ 126 - \quad 14B_2(T) - \quad 6B_6(T) + \quad 14B_8(T) &= 2^7 A_4(T) \end{aligned}$$

do not have a solution for which all $A_i(T)$, $B_i(T)$ are nonnegative.

V. APPENDIX

In this paper, a **multiset** S in a set X is understood to be an unordered sequence of elements of X in which each element is allowed to occur more than once. Thus a subset of X is a special kind of multiset. The notation $\mu_S(x)$ for the number of occurrences of $x \in X$ in S allows us to **identify** S with the mapping

$$\mu_S: X \longrightarrow \mathbb{N}, \; x \longmapsto \mu_S(x).$$

The symbols $\sqcup$, $\sqcap$, $\subset$ and $\setminus$ are used in the following sense:

Let S and T be two multisets in X. Then

1) $S \sqcup T$ is defined by $\mu_{S \sqcup T}(x) := \mu_S(x) + \mu_T(x)$ for all $x \in X$,

2) $S \sqcap T$ is defined by $\mu_{S \sqcap T}(x) := \mu_S(x)\mu_T(x)$ for all $x \in X$,

3) $S \subset T$ if and only if $\mu_S(x) \leq \mu_T(x)$ for all $x \in X$. If this is the case, then $T \setminus S$ is defined by $\mu_{T \setminus S}(x) := \mu_T(x) - \mu_S(x)$ for all $x \in X$.

References

1. R.A. Brualdi, V.S. Pless and R.M. Wilson, "Short Codes with a Given Covering Radius", to appear in *IEEE Trans. Inform. Theory.*

2. A.R. Calderbank and N.J.A. Sloane, "Inequalities for Covering Codes", to appear in *IEEE Trans. Inform. Theory.*

3. G.D. Cohen, M.G. Karpovsky, H.F. Mattson, Jr. and J.R. Schatz, "Covering Radius – Survey and Recent Results", *IEEE Trans. Inform. Theory*, vol. IT–31, pp. 328–343, 1985.

4. R.L. Graham and N.J.A. Sloane, "On the Covering Radius of Codes", *IEEE Trans. Inform. Theory*, vol. IT–31, pp. 385–401, 1985.

5. J. Simonis, "The Minimal Covering Radius t[15,6] of a 6–Dimensional Binary Linear Code of Length 15 is Equal to 4", to appear in *IEEE Trans. Inform. Theory.*

On Goldman's Algorithm for Solving First-Order Multinomial Autonomous Systems

William Y. Sit

Department of Mathematics
The City College of New York
New York, NY 10031

Introduction

In this article, a brief exposition of a method for finding first integrals for first order multinomial autonomous systems (FOMAS) of ordinary differential equations with constant coefficients will be given. The method is a simplified as well as a redesigned version based on a paper of Goldman (1987). We shall see how it can be applied to FOMAS with parametric coefficients. The algorithm is currently being implemented by the author, using the SCRATCHPAD II computer algebra language and system at the IBM T.J. Watson Research Center.

FOMAS occur and are of interest in many disciplines and their first integrals (or trajectories of motion) are generally difficult to find. Examples of FOMAS are too numerous to list, some well-known ones are the Riccati equation, the Lotka-Volterra equations for competing populations, Selkov's model for chemical reactions, the Lorenz system of the Rayleigh-Bernard problem, and Hamiltonian systems (where the Hamiltonian is a sum of monomial terms with constant coefficients).

Let $Y = (y_1, \dots, y_n)$ be n functions depending on the variable t. A monomial in Y is a product of the form $y_1^{k_1} \dots y_n^{k_n}$, where $k_1, \dots, k_n$ are constants. If $K = (k_1, \dots, k_n)$, we shall denote the monomial in Y by Y^K, and K is called the exponent vector for the monomial. By convention, exponent vectors are column vectors, but whenever convenient, we shall write exponent vectors as row vectors. We say that Y satisfies a first-order multinomial autonomous system (FOMAS) if for each i, $1 \le i \le n$, y_i satisfies a first order differential equation of the form:

$$y_i' = f_i(Y), \qquad (1).$$

where f_i is a linear combination of monomials in Y with coefficients which may be either constants or parametric constants. For example, the Lotka-Volterra equations for three competing species considered by Schwarz & Steeb (1984), form a FOMAS:

$$x_1' = x_1(1 + ax_2 + bx_3),$$

$$x_2' = x_2(1 - ax_1 + bx_3), \qquad (2).$$

$$x_3' = x_3(1 - bx_1 - cx_2).$$

When the exponent vectors occurring in f_i are all non-negative integers, as in the example above, a FOMAS reduces to a polynomial autonomous system (FOPAS).

A computer program was developed by Schwarz (1986) to compute first integrals of FOPAS's which are themselves polynomials in $y_1, \ldots, y_n$. Schwarz's algorithm literally takes a general polynomial of a fixed degree d in n variables and substitutes it into (1). This method does not work well on a FOMAS, because in a FOMAS, the exponent vectors need not have integral components. Also, it will not find integrals with exponent vectors that involve fractional or irrational numbers.

Goldman (1987) proved a theorem which gives necessary and sufficient conditions for the existence of a multinomial first integral for FOMAS. The proof also contained the outline of an algorithm for finding such integrals. In Goldman's paper, he introduced the notion of an integral array, which is a certain matrix satisfying some 10 conditions. He gave a few hints and several examples but did not elaborate on how such an integral array can be found in general (except in the case $q = 2$). Assuming such an array is found, he can compute the integral, in most cases, by solving systems of linear equations, or at worse in certain cases, by solving a system of algebraic equations. It was not clear when algebraic conditions are necessary.

In this brief exposition, Goldman's method will be expanded to a complete algorithm with a new simplified notation. The integral arrays are replaced by addition schemes (which is equivalent to integral arrays with some conditions removed). The generation of addition schemes is a combinatorial problem unrelated, in a sense, to FOMAS. When the first integral is a polynomial, the addition scheme is trivial to compute. We shall now begin by explaining some details of this theory.

Goldman's Theory

Goldman wrote (1) in the form:

$$y_i' = y_i \sum_{k=1}^{r} C_{ki} Y^{H_k} \quad (i = 1, \dots, n), \tag{3}$$

where H_k's are all the *distinct* exponent vectors that occur in the system. This notation is natural for Lokta-Volterra systems and was exploited by Mende (Peschel & Mende 1986). Let C_k be the row vector $(c_{k1}, \dots, c_{kn})$. For the Lotka-Volterra example (2) above, we have $r = 4$, $H_1 = (0,0,0)$, $H_2 = (0,0,1)$, $H_3 = (0,1,0)$, and $H_4 = (1,0,0)$; $C_1 = (1,1,1)$, $C_2 = (b,b,0)$, $C_3 = (a,0,-c)$, and $C_4 = (0,-a,-b)$. Goldman observed that the derivative of a monomial in Y with exponent vector K is given by:

$$\frac{d(Y^K)}{dt} = \sum_{k=1}^{r} (K, C_k) Y^{K+H_k} \tag{4}$$

Here (K, C) denotes the dot product of K and C_k. Let $I = \sum_{j=1}^{q} d_j Y^{B_j}$ be a multinomial first integral with q (distinct) terms. Applying (4), we have

$$I' = \sum_{k=1}^{r} \sum_{j=1}^{q} d_j (B_j, C_k) Y^{B_j + H_k} \equiv 0 \tag{5}$$

We can now equate the sums of the coefficients of like terms to zero. For $q = 1$, it is clear that I is an integral if $B = B_1$ is orthogonal to C_k for all k. Thus, all monomial first integrals can be found by solving a system of r linear equations in n unknowns:

$$(B, C_k) = w_k \quad (k = 1, \dots, r) \tag{6}$$

where all $w_k = 0$ for the present. Let C be the r x n matrix whose rows are $C_1, \dots, C_r$, and let rank C be c. Then there will be n - c linearly independent exponent vectors, say with corresponding monomial integrals, which form a basis for the space of all vectors orthogonal to $C_1, \dots, C_r$. We shall refer to this space as $C^{\perp}$.

Most of the non-linear computations of the algorithm lies in determining that rank $C = c$. If there were no parameters in C, this step presents no problem. In general, the difficulty depends on how many of the entries in C are parametric. Clearly $C^{\perp} \neq 0$ if and only if rank $C < \min(r,n)$. Suppose rank $C = c < \min(r,n)$. Then every $(c+1) \times (c+1)$ subdeterminant of C is zero and there exists a $c \times c$ submatrix C' of C such that det $C' \neq 0$. Thus, there is an affine algebraic variety V in the space of parameters for each c where rank $C = c$. For a given matrix C, a computer program has been written in SCRATCHPAD II by the author to generate algebraic conditions for rank $C = c$, for all possible c.

Thus the regimes in the space of parameters can be identified by algebraic varieties whose defining equations can be obtained by computing determinants. This is not a complete description yet, and there are additional linear and non-linear conditions defining finer partitions of the regimes above. However, all non-linear conditions come from determinants. Thus, contrary to Schwarz's algorithm, we can obtain algebraic systems defining the parametric regimes without reference to any undetermined coefficients in a first integral.

Addition Schemes

From now on, q will be assumed to be at least 2, and we shall also assume that I is *primitive*, that is, no proper subexpression of I is a first integral. In particular, this means B_j is not in $C^{\perp}$ and $d_j \neq 0$. According to (5), we let $E_1, \dots, E_s$ be an enumeration of the set of sums $B_j + H_k$ $(1 \leq j \leq q, 1 \leq k \leq r)$ such that $E_1, \dots, E_p$ $(p \leq s)$ correspond to the ones which equal at least two such sums. Let S_i be the set of index pairs (j,k) such that $B_j + H_k = E_i$. We shall record such information by a $q \times r$ matrix T called an *addition scheme* (more precisely, a representation of an addition scheme). Here $T_{jk} = i$ if (j,k) is in S_i for $i \leq p$ and $T_{jk} = 0$ otherwise. Let M be the $q \times r$ matrix defined by $M_{jk} = (B_j, C_k)$. When both B_j's and H_k's are known, T is completely determined, and M can be computed in terms of the parameters. (When the vectors B_j are not given and C involves parameters, our algorithm will try all possible addition schemes. While the number of addition schemes grows combinatorially as q and r increase, it is not difficult to generate them. For now, we assume a particular addition scheme is given.) It is clear then that in order to find the coefficients d_j, we would only have to solve the linear system:

$$\sum d_j M_{jk} = 0 \quad (1 \leq i \leq s) \qquad (7),$$

where the summation is over all (j,k) in S_i. Since I is primitive, B_j is not in $C^{\perp}$ and $d_j \neq 0$ for any j. Hence if $T_{jk} = 0$ then $M_{jk} = 0$.

Matching FOMAS to Addition Schemes

For any addition scheme T, we have the following properties:

T1: if $T_{jk} = 0$, then $(B_j, C_k) = 0$.

T2: if $T_{jk} = T_{j'k'}$ then $B_j + H_k = B_{j'} + H_{k'}$.

Treating T2 as a linear system of equations with indeterminates B_j, we can apply row reduction to obtain two important sets of conditions. Firstly, there exists a (finite dimensional) vector space R of linear integral relations on H_k's (which may be the trivial space). Each such relation is of the form

$$a_1 H_1 + \cdots + a_r H_r = 0, \quad (a_1 + \cdots + a_r = 0) \qquad (8),$$

where a_k's are integers. More precisely, they are sums of distinct differences $H_k - H_{k'}$. For example, if $T_{jk} = T_{j'k'}$ and $T_{jh} = T_{j'h'}$ where $(k,k') \neq (h,h')$, then we have $H_k - H_{k'} = H_h - H_{h'}$. A necessary condition for a given FOMAS to have a first integral with addition scheme T is that the vectors H_k satisfy the affine conditions (8) on H_k associated with T. Secondly, it can be shown that when I is primitive, the rank of the linear system T2 must be q-1. Thus there exist affine combinations L_{jq} over the integers of the exponent vectors H_k such that

$$B_j = B_{j'} + L_{jj'} \qquad (9),$$

where $j' = q$. By defining $L_{jj'} = L_{jq} - L_{j'q}$, (9) will hold for all j, j'. By property T1, any two zeros in column k of T give an orthogonal relation of the form

$$(L_{jj'}, C_k) = 0 \qquad (10)$$

thus giving additional *linear* conditions on the parameters of C. The conditions (8), (9) and (10) are easily computed, given T alone. Since the vectors H_k are distinct, and the vectors B_j are distinct, it follows that T can be the addition scheme for a given FOMAS only if:

(C1) no relations in R has the form $H_k - H_{k'}$,

(C2) the H_k's satisfy the inequalities $L_{jj'} \neq 0$ for all $j \neq j'$,

(C3) the H_k's satisfy all relations in R, and

(C4) the H_k's and C_k's satisfy all orthogonal relations (10).

Computing the First Integrals

When this is the case, equations (9) show that it suffices to find only one of the B_j's, say B_q. We now show how $B = B_q$ can be computed. Let $w_k = M_{qk} = (B_q, C_k)$. Then $T_{qk} = 0$ implies $w_k = 0$, and for any k, if $T_{jk} = 0$ for some $j \neq q$, then $(B_j, C_k) = 0$ so that by (9), we have $w_k = (-L_{jq}, C_k)$. If for some k, $T_{jk} \neq 0$ for all j, we cannot determine w and we let this be undetermined for the moment. Let w be the *column* vector $(w_1, \ldots, w_r)$.

For each c, assume rank C = c and solve the linear system (6). For efficiency, (6) can be solved for each c and for arbitrary w, under the assumption that rank (C, w) = c, so that the solution B can be written as a linear combination of r vectors with coefficients w_i and n - c vectors in $C^{\perp}$ with arbitrary coefficients. For the specific w we have, the requirement rank(C, w) = c gives us further conditions on the undetermined w_k, if any, or on the parameters of C, if none. When this cannot be satisfied, we must discard T. Assume the contrary, we simply substitute w into the general solution to obtain a particular solution B, where the arbitrary coefficients are all chosen as zeros. More generally, because of (9), we could compute M_{jk} for $j \neq q$ in terms of the possibly undetermined w_k and substitute to obtain B_j. If w = 0 (or $M_{jk} = 0$ for all k and some j), then B (respectively, B_j) is in $C^{\perp}$ and we may discard the current addition scheme. Thus, we have another condition for a FOMAS to be compatible with T:

(C5) the row vectors of the matrix M should not belong to the null space of C.

Assuming the addition scheme survives through the previous computations, we have computed M completely, in terms of perhaps some undetermined w_k. From this, the system on the coefficients d_j (that is, (7) with $1 \leq i \leq p$) is obtained. It is easy to see, as Goldman observed, that

(C6) any q - 1 columns of the coefficient matrix for this system should be linearly independent, for I to be primitive. This may yield further conditions on w or on C. Assume this. Normalize $d_1 = 1$ and solve uniquely for d_j, $2 \leq j \leq q$.

Main Steps of Goldman's Algorithm

We summarize here the essential steps of the complete algorithm as described in this article.

Addition Scheme Phase

Step 1. For given positive integers q and r, generate all addition schemes.

Step 2. For each addition scheme T, compute the relations (8) on H_k associated with T. Discard those for which (C1) is not satisfied. Discard also those that imply the B_j's are indistinct. For those satisfying (C1), compute the coefficients of $L_{jj'}$, as defined by (9), and triplets (j,j',k) for which the orthogonal condition (10) holds.

Step 3. Organize this information in a data base.

FOMAS Phase

Step 4. For a given FOMAS, determine all possible ranks c of the coefficient matrix and associated conditions on the parameters of C so that rank C = c, and determine the conditions on w so that rank(C, w) = c. Assuming these conditions, find C and solve the linear system (6) for B, with w arbitrary otherwise. Store such information in a table defining the regimes and solutions.

Step 5. Select from the database a compatible addition scheme T (verifying conditions C2, C3). Compute M and verify (C5). Identify those w_k that are undetermined (fixing j = q). Compute the coefficient matrix E of (6) from M and T, in terms of undetermined w_k's and the parameters. Specify (C6). If this is not possible, discard T. For each compatible T, perform the steps below. If none exists, stop.

Step 6. Select a regime from Step 4, combine the linear conditions C4 on the parameters of C with those possibly non-linear ones defining the regime; if these cannot be satisfied, discard this regime. For each accepted regime, perform the steps below. If no more regime exists, return to Step 5.

Step 7. Recheck (C5) for this regime. If not satisfied, return to Step 6. Otherwise, obtain $B_1, \ldots, B_q$ using results of Steps 4, 5, and 6.

Step 8. Recheck (C6) for this regime. If not satisfied, return to Step 6. Otherwise, normalize $d_1 = 1$ and solve uniquely for $d_2, \ldots, d_q$. Output the first integral and the conditions specified. Return to Step 6.

Computer implementation

The attractiveness of addition schemes, as compared to integral arrays, is that a library of addition schemes together with the associated linear conditons on H_k, the coefficients of L_{ij}, and the orthogonal conditions (10) can be generated in advance and then systematically applied to those FOMAS satisfying them. Thus while the generation of the addition schemes may have combinatorial complexity, the generation is done only once, and the application to a given FOMAS is simply by searching a library. At the time of this writing, the author has developed an efficient backtracking algorithm to generate addition schemes and compute the conditions (8), (9) and (10) associated with each. This algorithm was implemented in Turbo Pascal, and it generated for example, all 4×3 addition schemes and all 3×4 addition schemes as well as the associated relations, on a standard IBM PC within 5 minutes. Thus Steps 1 and 2 are fairly easy to implement.

Step 4 has also been implemented by the author on IBM's SCRATCHPAD II system, and it seems reasonable to predict that by the time of this conference, Steps 5, 6, 7, and 8 would be complete. What remains to be done is a classification of addition schemes and their associated conditions so that the data base can be organized for efficient searches (Step 3). A mathematical analysis of the complexity of the backtracking algorithm will also be desirable.

An Example

As an illustration for the power of the modified Goldman algorithm, I have used it to obtain *by hand computation* three new families of binomial integrals for the Lotka-Volterra equations (2): The three time-dependent integrals below are obtained by introducing a fourth equation $x_4' = x_4$ (that is, $x_4 = e^t$). Because they involve fractional exponents, they cannot be found by Schwarz'smethod.

Note. It was discovered after submitting this paper that there was a misprint in Schwarz's paper. The Lotka-Volterra equations he worked on has $x_2' = x_2(1 - ax_1 + cx_3)$ for the second variable. For this system, no new $(q \leq 3)$ first integral was found. The system considered in this paper seems more interesting for this reason.

(1) Assume $ab(b - c) \neq 0$. Then a first integral is

$$(x_1^c x_2^{-c} x_3^a e^{-at})^{1/(b-c)} + (x_1^b x_2^{-b} x_3^a e^{-at})^{1/(b-c)}.$$

(2) Assume $ab(b - c) \neq 0$ and $a = -c$. Then a first integral is

$$(x_1^{-1} x_2 x_3 e^{-t})^{b/(a+b)} + (x_1^a x_2^b x_3^{-a} e^{-bt})^{1/(a+b)}.$$

(3) Assume $ab(b - c) \neq 0$, $a = b$, $c \neq 0$. Then a first integral is

$$b(x_1 x_2^{-1} x_3 e^{-t})^{-c/(b-c)} + c(x_1^{-c} x_2^b x_3^{-b} e^{ct})^{1/(b-c)}.$$

Though I believe that these, together with (4) below, are all (up to a scalar multiple) the primitive binomial first integrals for the Lotka-Volterra equations (2), a more definitive statement can only be made after an exhaustive computer search when the full algorithm is implemented.

(4) Assume $b = 0$, $a \neq 0$. Then $(x_1 + x_2)e^{-t}$ is a first integral.

Bibliography

Goldman, L. (1987). Integrals of multinomial systems of ordinary differential equations. *J. of Pure and Applied Algebra* **45**, 225-240.

Schwarz, F., Steeb, W. H. (1984). Symmetries and first integrals for dissipative systems. *J. Phys. A: Math. Gen.* **17**, L819-L823.

Peschel, M., Mende, W. (1986). *The predator-prey model: do we live in a Volterra world?* New York: Springer.

Schwarz, F. (1986) A REDUCE package for determinining first integrals of autonomous systems of ordinary differential equations. *Computer Physics Communications* **39**, 285-296.

Acknowledgement. The author would like to thank Dr. Richard Jenks, for allowing him to use the SCRATCHPAD II facility at IBM T.J. Watson Research Center, and to members of his Computer Algebra Group for their generous assistance in this project.

A Covering Problem in the Odd Graphs.

Patrick Solé,

School of Computer and Information Science ,

Syracuse University, Syracuse , NY 13244-1240

Arif Ghafoor

Dept. of Electrical and Computer Engr.,

Syracuse University, Syracuse, NY 13244-1240

Abstract

The following problem originated in the design of interconnection networks: what is the graphical covering radius of an Hadamard code of length $2k-1$ and size $2k-1$ in the Odd graph O_k? Of particular interest is the case of $k = 2^{m-1}$, where we can choose this Hadamard code to be a subcode of the punctured first order Reed-Muller code $RM(1,m)$. We define the w-covering radius of a binary code as the largest Hamming distance from a binary word of Hamming weight w to the code. The above problem amounts to finding the k-covering radius of a $(2k, 4k)$ Hadamard code. We find upper and lower bounds on this integer, and determine it for small values of k.

Key words: Interconnection Networks, Coding Theory, Covering Radius, Odd Graphs, Hadamard Matrices, First order Reed Muller Code, Coset weight distribution.

1 Introduction

An interconnection network can be modeled as a finite, simple, undirected graph, Γ, where vertices represent nodes of the network, and the edges the physical links between the nodes. Suppose we want to expand the network Γ by interconnecting several of its copies. There are two extreme ways of interconnecting the copies:

- connect every node in a copy to all its images belonging to the rest of the copies.
- connect one node in each copy to all its images.

The first solution is costly since it requires many links, while the second is very fragile: if one particular node in one particular copy fails then the whole network is disconnected. The most economical solution is to use a semi-distributed approach by selecting a set of *centers* in Γ which can cover each copy of Γ by spheres for the graphical distance, and interconnect these centers of the spheres in some desired fashion. The centers in a given copy of Γ can act as *gateway nodes*, to handle message traffic going in and out of that copy of the network. In any case the parameter of interest in the choice of the centers is the radius of the spheres, as it affects the overall diameter of the expanded network. We are thus led to a *covering problem* in the graph Γ. From the preceding discussion, it can be noted that the number of centers cannot be less than the connectivity of Γ, without decreasing the fault-tolerance capability of the network.

2 The Odd graphs

In this paper we take Γ to be the *Odd graph* O_k [6]. An *Odd graph* is constructed using binary vectors with constant *Hamming weight*. A weight of a binary vector x of F_2^n is the number of its nonzero coordinates and is denoted by $w(x)$. The *Hamming distance* of two vectors x, y is the weight of their difference and is denoted by $d(x, y)$. The *Odd graph* O_k has for vertex set the binary words of length $2k-1$ and Hamming weight $k-1$. Two vertices are connected if and only if they have disjoint support or, equivalently, if they are at Hamming distance $2k-2$. The graph O_3 is the celebrated Petersen graph [3]. Odd graphs are selected due to their higher density than various other interconnection networks: they have degree k , diameter $k-1$, and $\binom{2k-1}{k-1}$ nodes [3]. Specifically, the following holds:

Theorem 1 : *The degree and the diameter of O_k are of asymptotic order $log\sqrt{N}$, where $N = \binom{2k-1}{k-1}$ is the number of nodes in O_k, and log is binary.*

<u>Proof:</u> We provide the proof by using the following classical Stirling approximation [8] to the binomial coefficient $\binom{2k-1}{k-1}$, which counts the number of nodes, N, as a function of the degree k:

$$N = \binom{X}{\lambda X} \approx 2^{X\theta(\lambda)}/\sqrt{2\pi\lambda(1-\lambda)} \tag{1}$$

where θ is the binary entropy function whose asymptotic expansion of the inverse (for x near zero) can be given as [8]:

$$\theta^{-1}(1-x) = 1/2 + 0.589\sqrt{x} + \ldots \tag{2}$$

(We quote this result for future use; for now, we only need the value $\theta(1/2) = 1$). Putting $X = 2k-1$, $x = 0$, and $\lambda = 1/2$, we get the desired result for the degree k, and the diameter $k-1$. Q.E.D.

It is clear from Theorem 1 that O_k is denser than many known interconnection networks such as mesh, star, ring, etc [10]. It is denser than the binary Hypercube graph (or its generalization

[11]) for which the diameter is proportional to the logarithm of the total number of nodes in the network. For the same reason, it is superior to the Chordal ring [12], for which the diameter is proportional to the square root of the total number of nodes. A high density is a desirable property for interconnection networks, because it ensures that not only communication delays in the network are smaller but also a large number of network units can be tightly packed together for microminiaturization [12][10].

The next theorem provides an important relationship between the Hamming distance and the *graphical distance* of two vertices in O_k [13]. We recall that the graphical distance between two vertices of a finite connected graph is the length of a shortest path between these two vertices. This simple but crucial result is used in Sections 5 and 6, where the problem of upperbounding the covering radius of the gateway nodes in O_k is solved by means of coding theory.

Theorem 2 : *The graphical distance $d_O(x,y)$ and the Hamming distance $d_H(x,y)$ between any two vertices x and y in O_k are related by:*

$$d_O(x,y) = min(d_H(x,y), d_H(x, \mathbf{1}+y)) \tag{3}$$

where $\mathbf{1}$ is the all-one vector and $+$ is the addition law in Z_2^{2k-1}.

Proof: When the graphical distance takes the values:

$$d_O(x,y) = 1,2,3,4,\ldots,k-1 \tag{4}$$

then the Hamming distance takes the values:

$$d_H(x,y) = 2k-2, 2, 2k-4, 4, 2k-6, 6, .., J \tag{5}$$

with

$$d_H(x, \mathbf{1}+y) = 1, 2k-3, 3, 2k-5, 5, ..., (2k-1-J) \tag{6}$$

where

$$J = \begin{cases} k, & \text{if } k \text{ is even} \\ k+1, & \text{if } k \text{ is odd} \end{cases}$$

This is straightforward to check by induction, and also well-known ([1] p.239, [13]). From equations (4), (5) and (6) we note that for the even graphical distances $d_O(x,y) = d_H(x,y)$, otherwise $d_O(x,y) = d_H(x, \mathbf{1}+y)$. Furthermore, it is also obvious that for even graphical distances $d_H(x,y) < d_H(x, \mathbf{1}+y)$, while the reverse is true for odd distances. Therefore, $d_O(x,y) = min(d_H(x,y), d_H(x, \mathbf{1}+y))$. Q.E.D.

3 The Graphical Covering Radius of a Hadamard Code in O_k

An Hadamard matrix of order n is a $n \times n$ real matrix with ± 1 entries such that its rows are pairwise orthogonal for the usual Euclidean scalar product. Hadamard matrices of order n, $n \geq 3$, can exist only if n is a multiple of 4, and the converse is widely believed [8]. If the first row and the first column are the all-one vector, the matrix is said to be *normalized.*

Let us suppose that k is even, and that a normalized Hadamard matrix M ([8] p.44) of order $2k$ exists. Removing the first row and the first column of $-M$, we are left with a set H of $k-1$ rows of Hamming weight $k-1$ (in binary notation, mapping 1 to 0, and -1 to 1), that can be identified with nodes of the graph. The nodes associated with H can be shown to be at graphical distance $k-1$ from each other [13] and can be effectively used to act as gateway nodes for the system expansion. We define the *graphical distance of a node to a code* C as the smallest graphical distance of an element of C to this node. We can then define the *graphical covering radius (r)* of a code C as the largest graphical distance of a node to C. In this paper we find upper and lower bounds on r in the case of $C = H$, and accordingly describe an efficient system expansion scheme.

In the following *log* is binary, and *Log* is decimal. All codes are binary but not necessarily linear. An (n, K, w) code is a code of length n, minimum weight w, and K codewords.

4 Lower bounds

Let w_i count the number of nodes at graphical distance i from a given node in O_k. Using the classical sphere-covering argument in O_k, we have the bound:

$$\sum_{i=0}^{r} w_i \geq \binom{2k-1}{k-1}/(2k-1) \tag{7}$$

where the w_i are given by:

$$v_0, v_{k-1}, v_1, v_{k-2}, v_2, \ldots \tag{8}$$

with the $v_i = \binom{k}{i}\binom{k-1}{i}$ (see, for example, the equation on p.219 of [1], and problem 10 of chapter 21 of [8]). This bound is most useful for small values of r (Cf. section 8). We now derive an asymptotic lower bound for large values of r.

Letting $r' = \lfloor r/2 \rfloor$, we note that

$$\sum_{i=0}^{r} w_i = \sum_{i=0}^{r'} v_i + \sum_{i=k-1-r'}^{k-1} v_i \tag{9}$$

Bounding a sum of products by a product of sums, we obtain:

$$(\sum_{i=0}^{r'} v_i) \leq (\sum_{i=0}^{r'} \binom{k-1}{i})(\sum_{i=0}^{r'} \binom{k}{i}). \tag{10}$$

The same approach can be used to evaluate the second sum in the RHS of equation 9. Using classical estimates like

$$\sum_{i=0}^{r'} \binom{k}{i} \leq 2^{(k)\theta(r'/(k))} \tag{11}$$

for sums of binomial coefficients ([8] p.310), the covering bound (equation 7) reduces to

$$\binom{2k-1}{k-1} \leq (2k-1)2^{2k\theta(r/2k)+1}. \tag{12}$$

Using the estimate:

$$log(\binom{2k-1}{k-1}) = 2k - 1 + O(log(k)), \tag{13}$$

(an immediate consequence of equation 1), and taking logs of both sides of equation 12, we get, after rearranging, an asymptotic lower bound on r:

$$r \geq 2k\theta^{-1}(1 + O(log(k)/k)), \tag{14}$$

where θ is the binary entropy function. For large k , we can use the asymptotic expansion of equation 2, and we obtain:

$$r \geq k - c\sqrt{klog(k)} + O(log(k)), \tag{15}$$

where c is an universal constant.

Bounds like equation 10 may seem quite crude, but they are sufficient for our asymptotic purposes.

5 Equivalent formulations

We define the *w-covering radius* of a binary code as the farthest possible Hamming distance of a binary word of weight w to the code. This is at most equal to the usual covering radius ([8] p.172). Then we have the characterization:

Theorem 3 : *The graphical covering radius of* C *in* O_k *is equal to the* $k-1$*-covering radius of the code* $D = C \bigcup \mathbf{1} + C$ *of length* $2k-1$.

Proof: Proof directly follows from Theorem 2. Q.E.D.

Let us suppose that D is linear. Let $A_j(x)$ denotes the number of words of weight j in the coset $x + D$ of D. The *weight of the coset* $x + D$ is the smallest j such that $A_j(x) \neq 0$. We are in a position to state the following obvious but useful Lemma:

Lemma 1 : *Let* T *be the coset of the largest weight among all the cosets* $x + D$ *with the property* $A_{k-1}(x) \neq 0$. *Then the* $k-1$*-covering radius of* D *is the weight of* T.

This simple property allows us to use known facts on coset weight distribution of binary codes ([9],[2]). The numerical values for r are given in the table in Section 8.

6 Upper bounds

The problem of finding an upper bound amounts to majorizing the $(k-1)$-covering radius of an Hadamard code of length $2k-1$ and size $4k$. Since all its codewords have weight k or $k-1$, which is more than the usual covering radius, the presence or absence of the zero or all-one vector are clearly immaterial. When k is a power of 2, we assume that tnis code is the punctured first order Reed-Muller. We recall that the *strength* of a code is the largest integer t such that every binary t-tuple appears the same number of times amongst any t-subset of its coordinates. We refer the interested reader to p.139 of [8], or to the fundamental reference [5].

Theorem 4 : *The* $k-1$*-covering radius of a* $(2k-1, 4k, k-1)$ *Hadamard code* H *is at most* $2\delta+1$, *where* δ *is the largest minimum weight of a self-complementary (k-1,4k) binary code of strength* 2.

Proof: The proof uses a suitable generalization of the concept of *leader* code [9]. We define the leader code associated to a binary code C and a binary vector y as the restriction of C to the non-zero coordinates of y. We denote it by C_y. Clearly the strength of C_y is at least the strength of C. For more information on this important concept, we refer to p.139 of [8], or to [5].

Let y be any vector of weight $k-1$. H is of strength 2, since the kernel of an Hadamard matrix of size $2k$ is a $2-(2k-1, k-1, k/2-1)$ design [15]. Moreover H_y is a self-complementary $(k-1, 4k, w)$ code. Let us show that the distance of y to H, $d(y, H)$ is at most $2w+1$.

By permutation of the coordinate places, we can write any codeword of H as (l/r) (where / stands for juxtaposition) with l in H_y, and r a binary vector of length k . Its weight is at most $k - w(l)$, since the codewords of H have weight k or $k-1$. The distance of y to this codeword is

$(k-1-w(l))+w(r)$, which is at most $2(k-1-w(l))+1$. But H_y is self-complementary, and we come up with a bound of $2w+1$.Q.E.D.

Now we can use a second moment argument analogous to that of [7] to prove:

Theorem 5 : *The $k-1$-covering radius of a $(2k-1, 4k)$ Hadamard code is at most $k-\sqrt{k-1}$.*
Proof: By [7] the covering radius of a self-complementary binary code of strength 2 and length n is at most $(n-\sqrt{n})/2$, hence a fortiori its minimum weight, which is the distance of the origin to the code. Using the previous theorem, the result follows.Q.E.D.

7 The merit factor of Hadamard codes

Despite the fact that their covering radius is asymptotically equivalent to the diameter of the network, the Hadamard codes are of practical interest. Indeed, if we define the *merit factor* F of a covering code as the ratio of the volume of the whole space to the volume of a sphere of radius the covering radius of the code, we have, asymptotically, by Stirling's approximation (equation 1) :

$$F \geq c'\sqrt{k},$$

where c' is an absolute constant. This means that we cut the number of links necessary for expanding the system, as compared to a a complete node to node interconnection of each copy of O_k by a factor of order $\sqrt{k}$ (see Section 9). Note that, since the connectivity of O_k is k, we cannot expect to decrease the number of centers by more than 2.

8 Numerical values

The few known values of r show that the true value of r is closer to the lower than to the upper bound. The lower bounds are obtained by computing numerically both sides of equation 7 (the covering bound).

k	r	Comment
4	1	a perfect code in O_4.[6]
6	3	from theorems 5 and equation 7.
8	4	from lemma 1 and [9].
10	5-7	from theorems 5 and equation 7
12	6-8	from theorems 5 and equation 7
16	10	from lemma 1 and [2].

9 An Efficient Expansion Scheme

For interconnection network an important design consideration is a provision for its future expansion, which does not require any change in the original topology of the network. In this section we present a simple but highly flexible expansion mechanism for networks using O_k graphs. In this technique we connect a given number of copies of O_k in form of a ring structure. Given Q copies, labeled from 0 through $Q-1$, let Γ_q be the set of centers, generated from the Hadamard code C, for the qth copy. A center, say $x_{iq} \in \Gamma_q$ is connected to the centers $x_{i(q+1)}$ of $\Gamma_{(q+1) mod\ Q}$ and $x_{i(q-1)} \in \Gamma_{(q-1) mod\ Q}$. Fig. 1 illustrate this ring structure. There are three major advantages to this approach. First, the scheme is fault-tolerant, since there are $| C |$ parallel paths between each pair of interconnected copies. Second, as mentioned in section 7, the total number of centers given as $| \Gamma |$ are considerably small as compared to the size of an O_k network. Therefore, the additional requirements imposed on these centers for the purpose of expansion are quite small.

Third, a practical advantage of this scheme is its flexibility to interconnect an arbitrary number of copies of the network. Each time a new copy needs to be added, we merely need to open a connection between two adjacent copies and connect the new copy in between them. A preferable place in the network for the new copy may be between copy 0 and $Q-1$. This will avoid the problem of relabeling the existing copies.

Lastly, it is worth mentioning that the overall diameter of a network is an important invariant since it greatly effects the communication delay in the network. For the proposed expansion scheme it can be noted that the overall diameter of the expanded system is $\lceil \frac{Q}{2} \rceil + r$, and directly depends on the graphical covering radius r. For Q much larger than $2r$, more intricated schemes need to be devised, and are under study.

10 The unicity problem

So far, we have not been concerned with the fact that for a given length there may exist several nonequivalent Hadamard matrices, hence several possible covering radii in O_k. Since codes with the same weight distribution may have a different coset weight distribution [2], this phenomenon could be expected. For small values of k, however, this cannot happen, since then, the general bounds we have developed for *any* Hadamard code are sufficient. $k = 8$ is the first value for which such a phenomenon may happen. It is known that there are 5 non-equivalent Hadamard matrices of order 16 ([8] p.48).

11 Conclusion

The problem we considered in this paper is to find the graphical covering radius of the Hadamard Code in the graph O_k . This determination is essential as we propose the use of a Hadamard code set C for expanding interconnection networks based on the graphs O_k.

The overall diameter of the proposed expansion scheme is shown to be dependent on the covering radius of C. The approach is expected to be applicable to a large class of distance-transitive graphs. The covering radius problem is shown to be related to the coset weight enumeration of the first order Reed-Muller and more generally, of Hadamard codes. Even if the bounds we obtained are susceptible of improvement, they are sufficient to ensure a good covering of the graph O_k with a high merit factor.

Acknowledgements

Discussion with Prof. H.F. Mattson Jr. led us to use the concept of leader code (see [4]). Prof. J.H. van Lint was kind enough to present this paper at the conference, and helped us to improve its readability. We are grateful to the anonymous referees for pointing out several obscure points in the proofs.

References

[1] E. Bannai, T. Ito. *Algebraic Combinatorics I: Association Schemes.* Benjamin-Cummings (1984).

[2] E.R. Berlekamp, L.R. Welch. "Weights distributions of the cosets of the (32,6) Reed-Muller code." IEEE Trans. on Inf. Theory, IT- 18,1,Jan.(1972).

[3] N.L. Biggs, "Some Odd Graph Theory", in Second International Conference on Combinatorial Mathematics, Annals of the New York Academy of Science Vol. 319. 1979. pp. 71-85.

[4] G. Cohen, M.G. Karpovsky, H.F. Mattson, J. R. Schatz, "Covering Radius-Survey and Recent Results", IEEE Trans. on Inf. Theory, IT-31,3, (1985), p.328.

[5] P. Delsarte, "Four Fundamental Parameters of a Code", Inf. and Contr., 23, pp. 407-438, (1973).

[6] P.Hammond, D.H. Smith, "Perfect codes in the graphs O_k." Journ. of Comb. Theory, Ser. B, 19 (1975), p. 239-255.

[7] T. Helleseth,T. Klove, J. Mykkelveit. "On the covering radius of binary codes."IEEE Trans. on Inf. Theory, IT,24,5,Sept.(1978).

[8] F.J. MacWilliams, N.J.A. Sloane *The theory of error correcting codes.* North Holland (1981).

[9] N.J.A. Sloane, R.J. Dick. "On the enumeration of cosets of first order Reed-Muller codes." IEEE Int. Conf. on Comm., Montreal June 1971, vol. 7.

[10] L.D. Wittie, "Communication Structures for large networks of microcomputers", IEEE Trans. on Compu., Vol. C-30(4), April 1981. pp. 264-273.

[11] J.R. Armstrong and F.G. Gray, "Fault diagnosis in a Boolean n-cube array of microprocessors," IEEE Trans. on Compu., Vol. C-30, pp. 581-590, Aug. 1981.

[12] B.W. Arden and H. Lee, "Analysis of Chordal Ring Network", IEEE Trans. on Compu. Vol. C-30(4), April 1981. pp. 291-295.

[13] A. Ghafoor, "Some classes of fault-tolerant communication architecture for distributed systems", Tech. Rep. TR-85-4, Dept. of Elect. and Comp. Engr. Syracuse University, 1985.

[14] A. Ghafoor, T.R. Bashkow, and I. Ghafoor, "Fault-tolerance and Diagnosability of a class of Binomial interconnection networks", Proc. of AFIPS, vol. 55, NCC, 1986.

[15] T. Beth, D. Lenz, M. Jungnickel, *Design Theory*, BI Institute, Mannheim, (1985).

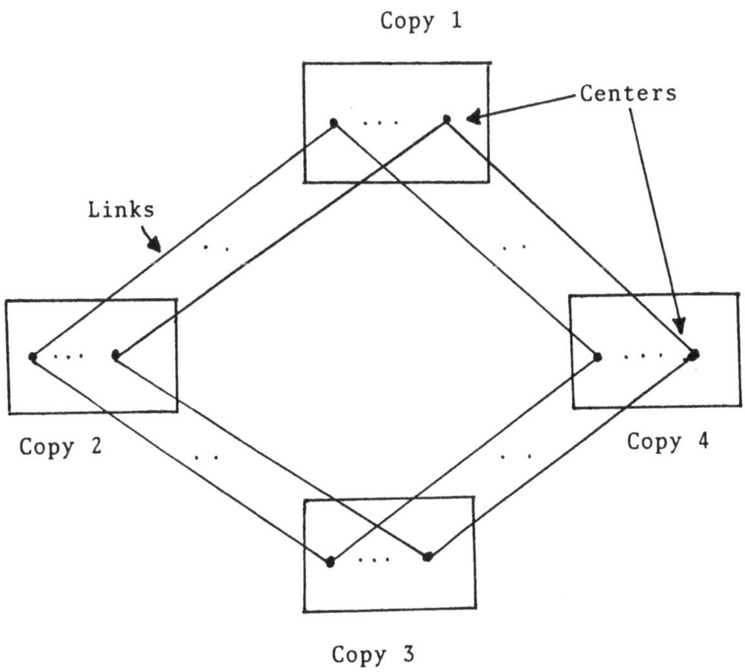

Fig. 1. Ring structure for the expanded system.

ON THE COMPUTATION OF HILBERT-SAMUEL SERIES AND MULTIPLICITY

Walter Spangher

Dipartimento di Scienze Matematiche
Università di Trieste
I-34100-Trieste (Italy)

INTRODUCTION In this note we give an algorithm for the computation of Hilbert-Samuel series $H_{A,\mathfrak{q}}$ where $(A,\mathfrak{m})$ is a local ring in the form $k[\xi]_{\mathfrak{p}}$ ($k[\xi]$ denotes a residue class ring of a polynomial ring $k[X]=k[X_1,\ldots,X_r]$ where k is a field and $\mathfrak{p}$ is a prime ideal of $k[\xi]$) and $\mathfrak{q}$ is an $\mathfrak{m}$-primary ideal of A. The algorithm permits to compute the *multiplicity* $e(\mathfrak{q},A)$ of $\mathfrak{q}$ (see (2.6) and (3.3)) too.

When $\mathfrak{q}$ is a parameter ideal, we give two other methods for the computation of $e(\mathfrak{q},A)$ (see (3.4) and (3.5)): the former makes use of the primary decomposition and the Boda -Vogel method [2]. The latter employs some results of Northcott [9] and it enables to compute the multiplicity without the primary decomposition (3.5).

In this way, we get two methods for the computation of the intersection multiplicity along a proper component of two varieties in an affine space.

In § 1, we introduce the basic notions and the notations relative to Gröbner bases and characterize the index of regularity of a Poincaré series, which will be used extensively in the paper. In § 2, we compute the *Poincaré series* $P(A[X]/I,t)$ where A is an Artinian local ring and I is a homogeneous ideal of the polynomial ring $A[X]$.

Finally, in § 3 we consider the multiplicity of a parameter ideal. In the sequel all the rings considered will be commutative and Noetherian.

§ 1 - PRELIMINARY DEFINITIONS, CONVENTIONS AND NOTATIONS

(1.1) Let A be an Artinian ring, and let B a polynomial ring $A[X_1,\ldots,X_r]$ with the $\mathbb{Z}$-graduation by defining $\deg(X_i)=1$ $(i=1,\ldots,r)$ and $\deg(a)=0$ for $a\in A$.

If $M=\bigoplus_{n\in\mathbb{Z}} M_n$ is a finitely generated graded B-module, we define the *Hilbert function of* M, by: $H(M,n)=l_A(M_n)$ $(n\in\mathbb{Z})$ ($l_A(\ )$ denotes the length of an A-module) and *the Poincaré series* P(M,t) of M by the formula: $P(M,t)=\sum_{n\in\mathbb{Z}} H(M,n)t^n$.

(1.2) We shall review further basic facts (see [3] , § 4):

(i) P(M,t) is a rational function of t, and can be written: $P(M,t)= Q(t)/t^p(1-t)^r$ where $p\in\mathbb{N}$ and $Q(t)\in\mathbb{Z}[t]$.

(ii) there is a polynomial $h\in\mathbb{Q}[T]$ such that $H(M,n)=h(n)$ for $n \gg 0$.

(1.3) We define *the index of regularity* a(M) of M, by the following property: $a(M)\in\mathbb{Z}$, $H(M,n)=h(n)$ if $n>a(M)$, and $H(M,a(M))\neq h(a(M))$.

By a direct computation,we can prove that a(M) is the degree of the rational function P(M,t).

If $t^{-p}Q(t)=\sum a_i t^i$ and $P(M,t)=\sum \alpha_n t^n$,we have (where $\begin{bmatrix} n\\ p\end{bmatrix}=0$ if $p<0$ or $p>n$,and $\begin{bmatrix} n\\ p\end{bmatrix}=\binom{n}{p}$ if $0\leq p\leq n$) : $\alpha_n=\sum_{i\in\mathbb{Z}} a_i \begin{bmatrix} n-i+r-1\\ r-1\end{bmatrix}=\sum_{i\leq n} a_i\binom{n-i+r-1}{r-1}$.

Let $n_1=\sup\{i\ ;i\in\mathbb{Z}, a_i\neq 0\}$ (i.e. $n_1=\deg(Q(t))-p$) ; we can show that $\alpha_n=h(n)$ for $n>n_1-r$ where h(T) is the following polynomial of $\mathbb{Q}[T]$:

$$h(T)=\frac{1}{(r-1)!}\cdot\sum_{n\in\mathbb{Z}} a_i \prod_{j=1}^{r-1}(T-i+j)$$

(by the computation of the difference between the polynomial function corresponding to the binomial coefficient $\binom{n}{p}$ and the function $\begin{bmatrix} n\\ p\end{bmatrix}$).

(1.4) If Q(t) has (1-t) as a factor, we can cancel to get P(M,t) in the form: $P(M,t)=R(t)/(1-t)^d$ with $R(t)\in\mathbb{Z}[t,t^{-1}]$,$R(1)>0$ and $d\geq 0$. If A is an Artinian local ring, then $d=\dim_B M$ (see [3], § 6,Proposition 5); the positif integer R(1) is called the degree of M.

(1.5) In particular, we can consider P(I,t) and P(B/I,t) where I is an homogeneous ideal of B ; since I and B/I are $\mathbb{N}$-graded B-modules, then we have p=0 in (1.2 (i)) and $R(t)\in\mathbb{Z}[t]$ in (1.4).

(1.6) *Hilbert-Samuel series.*

Let A be a Noetherian semilocal ring, q an ideal of A such that A/q is an Artinian ring,and let M be a finitely generated A-module. Then:

-- $q^nM/q^{n+1}M$ and $M/q^{n+1}M$ are A-modules of finite length,for every $n\in\mathbb{N}$.

-- $Gr(q,M)=\bigoplus_{n\geq 0} q^nM/q^{n+1}M$ is a finitely generated G(q,A)-module.

-- We have a surjective homomorphism $A/q[X_1,\dots,X_r]\longrightarrow Gr(q,A)$ where $X_i\longrightarrow\xi_i$ and $\xi_1,\dots\dots,\xi_r$ is a generating system of q/q^2 on A/q and then Gr(q,M) is a finitely generated graded $A/q[X_1,\dots,X_r]$ -module.

-- The series $H_{M,q}=\sum_{n\geq 0} l_A(q^nM/q^{n+1}M)t^n$ is called *the Hilbert-Samuel series of the A -module* M *,w.r.t. the ideal* q .

-- The series $H_{M,q}$ coincides with the Poincaré series $P(Gr(q,M),t)$ where $Gr(q,M)$ can be viewed as a graded $A/q[X_1,\dots,X_r]$ -module.

— We can consider the following series: $H^{(1)}_{M,q} = (1-t)^{-1} H_{M,q} = \sum_{n\geq 0} l_A(M/q^{n+1}M)\, t^n$.

-- Let A be a d-dimensional local ring; then we can write $H_{M,q}$ in the following forms

$$H_{M,q} = \frac{R(t)}{(1-t)^{\dim_A M}} \qquad \text{where } R(1)>0 \text{ and } R(t) \in \mathbb{Z}[t,t^{-1}]$$

$$H_{M,q} = \frac{R'(t)}{(1-t)^{d}} \qquad \text{where } d=\dim A,\ R'(t) \in \mathbb{Z}[t,t^{-1}]$$

The integer $e(q,M)=R'(1)$ is called *the multiplicity of* M *w.r.t.* q . By definition, we have the following property: $e(q,M) \geq 0$ and $e(q,M)=0$ iff $\dim_A M < \dim A$.

REMARK : The multiplicity as above defined is equivalent to the multiplicity as defined in [7](p.107) , but it is different from the multiplicity as defined in [3](p.72).

(1.7) *Gröbner bases.*

Let A be a Noetherian ring, $B=A[X_1,\dots,X_r]$ a polynomial ring, and $>$ a term-ordering on B (i.e. a total order on $\mathbb{N}^r$ compatible with the monoid structure, or on the monoid T(B) of monic monomials of B). If $f\in B$, is written as $f=\sum a_i m_i$ with $a_i \in A-\{0\}$, $m_i \in T(B)$ and $m_1 > m_2 > \dots\dots$, then we set $M(f)=a_1m_1$; if I is an ideal of B , we denote by M(I) the (monomial) ideal of B generated by $\{ M(f) ; f\in I \}$. If $f_1,\dots,f_s \in I$ are such that $M(I)=(M(f_1),\dots,M(f_s))$, then $I=(f_1,\dots,f_s)$ and $\{f_1,\dots,f_s\}$ is said to be a Gröbner basis of I.

PROPOSITION (1.8) (Buchberger,[4]) *With the above notation* (1.7), *let* A *be a field and let* $\{ f_1,\dots,f_s \}$ *be a Gröbner basis of* I *where* $M(f_i)$ *are monics* $(i=1,\dots,s)$. *Let* N *be the ideal of* T(B) *generated by* $M(f_1),\dots\dots,M(f_s)$ *, then the image of* T(B)-N *is an* A *-basis of the* A*-vector space* B/I *and, in particular, the two vector spaces* B/I *and* B/M(I) *have the same dimension over* A.

§ 2. COMPUTATION OF THE POINCARE SERIES.

Let k be a field, and let A be an Artinian local ring in the form: $A=k[Y]_p/q_p$ where $k[Y]=k[Y_1,\dots,Y_m]$, $p \in \mathrm{Spec}(k[Y])$ and q is a p-primary ideal.

Let I be an homogeneous ideal of $B=A[X_1,\dots,X_r]$; the main result of this section is the theorem (2.6) where we give an algorithm for the computation of the Poincaré series P(I,t) and P(B/I,t). We start with some general facts.

THEOREM (2.1) : *Let* A *be an Artinian local ring,* I *an homogeneous ideal of* $B=A[X]=A[X_1,\dots,X_r]$ *and* $<$ *a term-ordering on* B . *Then,* $l_A(I_n)=l_A(M(I)_n)$ *for every* $n\in\mathbb{N}$.

Proof : We set $T_n=\{j=(j_1,\ldots,j_r) \mid \sum_{i=1}^{r} j_i=n\ ,\ j_i\in\mathbb{N}\}$,we know that the term-ordering $<$ give a total order on T_n.

Since $M(I)_n=\bigoplus_{j\in T_n} M(I)_j$ and $M(I)_j$ is just isomorphic to an ideal $\mathfrak{a}_j$ of A,we can consider, for every $M(I)_j$, a Jordan-Hölder series:

$$(0)\subset(a_{1j})\subset(a_{1j},a_{2j})\subset\ \ldots\ldots\ldots\ldots\ \subset(a_{1j},\ldots\ldots,a_{n_j j})=\mathfrak{a}_j$$

and so we can write a particular Jordan-Hölder series of $M(I)_n$:

$$(0)\subset(a_{1i})X^i\subset\ \ldots\ldots\subset\mathfrak{a}_iX^i=M(I)_i\subset M(I)_i\oplus(a_{1k})X^k\subset\ \ldots\ldots\ldots\ \subset M(I)_i\oplus M(I)_k\subset\ldots\ldots$$

where $X^i=X_1^{i_1}\cdots X_r^{i_r}$, $i=(i_1,\ldots,i_r)$ is the first element of T_n, and $k=(k_1,\ldots,k_r)$ is the next element of i in T_n. Now we consider the following composition series of I_n:

$$(*)\quad (0)\subset(f_{1i})\subset(f_{1i},f_{2i})\subset\ \ldots\ldots\subset(f_{1i},\ldots,f_{n_i i})\subset(f_{1i},\ldots,f_{n_i i},f_{1k})\subset\ldots\ldots$$

where $f_{ij}\in I_n$ and $M(f_{ij})=a_{ij}X^j$ $(1\leq i\leq n_j)$.

By theorem 28 and Corollary 2,Ch.IV, § 14 of [12],the series (*)is a Jordan-Hölder series.

COROLLARY (2.2) : *Let* I *be an homogeneous ideal of* $B=A[X]$ *where* A *is an Artinian local ring. Then,for every term-ordering on* B : $P(I,t)=P(M(I),t)$ *and* $P(B/I,t)=$ $=P(B/M(I),t)$.

(2.3) *Computation of the index of regularity.*

Theorem (2.1) allows us to reduce the study of P(I,t) to the case that I is a monomial ideal; in particular we have: a(I)=a(M(I)) for every ideal I.

Let $c_{j1}X^j,\ldots\ldots\ldots,c_{jn_j}X^j,c_{k1}X^k,\ldots\ldots\ldots,c_{kn_k}X^k,\ldots\ldots\ldots$ $(c_{..}\in A)$ be a finite generating system of monomials of M(I); we set $S=\{j,k,..\}\subseteq\mathbb{N}^r$, and define this to be the "place-set" of this basis. Let $\mathfrak{a}_i$ be the ideal of A corresponding to $M(I)_i$ for $i=(i_1,\ldots,i_r)\in\mathbb{N}^r$;obviously we have: $\mathfrak{a}_i=\sum_{h\in S;h\mid i}\mathfrak{a}_h$ where $h\mid i$ denotes: $X_1^{h_1}\cdots X_r^{h_r}\mid X_1^{i_1}\cdots\cdots X_r^{i_r}$.

We obtain : $l_A(M(I)_n)=\sum_{i\in T_n} l_A\left(\sum_{h\in S;h\mid i}\mathfrak{a}_h\right)$, and the following upper bound for a(I) : $a(I)\leq\deg(\text{l.c.m.}\{j,k..\}=S)-r=\omega$.

Now,the Poincaré series $P(I,t)=Q(t)(1-t)^{-r}$ in the above situtation,is determined in one and only one way by: $Q(t)=\sum_{0\leq i\leq r+\omega} l_A(M(I)_i)t^i(1-t)^r$.

Therefore,for the computation of the Poincaré series P(I,t), it is enough to determine a monomial basis of M(I),the corresponding ideals $\mathfrak{a}_h$ $(h\in S)$ and $l_A(\mathfrak{a}_i)$ where only a finite number of $\mathfrak{a}_i$ $(i\in\mathbb{N}^r)$ are differents.

(2.4) Now, let A be an Artinian local ring in the form: $A=k[Y_1,\ldots,Y_m]_{\mathfrak{p}}/\mathfrak{q}_{\mathfrak{p}}$ where $\mathfrak{q}$ is a $\mathfrak{p}$-primary ideal. We note,first that A is determined by k,m and by a finitely generating set of $\mathfrak{q}$; to see this,if we refer to theorem 3.2 of [6],then we can prove that $\mathfrak{q}$ is primary and we can determine $\mathfrak{p}=\sqrt{\mathfrak{q}}$.

PROPOSITION (2.4) : *One can compute the length* $\lambda(q)$ *of a primary ideal* q .

Proof : Notice that $\lambda(q)=\lambda({}^hq)$ where ${}^hq \subseteq k[Y_o,\dots,Y_m]$ is the homogeneous ideal generated by the forms ${}^hf$,with $f\in q$. From $\mathrm{ord}_k({}^hq)=\lambda(q)\mathrm{ord}_k({}^h\sqrt{q})$ (see Satz 3 of [11]) where $\mathrm{ord}_k({}^hq)$ is the degree of $k[Y_o,Y]/{}^hq$,and since the Poincaré series of homogeneous ideals of a polynomial ring on a field,can be computed (see[1]), it follows that $\lambda(q)=l_A(A)$ can be computed.

(2.5) *Reduction to the case of a zero-dimensional primary ideal.*

Here we follow the method of[8],(2.1)-(2.3) (see also[6],Cor.2.6). With the notation of (2.4), if dim $q=d$,then (under a permutation on $\{Y_1,\dots,Y_m\}$)we have: $q \cap k[Y_1,\dots,Y_d]=(0)$. We set $k'=k(Y_1,\dots,Y_d)$; in the localization $k[Y_1,\dots,Y_m]\to$ $\to k'[Y_{d+1},\dots,Y_m]$, let p' be the prime ideal (resp. q' the primary ideal) corresponding to p (resp. to q), we have: $A=k'[Y_{d+1},\dots,Y_m]/q'$, dim $q'=0$.

PROPOSITION (2.5) : *Let* (A,m,K) *be an Artinian local ring in the form* $A=k[Y_1,\dots,Y_m]/q$ *where* k *is a field, and* q *an ideal (primary for a maximal ideal) of* $k[Y]$ *;then:*

(i) $l_k(A)=\mathrm{ord}({}^hq) = l_k((k[Y_o,Y]/{}^hq)_n)$ *for* $n \gg 0$.

(ii) $[K:k] = \mathrm{ord}({}^h\sqrt{q}) = l_k((k[Y_o,Y]/{}^h\sqrt{q})_n)$ *for* $n \gg 0$;

(iii) $l_k(A)$ *and* $[K:k]$ *can be computed.*

Proof: Since projdim ${}^hq=0$,then $l_k((k[Y_o,Y]/{}^hq)_n) = \mathrm{ord}({}^hq)$ for $n \gg 0$, By (1.8),for every term-ordering on $k[Y]$,we have that $l_k(A)=l_k(k[Y]/M(q))$; and so (i) follows since $l_k((k[Y_o,Y]/{}^hq)_n) = \sum_{1\le i\le n} l_k((k[Y]/M(q))_i)$. Notice that $K=k[Y]/\sqrt{q}$; (ii) follows immediately.

REMARKS: (1) We note that $l_k(A)=\lambda(q)\,[K:k]$ (see the "extension formula" as in [9], p. 168).

(2) We have also: $a(k[Y]/M(q)) = a(k[Y_o,Y]/{}^hq)+1$ which is the last $n\in\mathbb{N}$ such that $M(q)_n \neq (k[Y])_n$.

(2.6) We have now arrived at the main result:

THEOREM (2.6) : *Let* k *be a field,* p *a prime ideal of* $k[Y_1,\dots,Y_m]$, q *a* p*-primary ideal,* $A=k[Y]_p/q_p$ *,and let* I *be an homogeneous ideal of the polynomial ring* $B=A[X_1,\dots,X_r]$. *Then,the Poincaré series* $P(B/I,t)$ *can be computed.*

Proof: Step 1.- First of all,by (2.5) we can assume that dim $p=0$ and so $A=k[Y]/q$; we can compute $[K:k]$ where K is the residue field of A (see(2.5)).

Step 2.- Let $\phi : k[Y]\to A$ be the surjective homomorphism with $\ker\phi=q$, and $\bar\phi : k[Y,X]\to A[X]=B$ be the extension of ϕ ; if the ideal I is determined by a generating system of an ideal I' of $k[Y,X]$ with $\bar\phi(I')=I$, we set $J=\bar\phi^{-1}(I)=I'+q[X]$.

The ideal J is homogeneous w.r.t. the variables X.

Step 3.- Let $<$ be a term-ordering on $k[Y,X]$ defined by $X^iY^j < X^{i'}Y^{j'}$ if $X^i <_1 X^{i'}$ or $X^i = X^{i'}$ and $Y^j <_2 Y^{j'}$ where $<_1$ and $<_2$ are some term-orders on monomials in X and Y respectively. Therefore, we can consider $M(f), M_1(f), M_1(g)$ ($f \in k[Y,X]$, $g \in A[X]$) and we have: $M(M_1(J)) = M(J)$ and $M_1(J) = \bar{\phi}^{-1}(M_1(I))$.

Step 4.- By (2.1), observe that: $l_A((B/I)_n) = l_A((B/M_1(I))_n)$ ($n \in \mathbb{N}$); by (2.5) we have: $[K:k] \cdot l_A((B/M_1(I))_n) = l_k((B/M_1(I))_n) = l_k((k[Y,X]/M_1(J))_n)$ where n is the degree w.r.t. the variables X. By (1.8) we have: $l_k((k[Y,X]/M_1(J))_n) = l_k((k[Y,X]/M(J))_n)$.

Step 5.- The series $\sum_{n \in \mathbb{Z}} l_k((k[Y,X]/M(J))_n) t^n = [K:k] \cdot P(B/I,t)$ is a rational function of the form : $[K:k] \cdot Q(t) \cdot (1-t)^{-r}$ where $Q(t) \in \mathbb{Z}[t]$. In view of remark 2)-(2.5), we can compute $a(k[Y]/M_2(q)) = \mu$; let $\{f_1, \dots, f_p\}$ be a Gröbner basis of J w.r.t. the term-ordering $<$, with $M(f_i)$ monic, then by (2.3) we have:

$$a(I) \leq -r + \deg_X(\text{l.c.m.}\{M(f_1), \dots, M(f_p)\}) = \omega .$$

Now, for every $n = 0, 1, \dots, \omega + r$, one can compute:

$l_k((k[Y,X]/M(J))_n) = \#$ monic monomials ψ of $k[Y,X]$, with $\deg_X(\psi) = n, \deg_Y(\psi) < \mu + 1$ and

$$\psi \notin (M(f_1), \dots, M(f_p)) = M(J).$$

By (2.3), we have: $\sum_{0 \leq n \leq \omega + r} l_k((k[Y,X]/M(J))_n) t^n \cdot (1-t)^r = [K:k] \cdot Q(t)$.

REMARK: If we follow the method of [1] (pp.124-125), we can compute the Poincaré series $P(B/I,t)$, via the Möbius inversion function w.r.t. the characteristic function $j: S \to \{0,1\}$ where $S = \{(a_1, \dots, a_m, b_1, \dots, b_r); \sum_{1 \leq i \leq m} a_i \leq \mu ; a_i, b_j \in \mathbb{N}\}$, $j(a,b) = 1$ if $(a,b) \notin M(J)$ and $j(a,b) = 0$ if $(a,b) \in M(J)$.

§ 3. COMPUTATION OF THE MULTIPLICITY.

(3.1) (*Notations*) Let $k[Y] = k[Y_1, \dots, Y_m]$ be a polynomial ring where k is a field. We set: $k[\xi] = k[Y]/\mathfrak{a}$, $\mathfrak{p}$ a prime ideal of $k[Y]$ with $\mathfrak{a} \subseteq \mathfrak{p}$, $A = k[\xi]_{\bar{\mathfrak{p}}}$, $\mathfrak{q}$ a $\mathfrak{p}$-primary ideal with $\mathfrak{a} \subseteq \mathfrak{q}$ (where - denotes mod. $\mathfrak{a}$). In the local ring $(A, \mathfrak{m})$, we set $\mathfrak{q}'$ the $\mathfrak{m}$-primary ideal associated to $\mathfrak{q}$; we note $\phi : k[Y] \to A/\mathfrak{q}'$ the canonical homomorphism.

PROPOSITION (3.2): *The form ring* $G = Gr(\mathfrak{q}', A)$ *can be computed.*

Proof: First of all, note that we can assume that $\mathfrak{p}$ is a maximal ideal (see (2.5)) and so we have: $G \simeq Gr(\bar{\mathfrak{q}}, k[\xi])$. If $\bar{\mathfrak{q}} = (\bar{f}_1, \dots, \bar{f}_r)$ ($f_i \in k[Y]$), for the Rees algebra $R = R(\bar{\mathfrak{q}}, k[\xi]) = k[\xi][T\bar{f}_1, \dots, T\bar{f}_r, T^{-1}]$ we have: $R = k[Y,X,U]/\mathfrak{b}$ where,

$$\mathfrak{b} = (\mathfrak{a}, X_1 - f_1T, \dots, X_r - f_rT, TU-1) \cdot k[Y,X,T,U] \cap k[Y,X,U] .$$

Since $G = R/(T^{-1})$, we have: $G \simeq k[Y,X]/\mathfrak{c}$, where $\mathfrak{c} = (\mathfrak{b}, U)\, k[Y,X,U] \cap k[Y,X]$. Note that $\mathfrak{c} = \bar{\phi}^{-1}(\ker \psi)$ where $\bar{\phi} : k[Y,X] \to A/\mathfrak{q}'[X]$ is the polynomial extension of ϕ, and $\psi : A/\mathfrak{q}'[X] \to G$ is the homomorphism defined by $\psi(X_i)$ = the class of $\bar{f}_iT$ mod. (T^{-1}).

COROLLARY (3.3) : *The multiplicity* $e(q',A)$ *can be computed.*

Proof: Since $H_{A,q'}=P(G,t)$ (where G is considered as a graded $A/q'[X]$-module), by (3.2) and (2.6) the assertion is clear.

(3.4) It is know that the multiplicity of ideals generated by systems of parameters enjoy various nice properties. On the other hand, let V,W be two irreductible varieties of an affine space and let C be a proper component of $V\cap W$, then the intersection multiplicity $i(C,V\cdot W)$ of V and W along C, is given by the multiplicity of a primary parameter ideal (see[10], Ch. II, § 7, (7,a)).

In addition, let q be an $\mathfrak{m}$-primary ideal in a local ring $(A,\mathfrak{m})$, then there exist a parameter ideal $\tilde{q}$ such that $e(q,A)=e(\tilde{q},A)$; one can give $\tilde{q}$ via superficial elements (see[3], § 7,5) or by Northcott-Rees reduction of q .(see[7], Th. 14.13).

(3.4) *Boda-Vogel method.*

For an ideal I of A, we set $U(I)=\cap q$ where q runs through the primary ideals belonging to I, such that dim q = dim I.

Let $(A,\mathfrak{m})$ be a local d-dimensional ring($d\geq 1$) and $q=(\eta_1,\dots,\eta_d)$ an $\mathfrak{m}$-primary ideal of A; then if $a_o=(0)$, $a_k=(\eta_k)+U(a_{k-1})$ $(0<k\leq d)$ we have: $e(q,A)=l_A(A/a_d)$ (see [2], Proposition 1). Using the same notation as in (3.1), we set $A=k[\xi]_{\bar{p}}$ where p is a maximal ideal of k[Y]; now we follow the method of Boda-Vogel, but on k[Y], where there exist a primary decomposition algorithm (see[5]).

Let $z_1,\dots,z_d$ be elements of k[Y], corresponding to $\eta_1,\dots,\eta_d$; if $b_o=a$, $b_k=(z_k)+$ $+U(b_{k-1})$ (where $k[\xi]=k[Y]/a$), then we can give, via the primary decomposition, the p-primary component b of b_d and so we have: $e(q,A)=\lambda(b)$.

(3.5) *Computation of the multiplicity of a parameter ideal.*

Next we give another method for the computation of the multiplicity of a parameter ideal. Let $(A,\mathfrak{m})$ be a d-dimensional local ring and $q=(\eta_1,\dots,\eta_d)$ a parameter ideal; then we have the following properties:

(i) if η_1 is A-regular, then: $e(q,A)=e(q/(\eta_1),A/(\eta_1))$ (see[7], 14.11);

(ii) η_1 is $A/(0:\eta_1^\star)$-regular, where $(0:\eta_1^\star)=\bigcup_{n\geq 0}(0:\eta_1^n)$ (see[5], Cor. 4.2).

(iii) From the exact sequence : $0\longrightarrow(0:\eta_1^\star)\longrightarrow A\longrightarrow A/(0:\eta_1^\star)\longrightarrow 0$, where $\dim_A(0:\eta_1^\star)<\dim A$, we have: $e(q,A)=e(q/(\eta_1),A/(0:\eta_1^\star)+(\eta_1))$ (see[9], Ch. 7, Theorem 6).

(iv) If $a'_o=(0)$, $a_i=(a'_{i-1}:\eta^\star_i)$, $a'_i=a_i+(\eta_i)$ $(1\leq i\leq d)$, then we have: $e(q,A)=l_A(A/a'_d)$

Using the same notations as in (3.1), we set $A=k[\xi]_{\bar{p}}$, where p is a maximal ideal of k[Y]; we can assume that $a=\ker\chi$ where χ: $k[Y]\longrightarrow k[\xi]_{\bar{p}}$ is the canonical homomorphism. Let $z_1,\dots,z_d$ be elements of k[Y] corresponding to $\eta_1,\dots,\eta_d$. We can consider the following ideals of $k[Y]$: $b_0'=a$, $b_i=(b_{i-1}':z_i^*)$, $b_i'=b_i+(z_i)$ $(1 \leq i \leq d)$.

Let b be the p-primary component of b_d' ,then we have: $\chi(b_d')=a_d'$ and $e(q,A)=\lambda(b)$. We are going to determine b. Choosing $c_i=(b_d':z_i^*)(1 \leq i \leq d)$, we have that b is the only primary component of b_d' , which is not contained in every primary decomposition of c_i ; let $f=f_1 \cdots f_d$ where $f_i \in c_i \dot{-} p$; then $b=(b_d':f^*)$.

REFERENCES.

[1] Bayer D.A.,"The division algorithm and the Hilbert scheme",Ph.D.Harvard(1982).

[2] Boda E.-Vogel W.,"On system of parameters,local intersection multiplicity and Bézout theorem",Proc.Amer.Math.Soc.78,(1980),1-7.

[3] Bourbaki N.,"Algèbre commutative,Ch.8 et 9",Masson,Paris (1983).

[4] Buchberger B.,"Gröbner bases:an algorithmic method in polynomial ideal theory", in Recent trends in multidimensional systems theory,Bose N.K.Reidel (1985).

[5] Gianni P.-Trager B.-Zacharias G.,"Gröbner bases and primary decomposition of polynomial ideals",preprint (Dec.1984).

[6] Grieco M.-Zucchetti B.,"How to decide whether a polynomial ideal is primary or not",preprint (1987).

[7] Matsumura H.,"Commutative ring theory",Cambridge Univ.Press (1986).

[8] Mora F.,"An algorithmic approach to local rings",Proc.Eurocal'85,LNCS 204(1985).

[9] Northcott D.G.,"Lessons on rings,modules and multiplicities",Cambridge Univ. Press (1968).

[10] Samuel P.,"Méthodes d'Algèbre abstraite en géométrie algébrique",Springer,Berlin (1967).

[11] Vogel W.,"Zur Theorie der charakteristischen Hilbertfunktion in homogenen Ringen über Ringen mit Vielfachkettensatz",Math.Nachr.33,(1967),39-60.

[12] Zariski O.-Samuel P.,"Commutative Algebra Vol.I",Princeton (1958).

Succinct representations of counting problems

Jacobo Torán
Facultat d'Informàtica de Barcelona (UPC)
Pau Gargallo 5
08028 Barcelona
Spain

Abstract: We introduce the logarithmic time counting hierarchy (LCH) as a tool to classify certain combinatorial problems connected to the idea of counting, and consider the model of boolean circuits as a tool to encode in a succinct way instances of these problems. We observe that many natural problems, like "majority" are complete for the different classes of LCH; using this and the fact that as a general rule, the complexity of a problem increases exponentially when its succinct representation is considered, we obtain complete problems for the classes in the polynomial time counting hierarchy. With the help of the succinct encodings, we give sufficient conditions for a problem to be hard for the classes NP and PP. Finally we show another use of the succinct representations, proving translational results for the classes in the counting hierarchy.

The complexity of a problem is closely related with the way its instances are encoded. When the instances of the problem are finite sets, the usual way of encoding them is by describing or enumerating all its elements. If these sets have many regularities, it might not be necessary to describe them by a detailed enumeration, and a smaller instance representation can be considered. In [Ga,Wi,83], a succinct representation of graphs with regularities is studied, and in [Wa,86], different languages describing instances of combinatorial problems are investigated. These papers present many examples of problems that can be computed in polynomial time, but become complete in some higher class like NP, Σ_2^p or the first levels of the polynomial time counting hierarchy when their instances are given in a succinct way. (The polynomial time counting hierarchy (CH), defined below, was introduced in [Wa,86] as an extension of the polynomial time hierarchy, and contains the complexity classes obtained by alternating the existential, universal and counting or probabilistic quantifier, cf also [To,88a].) The explanation for the blow-up in the complexity of the succinct version of the problem is that the shrinking of the instances operates in the opposite way of a padding process, while the problem restricted to the instances that admit succinct representation, is not necessarily easier than when considered over all its instances.

In this work, we will consider succinct versions of combinatorial problems. In order to give a succinct version of a problem, we will use the model of boolean circuits computing functions $f : \Sigma^n \longrightarrow \Sigma^m, (n, m \in \mathbb{N})$. This model is very adequate for describing finite function problems, and it is a generalization of the model used in [Ga,Wi,83] for studying graph properties, where only one output bit was allowed. The boolean circuit model can be used also to study succinct versions of combinatorial problems about finite sets of numbers, trees, tiles, etc., by simply considering the sets induced by the range of the functions computed by boolean circuits. A succinct representation of a function $f : \Sigma^n \longrightarrow \Sigma^m, (n, m \in \mathbb{N})$ is a boolean circuit BC_f with n input and m output bits, such that for a given string $w \in \Sigma^n$, $BC_f(w)$ computes $f(w)$. If Q is a certain problem for functions with finite domain, sQ will denote the problem of deciding if the finite function computed by a given boolean circuit is in Q.

As we will see, there are many examples of function problems that can be computed in polynomial time, whose complexity increases when the succinct version of the problems is considered. This blow-up phenomenon has been described in [Pa,Ya,86], where it is shown that if a problem is hard for a complexity class with respect to a certain kind of projection reducibility, then its succinct version based on the boolean circuit model is hard for the exponentially larger complexity class. This result is proven for classes above LOGSPACE, showing that as a general rule the jump in the complexity of a problem is exponential when the succinct version is considered. In what complexity class are then the problems whose succinct version is in classes like NP or PP? We show that the result of [Pa,Ya,86] can be extended also for logarithmic time bounded complexity classes. These classes can be defined using machines running in logarithmic time, and have an indirect access mechanism to look-up a logarithmic part of the input string.

The logarithmic time bounded complexity classes have appeared before in the literature; in [Ch,Ko,St,81] the logarithmic time hierarchy is defined in a similar way as the polynomial time hierarchy, using logarithmic time bounded Turing machines with existential and universal states. In [Si,83] it is shown that this hierarchy is proper. In order to classify the complexity of many combinatorial problems associated with the idea of counting, we introduce the logarithmic time counting hierarchy (LCH) based on the exponentially larger hierarchy CH. LCH contains the logarithmic time hierarchy and it is known that its first level is strictly contained in the second [To,88a]. We will show that if a computational problem is complete in a class of LCH with respect to a certain kind of projection reduction, then its succinct version is complete in the corresponding class of the polynomial time counting hierarchy.

Using the above fact, we will give sufficient conditions for a problem to be hard for the classes NP or PP. As we will see, there are many natural examples of combinatorial problems that are complete for different levels of LCH; the succinct versions of these problems will provide us with complete problems for the corresponding classes in CH.

The succinct encodings help us also to prove translational lemmas betwen the classes of the two counting hierarchies, and we will show that if two classes in LCH coincide, then the corresponding classes in CH also coincide.

We start by defining the polynomial and logarithmic time counting quantifiers and hierarchies. For more information about these classes we refer to [To,88a], [To,88b].

Definition: The polynomial counting quantifier **C**, is defined in the following way; for a polynomial time computable function $f : \Sigma^* \longrightarrow \mathbb{N}$, a polynomial p and a two argument predicate P,

$$\mathbf{C}^p_{f(x)}\, y \;:\; P(x,y) \Longleftrightarrow ||\{y : \; |y| \leq p(|x|) \text{ and } P(x,y)\}|| \geq f(x).$$

If K is a language class, for any set A, $A \in \mathbf{C}K$ if there is a polynomial time computable function f, s.t. for every x, $f(x) > 0$, a polynomial p and a language $B \in K$ such that for any $x \in \Sigma^*$

$$x \in A \Longleftrightarrow \mathbf{C}^p_{f(x)}\, y \;:\langle x,y\rangle \in B$$

We alternate now the polynomial counting quantifier **C** with the existential and the universal quantifiers in order to define the counting hierarchy.

Definition: The polynomial counting hierarchy (CH) is the smallest family of language classes satisfying:

i/ P$\in$CH
ii/ If K $\in$CH then $\exists^p K$, $\forall^p K$ and $C^p K$ $\in$CH.

For simplicity, **C** will denote the class $\mathbf{C}^p$P, and the context will make clear when we talk about a quantifier and when about a language class. It is not hard to see that the threshold function of the **C** quantifier can be changed to strictly more than one half of the possible quantified strings. As a consequence, the class **C** is the same as the class PP of languages accepted by probabilistic Turing machines. This fact has been observed in [Sim,75], [Wa,86a].

CH is included in PSPACE, contains the polynomial time hierarchy, and all its classes have complete problems.

In order to define the logarithmic time counting hierarchy we need to define a model of a machine with a running time bounded by a logarithmic function.

Definition: A deterministic Turing machine with indirect access to the input is a Turing machine with the following elements:

- an input tape,
- k work tapes,
- a special tape to "look-up" the input,
- a special tape to write the aswers to the queries to the input,
- a "look-up" state q

The machine can write on the look-up tape the number of a position i of the input tape; whenever a configuration with state q is reached, in one computation step, the machine writes in the second special tape the content of the i-th position of the input tape. (If the input has length less than i, then the machine does not write anything). As we will see later, it is important that the content of the look-up tape (and the position of its head) remains untouched after each query to the input. We will suppose that in the initial configuration the input is written at the left end of the input tape without any blank spaces in between.

Observe that it does not make much sense to allow indirect access machines to work within a time bound greater or equal than linear, since then, the machines can be simulated by normal Turing machines, with roughly the same time bounds. The interest of indirect access machines arises when we consider sublinear time bounds. In particular, we will consider indirect access machines with computation time bounded by O(log(n)). Notice that an indirect access machine working in O(log(n)) steps can compute the length of its input using the following binary search technique: Query positions 2^i of the input tape until the first blank cell is found; every query can be done in constant time since in order to change the contents of the look-up tape from 2^i to 2^{i+1} only one new bit has to be added. Once an integer k is found such that $2^k \leq |x| < 2^{k+1}$, perform a binary search to find the last input position; again every query requieres only constant time since for the binary search only one bit of the look-up tape has to be changed every time. The whole procedure requires a logarithmic number of steps.

Definition : We will denote by LT the class of languages accepted by deterministic indirect access Turing machines with computation time bounded by O(log(n)), and FLT the class of functions $f : \Sigma^* \longrightarrow \mathbb{N}$ computed by deterministic indirect access Turing machines with computation time bounded by O(log(n)).

Observe that the size of the output of a function in FLT is logarithmic with respect to the size of its input. Based on the above definition, we define now the log-time counting hierarchy.

Definition: The logarithmic existential, universal and counting quantifiers are defined in the following way; for a function $f : \Sigma^* \longrightarrow \mathbb{N}, f \in$FLT, a constant c and a two argument predicate P,

$$\exists^{lc} y : P(x,y) \Longleftrightarrow \bigvee_{y} |y| \leq c\lceil \log(|x|) \rceil,\ P(x,y)$$

$$\forall^{lc} y : P(x,y) \Longleftrightarrow \bigwedge_{y} |y| \leq c\lceil \log(|x|) \rceil,\ P(x,y)$$

$$\mathrm{C}^{lc}_{f(x)} y : P(x,y) \Longleftrightarrow ||\{y : |y| \leq c\lceil \log(|x|) \rceil, \text{ and } P(x,y)\}|| \geq f(x)$$

If K is a language class, for any set A, $A \in \exists^l K$ if there is a language $B \in K$ and a constant c such that for any $x \in \Sigma^*$

$$x \in A \Longleftrightarrow \exists^{lc} y : \langle x, y \rangle \in B$$

and analogously for $A \in \forall^l K$. $A \in \mathrm{C}^l K$ if there is a language $B \in K$, a constant c and a function $f \in$FLT such that for any $x \in \Sigma^*$

$$x \in A \Longleftrightarrow \mathrm{C}^{lc}_{f(x)} y : \langle x, y \rangle \in B$$

Definition: The logarithmic time counting hierarchy (LCH) is the smallest family of language classes satisfying

i/ LT$\in$LCH

ii/ If $K \in$LCH then $\exists^l K, \forall^l K$ and $\mathrm{C}^l K \in$LCH

It is clear that LCH is included in LOGSPACE. In order to prove the completeness of some problems in the different classes of the log-time counting hierarchy, we cannot work with the usual polynomial time or logarithmic space reductions, since every problem in the hierarchy can be reduced to any other problem in the hierarchy via these reducibilities. We need a reducibility that works in logarithmic time, but this is somehow unnatural since big problem instances can only be reduced to small problem instances. We avoid these problems by using a definition of reducibility which works in polynomial time, but translates "locally" parts of problem instances in logarithmic time. The following definition is based on the definition of projection reducibility from [Sk,Va,85], and it is also used in [Pa,Ya,86].

Definition: Given two sets L_1, L_2, we will say that L_1 is polynomial projection reducible to L_2 if there is a function $\pi :\ \Sigma^* \longleftarrow\ \Sigma^*$ satisfying:

i/ For every $x \in \Sigma^*$ $\quad x \in L_1 \Longleftrightarrow \pi(x) \in L_2$

ii/ There is a polynomial p such that for every $x \in \Sigma^*$, $\quad |\pi(x)| \leq p(|x|)$

iii/ There is a function $\varphi : \Sigma^* \cdot \mathbb{N} \longrightarrow \{0,1\}$ computed by a deterministic log-time machine, such that for $x \in \Sigma^*, i \in \mathbb{N}, \varphi(x,i)$ is the i-th bit of $\pi(x)$, if $i \leq |\pi(x)|$, and is undefined otherwise.

The fact of L_1 being polynomial projection reducible to L_2 is denoted by $L_1 \leq_\pi L_2$.

The next two basic decision problems are complete in $\exists^l$LT and in C^lLT respectively, and will be used afterwards to classify some other problems in the hierarchy.

Example: Nonzero string. Let $L_1 = \{w \in \Sigma^* : w \text{ has at least one } 1\}$. L_1 is $\exists^l LT$-complete.

Example: Majority. Let $L_2 = \{w \in \Sigma^* : w \text{ has more 1's than 0's}\}$. L_2 is $\mathbf{C}^l LT$-complete. Observe that $w \in L_2$ if and only if more than half of the bits of w are 1.

We will study now the complexity of problems related with polynomial time computable functions of finite domain, $f : \Sigma^n \longrightarrow \Sigma^n$. A finite function defined over Σ^n, in Σ^n, will be given by a list of its values, each of them having the same length, n, (padded by leading 0's if necesary). We will only consider total functions, i.e. functions defined on all their input values.

The next theorems give sufficient conditions for a function problem to be hard for $\exists^l$LP and C^lLP. These conditions will be given by the existence of "$\exists$-critical" (resp. "C-critical) functions, which we define next. To simplify the notation in the definition of functions φ in the projections, for a string $w \in \Sigma^*$, we will represent by w_i the i-th bit of w.

Definition: Let Q be a set of functions $\xi : \Sigma^n \longrightarrow \Sigma^n, n \in \mathbb{N}$, and let f and g be a pair of polynomial time computable functions, $f, g : \Sigma^* \longrightarrow \Sigma^*$ such that for every $x \in \Sigma^*, |g(x)| = |f(x)| = |x|$. We will say that the pair (f, g) is $\exists$-critical for Q if the following conditions are true:

i/ For every $n \in \mathbb{N}$ $\quad f|_{\Sigma^n} \notin Q$.

ii/ For every $B \subseteq 0\Sigma^{n-1}, B \neq \emptyset$, $\quad h_B|_{\Sigma^n} \in Q$, where $h_B : \Sigma^* \longrightarrow \Sigma^*$ is defined by

$$h_B(x) = \begin{cases} f(x) & \text{if } x \notin B \\ g(x) & \text{if } x \in B \end{cases}$$

We will say that the pair (f, g) is C-critical for Q if the following condition is true:

For every $B \subseteq \Sigma^n, h_B|_{\Sigma^n} \in Q \iff ||B|| > 2^{n-1}$ being $h_B : \Sigma^* \longrightarrow \Sigma^*$ defined by

$$h_B(x) = \begin{cases} f(x) & \text{if } x \notin B \\ g(x) & \text{if } x \in B \end{cases}$$

The following powerful theorems yield easily many complete problems for $\exists^l$LT, and C^lLT as we shall see.

Theorem: Let Q be a certain set of functions with finite domain. If there exists a pair of functions (f, g) $\exists$-critical for Q, then Q is $\exists^l$LT-hard for the polynomial projection reducibility.

Proof: We show how to reduce L_1 to Q. Consider the following function φ defining the projection π from L_1 to Q (recall that for a string x, we represent by x_j the j-th bit of x. Since the projection has to be given bit by bit, the definition of φ is very hard to read. In the next figure we try to give an explanation of the function. For a string m, we will denote by $1^{|m|}$ the concatenation of $|m|$ 1's. For a given x, let $n+1 = 2^k$, for some k, be the minimum integer such that $|x| \leq 2^n$, (n can be computed in the following way: first, using the already mentioned binary search techniques, compute m where m is the minimum integer such that $|x| \leq 2^m$; n is then the string $1^{|m|}$). We need the fact of $n+1$ being a power of 2 in order to be able to divide by $n+1$; this will be explained in more detail after the definition of function φ. $\pi(x)$ will have size $(n+1)2^{(n+1)}$, and it will be a sequence of values of f and g. If the k-th bit of x is a 1, then $\pi(x)$ will have a value of g in the corresponding position, and otherwise it will have a value of f. Some values of f are added in order to represent in $\pi(x)$ the values of a complete function defined over Σ^{n+1}. Formally,

$$\varphi(x,i)=\begin{cases}(g(0r))_j & \text{if } r=\left\lceil\frac{i}{n+1}\right\rceil \text{ and } (i \bmod n+1)=j \\ & \text{and } i\leq(n+1)|x| \text{ and } x_r=1 \\ (f(0r))_j & \text{if } r=\left\lceil\frac{i}{n+1}\right\rceil \text{ and } (i \bmod n+1)=j \\ & \text{and } i\leq(n+1)|x| \text{ and } x_r=0 \\ (f(0r))_j & \text{if } r=\left\lceil\frac{i}{n+1}\right\rceil \text{ and } (i \bmod n+1)=j \\ & \text{and } (n+1)|x|<i\leq(n+1)2^n \\ (f(1r))_j & \text{if } r=\left\lceil\frac{i}{n+1}\right\rceil \text{ and } (i \bmod n+1)=j \\ & \text{and } (n+1)2^n<i\leq(n+1)2^{(n+1)} \\ \uparrow & \text{otherwise}\end{cases}$$

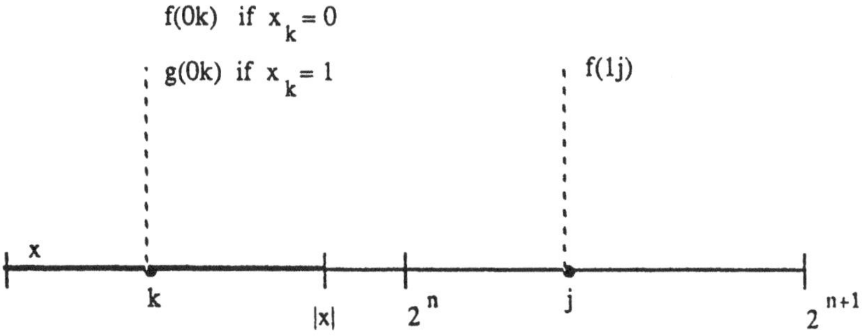

Figure 1

If $x \in L_1$ then there is a 1 in certain position i of x and the list of the function values will have one value of g. By condition ii/ in the above definition it follows that the function represented by $\pi(x)$ belongs to Q.

Conversely if $x \notin L_1$, the projection $\pi(x)$ will be the codification of $f|_{\Sigma^{n+1}}$ and by condition i/ in the definition of critical function, $\pi(x)$ is not in Q. The function can be computed in log-time, because n can be computed in log-time (already explained) and the divisions by $n+1$ are just shifts because $n+1$ is a power of 2. □

Theorem: Let Q be a set of functions with finite domain. If there exists a pair of functions (f,g) being C-critical for Q, then Q is C^{l}LT-hard for the polynomial projection reducibility.

The proof is similar to the one of the above theorem.

We will use these theorems for classifying the next problems. Observe that in order to show that a problem is $\exists^l$LT-hard (C^{l}LT-hard), we only need to give a pair of $\exists$-critical (C-critical) functions (f,g) for it. It can be proved that the next problems are complete for a class of LCH. For each problem we give a critical pair of functions (f,g). It is left to the reader to prove that these problems are in the claimed class in LCH and that the pairs of functions (f,g) are actually critical for the problems (or for its complements).

L_3:Constant value of the function.
$L_3=\{\xi:\Sigma^n\longrightarrow\Sigma^n:\xi \text{ is a constant function }\}$
$\exists$-critical functions (for $\overline{L_3}$) $f,g:\Sigma^*\longrightarrow\Sigma^*, f(x)=0^{|x|}, g(x)=1^{|x|}$.

L_4:Injectiveness
$L_4 = \{\xi : \Sigma^n \longrightarrow \Sigma^n : \xi \text{ is an injective function }\}$
$\exists$-critical functions (for $\overline{L_4}$) $f, g : \Sigma^* \longrightarrow \Sigma^*, f(x) = x, g(x) = 1^{|x|}$.

L_5:Increasing function.
$L_5 = \{\xi : \Sigma^n \longrightarrow \Sigma^n : \xi \text{ is strictly increasing }\}$
$\exists$-critical functions (for $\overline{L_5}$) $f, g : \Sigma^* \longrightarrow \Sigma^*, f(x) = x, g(x) = 1^{|x|}$.

L_6:Conservation of parity.
$L_6 = \{\xi : \Sigma^n \longrightarrow \Sigma^n$: for every $x \in \Sigma^n$, last digit of x = last digit of $\xi(x)\}$
$\exists$-critical functions (for $\overline{L_6}$) $f, g : \Sigma^* \longrightarrow \Sigma^*, f(x)$ =last digit of x, $g(x)$ = opposite of the last digit of x.

L_7:Large zero function restriction.
$L_7 = \{\xi : \Sigma^n \longrightarrow \Sigma^n : \xi(x) = 0 \text{ for more than one half of the inputs }\}$
C-critical functions $f, g : \Sigma^* \longrightarrow \Sigma^*, f(x) = (x), g(x) = 0$.

L_8:Large identity restriction.
$L_8 = \{\xi : \Sigma^n \longrightarrow \Sigma^n : \xi(x) = x \text{ for more than one half of the inputs }\}$
C-critical functions $f, g : \Sigma^* \longrightarrow \Sigma^*, f(x) = ((x+1)\text{mod}(2^n - 1)), g(x) = x$.

We give now some more examples of function problems that are complete in some other levels of LCH, without including the proof of their completeness.

L_9:Surjective function.
$L_9 = \{\xi : \Sigma^{2n} \longrightarrow \Sigma^{2n} : \xi \text{ is surjective on } 0^n\Sigma^n\}$
L_9 is complete in $\forall^l\exists^l$LT.

L_{10}:Large injective restriction.
$L_{10} = \{\xi : \Sigma^{2n} \longrightarrow \Sigma^{2n}$: There is an injective restriction of ξ of size$\geq 2^n\}$
L_{10} is complete in $\mathbf{C}^l\exists^l$LT.

We introduce now two combinatorial problems that with small modifications yield examples of complete problems for any of the levels of some subhierarchies of LCH. For simplicity we will call $\Sigma_k^l = \exists^l\forall^l\exists^l \dots Q^l$LT ($k-1$ alternations of quantifiers $\exists$ and $\forall$, where $Q^l = \forall^l$ if k is even and $\exists^l$ if k is odd), and $\mathbf{C}_k^l = \mathbf{C}^l\mathbf{C}^l \dots \mathbf{C}^l$LT ($k$ quantifiers $\mathbf{C}$).

A similar version of the first problem was independently defined by Grädel in [Grä,88].

a/ A tiling game: We will consider a finite set of strings $T \subseteq \Sigma^{2s}$ as the set of tiles of a one-dimensional domino game. Given a string $x \in T$, a second string $y \in T$ will match with x if the last s digits of x coincide with the first s digits of y. We will also consider a number $m \in \mathbb{N}$ and two players, Constructeur and Saboteur as in [Ch,86], building a tiled row of dominoes. Given a starting tile, alternately, each of the two players selects a domino that matches with the right part of the last selected tile, and adds it to the row. The aim of Constructeur is to build a tiled row of $m+1$ tiles, while the aim of Saboteur is to prevent Constructeur from reaching his aim. We investigate different versions of this tiling game.

R(m): *Tiling game of length m* (m odd). Consider a tiling game of 2^{mn} tiles of length $2mn$ given one after another in a string. The starting tile is given by the first $2mn$ digits of the string.

If Constructeur starts the game, can he complete a tiled row of $m+1$ tiles?

Theorem: $R(m)$ is Σ^l_m-complete.

i/ $R(m) \in \Sigma^l_m$. Given $w \in \Sigma^*$, $|w| = 2mn{\cdot}2^{mn}$ we can write the decision problem as a string of m quantifiers followed by a log-time predicate in the following way (an explanation is given below):

$$w \in R(m) \iff \exists i_1, i_1 \in \mathbb{N}, |i_1| = n \quad \forall i_2, i_2 \in \mathbb{N}, |i_2| = n \ldots Q i_m, |i_m| = n, i_n \in \mathbb{N}:$$

$$w_1 \ldots w_{2mn} w_{i_1 2mn} \ldots w_{(i_1+1)(2nm)-1} w_{i_2 2mn} \ldots w_{(i_2+1)(2mn)-1} \ldots w_{i_m 2mn} \ldots w_{(i_m+1)(2mn)-1}$$

is a correct row of the domino game, or for some i_j, (j even),

$$w_{i_j 2mn} \ldots w_{i_j 2(nm+1)-1} \text{does not match with } w_{i_{j-1} 2mn} \ldots w_{i_{j-1} 2(nm+1)-1}$$

This means that Constructeur has a way to select tiles in which Saboteur either plays unfair (placing tiles that do not match) or cannot prevent Constructeur from making the tiled row.

ii/ $R(m)$ is hard for Σ^l_m (m odd). Let L be a language in Σ^l_m, for a certain constant c and a log-time predicate P, for every $x \in \Sigma^*$,

$$x \in L \iff \exists u_1 |u_1| \le c\lceil\log(|x|)\rceil \forall u_2 |u_2| \le c\lceil\log(|x|)\rceil \ldots Q u_m |u_m| \le c\lceil\log(|x|)\rceil$$
$$P(x, u_1, \ldots, u_n)$$

Let n the smallest power of two greater than or equal to $c\lceil\log(|x|)\rceil$ and represent by $(k)_n$ the binary representation of length n of k, ($k < 2^n$). We can project L to $R(m)$. For that we will construct a domino game of $2^{2(m+1)n}$ tiles of size $2(m+1)n$ in such a way that in any row, the first m tiles can always be placed. The tile $m+1$, corresponding to a Constructeur is related to predicate P and to the tiles placed before, and if both players have played right, it can only be placed if $x \in L$. A more detailed explanation is given after the definition of function φ. Again, for clarification, we will describe the function in pieces, corresponding a piece to a tile of the domino game in the projection.

$\varphi(x, (u_1 u_2 \ldots u_m u_{m+1}) 2(m+1)n \ldots \varphi(x, (u_1 u_2 \ldots u_m u_{m+1}) + 1) 2(m+1)n - 1) =$

$$= \begin{cases} \overline{\underline{|(m)_n 0^n 0^n \ldots 0^n|(0)_n 0^n 0^n \ldots 0^n|}} & \text{if } u_j = (0)_n \ j \in \{1 \ldots m+1\} \\ \overline{\underline{|(0)_n 0^n 0^n \ldots 0^n|(1)_n u_1 0^n \ldots 0^n|}} & \text{if } u_j = (1)_n \ j \in \{2 \ldots m+1\} \\ \overline{\underline{|(1)_n u_1 0^n \ldots 0^n|(2)_n u_1 u_2 0^n \ldots 0^n|}} & \text{if } u_j = (2)_n \ j \in \{3 \ldots m+1\} \\ \overline{\underline{|(2)_n u_1 u_2 0^n \ldots 0^n|(3)_n u_1 u_2 u_3 0^n, \ldots 0^n|}} & \text{if } u_j = (3)_n \ j \in \{4 \ldots m+1\} \\ \ldots & \\ \ldots & \\ \overline{\underline{|(m-2)_n u_1 u_2 \ldots u_{m-2} 0^n 0^n|(m-1)_n u_1 \ldots u_{m-2} u_{m-1} 0^n|}} & \text{if } u_j = (m-1)_n \\ & j \in \{m, m+1\} \\ \overline{\underline{|(m-1)_n u_1 u_2 \ldots u_{m-2} u_{m-1} 0^n|(m)_n u_m 0^n \ldots 0^n|}} & \text{if } u_{m+1} = (m)_n \\ & \text{and } P(x, u_1, u_2 \ldots u_m) \\ \overline{\underline{|(m)_n 0^n \ldots 0^n|(0)_n 0^n \ldots 0^n|}} & \text{otherwise} \end{cases}$$

We will see that Constructeur can build a tiled row of $m+1$ tiles if and only if $x \in L$. Observe that the only tiles that depend on P have the form $\overline{|(m-1)_n u_1 \dots u_{m-1} 0^n|(m)_n u_m 0^n \dots 0^n|}$ and in these tiles is encoded the whole story of the game. The remaining tiles will always be in our set of dominoes: this means that the tiled row constructed in the game, will always have length at least m. The string of length n written at the left of every tile guarantees that the j-th tile in the row encodes an election u_{j-1} for the quantifier $j-1$. It will be possible to place the tile $m+1$ if and only if there is a choice of u_m such that $P(x, u_1, \dots u_m)$. If $x \in L$ then $\exists u_1 \forall u_2 \dots Q u_m P(x, u_1, \dots u_m)$. Considering Saboteur's selections, Constructeur only needs to select the tiles that codify the values corresponding to the existential quantifiers in the formula. Conversely, if $\forall u_1 \exists u_2 \dots \overline{Q} u_m \neg P(x, u_1 \dots u_m)$, Saboteur can then select tiles so that the m-th domino in the row is $\overline{|(m-2)_n u_1 \dots u_{m-2} 0^n 0^n|(m-1)_n u_1 \dots u_{m-2} u_{m-1} 0^n|}$ and for any u_m the tile $\overline{|(m-1)_n u_1 \dots u_{m-2} u_{m-1} 0^n|(m)_n u_m 0^n \dots 0^n|}$ is not in our set of dominoes. It follows that Constructeur cannot finish the row.

b/ The problem of the green tree. Consider a tree whose leaves can be labeled with either a 1 (the leaf is green) or a 0 (the leaf is brown). We will say that an interior node of the tree is green if more than half of its direct descendants are green. A tree is green if its root is green. We study the complexity of the problem of knowing if a given tree is green. We will restrict ourselves to complete trees having degree $2^n, n \in \mathbb{N}$ i.e. each interior node has 2^n direct descendants. A complete tree with its leaves labeled, having k levels and with every node having 2^n direct descendants will be represented by a list of 2^{kn} bits, each representing the label of one of the leaves.

$T(k)$: *Green tree of height k.* Given a list of 2^{kn} bits, does it encode a green tree of k levels?

Theorem: $T(k)$ is $\mathbf{C}_k^l$ complete.

i/ $T(k) \in \mathbf{C}_k$. Given $w \in \Sigma^*$, the length of w can be checked using standard techniques. Suppose $|w| = 2^{kn}, n \in \mathbb{N}$,

$$w \text{ is green} \iff \mathbf{C} i_1, i_1 \in \mathbb{N}\ |i_1| = n\ \mathbf{C} i_2, i_2 \in \mathbb{N}\ |i_2| = n \dots \mathbf{C} i_k, i_k \in \mathbb{N}\ |i_k| = n :$$

$$w_{(i_1 i_2 \dots i_k)} = 1$$

ii/$T(k)$ is hard for $\mathbf{C}_k$. Let $L \in \mathbf{C}_k$, there is a constant c and a polynomial time predicate P such that for every $x \in \Sigma^*$,

$$x \in L \iff \mathbf{C} u_1, |u_1| \le c\lceil \log(|x|) \rceil\, \mathbf{C} u_2, |u_2| \le c\lceil \log(|x|) \rceil \dots \mathbf{C} u_k, |u_k| \le c\lceil \log(|x|) \rceil :$$

$$P(u_1 \dots u_k)$$

We give the projection reducing L to $P(k)$. Let n be the smallest power of 2 greater than or equal to $c\lceil \log(|x|) \rceil$, $\pi(x)$ will have size 2^{kn}, and the function φ defining π is:

$$\varphi(x, (u_1 u_2 \dots u_k)) = \begin{cases} 1 & \text{if } P(u_1 u_2 \dots u_k) \\ 0 & \text{if } \neg P(u_1 u_2 \dots u_k) \end{cases}$$

The above problems are "bounded" in the sense that we only allow a tiling game to have a given number of moves, or a tree to have a given number of levels. It is not hard to see that if we do not consider these limitations, the the problems become complete for the class P of problems computed in deterministic polynomial time, with respect to the log-space reduction. For a complete proof we refer to [To,88a].

Although we have only shown the existence of "natural" problems that are complete in any of the levels of some of the subhierarchies in LCH, it is clear that using the same techniques, complete problems for any of the classes in LCH can be constructed.

We will now combine the obtained results with some known facts about succinct representations of functions to obtain complete problems for the polynomial time counting hierarchy.

In [Pa,Ya,86], a powerful result is presented which establishes, under certain restrictions, that if a certain problem is hard for a complexity class with respect to a certain kind of projection reducibility, then the succinct version of the problem is hard for the exponentially larger class. This result was presented for complexity classes above LOGSPACE, but it is not hard to extend it to our classes.

Theorem:[Pa,Ya,86] Let Q be a complete problem for the class K $\in$LCH, with respect to the polynomial time projection reducibility. The succinct version of the problem, sQ, is complete with respect to the polynomial time m-reducibility, for the corresponding (exponentially larger) class K' $\in$CH.

This result provides a lower bound for the complexity of problems over succinct instances, assuming a lower bound for the standard version of the problem. We will need also a dual lemma, providing upper bounds for the problem encoded in the standard way, assuming an upper bound for the succinct version of the problem.

Lemma: If Q is a problem in K, K $\in$LCH, then the succinct version of Q, sQ, is in K', where K' is the corresponding exponentially larger class in CH.

Proof: Let K $\in$LCH and $L \in K$. For a certain $k \in \mathbb{N}$, a constant c_1 and a log-time predicate P, for any $x \in \Sigma^*$,

$$x \in L \Longleftrightarrow Q_1^l i_1, |i_1| \leq c_1 \lceil \log(|x|) \rceil, Q_2^l i_2, |i_2| \leq c_1 \lceil \log(|x|) \rceil, \ldots Q_k^l i_1, |i_k| \leq c_1 \lceil \log(|x|) \rceil : P(x, i_1, i_2 \ldots i_k)$$

Let sL the succinct version of L, and $BC(x)$ the boolean circuit representation of an instance x. Supose that $BC(x)$ receives n bits as input, and produces m bits as output. $BC(x)$ encodes then a string x of $m2^n$ bits, and since $m + n \leq |BC(x)|$, $2^{|BC(x)|} \geq |x|$. It follows that for a certain constant c_2,

$$BC(x) \in sL \Longleftrightarrow Q_1 i_1, |i_1| \leq |BC(x)|^{c_2}, Q_2 i_2, |i_2| \leq |BC(x)|^{c_2}, \ldots Q_k i_1, |i_k| \leq |BC(x)|^{c_2} : P'(x, i_1, i_2 \ldots i_k)$$

where P' is the polynomial time predicate (respect to $BC(x)$) working like P, except that when P queries a bit in a position i of x, P' queries the bit indirectly from $BC(x)$ using the following process: Compute r, $r = \lceil \frac{i}{m} \rceil$ (input that produces the part of x that contains its i-th bit). Then "feed" the circuit with the r-th string of length n in lexicographical order, and query the bit of position (i mod m) from the output. This bit is the i-th bit of x. (The process can be done in polynomial time with respect to $BC(x)$ since $BC(x)$ is given as input.) It follows that $sL \in K'$. □

Combining these results with the results obtained in the previous part of this article, we will automatically find complete problems for various levels of CH.

Corollary: The succinct versions of problems $L_1 \dots L_{10}, R(m), T(k)$ are complete in the corresponding classes of CH with respect to the m-reducibility. A summary of the results can be seen in the next figure.

	Problem:	**Succinct version complete in the class:**
L_1	Nonzero string..................	NP
L_2	Majority........................	PP
L_3	Constant value of the function	co-NP
L_4	Injectiveness....................	co-NP
L_5	Increasing function..............	co-NP
L_6	Conservation of parity..........	co-NP
L_7	Large zero function restriction ..	PP
L_8	Large identity restriction........	PP
L_9	Surjective function..............	$\forall\exists$
L_{10}	Large injective restriction.......	$C\exists$
$R(m)$	Tiling game of length m (m odd)	Σ^p_m
$T(k)$	Green tree of height k...........	$\mathbf{C}_k$

Figure 2.

We end this article using the above results to show the strong relation between both hierarchies, LCH and CH, proving that if LCH colapses, then CH collapses too. Results of this type, called translational results, have been proven before in [Bo,76] and [Ib,72]. It should be noticed however that the classes of LCH characterized using only existential and universal quantifiers define a proper subhierarchy of LCH [Sip,83].

Theorem: Let K_1 and K_2 be two classes in LCH. If $K_1 \subseteq K_2$ then $K_1' \subseteq K_2'$, where K_1' and K_2' are the corresponding exponentially larger classes in CH.

Proof: Let L be complete in K_1 with respect to the polynomial projection reduction, and suppose $K_1 \subseteq K_2$. The succinct version of L, sL, is complete for K_1' with respect to the polynomial time many one reduction. Since $L \in K_2$, $sL \in K_2'$. It follows that $K_1' \subseteq K_2'$ by the closure of the classes in CH under the polynomial time m-reducibility. □

In the proof of the above theorem, the existence of a complete problem in K_1 with respect to the polynomial projection reducibility is used. As mentioned before, we have only shown the existence of some "natural" problems complete in some of the subhierarchies of LCH. However, using the same techniques, complete problems in any of the classes in LCH can be constructed.

As a final remark, we mention that the containments $\mathbf{C}^l \subset \exists^l\mathbf{C}^l$ and $\mathbf{C}^l \subset \forall^l\mathbf{C}^l$ are strict. These results have been proven in [To,88a]

Acknowledgement: The author would like to thank the unknown referee for some helpful comments.

References

[Bo,76] R.V. Book: Translational lemmas, polynomial time, and $(\log n)^j$ space. *Theoret. Comput. Sci.* 1 (1976), 215–226.

[Ch,Ko,St,81] A.K. Chandra, D.C. Kozen, L.J. Stockmeyer: Alternation. *Journal ACM* 28 (1981), 114–133.

[Ch,86] B.S. Chlebus: Domino-Tiling games. *Journal Comput. Syst. Sci.* 32 (1986), 374–392.

[Ga,Wi,83] H. Galperin, A. Wigderson: succinct representations of graphs. *Information and Control* 56 (1983), 183–198.

[Grä,88] E. Grädel: Domino Games and complexity. Manuscript Università di Pisa (1988).

[Ib,72] O. Ibarra: A note concerning nondeterministic tape complexities. *Journal ACM* 19 (1972), 608–612.

[Pa,Ya,86] C.H. Papadimitriou, M. Yannakakis: A note on succinct representations of graphs. *Information and Control* 71 (1986), 181–185.

[Sim,75] J. Simon: On some central problems in computational complexity. Ph.D. Thesis, Cornell University (1975).

[Sip,83] M. Sipser: Borel sets and circuit complexity. Proc. 15th STOC (1983), 61–69.

[Sk,Va,85] S. Skium, L.G. Valiant: A complexity theory based on boolean algebra. *Journal ACM* 32 (1985), 484–502.

[To,88a] J. Torán: Structural properties of the counting hierarchies. Ph. D. Thesis. Facultat d'Informàtica de Barcelona. (1988).

[To,88b] J. Torán: An oracle characterization of the counting hierarchy. Proceedings 3rd Structure in Complexity Theory Conference (1988), 213–224.

[Wa,86] K. Wagner: The complexity of combinatorial problems with succinct input representation. *Acta Informatica* 23 (1986), 325–356.

HOW TO GUESS ℓ-th ROOTS MODULO n BY REDUCING LATTICE BASES

Brigitte VALLÉE
Département de Mathématiques, Université de Caen, France

Marc GIRAULT
Service d'Etudes communes des Postes et Télécommunications
Caen, France

Philippe TOFFIN
Département de Mathématiques, Université de Caen, France

ABSTRACT. In numerous problems of computational number theory, there often arise polynomial equations or inequations modulo a number n. When n is a power of a prime number, polynomial-time algorithms, either deterministic or probabilistic, allow one to solve these problems. The same is true, via the Chinese remainder theorem, when the factorisation of n is known. A natural and important question is the following one:

Is the task of solving polynomial equations or inequations modulo n as difficult as the factorisation of n?

We show here that, even if the factorisation of n is unknown, we can solve in polynomial probabilistic time polynomial inequations or polynomial equations modulo n provided we are given a sufficiently good initial approximation of a solution.

Our main tool is lattices that we use after a linearisation of the problem; we study a particular kind of lattice, which generalize that of Frieze et al, and the solution of our problem relies on the geometrical regularity of these lattices.

Our results are both algorithmical and structural:

On the one hand, we exhibit an algorithm, based on lattice reduction ideas, which reconstructs truncated roots of polynomials, and we extend here some previous results, only obtained in the linear case by Frieze et al. This algorithm has numerous practical applications, since the security of many cryptographic schemes is based on the difficulty of solving polynomial equations or inequations modulo n. We first deduce that it is easy to break higher–degree versions of Okamoto's recent cryptosystem and we extend, in this way, previous attacks of Brickell and Shamir. We also obtain new results about the predictability of the RSA pseudo-random generator.

On the other hand, we establish, for any ℓ, new theoretical results about the comparative distribution of ℓ-th powers and their ℓ-th roots, and we can prove, in the case $\ell = 2$, a very natural property about this distribution. These results can be seen as extensions, in a slightly different way, of a previous theorem of Blum.

This work was supported in part by PRC "Mathématiques et Informatique" and in part by a convention between SEPT and University of Caen.

1. INTRODUCTION.

It is known that extracting ℓ-th roots modulo a composite number n is a hard problem. In fact, solving the congruence $x^2 \equiv y$ $[n]$ is equivalent to factoring n. In this paper, we study the problem of extracting ℓ-th roots, i.e solving the congruence

$$x^\ell \equiv y \quad [n] \tag{1}$$

when the factorisation of n is unknown. Our treatment applies to moduli n that are either square-free or almost square-free, and we assume that we start with initial approximations x_0 of x and y_0 of y.

There are several reasons to be interested in such a problem:

(i) Square-free moduli are likely to represent harder cases than highly composite numbers –which are easier to factor–

(ii) In cryptographic applications, one often must guess some bits on each of the variables x and y, solutions of the congruence (1) while the other bits are given in the problem: This is the general framework

- in the cryptographic proposals of Okamoto,
- and also in the problems posed by the predictability of the RSA pseudo-random generator.

(iii) As a particular case, the comparative distribution of the quadratic residues and their square roots play an important role in factorisation algorithms based on congruences of squares.

We exhibit here a general method, based on the geometrical study of some particular kind of lattices, and we build a new framework that allow us to extend and unify previous results of Frieze, Brickell, Shamir and Blum.

Our paper is organized as follows: First we recall the problems already posed, their partial solutions and describe how our results solve extensions of these problems. After this, we introduce our main tool, lattices, and explain how their geometrical properties fit in our subject. Finally, we come back to the numbers, deduce our results and describe their cryptographic applications. We finish by asking some open problems.

1.1. Some definitions.

Let $Z(n)$ denote the ring of the integers modulo n that we identify with the interval of the integers of length n centered at 0. We are given an integer $\ell \geq 2$ and we look for ℓ-th roots modulo n: We always assume that n and ℓ are coprime.

We shall often consider moduli n which are not too different from a squarefree one, and we define, for $\delta \in [0,1]$, a number n to be *δ-squarefree* number if

$$n = \prod_{i=1}^{j} p_i^{e_i}, \quad \text{and} \quad \prod_{i=1}^{j} p_i^{e_i - 1} \leq n^\delta \quad (p_i \text{ are the distinct prime factors of } n)$$

These numbers are generalizations of squarefree ones (a squarefree number is 0 squarefree) and have been already introduced by Frieze et al. [4].

A number n is said to be *δ-monosquarefree* if it can be written:

$n = p^2 q$ with a prime p equal to n^δ, coprime with the squarefree q.

When n is a δ-monosquarefree number and $\ell \geq 2$ an integer, some particular elements x_0 of $Z(n)$ play a particular role ; these are the x_0's such that x_0 modulo pq is less than $n^{(1-\delta)/2\ell}$. Such an element x_0 is called *easy* .

We use approximations of a number x_0 of $Z(n)$, and we consider various sorts of neighbourhoods of x_0. If $|u|$ denotes, for $u \in Z(n)$, the maximum of u and $-u$, it is usual to consider the intervals

$$I(a, x_0) = \{x \in Z(n) \;/\; x = x_0 + u, \;\; |u| \leq n^a\}.$$

More generally, if a, a_1, a_2 are three reals of $[0,1]$, we define the subsets

$$K(a_1, a_2, x_0) = \{x \in Z(n) \;/\; x = u_1 x_0 + u_2, \;\; |u_1| \leq n^{a_1} \;,\;\; |u_2| \leq n^{a_2}\}$$

formed with intervals centered at small multiples of x_0. We distinguish two particular important cases: $a_1 = 0$ and also $a_1 = a_2 = \frac{a}{2}$ and we set

$$H(a, x_0) = K(0, a, x_0) \;\text{ and }\; J(a, x_0) = K(\frac{a}{2}, \frac{a}{2}, x_0)$$

Remark that $H(a, x_0)$ is the union of the three intervals $I(a, x_0)$, $I(a, 0)$ and $I(a, -x_0)$. The intervals I, and more generally, the subsets H, define that we call *inhomogeneous approximations*. On the other side, the subsets J define *homogeneous approximations*.

1.2. Some natural problems.

Instead of solving exactly the equation $x^\ell \equiv y \;\; [n]$, which is surely a difficult problem when the factorisation of n is hidden, we allow approximations on one or the other of the two variables x or y. The following problems thus arise in a natural way:

Problem 1:

Given an interval $I(b, y_0)$ of $Z(n)$, Find x such that x^ℓ belongs to $I(b, y_0)$.

Problem 2:

Given

(i) y_0, an ℓ-th power in $Z(n)$ whose ℓ-th roots are unknown,

(ii) a subset $H(a, x_0)$ (resp $J(a, x_0)$) which is known to contain an ℓ-th root x of y_0,

Find x.

1.3. Our main results.

In fact, the problem we solve here is more general; it is a natural extension of both problems 1 and 2 since it allows approximations on both variables x and y.

Problem 3:

Given two subsets of $Z(n)$

(i) an interval $I(b, y_0)$

(ii) and a subset $K(a_1, a_2, x_0)$

Do there exist x in $K(a_1, a_2, x_0)$ and y in $I(b, y_0)$ such that $x^\ell \equiv y \; [n]$?

If yes, find them.

Before stating our main result, which essentially solves this last problem, we need two definitions:

Two subsets A and B of $Z(n)$ are said to be *ℓ-compatible* if and only if there exists a pair (x, y) of $A \times B$ such that $x^\ell \equiv y$ modulo n; any such pair is called a *compatibility pair*. We say that two compatibility pairs (x, y) and (x', y') are *twins* if and only if $x' = -x$ and $y' = y$; remark that, if A is symmetrical with respect to 0 and if ℓ is even, the compatibility of such a pair entails the compatibility of the other one.

The following definition allows us to state sufficient conditions under which subsets are compatible:

Let a_1, a_2, b, δ be four real numbers of $[0,1]$, $\epsilon > 0$ a real number and $\ell \geq 2$ an integer; the triple (a_1, a_2, b) satisfy the conditions $C(\ell, \delta, \epsilon)$ if and only if

$$\frac{\ell(\ell+1)}{2} a_1 + \frac{\ell(\ell-1)}{2} a_2 + b = 1 - \delta - \ell\epsilon \text{ and } b \geq \ell a_2 \tag{2}$$

We can give now our first main result, a theoretical one, which describes the relative spreading of ℓ-th roots and ℓ-th powers; it is a uniqueness result : If x and x_0 are two "close" elements of $Z(n)$, their ℓ-th powers are not often "close".

THEOREM 1.

For $\epsilon > 0$, for $\ell \geq 2$, for n δ-squarefree, $n \geq n_0(\ell, \epsilon)$, for a triple (a_1, a_2, b) satisfying $C(\ell, \delta, \epsilon)$, there exists an exceptional set $S(\epsilon)$ of $Z(n)$ such that the following is true:
(i) $|S(\epsilon)| \leq n^{1-\epsilon}$
(ii) For any x_0 not in $S(\epsilon)$, for any y_0, the two subsets $K(a_1, a_2, x_0)$ and $I(b, y_0)$ have at most one compatibility pair (for an odd ℓ) and at most two twin compatibility pairs (for an even ℓ).

And our second main result, which is constructive:

THEOREM 2.

For $\epsilon > 0$, for $\ell \geq 2$, for n δ-squarefree, $n \geq n_0(\ell, \epsilon)$, for a triple (a_1, a_2, b) satisfying $C(\ell, \delta, \epsilon)$, there exists a polynomial probabilistic algorithm that decides, except in $S(\epsilon)$, on the ℓ-compatibility of the two subsets $K(a_1, a_2, x_0)$ and $I(b, y_0)$.

1.4. The particular case $\ell = 2$.

The following result shows that the square roots of a sufficiently large interval have an almost regular distribution in the whole of $Z(n)$ and thus extends a previous result of Blum [2]. More precisely, we can obtain here an existence result:

THEOREM 3.

For any $\epsilon > 0$, for any $n \geq n_0(\ell, \epsilon)$, for any pair (a, b) satisfying the two conditions

$$a + b = 1 + 4\epsilon \text{ and } b \geq 2a$$

there exists an exceptional subset $S(\epsilon)$ of $Z(n)$ such that the following is true:
(i) $|S(\epsilon)| \leq n^{1-\epsilon}$
(ii) For any x_0 not in $S(\epsilon)$, for any y_0, the two intervals $I(a, x_0)$ and $I(b, y_0)$ are 2-compatible.

There is also a constructive version of this result.

1.5. The predictability of the RSA Pseudo-Random Generator.

Alice chooses an element x_0 in $Z(n)$ and calculates $y_0 \equiv x_0^\ell\ [n]$; then she hides the $[a \log_2 n]$ least significant bits of x_0 and the $[b \log_2 n]$ least significant bits of y_0. She asks Bob to guess them.
Can Bob do it ? What is the total proportion of bits that he can guess ?
Our answer is the following one: Bob can guess a total proportion of bits equal to

$$a + b = \frac{2}{\ell}(1 - \delta - \ell\epsilon) \text{ provided that } b \geq \ell a.$$

Thus, for $\ell = 2$, since δ can be neglected, Bob guesses almost as many bits as Alice shows him.

We give another application to Pseudo-Random Generators:

The RSA Pseudo-Random Generator of degree ℓ and of modulus n is very well known : we choose x_0 in $Z(n)$ and define the sequence (x_i) such that

$$x_{i+1} \equiv x_i^{\ell} \; [n]$$

By removing the $[a \log_2 n]$ least significant bits of x_i, we obtain a truncated sequence. An interesting question is:

How to choose a so that this sequence be predictable ?

We can answer:

The pseudo-random generator obtained by truncature of the $[a \log_2 n]$ least significant bits of the RSA generator of degree ℓ and of modulus n is right predictable provided that

$$a < 2\frac{1-\delta}{\ell(\ell+1)};$$

the degree of squarefreeness δ of n can be neglected if $\ell = 2$.

1.6. Application to the attack of cryptographic schemes.

The task of breaking Okamoto's schemes [6], [7], [8], relies upon the existence of algorithms which solve, in probabilistic polynomial time, even if the factorisation of the modulus n is hidden, the already posed problems; we don't wish to give here a precise description of these schemes and their attacks –this description, in the particular case $\ell = 2$ can be found in [10]– here, we only explain the relation between our results and the attacks of these cryptographic schemes.

In these schemes, the modulus n is supposed to be particular: $n = p^2 q$ where p and q are distinct primes ($p < q$); such a number is 1/3 monosquarefree.

Solving Problem 1: the attack of the signature scheme.

In [6], Okamoto and Shiraishi proposed a signature scheme:
Given a 'one-way' function h, a signature x is considered as valid for a message u if

$$h(u) \leq x^{\ell} \; [n] \leq h(u) + O(n^b) \quad \text{with } |x| \text{ not too small and } \; b \text{ much smaller than } \; 1$$

Therefore, Problem 1 is exactly the problem we must solve if we want to break this signature scheme. It has been already solved with $b = 2/3$ in the two particular cases:

(1) if $\ell = 2$, even if the factorisation of n is hidden (Brickell [3])

(2) for any ℓ, provided the factorisation of $n = p^2 q$ (p and q distinct primes) is known (Okamoto [6], Brickell [3]). It is easy to see that this result can be generalised when q is any number coprime with p.

We here improve this previous result in two directions without using the particular form of the modulus:

(1) in the case $\ell = 2$, we show that *one can find the square root x in almost any prescribed interval $I(a, x_0)$ provided that a is sufficiently large: $a > 1/3$*

(2) for any ℓ, for any n, we show *one can guess x, provided one knows an estimate of it (which depends on ℓ and δ, the degree of squarefreeness of n) even if the factorization of n is hidden and, thus, attack the signature scheme with some extra information.*

Solving Problem 2: the attacks of the cryptosystems.

In [7], Okamoto proposed a first public key cryptosystem:
The public key is the pair (n, x_0), where x_0 is an easy element of $Z(n)$. From a message u, which is small compared to n, the cipher text y is built as follows:

$$y \equiv (x_0 + u)^{\ell} \; [n]$$

As quoted in [8], Shamir [9] has two attacks to break this system: the first one works for any pair (n, x_0) while the second one uses the particular form of the public key.

Okamoto [8] then proposed a new cryptosystem: x_0 is the known quotient modulo n of two secret easy numbers of $Z(n)$. A message (u_1, u_2), where the u_i's are small compared to n, gives a cipher text y such that

$$y \equiv (u_1 x_0 + u_2)^\ell \ [n].$$

Therefore, the security of these cryptosystems are based on the difficulty of solving Problem 2, in the inhomogeneous case for the first version [7], in the homogeneous case for the second version [8].
Shamir [9] (1986) has already attacked the first -inhomogeneous- version of Okamoto's cryptosystems by solving the second problem in the case of an inhomogeneous approximation. He has two different attacks (explained in [8]) depending on whether one of the following conditions is realized:
(1) for any pair (x_0, n), when $\ell = 2$ and $a < 1/3$
(2) for any ℓ, for any δ-monosquarefree n, for any easy x_0.

The case of a homogeneous approximation, under the same conditions on the parameters, is an open question of Okamoto (1987); in this case, a positive answer to this question will break the second version of Okamoto's cryptosystem.

We can actually solve this problem under these conditions.

Moreover, we have a more general attack that extends the first attack of Shamir and allows us to solve Problem 2.

THEOREM 4.

For any ℓ, for the two kinds of approximation,there exists a probabilistic polynomial time algorithm which solves Problem 2 provided that n be any sufficiently large δ-squarefree number and

$$a < 2\,\frac{1-\delta}{\ell(\ell+1)}.$$

The corrective term, which depends on δ, can be neglected when $\ell = 2$.

We deduce the following result:

All the versions of Okamoto's cryptosystems can be broken.

We present also a second attack that uses the particular form of the public key and extends the second attack of Shamir.

2. THE IRRUPTION OF LATTICES.

Here, we show how lattices come in a natural way to our problem; we introduce a first lattice, then a second one in order to work with the norm sup. We recall some geometrical properties of a lattice linked with the length of the first minimum and finish by analyzing the size of the first minimum in our specific lattices.

2.1. The first lattice.

We are given an integer $\ell \geq 2$ and a pair (x_0, y_0) of two elements of $Z(n)$ and we consider the ℓ-th compatibility of the two subsets $K(a_1, a_2, x_0)$ and $I(b, y_0)$: we want to find a triplet (u_1, u_2, v) of "small" integers solving the equation

$$(u_1 x_0 + u_2)^\ell \equiv y_0 + v \ [n]$$

Through a binomial expansion, we obtain

$$x_0^\ell u_1^\ell + C_\ell^1 x_0^{\ell-1} u_1^{\ell-1} u_2 + \ldots\ldots + C_\ell^i x_0^{\ell-i} u_1^{\ell-i} u_2^i + \ldots. + C_\ell^1 x_0 u_1 u_2^{\ell-1} + u_2^\ell - v \equiv y_0 \ [n]$$

We consider the lattice $L(\ell, x_0)$ of the vectors $w = (w_0, w_1, \ldots., w_\ell)$ of $\mathbf{Z}^{\ell+1}$ such that

$$\sum_{i=0}^{\ell-1} C_\ell^i x_0^{\ell-i} \, w_i \, - w_\ell \equiv 0 \ [n]$$

and we must find, in $L(\ell, x_0)$, a point w which is -in the sense of an unusual norm- "near" to the point $(0, 0, \ldots., y_0)$. Lattice $L(\ell, x_0)$ has the following matrix $(\ell+1, \ell+1)$

$$\begin{pmatrix} 1 & 0 & 0 & \cdots & 0 & 0 \\ 0 & 1 & 0 & \cdots & 0 & 0 \\ 0 & 0 & 1 & \cdots & 0 & 0 \\ \vdots & \vdots & \vdots & \ddots & \vdots & \vdots \\ 0 & 0 & 0 & \cdots & 1 & 0 \\ x_0^\ell & C_\ell^1 x_0^{\ell-1} & C_\ell^2 x_0^{\ell-2} & \cdots & C_\ell^{\ell-1} x_0 & n \end{pmatrix}$$

For each component, the prescribed approximation, linked to the choice of the neighbourhoods, is the following one:

$$|w_i| \leq n^{(\ell-i)a_1 + ia_2} \quad \text{for all } i: \ 0 \leq i \leq \ell - 1 \ \text{ and also } \ |w_\ell - y_0| \leq 2n^b$$

(For the last condition, we use the hypothesis: $b \geq \ell a_2$)

2.2. The second lattice.

Except in the case where $a_1 = a_2 = \dfrac{b}{\ell}$, these approximations are not equal, and we are going to use a system of multipliers

$$k = (k_0, k_1, \ldots., k_\ell)$$

in order to "expand-contract" the lattice $L(\ell, x_0)$ into another one, the lattice $M(\ell, x_0)$, in which the approximations are made equal: we can now, in this lattice, use the norm sup. The lattice $M(\ell, x_0)$ is the set of the vectors $t = (t_0, t_1, \ldots., t_\ell)$ which satisfy

$$t_i = k_i \, w_i \quad \text{for all } i: \ 0 \leq i \leq \ell \ \text{ with } \ w = (w_0, w_1, \ldots., w_\ell) \in L(\ell, x_0)$$

To preserve the determinant of the lattice, we can suppose, after a possible homothety, that

$$\prod_{i=0}^{\ell} k_i = 1$$

Then, the system of multipliers k and the associated lattice $M(\ell, x_0)$ will be said simple. If we now look for rationals k_i with the form $k_i = n^{c_i}$ for all $i : 0 \leq i \leq \ell$, we have the conditions of approximation equalizations:

$$c_i + (\ell - i)a_1 + ia_2 \text{ independent of } i : 0 \leq i \leq \ell - 1 \text{ and equal to } c_\ell + b \quad (3)$$

and the condition of simplicity

$$\sum_{i=0}^{\ell} c_i = 0. \quad (4)$$

Now, we transport our problem into the simple lattice $M(\ell, x_0)$ which has the following matrix:

$$\begin{pmatrix} k_0 & 0 & 0 & \dots & 0 & 0 \\ 0 & k_1 & 0 & \dots & 0 & 0 \\ 0 & 0 & k_2 & \dots & 0 & 0 \\ \vdots & \vdots & \vdots & \ddots & \vdots & \vdots \\ 0 & 0 & 0 & \dots & k_{\ell-1} & 0 \\ k_\ell x_0^\ell & k_\ell C_\ell^1 x_0^{\ell-1} & k_\ell C_\ell^2 x_0^{\ell-2} & \dots & k_\ell C_\ell^{\ell-1} x_0 & k_\ell n \end{pmatrix}$$

and we look for a point $t = (t_0, t_1, ..., t_\ell)$ of this lattice near to the point $m = (0, .., k_\ell\, y_0)$ for the norm sup.

2.3. Some questions.

We are thus led to some important questions:

1) How to find this point? Will it be a convenient one ? More precisely,
2) Will this point t be sufficiently close to $m = (0, 0, ..., 0, k_\ell\, y_0)$ so that the prescribed approximations can be satisfied ?
3) Does it come from our problem: is it a "power" point, i.e does there exist a triplet (u_1, u_2, v) such that

$$t_i = k_i\ u_2^i u_1^{\ell-i} \text{ for all } i : 0 \leq i \leq \ell - 1 \text{ and also } t_\ell = k_\ell(y_0 + v - u_2^\ell) \quad (5)$$

2.4. The ClosePoint Algorithm.

Given in $\mathbf{Q}^{\ell+1}$ a lattice L of rank $\ell + 1$ and a point m, finding the lattice point nearest to m is a NP- hard problem. But there are many polynomial time algorithms, using the LLL lattice reduction, that give a lattice point t nearest to m

i) within a factor $F(\ell)$
ii) with respect to the euclidean norm.

These two restrictions will pose some questions.

We use here the simplest algorithm [1], which also gives the worst estimate for $F(\ell)$

$$F(\ell) = O(\ell 2^\ell)$$

but it will be adequate, since we must choose ℓ small compared to the expected distance so that $F(\ell) = n^\epsilon$. We describe it:

1) Reduce, by the LLL algorithm, the given basis of L into a basis $e = (e_0, e_1, ..., e_\ell)$
2) Express m in the basis e as $m = \sum_{i=0}^{\ell} m_i e_i$
3) $t = \sum_{i=0}^{\ell} t_i e_i$ where t_i is the integer nearest to m_i.

Note that the gcd d of $(w_0, w_1, ..., w_\ell)$ divides n. If we let

$$s_i = C_\ell^i w_i \quad \text{for } i : 0 \leq i \leq \ell - 1 \quad \text{and} \quad s_\ell = -w_\ell$$

we have also

$$d = \gcd(s_0, s_1, ..., s_\ell)$$

and

$$x_0 \text{ is a root modulo } n \text{ of the polynomial } Q(X) = \sum_{i=0}^{\ell} s_i X^{\ell - i}$$

For a fixed divisor d of n there are at most

$$\prod_{i=0}^{\ell} \frac{3R}{k_i d} = \left(\frac{3R}{d}\right)^{\ell+1}$$

such polynomials.

We now estimate the number of the roots x_0 modulo n of such a polynomial. If we replace n by n' defined by the quotient of n by d and s_i by s'_i, the quotient of s_i by d, we obtain

$$\sum_{i=0}^{\ell} s'_i \, x_0^{\ell-i} \equiv 0 \ [n']$$

Each solution modulo n' lifts to at most d solutions modulo n and, now, by the Chinese theorem, there are at most [4]

$$\ell^f n^\delta \ \text{ solutions modulo } \ n'$$

except for $\ell = 2$, when we can neglect the term n^δ.

We finally deduce :

$$|U(R)| \leq (3R)^{\ell+1} \ \ell^f \ n^\delta \sum_{d|n} \frac{1}{d^\ell}$$

To simplify this further, we use the two estimates: first,

$$\sum_{d|n} \frac{1}{d^\ell} \leq 2 \ \text{ for } \ \ell \geq 2$$

and the well-known theoretic estimate for the number f of different prime factors of n that shows there is a constant c_0 such that for n sufficiently large, one has:

$$f \leq c_0 \frac{\log n}{\log \log n}$$

And thus, we deduce :

$$\ell^f \leq n^\epsilon \ \text{ provided } \ \log n \geq \ \ell^{c_0/\epsilon}$$

and also:

$$|U(R)| \leq (3R)^{\ell+1} \ n^{\delta+\epsilon}$$

If we choose $3R = n^{(1-\delta-2\epsilon)/(\ell+1)}$, we have $|U(R)| \leq n^{1-\epsilon}$. We define now $S(\epsilon)$ to be equal to $U(R)$ with this particular value of R and this ends the proof.

2.5. The ϵ-regular lattices.

By definition, a lattice L of rank $\ell + 1$, whose determinant n is δ-squarefree, is said to be ϵ-regular if and only if its first minimum, denoted by $\lambda_1(L)$ satisfies

$$\|\lambda_1(L)\|_\infty \geq n^{(1-\delta-2\epsilon)/(\ell+1)}$$

By comparing this bound with the geometrical mean of the minima's euclidean lengths -which is of the order of $n^{1/(\ell+1)}$-, and using Minkowski's inequality, we remark that all the successive minima of a regular lattice are almost equal. All this explains the name of such a lattice and allow to get the following geometrical facts, both theoretical and practical:
Let $B(m,r)$ be an euclidean ball of center m and radius r in $\mathbf{Q}^{\ell+1}$. We let

$$r_0 = n^{(1-\delta-2\epsilon)/(\ell+1)} \qquad r_1 = \frac{r_0}{F(\ell)} = n^{(1-\delta-\ell\epsilon)/(\ell+1)}$$

$$r_2 = (2\gamma_\ell)^{\ell/2} \ n^{1/(\ell+1)+\delta+2\epsilon} \ (\gamma_\ell \text{ is Hermite's constant}) \qquad r_3 = F(\ell) r_2$$

and we have:

Uniqueness: if $r < r_0/2$, this ball contains **at most** one point of the lattice,

Constructive Uniqueness: and if, moreover $r < r_1$, the algorithm **ClosePoint** gives the following results: it outputs this point, if it exists or answer "none", if it does not exist.

Existence: if $r > r_2$, this ball contains **at least** one point of the lattice,

Constructive Existence: and if, moreover $r > r_3$, the algorithm **ClosePoint** finds a lattice point.

2.6. The ϵ-regularity of the lattices $M(\ell, x_0)$.

We shall prove here that most of the lattices $M(\ell, x_0)$ are ϵ-regular. More precisely, we state the following result:

Proposition:

For $\epsilon > 0$, for $\ell \geq 2$, for n δ-squarefree, $n \geq n_0(\ell, \epsilon)$, for any simple system of multipliers k, there exists an exceptional subset $S(\epsilon)$ of $Z(n)$, depending on k, such that the following is true:

(i) $|S(\epsilon)| \leq n^{1-\epsilon}$

(ii) For any x_0 not in $S(\epsilon)$, the simple lattice $M(\ell, x_0)$ is ϵ-regular.

We give now a sketch of the proof, whose outline follows FHKFS [4]. However, we don't need the reciprocal lattice.

For a fixed R, we are interested in estimating the size of the set

$$U(R) = \{x_0 \in Z(n) \ / \ \exists t \in M(\ell, x_0), \ \|t\|_\infty \leq R\}$$

If $x_0 \in U(R)$, then there exists $w \in L(\ell, x_0)$ such that

$$|w_i| \leq \frac{R}{k_i} \quad \text{for all} \quad i : 0 \leq i \leq \ell$$

and also

$$w_\ell \equiv \sum_{i=0}^{\ell-1} w_i C_\ell^i \, x_0^{\ell-i} \ \ [n]$$

2.7. Another exceptional set $T(\epsilon)$.

Finding the shortest vector of the lattice $M(\ell, x_0)$ is almost certainly an NP hard problem, and thus we can't know whether we are in the exceptional subset $S(\epsilon)$: we are going to consider another exceptional set $T(\epsilon)$ with a more algorithmical definition.
We define $e_0(x_0)$ to be the first vector of the LLL reduced basis of $L(\ell, x_0)$. We know that [5]

$$|e_0(x_0)| \leq 2^{\ell/2}\ |\lambda_1(x_0)|$$

If we let

$$T(\epsilon) = \{x_0 \in Z(n)\ / |e_0(x_0)| \leq 2^{\ell/2} n^{(1-\delta-2\epsilon)/(\ell+1)}\}$$

we can work with $T(\epsilon)$ in the same way as with $S(\epsilon)$ provided

$$2^{\ell(\ell+1)/2} \text{ is small compared to } n^{\epsilon}.$$

3. FROM LATTICES TO NUMBERS.

We come back to our original problem. First, we choose a convenient lattice where we can apply the results of the previous section : we prove our main results (Theorems 1 and 2). Then, we make precise some conditions on the parameters and state an optimal choice on them, which we use in the proof of Theorem 4. After this, we describe the pecularities of the case $\ell = 2$ which allows us to prove Theorem 3. We finish by giving another attack of Okamoto's second cryptosystem that extends the second attack of Shamir.

3.1. Choosing the expansion-contraction $M(\ell, x_0)$ of the lattice $L(\ell, x_0)$.

We can choose now the expansion-contraction in order to use the second consequence of the ϵ-regularity of the lattices $M(\ell, x_0)$: we want a compatibility pair to build a vector t of $M(\ell, x_0)$ in the ball $B(m, r_1)$.
If we come back to the conditions of approximation equalizations (3), we require

$$c_i + (\ell - i)a_1 + ia_2 = \frac{1-\delta-\ell\epsilon}{\ell+1} \text{ for all } i : 0 \leq i \leq \ell - 1 \text{ and also } c_\ell + b = \frac{1-\delta-\ell\epsilon}{\ell+1}.$$

Since the system of multipliers is simple, we have (4)

$$\sum_{i=0}^{\ell} c_i = 0 = 1 - \delta - \ell\epsilon - \left(a_1\frac{\ell(\ell+1)}{2} + a_2\frac{\ell(\ell-1)}{2} + b\right).$$

Our requirement is thus satisfiable if and only if the triple (a_1, a_2, b) satisfies $C(\ell, \delta, \epsilon)$ defined in (2).

3.2. The proofs of the main results, Theorem 1 and Theorem 2.

We return now to our problem.
We fix $\epsilon > 0$ and a triple (a_1, a_2, b) which satisfies $C(\ell, \delta, \epsilon)$; we consider $x_0 \notin S(\epsilon)$ and (x, y) a possible compatibility pair for the two subsets $K(a_1, a_2, x_0)$ and $I(b, y_0)$.
We associate to this pair the "power" point w of $L(\ell, x_0)$, and then the corresponding point t of our particular lattice $M(\ell, x_0)$ defined in (5).

This power point t is in the euclidean ball $B(m, r_0/2)$ which contains at most one point of $M(\ell, x_0)$, and *a fortiori* at most one power point of $M(\ell, x_0)$. Each power point comes from two twin triples (u_1, u_2, v) (if ℓ is even) and one triple (if ℓ is odd) and we easily get these triples by extracting ℓ-th roots in the integers in the equations (5). After this, the verification of the prescribed approximations on the triple (u_1, u_2, v) remains to be done. All this shows our main theoretical result (Theorem 1)

Now, we adopt a constructive point of view for the proof of Theorem 2:
Under the same hypotheses, the power point t, if it exists, is in the euclidean ball $B(m, r_1)$ and there is no other lattice point in the ball $B(m, r_0/2)$. The **ClosePoint** algorithm allows us to find a lattice point t' and clearly, we have $t = t'$: we thus find t.
If $B(m, r_1)$ does not contain any power point, the algorithm **ClosePoint** may find any other lattice point, which is checked not to be suitable.

3.3. Some precisions about the choices of n, ℓ, ϵ.

There are three occasions where we needed some conditions on this triple. We required:

1) $c_0 \log \ell \leq \epsilon \log \log n$ (in the proof 2.6)
2) $\ell(\ell+1) \leq \epsilon \log n$ (in the remark 2.7)
3) $\log F(\ell) \leq \epsilon \log n$ with $F(\ell) = O(\ell 2^\ell)$ (in the description 2.4)

We will thus choose

$$\log \ell \leq K\epsilon \log \log n \quad \text{i.e.} \quad n \geq n_0(\ell, \epsilon)$$

where K is a constant to be determined as a function of c_0.

3.4. The optimal choice of the parameters.

We now return to the particular subsets $H(a, x_0), I(a, x_0)$ or $J(a, x_0)$, and our conditions $C(\ell, \delta, \epsilon)$ can be written in each of these cases :

$$\text{Inhomogeneous case : } \frac{\ell(\ell-1)}{2} a + b = 1 - \delta - \ell\epsilon \text{ and } b \geq \ell a$$

$$\text{Homogeneous case : } \frac{\ell^2}{2} a + b = 1 - \delta - \ell\epsilon \text{ and } b \geq \ell \frac{a}{2}$$

In each of these cases, the optimal choice (a_0, b_0) for the pair (a, b) is the following

$$\text{For both cases : } a_0 = 2 \frac{1 - \delta - \ell\epsilon}{\ell(\ell+1)}$$

$$\text{In the first case: } b_0 = 2 \frac{1 - \delta - \ell\epsilon}{\ell+1}. \quad \text{In the second case: } b_0 = \frac{1 - \delta - \ell\epsilon}{\ell+1}.$$

3.5. A precise description of the GuessRoot Algorithm.

Input: Two subsets $K(a_1, a_2, x_0)$ and $I(b, y_0)$ of $Z(n)$, an integer $\ell \geq 2$ and an estimate of δ.

Output: two booleans, *careful* and *compatible* and, if it exists, a compatibility pair (x, y)

1) Determine $\epsilon > 0$ such that the triple (a_1, a_2, b) "satisfies" the $C(\ell, \delta, \epsilon)$ conditions and verify that the triple (n, ℓ, ϵ) satisfies the conditions 3.3. If not, answer "Bad choices of the parameters" and go to the end.
2) Choose the system k of the multipliers as in 3.1, calculate the point m and reduce the associated lattice $M(\ell, x_0)$; *careful*:= false; *compatible* := false;
3) If x_0 is in $T(\epsilon)$ then *careful*:= true.

4) $t :=$ **ClosePoint**$(m, M(\ell, x_0))$
   if t is a power point then
      calculate the triple (u_1, u_2, v)
      if the prescribed approximations are satisfied then *compatible*:= true
5) If not *compatible* and *careful* then answer
      "Failure of the algorithm";
   If not *compatible* and not *careful* then write
      "no compatibility pair";
   If *compatible* then output $x = u_1 x_0 + u_2, y = y_0 + v$,

3.6. The comparative distribution of ℓ-th roots and ℓ-th powers

We first recall Blum's result [2] :

Let $n \geq 5$ be a squarefree number. The probability that $I(1/2, x_0)$ and $\{x_0^2\}$ have a compatibility pair different from the trivial one is less than $48n^{-1/2} \log\log n$.

to be compared with our result:

Let n be a δ-squarefree number, $\ell \geq 2$ be an integer and $(0, a, b)$ a triple which verifies $C(\ell, \delta, \epsilon)$. The probability that $I(a, x_0)$ and $I(b, x_0^\ell)$ have a compatibility pair different from the trivial one is less than $n^{-\epsilon}$.

We can interpret the exceptional subset $S(\epsilon)$:
When, as usual, $(0, a, b)$ verifies $C(\ell, \delta, \epsilon)$, we notice the following fact:
If (x_1, y_1) is a compatibility pair of $I(a, x_0) \times I(b, x_0^\ell)$ and if x_0 is an element of $S(\epsilon)$, then the whole interval $[x_0, x_1]$ is contained in $S(\epsilon)$.

$S(\epsilon)$ prevents us from splitting roots, because it contains some intervals where the roots are piling up.

3.7. The particular case $\ell = 2$; proof of Theorem 3.

Here, we consider the case of two intervals $I(a, x_0)$ and $I(b, y_0)$. The study of their compatibility takes us into the section of $L(\ell, x_0)$ by the hyperplane $w_0 = 1$ that we call $L'(\ell, x_0)$. The analysis of the ϵ-regularity of the associated lattice $M'(\ell, x_0)$, which is of rank ℓ, is the same as in 2.6 and we can neglect the term n^δ if $\ell = 2$. In this particular case, we must find "small" solutions (u, v) of the equation:

$$2x_0\, u - (v - u^2) \equiv y_0 - x_0^2 \ \ [n]$$

The lattice $M'(\ell, x_0)$ has the following matrix

$$\begin{pmatrix} k_1 & 0 \\ 2k_2 x_0 & k_2\, n \end{pmatrix}$$

and we must find, in $M'(\ell, x_0)$, a point "near" to $m = (0, k_2(y_0 - x_0^2))$.
We first observe a very particular fact due to the case $\ell = 2$. Given $w = (w_1, w_2)$ and (x_0, y_0), the following system has always a solution (u, v): $w_1 = u, \ \ w_2 = y_0 - x_0^2 + v - u^2$.

All the points of $M'(\ell, x_0)$ come from our problem : they are power points !

On the other side, we can choose the multipliers (k_1, k_2) in order to use the third and fourth consequence of the ϵ-regularity of $M'(\ell, x_0)$: we require

$$c_1 = \frac{1}{2} - a + 2\epsilon, \quad c_2 = \frac{1}{2} - b + 2\epsilon, \quad \text{and} \quad c_1 + c_2 = 0,$$

which is possible by hypothesis.

We thus obtain a point t of $M'(\ell, x_0)$, then a point w of $L'(\ell, x_0)$ and solve the previous system to obtain our compatibility pair. This ends the proof of Theorem 3.

3.8. An extension of Shamir's second attack.

Okamoto thought that the first attack of Shamir [9] could not be applied to the homogeneous case. We have shown here how an extension of the ideas of Shamir could generalize this first attack. Moreover, we now show that the second attack of Shamir [9] can be extended to the homogeneous case.

We recall the general framework of the second version of Okamoto's cryptosystems [8]: We consider a δ-monosquarefree number and two easy numbers x_1 and x_2; we let $x_0 = x_1/x_2$ and we want to recover the solution x in $J(a, x_0)$ of the equation: $x^\ell \equiv y_0$ $[n]$. More precisely, we try to find an x of the form $x = u_1 x_0 + u_2$, under the conditions $0 < u_1, u_2 < n^c$ with $c < (1-\delta)/2\ell$.
We know the quantities n, y_0, $x_0 = x_1/x_2$. The factors p and q of n and the pair (x_1, x_2) are hidden. Moreover, since the x_i are easy, we have

$$x_i = y_i + z_i pq \; [n] \quad \text{with} \quad 0 < y_i < \frac{1}{2} n^c \quad \text{and also} \quad 0 < z_i < p$$

The secret key is $(y_1, y_2, z_1, z_2, p, q)$.

We first show that we can easily recover the triple (y_1, y_2, p). We have in effect

$$(x_2 p)\, x_0 \equiv (x_1 p) \;\; [n]$$

and, since x_i is easy, all the multiples $(x_i k)$ $[n]$ of x_i (for $0 < k < p$) are in the order of $n^{1-\delta}$ while the particular multiple $x_i p$ $[n] = y_i p$ is less than $n^{c+\delta}$.
Provided $c + \delta$ is less than $(3/4)(1-\delta)$ and less than $1/2$, this last multiple is less than the others and sufficiently small to be found by variants of Euclid's algorithm. These conditions are equivalent to

$$\delta < \min\left(\frac{3\ell - 2}{7\ell - 2}, \frac{\ell - 1}{2\ell - 1}\right),$$

the latter inequality being always satisfied in the usual case where $\delta \leq \frac{1}{3}$.
We thus determine $y_1 p$ and $y_2 p$ and, after a gcd computation, each of the three terms y_1, y_2, p.

Since $pq = n/p$ is now known, we consider our equation modulo pq and obtain

$$(u_1 y_1 + u_2 y_2)^\ell \equiv y_0 y_1^\ell \;\; [pq]$$

But, by Okamoto's condition, we have

$$0 < u_1 y_1 + u_2 y_2 < n^{(1-\delta)/\ell}$$

and, by extracting an ℓ-th root in the integers, we recover the value A of $u_1 y_1 + u_2 y_2$. It remains to find small solutions to this linear equation. This can be done also by Euclid's algorithm because of the expected values of u_1 and u_2.

4. SOME EXTENSIONS AND OPEN PROBLEMS.

4.1. An easy extension.

It is clear that our method makes it possible to find a root x of any polynomial P of degree ℓ, provided we know a sufficiently good approximation x_0 of this root. We must solve

$$P(u_1 x_0 + u_2) \equiv \; y_0 + v \;\; [n]$$

Instead of a binomial expansion, we use Taylor's formula, and we obtain

$$x_0^\ell u_1^\ell \frac{P^{(\ell)}(u_2)}{\ell!} + x_0^{\ell-1} u_1^{\ell-1} \frac{P^{(\ell-1)}(u_2)}{(\ell-1)!} + \ldots\ldots + x_0^{\ell-i} u_1^{\ell-i} \frac{P^{(\ell-i)}(u_2)}{(\ell-i)!} + \ldots . + P(u_2) - v \equiv y_0 \;\; [n]$$

We replace the condition (5) by the following one

$$t_i = k_i \, \frac{P^{(\ell-i)}(u_2)}{(\ell-i)!} \, u_1^{\ell-i} \quad \text{for all } i : 0 \le i \le \ell - 1 \quad \text{and also} \quad t_\ell = k_\ell(y_0 + v - P(u_2)\,)$$

and we have the same kind of approximations equalizations.

In the case of an homogeneous approximation, we can also work in another lattice; Instead of $L(\ell, x_0)$, we work in a lattice whose matrix contains as a last row:

$$(P(x_0), \; P'(x_0), \; \frac{P''(x_0)}{2!}, \; \ldots, \; \frac{P^{(i)}(x_0)}{i!}, \; \ldots \; , \; \frac{P^{(\ell-1)}(x_0)}{(\ell-1)!}, \; n)$$

The analysis of the geometrical properties of these lattices, when P is fixed, and when x_0 varies, is the same as in 2.6 ; instead of the previous polynomials Q, we study polynomials Q that can be written as small linear combinations of

$$P(x_0), \; P'(x_0), \; \frac{P''(x_0)}{2!}, \; \ldots, \; \frac{P^{(i)}(x_0)}{i!}, \; \ldots \; , \; \frac{P^{(\ell-1)}(x_0)}{(\ell-1)!}.$$

4.2. Quasi-uniform algorithms for finding small quadratic residues and application to integer factorisation.

The problem of finding small quadratic residues mod n, when n is a large composite number of unknown factorisation, is almost surely a hard problem. We exhibit in [11] polynomial-time algorithms, based on similar ideas, which find, in a quasi-uniform way, elements x of $Z(n)$ such that the quadratic residues x^2 $[n]$ are smaller than $n^{2/3}$. We apply such an algorithm to integer factorisation and exhibit a new algorithm which appears to have the best rigourous bound known to date for integer factorisation algorithms.

4.3. An open question about $S(\epsilon)$.

We now pose an open question:

Can we describe precisely the exceptional set $S(\epsilon)$? This description, in the framework of 3.7, is made in [11]. How to generalize it? Can the study of this set bring any information on the localization of the factors of n ?

4.4. An open question about the compatibility probability.

Let a et b be two reals of $[0,1]$. Define

$$Q(a,b,\ell) = \{(x_0,y_0) \in Z(n)^2 \ / \ I(a,x_0) \text{ and } I(b,y_0) \text{ are } \ell\text{-compatible } \}$$

What is the probability $q(a,b,\ell)$ of the event $Q(a,b,\ell)$?
In the particular case $\ell = 2$, we have shown, that

$$q(a,b,2) \geq 1 - \frac{1}{n^\epsilon}$$

provided

$$a + b = 1 + 4\epsilon \quad \text{and} \quad b \geq 2a$$

There is a simple heuristic argument which could generalize this result in the case when the function $x \mapsto x^\ell$ is ℓ–to–one for each p_i. (This condition is equivalent to $p_i \equiv 1 \ [\ell]$).
We thus pose an open question:

Is it true that $q(a,b,\ell)$ is almost equal to 1 if $a+b$ is near to 1?

Acknowledgements.

We are happy to thank Jacques Stern for his encouragements and useful advices for the organization of this paper. We also wish to thank Philippe Flajolet, our English and TEX adviser.

5. BIBLIOGRAPHIC REFERENCES

[1] L. Babai: On Lovasz's lattice reduction and the nearest lattice point problem, *Combinatorica* 6, pp 1-14.

[2] M. Blum: How to exchange (secret) keys, *ACM transactions on Computer systems*, 1, 2, may 83, pp 175-193.

[3] E. Brickell, J. Delaurentis: An attack on a signature scheme proposed by Okamoto and Shiraishi, *Proc of Crypto'85*, pp 1-4.

[4] A. Frieze, J. Hastad, R. Kannan, J.C. Lagarias, A. Shamir: Reconstructing truncated variables satisfying linear congruences, to appear in *SIAM Journal of Computing*

[5] A.K. Lenstra, H.W. Lenstra, L. Lovasz : Factoring polynomials with integer coefficients, *Mathematische Annalen*, 261, (1982) pp 513-534

[6] T. Okamoto, A. Shiraishi: A fast signature scheme based on quadratic inequalities, *Proc of the 1985 Symposium on Security and Privacy*, April 1985, Oakland, CA.

[7] T. Okamoto: Fast public-key cryptosystem using congruent polynomial equations, *Electronics Letters*, 1986, 22, pp 581-582.

[8] T. Okamoto: Modification of a public-key cryptosystem, *Electronics Letters*, 1987, 23, pp 814-815.

[9] A. Shamir: Private communications to Okamoto, August and October 1986, (quoted in Okamoto [8]).

[10] B. Vallée, M. Girault, Ph. Toffin: How to break Okamoto's cryptosystems by reducing lattice bases, *Proceedings of Eurocrypt'87*, Lecture notes in Computer Science.

[11] B. Vallée: Quasi-uniform algorithms for finding small quadratic residues and application to integer factorisation, or
Factorisation entière par génération quasi-uniforme de petits résidus quadratiques, preprints of Département de Mathématiques de l'Université de Caen.

A Study on Imai-Hirakawa Trellis-Coded Modulation Schemes

Kazuhiko YAMAGUCHI and Hideki IMAI
Division of Electrical and Computer Engineering
Faculty of Engineering
Yokohama National University
156, Tokiwadai, Hodogaya-ku, Yokohama, 240, Japan

1. Introduction

Combined coding and modulation schemes have been studied to realize high speed and highly reliable data transmission over band-limited channels. Imai and Hirakawa[1][3] proposed a coded multilevel signal modulation scheme in 1975. The scheme uses several classes of binary error-correcting codes, symbols of which are combined to set up the transmission signal. We call this Imai-Hirakawa (IH) scheme.

Another important combined coding and modulation scheme was presented by Ungerboeck[4]. This scheme is a Trellis-Coded Modulation (TCM) scheme using a convolutional code designed on the basis of Euclidean distance (ED). We abbreviate it to U-TCM scheme.

IH scheme using several classes of convolutional codes can be considered as a different type of TCM from U-TCM. This scheme, which is abbreviated to IH-TCM scheme, was suggested by Imai and Hirakawa[1], but it was not well studied until Yamaguchi and Imai[2] have recently shown that it has excellent Euclidean distance property and easy implementation.

In this paper, we analyze error-correcting capability and complexity of implementation for the TCM scheme in detail. First, we derive the free ED of IH-TCM and define a criterion for complexity of implementation. IH-TCM systems which have large free ED and simple implementation are found. Second, the complexity of implementation are theoretically analyzed by using the random coding theorem. It shows that IH-TCM is easier to implement than the optimum multilevel coding scheme with the same decoding error probability. Finally, the bit error performance of the decoded data is evaluated by computer simulation. From the results we see that IH-TCM system, which optimize asymptotic coding gain and complexity of implementation, is not always good. It is shown, however, that properly designed IH-TCM has good bit error performance and simple implementation.

2. Coding and Decoding of IH Scheme

For IH scheme, each 2^M-ary signal is mapped to binary M-tuple $y_0, y_1, \cdots, y_{M-1}$. The method of mapping, which is given by Imai and Hirakawa[1], is essentially equivalent to the set partitioning method of Ungerboeck[4].

Encoding procedure of IH scheme is described as follows (See Fig. 1). We use binary error-correcting codes $C_0, C_1, \cdots, C_{M-1}$ which have rates $R_0, R_1, \cdots, R_{M-1}$, respectively. First the information sequence is partitioned into M component information sequences. The i-th component information sequence is fed to the encoder of C_i, which is denoted as E_i. Let $y_i(t)$

be the output of E_i at time t. Then M-dimensional vector $Y(t) = (y_0, y_1, \cdots, y_{M-1})$ is the encoded signal at time t.

$D_0, D_1, \cdots, D_{M-1}$ are decoders of $C_0, C_1, \cdots, C_{M-1}$, respectively. D_i estimates the output sequence of E_i by using the received signal sequence and the estimated sequences yielded from $D_0, D_1, \cdots, D_{i-1}$.

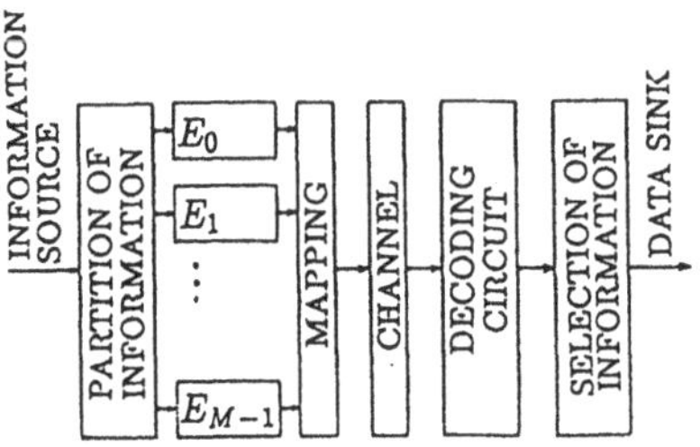

Fig. 1. Block diagram of IH scheme.

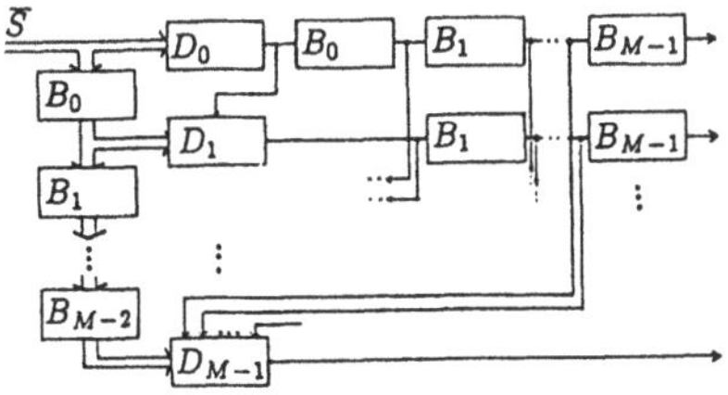

Fig. 2. Decoding circuit of IH scheme.

The decoding circuit of IH scheme is illustrated in Fig. 2, where B_i is a delay circuit. The delay time of B_i is the same as the decoding time of D_i.

This decoding is a little inferior to the maximum likelihood decoding (MLD) with respect to the error-correcting capability, but its implementation is much easier than that of the MLD.

3. Theoretical Studies of Performance

3.1 Euclidean distance of IH-TCM

First we show the relation between the free ED of IH-TCM and free Hamming distances (HD) of C_is. Let us denote the free HD for C_i as $d_{free}(i)$ and the free ED for the IH scheme as d_{freeED}. Then the following equation can be easily derived:

$$d_{free\ ED} = \min_i \left[\sqrt{d_{free}(i)} \cdot \Delta_i\right] \tag{1}$$

where Δ_0, Δ_1, $\cdots$, Δ_{M-1} are the minimum subset distances[4] given by the set partitioning method. The asymptotic coding gain of this scheme over the uncoded modulation scheme of the same rate is given by

$$G = log_{10}\ (d^2{}_{freeED}\ /\ \Delta_{un}) \tag{2}$$

where Δ_{un} is the minimum distance of the uncoded modulation scheme.

For example, in order to construct a rate 2/3 coded 8-PSK system, we can use three codes: a rate 1/3 convolutional code as C_0, a rate 2/3 code as C_1 and a rate one code (i.e, uncoded information) as C_2. The $d_{free\ ED}$ of the system is less than 2.0, because C_2 is not encoded. For another example of rate 2/3 coded 8-PSK systems, we choose the optimum convolutional codes of rate R_0=1/2, R_1=3/4 and R_2=3/4 as C_0, C_1 and C_2, encoders of which have 10, 3 and 3 memory cells, respectively. Here the optimum convolutional code means the non-catastrophic code that has the largest free HD with given encoder memory cells. In this case, the free HD of

C_0 is 14 and that of C_1 and C_2 is 4. $d_{free\ ED}$ of the system is 4.0 which gives IH system with 6.0dB coding gain over uncoded 4-PSK.

3.2 Complexity of Implementation for IH-TCM

We show an example of IH-TCM system which has large free ED, but it uses high rate (3/4) convolutional codes and the number of total encoder memory cells is up to 6.

In order to evaluate the complexity of implementation for TCM, we usually use the number of encoder memory cells, constraint length or number of encoder states. However, such parameters are not suitable for evaluating IH TCM, because the complexity depends on the decoding method and the simple decoding method described in *2* can be used for IH-TCM.

In a TCM scheme, complexity of implementation mainly depends on the Viterbi decoder. Therefore, we define a criterion for the complexity as the number of comparisons in ACS (Add Compare Select) circuit during a period of receiving one channel symbol. This number is denoted by L in the sequel.

We can carry out Viterbi algorithm (VA) by using either hardware or software logic. In order to compare two TCM systems, the same type of decoders must be assumed. Under this assumption, the complexity is roughly estimated by the criterion.

If a rate k/n convolutional code with ν encoder memory cells is employed, the number of comparisons per channel symbol L is given by

$$L = 2^{\nu} \cdot (2^k - 1) \cdot r/n \tag{3}$$

where r is the number of transmission bits per channel symbol ($r = n$ for Ungerboeck's scheme and $r = 1$ for each C_i of IH scheme). In case of a rate $(n-1)/n$ binary convolutional code, we can use the algorithm which satisfies

$$L \leq 2^{(\nu+1)} \cdot (n-1) \cdot r/n \tag{4}$$

with the same decoding error probability as that of VA[6]. Total L of IH-TCM system is given by the sum of L for C_i $(i = 0, 1, \cdots M-1)$.

In particular, when a single-error-detecting block code of length n is used and MLD is applied, L is given by

$$L = (n-1)/n \tag{5}$$

3.3 Examples of IH-TCM system

In order to compare IH-TCM with U-TCM, we employ 8-PSK systems with rate 2/3, and 16-QAM systems with rate 3/4. The results are shown in Tables 1, 2, Figs. 3 and 4. In the figures, the criterion L is written as 'number of comparisons.'

In Tables 1 and 2, τ, which is the decoding delay of IH-TCM, is the sum of the delay of $D_0, D_1, \cdot, D_{M-1}$. If D_i uses a path memory of length $\nu_i \cdot \lambda \cdot n_i$ (bits), the whole delay of IH-TCM is given by the sum of the lengths. Usually λ is a small positive integer like 5 or 6. We assume that every λ of D_i (i=0,1,$\cdots$, M-1) is the same.

Figs. 3 and 4 show that IH-TCM has large coding gain and easy implementation. For example, we can obtain an 8-PSK system which has 6.0dB asymptotic coding gain over the uncoded 4-PSK, and a 16-QAM system which achieves 9.8dB gain over the uncoded 8-PSK. By

Table 1. Examples of rate 2/3 coded IH-TCM (8-PSK).

No		R_i	ν_i	d_{free}	d_{ED}^2/Δ_1^2	G (dB)	L	τ
1	C_0 C_1 C_2	1/3 2/3 (1)	2 (0) -	8 (2)* 1	2.00	3.00	2	6λ
2	C_0 C_1 C_2	1/3 2/3 (1)	2 1 -	8 2 1	2.00	3.00	2.33	9λ
3	C_0 C_1 C_2	1/3 2/3 (1)	2 2 -	8 3 1	2.00	3.00	5.33	12λ
4	C_0 C_1 C_2	1/3 5/6 5/6	4 3 (0)	11 3 (2)*	2.05	4.80	19.5	30λ
5	C_0 C_1 C_2	1/3 5/6 5/6	6 4 (0)	15 4 (2)*	4.00	6.00	48.8	42λ
6	C_0 C_1 C_2	1/2 3/4 3/4	4 3 3	7 4 4	2.05	3.12	32	32λ
7	C_0 C_1 C_2	1/2 3/4 3/4	5 3 3	8 4 4	2.34	3.70	40	34λ
8	C_0 C_1 C_2	1/2 3/4 3/4	6 3 3	10 4 4	2.93	4.66	56	36λ
9	C_0 C_1 C_2	1/2 3/4 3/4	8 3 3	12 4 4	3.51	5.46	152	40λ
10	C_0 C_1 C_2	1/2 3/4 3/4	10 3 3	14 4 4	4.00	6.02	536	44λ
11	C_0 C_1 C_2	1/2 3/4 3/4	10 5 3	14 5 4	4.10	6.13	572	52λ
12	C_0 C_1 C_2	1/2 3/4 3/4	11 5 3	15 5 4	4.39	6.43	1084	54λ
13	C_0 C_1 C_2	1/4 7/8 7/8	2 3 3	10 3 3	2.93	4.66	29	56λ
14	C_0 C_1 C_2	1/4 7/8 7/8	3 3 3	13 3 3	3.00	4.77	30	60λ
15	C_0 C_1 C_2	1/4 7/8 7/8	3 4 3	13 4 3	3.81	5.80	44	68λ
16	C_0 C_1 C_2	1/4 7/8 7/8	4 4 3	16 4 3	4.00	6.02	46	72λ

d_{free} is free Hamming distance of C_i.
d_{ED} is free Euclidean distance of IH-TCM system. ($=d_{freeED}$)
Coding gain G is calculated relative to uncoded 4-PSK.
* single-error-detecting block code

Table 2. Examples of rate 3/4 coded IH-TCM (16-QAM).

No		R_i	ν_i	d_{free}	d_{ED}^2/Δ_1^2	G (dB)	L	τ
1	C_0 C_1 C_2 C_3	1/4 7/8 7/8 (1)	2 3 3 -	10 3 3 1	4.10	6.13	29	56λ
2	C_0 C_1 C_2 C_3	1/4 7/8 7/8 (1)	2 4 3 -	10 4 3 1	5.46	7.38	43	64λ
3	C_0 C_1 C_2 C_3	1/2 3/4 7/8 7/8	2 3 3 3	5 4 3 3	3.42	5.34	42	64λ
4	C_0 C_1 C_2 C_3	1/2 3/4 7/8 7/8	3 3 3 3	6 4 3 3	4.10	6.13	44	66λ
5	C_0 C_1 C_2 C_3	1/2 3/4 7/8 7/8	4 5 3 3	7 5 3 3	4.78	6.80	84	76λ
6	C_0 C_1 C_2 C_3	1/2 3/4 7/8 7/8	5 5 3 3	8 5 3 3	5.47	7.38	92	78λ
7	C_0 C_1 C_2 C_3	1/2 3/4 7/8 7/8	6 5 3 3	10 5 3 3	6.83	8.35	108	80λ
8	C_0 C_1 C_2 C_3	1/2 3/4 7/8 7/8	8 6 4 3	12 6 4 3	8.20	9.14	266	96λ
9	C_0 C_1 C_2 C_3	1/2 3/4 7/8 7/8	10 8 4 3	14 7 4 3	9.54	9.81	938	108λ

Coding gain G is culculated relative to uncoded 8-PSK.

using the proposed criterion for these modulation systems, we see that some IH-TCM systems are easier to implement than U-TCM with the same coding gain. On the other hand, some IH-TCM systems can achieve higher coding gain than U-TCM with the same complexity of the decoders.

As an example, the 8-PSK system having 6.0 dB coding gain, which was previously mentiond, is constructed from three binary convolutional codes: a (1,4,4) code, a (7,8,4) code and a (7,8,3) code, where a (k,n,M) code means a rate k/n convolutional code that has the maximum free Hamming distance and is encoded by a 2^M-state encoder. This system is implemented as easily as the U-TCM system having 4.1dB gain whose encoder has sixteen states.

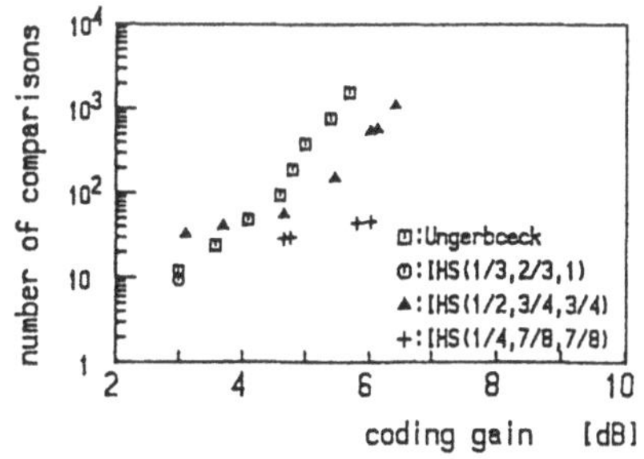

Fig. 3. Hardware complexity vs. coding gain. (rate 2/3 coded 8PSK)

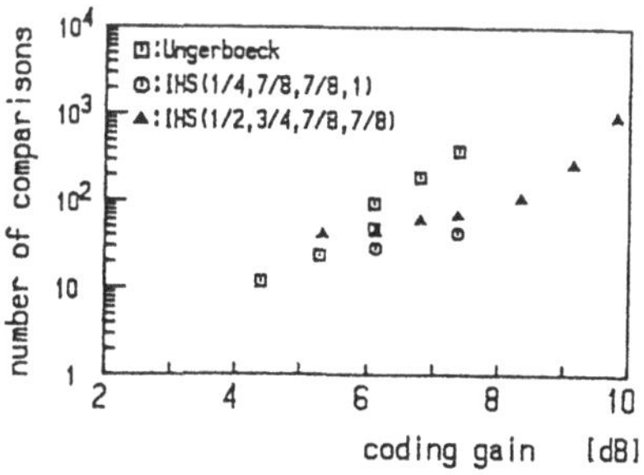

Fig. 4. Hardware complexity vs. coding gain. (rate 3/4 coded 16QAM)

From Tables 1 and 2 we see that the decoding delay of IH-TCM is long. This may be a problem for some applications such as voice data transmission. The decoding delay of U-TCM systems is relatively small. For example, the delay of the rate 3/4 U-TCM for 16-QAM is only 32 λ. However, IH-TCM systems are useful, unless small decoding delay is required.

4. Evaluation IH-TCM Using Random Coding Theorem

In *3.1* and *3.2* , we have shown some examples of IH-TCM systems and discussed the performance of hardware complexity and coding gain.

In this chapter, we analyze the complexity for implementation of these schemes theoretically. We apply the random coding theorem[7] to IH-TCM with much larger constraint length than that mentioned in *3* in order to analyze the complexity of implementations for very highly reliable communication.

IH scheme using M error-correcting codes is expressed by an equivalent model which has M parallel communication channels shown in Fig. 5.

Suppose that we communicate over a white Gaussian channel using 2^M-ary level modulation such as 2^M-ary PSK or QAM. Then, the parallel communication channels in the equivalent model are considerd to be binary input and Q-ary output channels. Strictly speaking, these channels are not independent each other, because decoding errors that occur in D_i affects decoding in $D_{i+1}, D_{i+2} \cdots D_M$. However such propagation of decoding errors has only additive effect in the error probability of each decoder output sequence. Therefore we can neglect the effect of the propagation if we discuss the probability by its order.

Let us derive the optimum IH-TCM for this communication channel by using the random coding theorem. We assume that the original 2^M-ary level communication channel has quantized output. We take the optimum 2^M level TCM as the target of comparison, and compare its hardware complexity with that of the optimum IH-TCM at the same decoding error probability. We simply call the optimum IH-TCM as IH-TCM and the optimum 2^M level TCM as M-TCM.

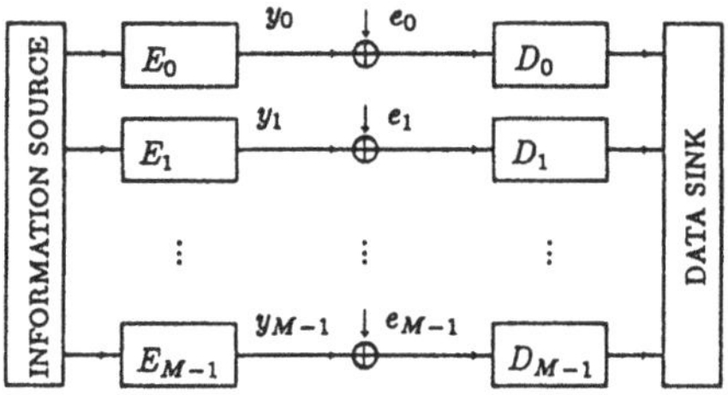

Fig. 5. Equivalent channel model of IH scheme.

The M-TCM is a larger class of TCM than U-TCM, because M-TCM employs a time-varying convolutional code, as well as a time-invariant convolutional code which is used for U-TCM. The usage of time-varying code means that the class of M-TCM also includes multi-dimensional TCMs[8]etc..

In this chapter, the decoding error probability is defined as the probability that the B consecutive channel symbols are not correctly decoded.

Let P_{BMul} denote the decoding error probability of the M-TCM and P_{BIH} that of the optimum IH-TCM. Let $P_B(i)$ be the error probability of D_i for decoding B channel symbols, when $P_B(j)(0 \leq j < i)$ is 0. Then P_{BIH} is written as

$$P_{BIH} = 1 - \prod_{i=0}^{M-1} (1 - P_B(i)) \sim \sum_{i=0}^{M-1} P_B(i) \quad . \tag{6}$$

We compare hardware complexity of IH-TCM with M-TCM under the following condition:

$$P_{BMul} = P_{BIH} \quad . \tag{7}$$

For the sake of simplicity, we replace (7) by the following equations:

$$\frac{1}{M} P_{BMul} = P_B(0) = P_B(1) = \cdots = P_B(M-1) \quad . \tag{8}$$

The condition (8) means that each C_i is required to have the same decoding error performance. This requirement seems reasonable for constructing reliable IH-TCM systems.

Computing P_{BMul} and $P_B(i)$ is quite difficult, because we cannot specify the optimum TCMs. Therefore we investigate the upper bound and lower bound for these probabilities.

From the random coding theorem, there is a rate $R = k/n$ time-varying convolutional code satisfying the following inequality for the binary input memoryless channel:

$$P_B < \frac{B(2^k - 1)\, exp[-\nu \cdot k \cdot E_{cex}/\{R(1+\epsilon)\}]}{(1 - 2^{\epsilon k2})} \tag{9}$$

where ϵ is a positive number and ν is the constraint length of the code. E_{cex} is called expurgated upper bound and is derived from

$$Ex = \left\{ -\rho\ ln \sum_{x} \sum_{x'} q(x)q(x') \cdot \left[\sum_{y} \sqrt{p(y \mid x)p(y \mid x')} \right]^{1/\rho} \right\} \tag{10}$$

$$0 < R\ \ log_e\, 2 < C\ \ /\ \ 2(1+\epsilon)$$

where C is the capacity of the channel. Then, E_{cex} is given by

$$E_{cex} = \max_{q(x)} [E_x] \qquad (1 \leq p < \infty) \quad . \tag{11}$$

where x is an input of the channel and y is an output. $q(x)$ is the probability of x and $p(y \mid x)$ is the transition probability of the channel. ρ can be determined by

$$R \log_e 2 = \max[E_x/\rho(1+\epsilon)] \quad . \tag{12}$$

In (9), if we choose large ν, ϵ can be negligibly small. Hence the characteristics of the bound is mainly decided by E_{cex}. We then evaluate $P_{BIH}(i)$ by using E_{cex}. In the similar way, we use E_c, which is the upper bound for discrete-input memoryless channel, in order to obtain P_{BMul} and E_x, which is the upper bound of discrete-input memoryless channel, in order to obtain P_{BMul} and $P_{BIH}(i)$.

On caluculating these decoding error probabilities, there are several parameters to be specified such as rates R, R_i, constraint lengths ν, ν_i, modulation scheme and its SNR.

Now, we use rate 2/3 coded 8-PSK with 64 level output quantized channel. For IH-TCM, we must minimize the complexity in every combination of R_i and given P_{BIH}. For simplicity, however, we a priori decide $R_0 = 1/4$, $R_1 = R_2 = 7/8$ here, because it is seen from Fig. 3 that this combinataion of code rates gives good performance of coding gain and complexity of implementation. Naturally, this decision may put some limit to the potential of IH-TCM.

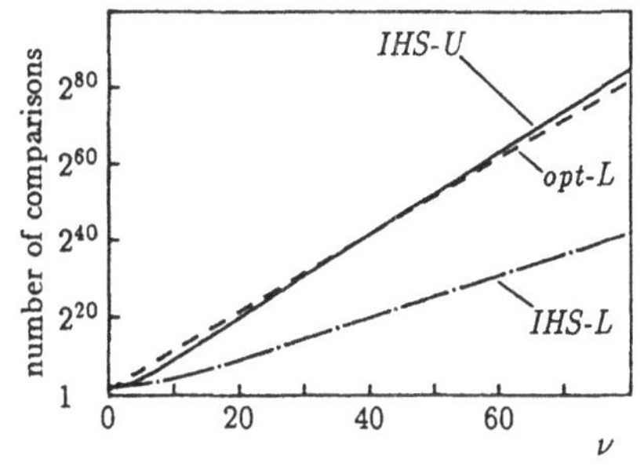

Fig. 6. Hardware complexity analyzed by error exponent.

Fig. 6 shows a result concerning to the complexity of implementation. In this figure IHS-U and IHS-L, which are obtained from E_x,E_{cex}, are the upper and lower bounds of complexity for IH-TCM, and opt-L, which is obtaind from E_c, is the lower bound for the M-TCM. This figure shows the result in case of 10 dB SNR, but the relasion between the curves is similar for different SNR. The horizontal scale ν of Fig.6 is the constraint length of M-TCM. IH-TCM at the point ν is chosen to have the same decoding error probability as M-TCM of constraint length ν. Therefore the horizontal scale indicates the decoding error probability of both TCMs.

The actual curve of complexity for IH-TCM exists between IHS-U and IHS-L, and the curve for M-TCM exists above opt-L. It is known from the random coding theorm that IHS-L is a tight bound and opt-L is not. Therefore the result shows that IH-TCM is much easier to implement than the optimum multilevel coding scheme with the same decoding error probability for large constraint length.

5. Computer Simulation

We have derived the asymptotic coding gain of IH-TCMs in *3*, but the real coding gain at a practical signal-to-noise-ratio (SNR) is not given. At a practical SNR, the real coding gain of a TCM scheme decreases from the asymptotic one.

Since the decoding of IH-TCM is not MLD, the decrease of coding gain of IH-TCM is larger than that of U-TCM. However, the complexity of implementation for IH-TCM is expected to be still smaller than that of U-TCM having the similar coding gain.

In this chapter, we show the bit error probability performance of IH-TCM and U-TCM obtained by computer simulations in order to evaluate IH-TCM scheme totally. Fig. 7 shows the bit error probability performance in case of rate 2/3 coded 8-PSK TCM systems. The demodulation of the system is assumed to employ soft decison with 64 level quantization. The specification of the simulation is shown in Table 3. In the simulation, we suppose the additive white gaussian noise channel. It is assumed that the synchronization of demodulator is always perfect. The length of the path memory in VA is $L = 6\ \nu \cdot n$ (bits). The encoders of all TCM systems are systematic encoders with feedback. Bit error probability of Fig. 7 is calculated for the information bit.

Table 3. Condition of Fig. 6.

SCHEME	CODING GAIN	NAME	RATE ν_i d_{freeHD}	L
IH-TCM	3.0	IH-1	C_0 1/3 2 8 C_1 2/3 0 2 C_2 1(UNCODED)	2.0
		IH-1MLD	THE SAME AS ABOVE	84.0
		IH-2	C_0 1/3 4 12 C_1 2/3 1 2 C_2 1(UNCODED)	7.3
U-TCM	3.0	U-1	2/3 2 -	12.0

In comparing IH-1 with U-1 whose asymptotic coding gain is 3.0 dB, the real coding gain of IH-1 is 0.75 dB inferior to that of U-1 at the bit error rate of 10^{-5}. However, the number L of IH-1 is only 2 while that of U-1 is 12. As for IH-1MLD which uses maximum likelihood decoding

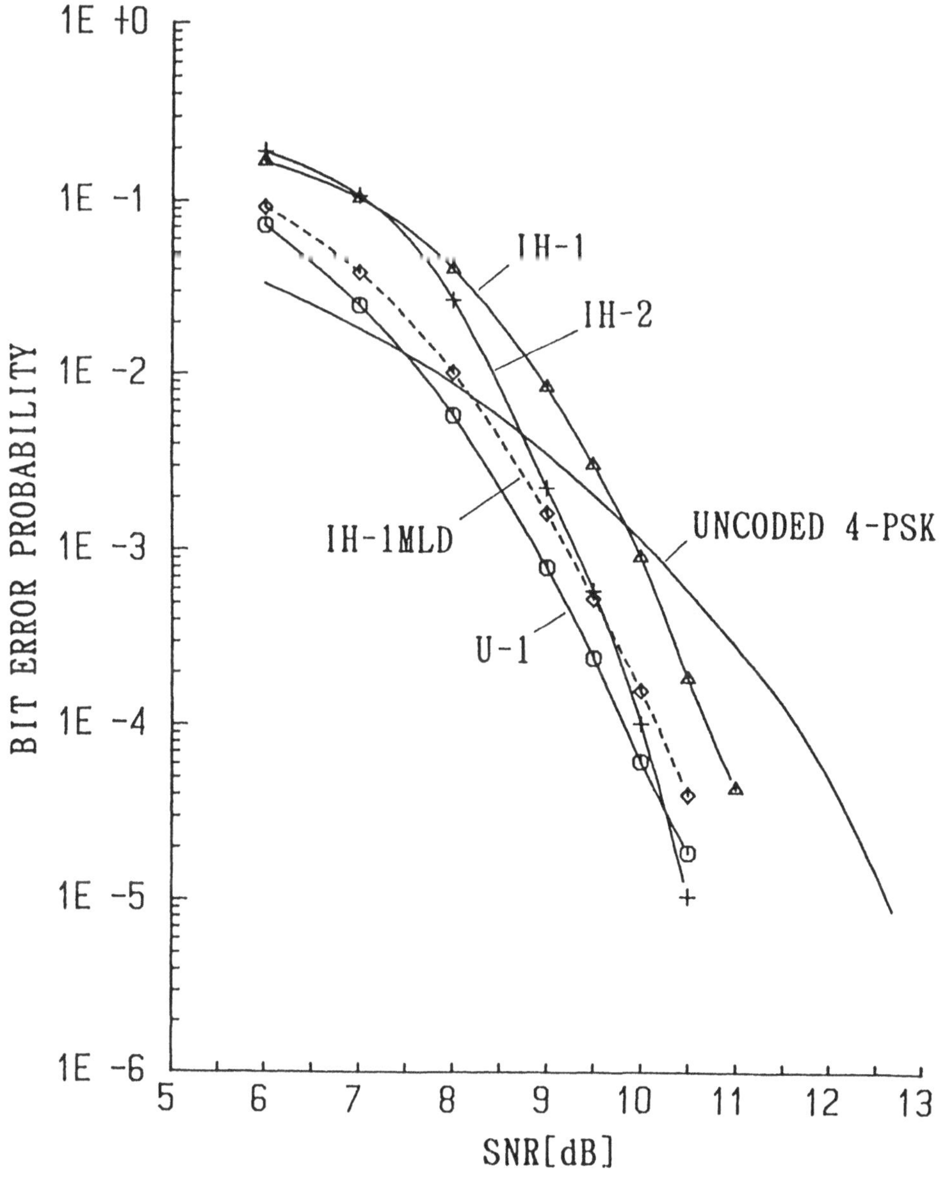

Fig. 7. Bit error performance of IH-TCM and U-TCM.

for IH-1, the coding gain of IH-1MLD is close to that of U-30. The number L of IH-1MLD is 9.3, which is less than that of U-1.

IH-2 has 3.0 dB asymptotic coding gain too, and its L is 7.3. From the results we see that IH-2 has better bit error performance than IH-1, IH-1MLD and U-1 if the bit error probability is less than 10^{-5}.

6. Conclusion

Although IH-TCM has a defect that the decoding delay is long, the results obtained in *3* have shown that IH-TCM acheives large asymptotic coding gain and is easy to emplement. For constructing IH-TCM system, use of the asymptotic coding gain is a convenient way of estimation for the reliability. However, it must be used carefully. In some cases, IH-TCM which shows high reliability and small complexity in computer simulation is not the same as that obtaind by maximizing asymptotic coding gain.

We have shown some bit error performances of IH-TCM and U-TCM obtaind by computer simlation. It follows from the result that there are IH-TCM systems with smaller complexity of implementation than U-TCM systems having the same decoding error probability at a practical SNR.

The results of computer simulation have been given for only a few cases of IH-TCM systems. However, they suggest that IH-TCM has the possibility of being a practical way to realize highly reliable and efficient communication systems.

Acknowledgement

The authors wishes to thank Prof. Ryuji Kohno of the Yokohama National University for his helpful suggestion. They are also indebted to Mr. Yuichi Saito for contribution to computer simulations.

References:

[1] IMAI,H. and HIRAKAWA, S.: 'A new multilevel coding method using error-correcting codes', *IEEE Trans. Inf. Theory*, **IT-23**, May, 1977.

[2] YAMAGUCHI, K. and IMAI, H.: 'Highly reliable multilevel channel coding system using binary convolutional codes,' *Electronics Letters*, **23**, No. 18, pp. 939-941, August 1987.

[3] YAMAGUCHI, K. and IMAI, H.: 'A highly reliable trellis coded modulation using binary convolutional codes,' *Trans. of the Institute of Electronic, Information and Communication Engineers*, **J71-A**, No. 4, pp. 1018-1025, Aprir 1988 (in Japanese).

[4] IMAI, H. and HIRAKAWA, S.: 'A new system of multilevel signal transmission with error control', *Paper of Technical Group on Algorithm and Language, Institute of Electronics and Communication Engineers*, **AL75-30**, pp.1-10, Oct., 1975, (in Japanese)

[5] UNGERBOECK, G.: 'Channel coding with multilevel/phase signals', *IEEE Trans. Inf. Theory*, **IT-28**, pp.55-67, Jan. 1982.

[6] YAMADA,T. HARASHIMA H. and MIYAKAWA,H.: 'A new maximum likelihood decoding of high rate convolutional codes using a trellis,' Trans. Institute of Electronics and Communication Engineers of Japan(A), **66-A**, No. 7, pp.611-616, July 1977 (in Japanese).

[7] Viterbi A. J, Omura J. K.: 'Principles of digital communication and coding', McGraw-Hill, 1979.

[8] A. R. Calderbank and N. J. A. Sloane : 'New trellis codes based on lattices andcosets', *IEEE Trans. Inf. Theory*, **IT-33**, pp.177-195, March. 1987.

EMBEDDING FLEXIBLE CONTROL STRATEGIES INTO OBJECT ORIENTED LANGUAGES

Sabina Bonamico

Dipartimento di Matematica
"Guido Castelnuovo"
Università di Roma "La Sapienza"
P.le Aldo Moro 5
00184 Roma, Italy

Gianna Cioni

Istituto di Analisi dei Sistemi
ed Informatica
Consiglio Nazionale delle Ricerche
Viale Manzoni 30
00185 Roma, Italy

EXTENDED ABSTRACT

The original solicitation for the work presented in this paper is the observation that scientists and engineers spend most of their working time in manipulating mathematical objects, such as relations, functions and models. This manipulation consists not only in computing, but mainly in analyzing them in order to verify or deduce specific properties. While the classical computing tools, as numerical software libraries and systems for symbolic and algebraic computation, are widely used, the availability of powerful automatic deduction tools still represents at this moment an interesting research topic.

The reason of this interest is twofold: the need for effective automatic deduction capability, based on the natural formal definition of the mathematical objects (in terms of axioms, properties, types and constraints), and the requirement of a uniform semantic of this deductive and of the computing mechanism in order to achieve a completely integrated software environment.

Traditionally problems which need automatic deduction are treated by logic programming, which being based on a declarative style, appears the most natural approach expecially from the point of view of the readability of the programs. Unfortunately most of the actual implementations of logic programming systems offer only a fixed and predefined strategy of the deduction mechanism. In order to improve the flexibility of these systems many research lines have been followed to add meta-levels on the top of standard resolution mechanism whithout modifying the basic level. Another interesting research direction is that of offering to the user the possibility of choosing among different strategies on the base, for instance, of the knowledge he owns on his specific problem. For this purpose of offering flexible control strategies, we have been considering an alternative set of deduction methods to be used in the different situations, according to the characteristics of the given mathematical problem.

The second goal of the present work, namely the integration of deduction capability into a computing environment with uniform semantic, has been pursued by an object oriented language which also guarantees the correctness of the specification and the maximum degree of flexibility.

Let us consider logic problems, because in this case it is possible and easy to build a set of alternative methods to perform automatic deduction and to define the exact requirements from the point of view of the programming methodologies.

Logic problems can be completely represented by a set of states S, a set of initial states $S_I \subset S$, a set of final states $S_F \subset S$ and a transition function $T: S \rightarrow S$, which describe the evolution of the system from the current state to the next one. The transition function is generally not deterministic, because different choices can be selected and legally applied to the current state. Then a strategy to choose among different states is necessary in order to obtain the best resolution path.

Two different approaches can be followed on the basis of the knowledge we have on the problem and on the related solution (e.g. the best one, the first one, all) [NIL82]: "Uninformed" search strategies and "Heuristic" search strategies.

The first kind of strategies (Depth First Search, DFS; Breadth First Search, BFS) arbitrarily select the successive states or, at least, independently of the achieved path. For this reason they can cause a waste of storage and time. The second kind of strategies (Best First Search, BS; AND Search, A*; AND/OR Search, AO*) exploit the problem knowledge to improve the efficiency of the system.

To direct this search and to keep track of the rules which have been applied, we need a Tentative Control, as Backtracking, which is able to repair the errors and to select the most appropriate rule to be considered after a failure.

These two elements, Searching and Control Strategies, are the crucial blocks to obtain a good resolution mechanism in respect to time and memory space, for all the problems which can be expressed following a logical approach. If user has some knowledge about the problem he tries to solve, these informations can be used in order to dry the Search Strategy and also to effectively reduce the use of Backtracking. When Backtracking cannot be completely avoided, a good knowledge of problem allows to employ one of the so-called Intelligent Backtrackings which establish the tree level to be reached to perform an efficient choice [BRU81]. Logical programming paradigm is classically the most used to solve these kinds of problems, but it is commonly known that the Resolution of the standard PROLOG interpreters is not flexible, because it uses a DFS strategy and a non-intelligent Backtracking.

In spite of the fact that some researches have been developed in the context of logic programming in order to obtain a more sophisticated Search Strategy, and also a different way to organize Backtracking [COD84], [BRU84], in many cases this is paid by a reduction of readability and of expressivity clearness.

In our case, we have the advantage of treating with problems whose characteristics are well defined, whose typical language is the mathematical or algebraic one, and these features give us the idea of predefining different strategies, organized in a structured way by many levels which user can choose for the resolution of his specific problem, embedding all that in an Algorithmic Programming Language.

Therefore, we have chosen to use the imperative approach of a very high-level Object Oriented Language, also for obtaining flexibility and correctness in a unique and natural computing environment. Then, we decided to use the same approach for the Logical Deduction, as well.

Our specification and implementation language, which is actually the LOGLAN [LOG84], and will be an extension of that language in an immediate future, owns a high degree of modularity and gives us the possibility of decomposing a problem in many subproblems, which can be treated with different methods encapsulated in different units. That is obtained by well defined and correctely implemented units, stressing mainly the visibility rules. Moreover, LOGLAN offers a well structured hierarchization, in the sense that a unit can inherite properties from another one and add them to his attributes in a dynamic way. The main elements of the language are the classes and, during the execution, the objects which can be prefixed at multi-levels and moreover also nested, offering to the user a very flexible and secure programming environment. Also the redefinition of the operations can be made in a dynamic and flexible way using virtual definitions and types as formal.

But at the main level, the first characteristic of our choice is the possibility, given by the methodology of this kind of language, of building all our computation and deduction steps in a unique and correct approach. The correspondence between the formal specification and the coding [MIR87] is in this approach very strictly and the guarantee of correctness is passed from the definition of properties to the execution of real programs.

By using that language we started to build a system to make Automatic Deduction through different kinds of strategies and Backtracking. In a first time we have verified this approach in the games theory, taking into account the most classical and known problems and later on we approached the Formal Integration problem, in order to give a heuristic solution.

With this approach user has only the duty to specify the characteristics of his problem, using the same semantics of the computation.

The system is organized in three levels. The first one is depending on the problem and here the user must specify all the characteristics on states, and transitions, following the menu given in the block. The second one is concerning the way to present on the monitor the structure, the evolution and the results of the problem, and it is a totally general schema of user interface. The third, and more important, contains the alternative ways to do Backtracking (Fixed and Intelligent) and many different Search Strategies. Another block is concerning with the specification of the kind of solution asked by the user (the first, the best or all). Also this block has a general structure.

All the levels are prefixed together and user, choosing the best solution method, obtains the real structure of the program which will be executed.

If the case of Formal Integration, following the line of the fundamental work of far 1963 [SLA63], the solution is reached in all the cases using the Intelligent Bactracking in the AND/OR tree. The solution can be also reached with a good degree of efficiency by a modification of the AO* strategy which takes into account the cost of the different yet performed computations.

References

[**BRU81**] M. BRUYNOOGHE: *Intelligent Backtracking for an interpreter of Horn clause logic programs*, in Proc. on Coll. on Math. Logic in Programming, Salgotarjan, North - Holland, (1981).

[**BRU84**] M. BRUYNOOGHE, L.M. PEREIRA: *Deduction revision by intelligent backtracking*, in Implementation of PROLOG, ed. by J.A. Campbell, J.Wiley, (1984).

[**COD84**] C. CODOGNET, P. CODOGNET, G. FILÉ: *Yet another Intelligent Backtracking Method*, in Implementation of PROLOG, ed. by J.A. Campbell, J.Wiley, (1984).

[**LOG84**] LOGLAN 82, Report Warsaw (1984).

[**MIR87**] G. MIRKOWSKA, A. SALWICKI: *Algorithmic Logic*, Reidel Publ. Co., (1987).

[**NIL82**] N.J. NILLSON: *Principles of Artificial Intelligence*, II ed., Springer Verlag, (1982).

[**SLA63**] J. R. SLAGLE: *Heuristic Program that solves Symbolic Integration Problems in Freshman Calculus*, in Computer and Thought, Feigenbaum ed., Mc Graw Hill, (1963).

MAJORITY DECODING OF LARGE REPETITION CODES FOR THE R-ARY SYMMETRIC CHANNEL

Paul CAMION

INRIA Domaine de Voluceau
Rocquencourt BP 105
78153 LE CHESNAY Cedex - FRANCE

ABSTRACT

A r-ary symmetric channel has as transition probability matrix the r x r matrix $q_{xy} = p$ if $x \neq y$ and $q_{xy} = 1-(r-1)p = q$ if $x=y$. Given a set Y of r symbols, the code here consists of r codewords, each one of them is made up of n identical symbols. Whenever q is larger than p, maximum likelihood decoding amounts to find out in the received vector which symbol is repeated most. Thus MLD here reduces to majority decoding.

A generating function for the error probability as well as the probability of decoding failure for the system is obtained. Also recurrence relations are given for computing those probabilities.

We more generally consider a DMC (Discrete Memoryless Channel) which we call *transitive*. A r-ary transitive DMC is a DMC such that there exists a transitive permutation group G on the set **Y** of symbols such that

$$\forall (x,y) \in \mathbf{Y} \times \mathbf{Y}, \quad \forall \sigma \in G \ : \ q_{xy} = q_{\sigma x, \sigma y}.$$

The results corresponding to those announced for the r-ary symmetric channel are obtained for the majority decoding repetition codes over r-ary transitive DMC.

INTRODUCTION

A r-ary repetition codes are class one candidates as first outer codes for the remarkable generalized concatenated codes introduced by ZINOVIEV in 1976 (cfr 4). A class of examples of those codes needing the here introduced algorithms will be presented in a further paper.

When coding for transmitting over a very noisy channel, they also may be used as inner codes in a classical concatenation. They first are the most easily decodable among codes ensuring a high protection.

Moreover, as observed in the example of section 3, to the somewhat high probability of decoding failure corresponds erasures to be decoded by the outer code. This provides concatenated codes easily decodable and protecting against the most harmful disturbances.The example dealt with in section 3 is doubtless introduced to give some idea of the security provided, but especially to show that computing the concerned probabilities needs a more powerful tool than just expanding a polynomial in r variables even for small values such as r=8 and n=16.

Recurrence relation (2) in section 5 provides a tool which makes possible the computation of the concerned probabilities with the help of a computer. The classical use of exponential generating functions for handling the shuffle product (cfr [2]) leads to relations (4) and (5) which brings a complete solution to the problem with the help of symbolic calculus.

Unlike binary channels, r-ary channels may generally not be considered symmetric. However, in case of phase modulation, for example, there exists a cyclic group of permutations acting on the set of signals which preserves the transition probabilities. This allows to consider a transitive DMC and we here obtain expressions for the probabilities to be computed ((8) and the previous one) involving quantities which are to be obtained by recurrence (2).

If on the one hand some symmetry is helpful for computing purposes, in particular for computing the probability of decoding failure, on the other hand there is no possible decoding failure in some specific DMC that are investigated in section 4 where dominance replaces symmetry.Those DMC are called "perfect " in the sense that every word of the ambient space is decodable for maximum likelihood as in a perfect block code.

1. The Binary repetition code

A repetition block code with odd length for the 2-ary symmetric channel is perfect. A received vector is decoded as 0 if its Hamming distance to the 00...0 n-tuple is shorter than its distance to the 11...1 n-tuple, otherwise it is decoded as 1.

For n = 2s+1, the *bit error probability* is

$$\sum_{s+1\leq i\leq n} \binom{n}{i} p^i q^{n-i}$$

If p is smaller than 1/2 then the bit error probability $P_e^{(2s+1)}$ does approach zero as $s \rightarrow \infty$. This is a well know fact. (McEliece [3]).

2. The r-ary repetition code

Here the codewords are the n-tuples $\mathbf{x_i} = a_i\, a_i \ldots a_i$, i=1, ... , r. Maximum likelihood decoding (MLD), sets as codeword $\hat{x}$ when decoding $\mathbf{x} = a_{i_1}\, a_{i_2} \ldots a_{i_n}$ the codeword $\mathbf{x_i}$ for which $P(\mathbf{x}|\mathbf{x_i})$ is largest. Clearly, if q/p is larger than one for the r-ary symmetric channel defined in the abstract, then $\mathbf{x}$ is decoded by MLD as the codeword $\mathbf{x_i}$ for which the number of indexes i_j that equal i is largest. Notice that for r larger than two then MLD fails to decode some vectors. For instance if r=3, then P(aabbc|aaaaa) = P(aabbc|bbbbb). Also notice that the set of vectors decoded as $\mathbf{x_i}$ by MLD is no longer the Hamming bowl $B_e(\mathbf{x_i})$ consisting of all vectors at Hamming distance at most e from $\mathbf{x_i}$. For instance, aabcd and aabbb both are at distance three from aaaaa.

3. Solving by inspection a small sized problem

Let us consider the 3-ary symmetric channel. The set of signals **Y** is {a,b,c} and we are given the transition probabilities : $q_{xx} = 0.9$ and $q_{xy} = 0.05$ for x≠y. We set n=4. Expanding $(a+b+c)^4$ in $\mathbb{Z}[a,b,c]$, we group the monomials corresponding to n-tuples decodable as aaaa, then those corresponding to n-tuples undecodable and finally those corresponding to n-tuples decodable as bbbb or cccc :

$$M_{cd} = a^4 + 4\,a^3\,b + 4\,a^3\,c + 12\,a^2\,bc$$
$$M_{flr} = 6\,a^2\,b^2 + 6\,a^2\,c^2 + 6\,b^2\,c^2$$
$$M_{err} = 4\,a\,b^3 + 4\,a\,c^3 + 4\,bc^3 + 4\,b^3\,c + 12\,ab^2c + 12\,abc^2 + b^4 + c^4.$$

From which we respectively compute the probability P_{cd} of correct decoding, the probability P_{flr} of decoding failure and finally the probability of decoding error :

$$P_{cd} = q^4 + 8\,q^3p + 12\,q^2p^2$$
$$P_{flr} = 12\,q^2\,p^2 + 6\,p^4$$
$$P_{err} = 32\,qp^3 + 10\,p^4$$

Here $P_{cd} = 0{,}972$

$P_{flr} = 2{,}43375\ 10^{-2}$

$P_{err} = 3{,}6625\ 10^{-3}$

The fact that we get a relatively high probability of decoding failure suggests using an outer code for decoding erasure. Indeed observable failures for the inner code may be interpreted as erasure for the outer code.

As outer code, we can use a Hamming (4,2) code over GF(3). It will correct every single error, every single erasures or (exclusively) every pair of erasures. Thus the probability of correct decoding by the outer code is

$$P_{cd}^{4} + 4\,(P_{flr} + P_{err})\,P_{cd}^{3} + 6\,P_{flr}^{2}\,P_{cd}^{2},$$

which amounts to 0.9988273.

Finally the whole scheme transmits every pair of ternary symbols with that security. This amounts to a security of 0.9994134 for every single symbol. Thus we finally have the situation of a channel which would transmit every ternary signal with error probability less than 6/10,000 in place of 1/10. However the transmission rate goes down to 1/8.

4. Perfect MLD repetition codes for the general r-ary Memoryless Discrete Channel

As a result of the investigation in section 3, for any given n there exist $q_{aa} > q_{bb} > q_{cc}$ and moreover q_{cc} larger enough than any other transition probability, such that every word of length n is MLD decodable for the ternary DMC.

This is the fact for the following datas :

$q+2p = 1\ ,\ 0 < p < q,$

$q_{aa} = q-\varepsilon,\ q_{bb} = q-2\varepsilon,\ q_{cc} = q-3\varepsilon,$

$q_{ab} = q_{ba} = p,\ q_{ac} = q_{ca} = p+\varepsilon,\ \ q_{bc} = q_{cb} = p+2\varepsilon.$

We need that $\varepsilon > 0$ and $q-3\varepsilon > p+2\varepsilon$, i.e $0 < \varepsilon < (1-3p)/5$. Moreover, majority decoding agrees with MLD if

$$\frac{q-3\varepsilon}{p+2\varepsilon} > \left(\frac{p+2\varepsilon}{p}\right)^{n-1}.$$

For $q = 0.5 + 4\varepsilon$, we have that $p = 0{,}25-2\varepsilon$ and the LHS is then worth $2+4\varepsilon$. We then choose ε sucht that

$$\log\left(1+\frac{2\varepsilon}{p}\right) < \frac{\log 2}{n-1}$$

We more generally have with the definition of perfect in the abstract

Theorem 1

For every r, n there exists r-ary DMC's for which the repetition code of length n is perfect.

Let $Y = \{a_0,\ldots,a_{r-1}\}$, $r > 3$.
Set $q_{a_i a_i} = q - i\varepsilon$, for $i \leq r-2$, $p=(1-q)/(r-1)$, with $p<q$.
$q_{a_{r-1}a_{r-1}} = q - \binom{r-1}{2}\varepsilon$, $q_{a_j a_i} = q_{a_i a_j} = p$ for $i<j<r-1$, $q_{a_{r-1}a_i} = q_{a_i a_{r-1}} = p + i\varepsilon$ for $i < r-1$.

Thus

$$\sum_{y \neq a_i} q_{a_i\,y} = (r-1)\,p + i\varepsilon\ , \text{ for } i<r-1, \text{ else } (r-1)p+\binom{r-1}{2}\varepsilon.$$

Next for $i<j$, $q_{a_i\,a_i} > q_{a_j\,a_j}$.

We now have to show that the smallest q_{xx} is larger that the greatest q_{xy}, $y\neq x$, for a suitable choice of ε.

$$q_{a_{r-1}\,a_{r-1}} = q - \binom{r-1}{2}\varepsilon\,;\ \left(\binom{r-1}{2} + r-2\right)\varepsilon < q-p\ \ ;\ \varepsilon < 2(1-rp)/(r+1)(r-2)\,.$$

Moreover, it is possible to adjust ε in order that

$$\frac{q-\binom{r-1}{2}\varepsilon}{p+(r-2)\varepsilon} > \rho^{n-1},\ \text{where } \rho = \max\left(\frac{p+(r-2)\varepsilon}{p}\ ,\ \frac{q}{q-\binom{r-1}{2}\varepsilon}\right),$$

if we want that MLD agrees with majority decoding.

Finally for every received word **x** for which the number of occurences of some letter a is largest and equal the number of occurences of an other letter b, we will ascertain that

$$P(\mathbf{x}|\mathbf{a}) - P(\mathbf{x}|\mathbf{b}) \neq 0, \qquad (1)$$

where **a** and **b** are the repeated a and repeated b codewords respectively.

But (1) is a non-zero polynomial in ε and there is a finite set of such polynomials. Thus there exist a value for ε in the required interval for which none of these polynomials cancel.

5. Recurrence relations and generating functions for the probability of correct decoding and the probability of decoding failure in majority decoding repetition codes over a r-ary DMC.

The $\mathbb{Z}$-algebra over the monoïd **Y*** generated by the set **Y** is denoted by $\mathbb{Z}\langle Y\rangle$. The shuffle product as used in Foata and Schützenberger [2] of two monomials u and v, $u \in \mathbb{Z}\langle Y_0\rangle$, $v \in \mathbb{Z}\langle Y_1\rangle$ is $u * v \in \mathbb{Z}\langle Y_0 \cup Y_1\rangle$, $u*v = \sum w\{w|_{Y_0}=u \text{ and } w|_{Y_1}=v\}$, where $w|_{Y_i}$ is the monomial in $\mathbb{Z}\langle Y_i\rangle$ obtained by setting to 1 every symbol of w lying in $Y_{1-i}, i=0,1$. Then the shuffle product is extended linearly to any two elements in $\mathbb{Z}\langle Y_0\rangle$ and $\mathbb{Z}\langle Y_1\rangle$ respectively.

Denoting by M(**Y**,s,n) the sum of all monomials of degree n of $\mathbb{Z}\langle Y\rangle$ with degree at most s in every symbol of **Y**, we straightforward have that :

$$M(\mathbf{Y},s,n+1) = Y\, M(\mathbf{Y},s,n) - \sum_{x \in \mathbf{Y}} x(x^s * M(\mathbf{Y}\backslash x,s,n-s)) \tag{1}$$

Where $$Y = \sum_{x \in \mathbf{Y}} x.$$

Now let **Y** be the set of symbols for a r-ary DMC, $|\mathbf{Y}| = r$, with transition probabilities q_{xy}. We consider the repetition code over **Y**. The codeword $\mathbf{a}_n$ = aa...a of length n is transmitted through the channel. We write $\mathbf{X}_a$ for **Y**\a. The morphism $\mathbb{Z}\langle X_a\rangle \to \mathbb{R}$ defined by replacing every symbol x of $\mathbf{X}_a$ by q_{ax} maps M($\mathbf{X}_a$,s,n) onto P($\mathbf{X}_a$,s,n) which is the probability that in the transmitted word $\mathbf{a}_n$ every symbol a is transformed into a symbol of $\mathbf{X}_a$ and that this one appears at most s times repeated in the received word.
From (1), it follows that

$$P(\mathbf{X}_a,s,n+1) = \sum_{x \in \mathbf{X}_a} q_{ax}\, P(\mathbf{X}_a,s,n) - \sum_{x \in \mathbf{X}_a} q_{ax}^{s+1} \binom{n}{s} P(\mathbf{X}_a\backslash x,s,n-s) \tag{2}$$

We now denote by $g_s(t)$ the polynomial

$$1 + t + \frac{t^2}{2!} + \dots \frac{t^s}{s!}.$$

We have that

$$\prod_{x \in \mathbf{X}_a} g_s(q_{ax}\, t) = \sum_{n \geq 0} P(\mathbf{X}_a,s,n)\, \frac{t^n}{n!}. \tag{3}$$

The generating function for the probability that a received word be decoded as $\mathbf{a}_n$ given that $\mathbf{a}_n$ is transmitted is

$$P_{cd}(a,t) = \sum_{s\geq 0} \prod_{x\in \mathbf{X}_a} g_s(q_{ax}t) \frac{(q_{aa}t)^{s+1}}{(s+1)!}. \tag{4}$$

The generating function for the probability that a received word be decoded as another codeword than $\mathbf{a}_n$ given that $\mathbf{a}_n$ is transmitted is

$$P_{err}(a,t) = \sum_{s\geq 0} g_s(q_{aa}t) \sum_{x\in \mathbf{X}_a} \prod_{y\in \mathbf{X}_a\backslash x} g_s(q_{ay}t) \frac{(q_{ax}t)^{s+1}}{(s+1)!} \tag{5}$$

Finally we have that

$$e^t = P_{cd}(a,t) + P_{err}(a,t) + P_{flr}(a,t). \tag{6}$$

Under our hypothesis on a transitive DMC, that function remains unchanged when replacing a by any other symbol y of $\mathbf{Y}$ and $\mathbf{X}_a$ by $\mathbf{Y}\backslash y = \mathbf{X}_y$.
We denote by $P(\mathbf{X},s,n)$ the common value

$$P(\mathbf{X}_y,s,n),\ y\in Y. \tag{7}$$

From (3), (4), (7), we deduce that the probability of correct decoding $P_{cd}(n)$ for a repetition code on the considered MDC is, for $q_{xx} = q$, $\forall x\in \mathbf{Y}$,

$$P_{cd}(n) = \sum_{0\leq s\leq n-1} \binom{n}{s+1} q^{s+1}\, P(\mathbf{X},s,n-s-1)$$

where $P(\mathbf{X},s,i)$ is computed recursively by (2), $i=0, \dots, n-1$. Notice that $P(\mathbf{X},0,n)=0$ for $n > 0$. Also

$$P_{err}(n) = \sum_{x\in \mathbf{X}_a} \sum_{0\leq s\leq n-1} \binom{n}{s+1} q_{ax}^{s+1} \sum_{0\leq i\leq s} \binom{n-s-1}{i} q^i\, P(\mathbf{X}_a\backslash x,s,n-s-i-1), \tag{8}$$

which is invariant under the action of the transitive group G.

Finally a received word is not decodable iff at least two symbols appear the same largest number of times. Thus the probability of decoding failure $P_{flr}(n)$ is

$$P_{flr}(n) = 1- P_{cd}(n) - P_{err}(n).$$

In the particular case where the channel is r-ary symmetric, i.e. when $q_{xy} = p = (1-q)/(r-1)$ for $x\neq y$, relation (2) simplifies to

$$P(m,s,n+1) = mp\, P(m,s,n) - mp^{s+1} \binom{n}{s} P(m-1,s,n-s), \tag{9}$$

by writing m for r-1. In (9), p is thus an arbitrary fixed positive number such that mp≤1.

6. Combinatorial identities raised by the present investigation

Theorem 2

For every integer m, we have the identity

$$e^t = 1 + \sum_{s\geq 0} \sum_{1\leq i\leq m} \binom{m}{i} g_s^{m-i}\left(\frac{t}{m}\right)\left(\frac{1}{(s+1)!}\left(\frac{t}{m}\right)^{s+1}\right)^i$$

in which every term under $\sum_{s\geq 0}$ *in the RHS is an exponential generating function in which the coefficient of* $t^n/n\,!$ *is the probability that a random word of length n in m symbol, each of them equally probable, has some symbols occuring exactly s+1 times, the others occuring at most s times.*

This follows from the fact that every word of length n in m symbols has exactly i letters occuring the same largest number of times, i=1,...,m.

Relation (6) gives a more general identity. There the formal series are in A[[t]] where A is the ring $\mathbb{Q}\,[\{q_{xy}\}_{x,y\in Y}]$.

7. Relations with the birthday paradox

Relation (9) also allows exact computations of the average number of people sampled up to the moment where the sample contains s+1 different people with the same birthday for m possible birthdays. This is indeed

$$\sum_{n\geq 0} P(m,s,n).$$

Here, when computing P(k,s,t) by (9) for integers k and t smaller that m and n+1 respectively, p remains fixed at value 1/m.

For m = 52 and s = 1,2,3,4,5,6,7,8, we then obtain from (9) for (11) the respective values 8.7183, 24.9400, 46.1059, 70.6872, 97.7810,126.8144, 157.4009, 189.2657. This means for example that the average number of drawings with replacement from a deck of 52 cards to get some card exactly 6 times drawed is 97.7810.

REFERENCES

[1] W. FELLER : *An Introduction to Probability Theory and Its Applications.* 1968 John Wiley & Sons.

[2] D. FOATA, M. SCHÜTZENBERGER : *Théorie Géométrique des Polynômes Eulériens.* 1970 Lecture Notes in Mathematics # 138 Springer Verlag.

[3] R.McELIECE : *The Theory of information and coding*. 1977 Addison-Wesley.

[4] T. ERICSON : *Concatenated Codes - Principles and possibilities* -AAECC- 4, Karlsruhe 1986.

ON BINARY CODES OF ORDER 3

B. Courteau and A. Montpetit
Département de mathématiques et d'informatique
Université de Sherbrooke, Sherbrooke, Canada, J1K 2R1

ABSTRACT We present a class of binary (not necessarily linear) codes containing perfect codes, Preparata codes, BCH codes of lenght 2^{2m+1}-1, Hadamard code of length 11, l-error-correcting uniformly packed codes and also some other classes presenting remarkable regularity properties. The points of a code C of order 3 are scattered in the ambiant Hamming space so that for any point x the number of paths of length 3 joining x to the code only depends on the fact that the Hamming distance d(x, C) from x to C is 0, 1 or is greater than 1.

In the linear case the set of columns of a parity check matrix of a code of order 3 is a triple-sum-set [5], which is a natural extension of partial difference sets [7, 15, 2] and the orthogonal of such a code admits at most three non-zero weights.

The aim of this communication is to introduce and characterize codes of order 3 and of order 3-star in the binary case and to give some examples. The general q-any case is treated in [4] and the proofs of the results are to be found there

1. Hamming graph and the combinatorial matrix of a code

Let $GF(2)^n$ the binary Hamming space equipped with the Hamming distance d: d(x, y) is the number of components in which the vectors x and y in $GF(2)^n$ differ. The Hamming graph H(n) has $GF(2)^n$ as vertex-set and $E=\{(x,y) \mid x, y \in GF(2)^n, d(x, y) = 1\}$ as edge-set.

Let $C \subset GF(2)^n$ an arbitrary binary code of length n.

Définition 1.1: The combinatorial matrix of the code C is the matrix A of dimension $2^n \times \infty$ whose element in position $(x, j) \in GF(2)^n \times N$ is A(x, j) = number of paths of length j joining x to the code C in the Hamming graph H(n).

The combinatorial matrix is closely related to the external distance s' and dual distances $d'_1, \ldots, d'_{s'}$ which are defined [8] as follows: s' is such that s'+1 is the number of non zero components of the MacWilliams transform of the distance distribution of C and $d'_0 = 0, d'_1, \ldots, d'_{s'}$ are the indices of these non-zero components. We need the following result proved in [3].

Theorem 1.1 The external distance of C is s' $\Leftrightarrow$ s' is the minimum among the numbers t for which there exists an t+1 homogeneous linear recurrence

$$(1) \quad \sum_{j=1}^{t+1} c_j A(x, j+m) = 0, \ x \in GF(2)^n, m \in N$$

where $c_0, \ldots, c_{t+1}$ are integers and $c_{t+1} \neq 0$.

Moreover, the recurrence of minimum order $s'+1$ is unique if we impose the condition $c_{s'+1} = 1$ and the coefficients verify

$$(2) \quad \sum_{j=0}^{s'+1} c_j z^j = \prod_{r \in J} (z - P_1(r))$$

where $J = \{0, d'_1, ..., d'_{s'}\}$ and $P_1(r) = n - 2r$

Remark 1.2: There is a non homoneneous version of the preceding theorem in which the recurrence (1) is replaced by

$$(3) \quad \sum_{j=0}^{t} b_j \, A(x, j) = 1, \qquad x \in GF(2)^n$$

and the coefficients b_j of the minimum order relation verify

$$(4) \quad \sum_{j=0}^{s'} b_j z^j = 2^n \, |c|^{-1} \prod_{r \in J_0} (1 - z/P_1(r))$$

with $J_0 = J - \{0\}$.

2. Codes of order 3 or 3-star

Definition 2.1 The star-extension of the code C is the code $C^* = \{(\lambda, x) \mid \lambda \in GF(2), x \in C\}$.

Let A and A^* be the combinatorial matrices of C and C^* respectively.

Lemma 2.2 For all $j \in N$ and $(\lambda, x) \in GF(2)^{n+1}$, we have

$$A^*((\lambda, x), j) = \sum_{k=0}^{j} \binom{j}{k} A(x, k)$$

Definition 2.3 Let $C \subseteq GF(2)^n$ be a binary code of length n. We shall say that <u>C is of order 3</u> with parameters μ_0, μ_1, μ_2 if

$$A(x, 3) = \begin{cases} \mu_0 & \text{if } d(x, C) = 0 \\ \mu_1 & \text{if } d(x, C) = 1 \\ \mu_2 & \text{if } d(x, C) > 1. \end{cases}$$

C shall be said <u>of order 3-star</u> if C^* is of order 3.

Remark 2.4: Let $C = \ker H$ be a linear 1-error-correcting (n, n-k)- code with parity control matrix H and $\Omega \subseteq GF(2)^k$ be the set of columns of H with a fixed ordering. We say that $\Omega \subseteq GF(2)^k$ is a <u>triple-sum-set</u> with parameters λ_1, λ_2, if

$$\text{card}(\{(a_1, a_2, a_3) \in \Omega^3 / x = a_1 + a_2 + a_3\}) = \begin{cases} \lambda_1 & \text{if } x \in \Omega \setminus \{0\} \\ \lambda_2 & \text{if } x \in \Omega^c \setminus \{0\} \end{cases}$$

where Ω^c is the complement of Ω in $GF(2)^k$. This is a natural generalization of the notion of partial difference set [7]. In the linear case we have the following equivalences:

C is of order 3 $\Leftrightarrow \Omega$ is a triple-sum-set

C is of order 3-star $\Leftrightarrow \Omega \cup \{0\}$ is a triple-sum-set

3. Characterization of codes of order 3 or 3-star

Using theorem 1.1, remark 1.2 and lemma 2.2, we obtain the following results.

Theorem 3.1 If C is a code of order 3 or 3-star with parameters μ_0, μ_1, μ_2 then $s' \le 3$. Moreover, if $\mu_2 = 0$ then $s' \le 2$.

Theorem 3.2 Let C be a code with minimum distance $d \ge 3$ such that $s' = 3$ and let d'_1, d'_2, d'_3 be its dual distances. Then
(i) C is of order 3 $\Leftrightarrow d'_1 + d'_2 + d'_3 = 3n/2$
(ii) C is of order 3-star $\Leftrightarrow d'_1 + d'_2 + d'_3 = 3(n+1)/2$.

Theorem 3.3 Let C be a code with minimum distance $d \ge 3$. Then, $s' \in \{1, 2\} \Leftrightarrow$ C is of order 3 and of order 3-star.

Let e(C) denote the error-correcting capability of the code C. The following result determine all codes C of order 3 or 3-star such that $e(C) \ge 2$.

Theorem 3.4
(i) There do not exist codes C of order 3 for $e(C) \ge 3$.
(ii) There do not exist codes C of order 3-star for $e(C) > 3$. The only codes of order 3-star with $e(C) = 3$ are the perfect Golay code of length 23 and the repetition code of length 7.
(iii) C is of order 3 with $e(C) = 2 \Leftrightarrow$ C is 2-error correcting (λ, μ)-uniformly packed [10] with $\lambda = \mu$.
(iv) C is of order 3-star with $e(C) = 2 \Leftrightarrow$ C is 2-error correcting strongly (λ, μ)-uniformly packed code (i.e. $\mu = \lambda + 1$).

Remark 3.5 By a result of Goethals-Van Tilborg [10], the only codes of order 3-star such that $e(C) = 2$ are the BCH codes of length $2^{2m+1}-1$, the Preparata codes of length $2^{2m}-1$ and the Hadamard code B_{12} of length 11, these two last codes being non linear.

Problem 3.6 Determine all codes C of order 3 or 3-star such that $e(C) = 1$.

4. Some examples in the case $e(C) = 1$

4.1 All perfect or uniformly packed single-error correcting codes, linear or not, are codes of order 3 and of order 3-star in virtue of theorem 4.3 since by a theorem of Mac Williams [11] and a theorem of Van Tilborg [12], $s'=1$ and $s'=2$ respectively for these codes.

4.2 Calderbank and Goethals [1] have given two classes of non-uniformly packed codes of length $n=2^m-1$ for $m \ge 4$. These two classes admits three dual distances satisfying the condition of theorem 4.2 (ii). They are linear codes of order 3-star but not of order 3.

4.3 Orthogonals of hyperquadric codes considered by Wolfmann [13] in odd dimension have three dual distances satisfying theorem 4.2 (ii). Thus they are (linear) codes of order 3-star but not of order 3.

4.4 A new class of codes of order 3-star.

Let r be a natural number. In $GF(2)^{3r} \cong GF(2^r)^3$ consider the partial spread $P = \{V_1, ..., V_m\}$ deduces from an oval of the projective plane $PG(2^r, 2)$. This means that $m=2^r+1$ [9] and that $V_1, ..., V_m$ are r-dimensional vector subspaces of $GF(2)^{3r}$ such that for all triplets of distinct indices $i, j, k \in \{1, ..., m\}$

$$V_i \cap (V_j + V_k) = \{0\}.$$

Let $\Omega = \bigcup_{i=1}^{m} V_i \setminus \{0\}$ and $C = C(\Omega)$ be the code (up to equivqlence) whose parity check matrix admits Ω as set of columns . The length of C is $n = m(2^r-1) = 2^{2r} - 1$. Applying proposition 4 of [6] we see that C admits the three dual distances $d'_1=(m-2)2^{r-1}$, $d'_2=(m-1)2^{r-1}$ and $d'_3 = m2^{r-1}$. Thus

$$\sum_{i=1}^{3} d'_i = (3m-3)2^{r-1} = 3 \times 2^r \times 2^{r-1} = \frac{3}{2}\, 2^{2r} = \frac{3}{2}\,(n+1).$$

Hence, by theorem 4.2(ii), C is a code of order 3-star. When r is even these codes are not orthogonal to any hyperquadric code because otherwise they should admit only two dual distances [14].

References

[1] A.R. Calderbank and J.-M. Goethals, Three-Weight Codes and Association Schemes, Philips J. Res. 39 (1984) 143-152.

[2] P. Camion, Difference Sets in Elementary Abelian Groups (Les Presses de l'Université de Montréal, Montréal, 1979).

[3 P. Camion, B. Courteau, P. Delsarte, On r-partition Designs in Hamming Spaces, Rapport de recherche no 626. INRIA, Rocquencourt, 78153 Le Chesnay, France, février 1987.

[4] B. Courteau, A. Montpetit, On a class of codes admitting at most three non-zero dual distances, submitted to Discrete Math.

[5] B. Courteau and J. Wolfmann, On Triple-sum-sets and two or three weights codes, Discrete Math. 50 (1984) 179-191.

[6] B. Courteau, G. Fournier, R. Fournier, A Characterization of N-weights projective codes, IEEE Trans. Inform. Theory 27 (1981) 808-812.

[7] I.M. Chakravarti and K.V. Suryanarayana, Partial difference sets and partially balanced weighting desings for calibration and tournaments, J. Combinatorial Theory (A) 13 (1972) 426-431.

[8] P. Delsarte, Four Fundamental Parameters of a Code and Their Combinatorial Significance, Inform. and control 23 (1973) 407-438.

[9] P. Dembowski, Finite Geometries (Springer-Verlag, New York, 1968).

[10] J.-M. Goethals and H.C.A. Van Tilborg, Uniformly packed codes, Philips Research Reports 30 (1975) 9-36.

[12] H.C.A. Van Tilborg, Uniformly packed codes, Thesis, Tech. Univ. Eindhoven, 1976.

[13] J. Wolfmann, Codes projectifs à deux ou trois poids associés aux hyperquadriques d'une géométrie finie, Discrete Math. 13 (1975), 185-211.

[14] J. Wolfmann, Codes projectifs à deux poids, 'caps' complets et ensembles de différences, J. Combinatorial Theory (A) 23 (1977).

POLYNOMIAL DECOMPOSITION ALGORITHM OF ALMOST QUADRATIC COMPLEXITY

J. Gutiérrez, T. Recio and C. Ruiz de Velasco
Departamento de Matemáticas, Estadística y Computación,
Universidad de Cantabria,39071- Santander, Spain

§ 0

In Barton & Zippel (" Polynomial decomposition algorithms " J. Symbolic Computation (1985) V.1) the authors give a polynomial decomposition algorithm for an univariate polynomial f(X) over a field of characteristic zero with roughly estimated complexity which is in the worst case exponential in the degree of the given polynomial.

This is due to the fact that they use essentially factorization procedures of polynomials in their algorithm, recombining in all the different ways the obtained factors. Moreover, they do not compute explicitly the complexity but for one of the steps, precisely the one that requires the computation of a polynomial g(X) such that f(X)= g(X)oh(X) once h(X) is finded. This step needs $O(deg(f(X)^2)$ field operations according to the authors .

In this communication we evaluate the complexity of a different algorithm by Gutierrez & Ruiz de Velasco obtaining a $n^{2+\delta}$ upper bound complexity, where d is an arbitrary small positive constant, improving therefore greatly the time of the previous algorithm.

§ 1

Let us survey briefly the algorithm of [G-R] (where the reader can find supplementary details).

First of all, given a polynomial f(X) over any field, with n=deg(f(X)) prime to the characteristic (therefore without any restriction if the characteristic is zero), in order to find a decomposition of f(X) we can assume that $n > 1$.

Then we are going to find polynomials $g_1(X)$, $g_2(X)$,....., $g_r(X)$ of degree > 1 such that

(a) $f(X) = g_r(X) \circ g_{r-1}(X) \circ \ldots\ldots\ldots \circ g_1(X)$

(b) $g_i(0) = 0$ for $i = 1, \ldots\ldots, r-1$.

(c) $g_i(X)$ are monic for $i = 1, \ldots\ldots, r-1$.

(d) $g_i(X)$ are indecomposable polynomials for all i.

As in [B-Z] the main step is to find a polynomial h(X) such that h(0)=0, of degree > 1, and verifiying $f(X) = g(X) \circ h(X)$ for some polynomial g(X). Then we compute g(X) and proceed recursively.

The main point in [G-R] eviting factorization is the fact that a precise algorithm is given that for a given degree m (dividing n) determines uniquely the coefficients of h(X) (with deg(h(X)) = m) or, in the other hand, decides that such a h(X) does not exist.

If such that h(X) exists we compute g(X) as in [B-Z] at cost n^2 and continue with g(X). If such h(X) does not exist we try with another divisor of n, eventually concluding that f(X) itself is indecomposable. If the divisors of n are tested in an ordered form, the computed h(X) are then indecomposable.

§ 2

Here we compute the cost of finding the coefficients of h(X) or the cost of rejecting the choosen divisor of deg(f(X)).

Let m be a strict divisor of n and let h(X),

$h(X) = X^m + b_1X^{m-1} + \dots\dots\dots\dots + b_{m-1}X$. Then we make the following symbolic divisions obtaining only the first m terms of each quotient and the corresponding of each remainder:

$$f(X) = q_1(X)\ h(X) + r_0(X)$$

$$q_1(X) = q_2(X)\ h(X) + r_1(X)$$

$$\vdots$$

$$q_{t-1}(X) = q_t(X)\ h(X) + r_{t-1}(X)$$

where tm = n and therefore $q_t(X) = 1$. Now if h(X) is a good candidate for the decomposition of f(X) then $r_{t-1}(X)$ must be constant. Imposing this condition we obtain a system of equations in the b_i´s and the m first coefficients of all the $q_i(X)$´s that determine exactly the b_i´s. Actually (see the proof [G-R]) if

$$q_j(X) = X^{n-jm} + A^j{}_1X^{n-jm-1} + \dots + A^j{}_iX^{n-jm-i} + \dots + A^j{}_{m-1}X^{n-jm-(m-1)} + \dots$$

for all j and

$$f(X) = X + A^0{}_1X^{n-1} + A^0{}_2X^{n-2} + \dots\dots + A^0{}_iX^{n-i} + \dots\dots + A^0{}_{n-1}X + A^0{}_n,$$

then

$$b_1 = A^0{}_1 / t \quad \text{and}$$

$$A^j{}_1 = A^{j-1}{}_1 - b_1 \quad \text{for all } j \quad \text{and}$$

$$C_1 = A^1{}_1 + A^2{}_1 + \ldots\ldots + A^{t-1}{}_1$$

Moreover

$$b_2 = (A^0{}_2 - b_1 C_1) / t$$

$$A^j{}_2 = A^{j-1}{}_2 - b_1 A^j{}_1 - b_2 \quad \text{for all } j$$

$$C_2 = A^1{}_2 + A^2{}_2 + \ldots\ldots + A^{t-1}{}_2$$

and so on until b_{m-1} is computed.

Clearly the number of field operations to compute the b_i´s is at most $O(n^2)$.

§ 3

With this h(X) we can compute the rest of the remainders - as we compute g(X) - in time $O(n^2)$ [B-Z]. If all the remainders are constant we can proceed recursively with g(X). If some remainder is not constant then there exist not a h(X) for degree m.

Therefore if we adopt the hypothesis that this algorithm runs in time $O(n^{2+\delta})$ then the recursive step will take $O(sn^2)$ where s is the number of divisors of n that we have rejected before finding a good candidate.

Then the problem is solved in time $O(n^{2+\delta})$ as the remaining polynomials have degree smaller than n. Therefore total time is $O(sn^2)+O(n^{2+\delta})$.

It only rests to know how many divisors has n. Now (see [V]) $s = O(n^\delta)$ and thus $O(sn^2) + O(n^{2+\delta}) = O(n^{2+\delta})$.

Added after the Conference.

The referee has pointed to us that in the 28th Ann. IEEE Symp. Foundations of Computer Science, Los Angeles,1987, J. von zur Gathen, D. Kozen and S, Landau in " Functional decomposition of polynomials" have given a positive answer to the polynomial decomposition problem. Also he/she mentions differents previous technical reports by part of these authors concerning the same problem. In particular he/she mentions a bound $O(n \log^2 n \log\log n)$ which is more precise that the one given in our paper, although our aim was to lessen the exponential bound by D. Barton and R. Zippel.

We are grateful to the referee for this information.

References

[B-Z] Barton,D & Zippel,R. " Polynomial decomposition algorithms " J. Symbolic Comp. (1985) **1**,159-168.

[G-R] Gutiérrez, J & Ruiz de Velasco,C. " A polynomial decomposition algorithm" J.Symbolic Comp.(submitted, February 1988)

[V] Vinogradov,I. " Fundamentos de la teoria de los numeros " Ed. Mir. Moscu (1977).

This paper is partially supported by D.G.I.C.Y.T PA 86 - 0471

THE WEIGHTS OF THE ORTHOGONALS OF CERTAIN CYCLIC CODES OR EXTENDED GOPPA CODES

J. WOLFMANN
G.E.C.T
Université de Toulon
83130 LA GARDE , FRANCE

Abstract : Starting from results on elliptic curves and KLOOSTERMAN sums over the finite field $GF(2^t)$, we determine the weights of the orthogonals of the following binary linear codes : the MELAS code of length $2^t - 1$, the irreducible cyclic binary code of length $2^t + 1$, and the extended binary GOPPA codes defined by polynomials of degree two.

Extended abstract

0. Introduction

In this paper we present a deep link between elliptic curves and KLOOSTERMAN sums over the finite field $\mathbf{F} = GF(2^t)$ and,on the other hand, special families of cyclic and GOPPA codes defined by polynomials of degree two (quadratic GOPPA codes). The references on coding theory are to be found in [3].

Notations. $\mathbf{F} = GF(2^t)$ denotes the finite field with 2^t elements, where $t \geq 3$, and $\mathbf{F}^\times = \mathbf{F} - \{0\}$; the **trace** of $z \in \mathbf{F}$ over GF(2) is

$$\mathrm{tr}\, z = z + z^2 + \ldots + z^{2^i} + \ldots + z^{2^{t-1}}.$$

If γ is in an extension field of GF(2), then $m_\gamma(x)$ is the **minimal polynomial** of γ over GF(2).

If $x = (x_1, x_2, \ldots, x_n) \in GF(2)^n$, the **weight** of x is the number

$$w(x) = \#\{i \mid x_i \neq 0\},$$

1.KLOOSTERMAN sums

we consider the KLOOSTERMAN sum defined by

$$K(a) = \sum_{x \in \mathbf{F}^\times} (-1)^{\mathrm{tr}(x^{-1} + ax)} \quad (a \in \mathbf{F}^\times),$$

Theorem 1.(LACHAUD [2])

The set of the $K(a)$, $a \in \mathbf{F}^\times$, is the set of all the integers $s \equiv -1 \pmod 4$ in the range

$$[-2^{(t/2)+1}, 2^{(t/2)+1}]$$

In order to apply the results of section 2 and 3 to cyclic and GOPPA codes we will give a descriptive result in lemma below. Let α be a primitive root of **F**, and let β be a primitive root of unity of order $2^t + 1$ in $GF(2^{2t})$. We introduce the following two sets :

$$\Omega = \{ x \in \mathbf{F}^{\times} \mid tr\, x^{-1} = 0 \}, \qquad \bar{\Omega} = \{ x \in \mathbf{F}^{\times} \mid tr\, x^{-1} = 1 \}.$$

Lemma 1.2. *We have*

$$\Omega = \{ x \in F^{\times} / x = \alpha^{i} + \alpha^{-i}, i = 1, 2, \ldots, 2^{t-1} - 1 \},$$

and $$\bar{\Omega} = \{ x \in F^{\times} / x = \beta^{i} + \beta^{-i}, i = 1, 2, \ldots, 2^{t-1} \}.$$

2. Two linear codes

The elements α and β are as above. Let $n = 2^{t-1}$. Define two mappings

$$m : \mathbf{F} \to GF(2)^{n-1} \quad \text{and} \quad \bar{m} : \mathbf{F} \to GF(2)^{n}$$

by

$$m(a) = (tr\,(a\omega_1), \ldots, tr\,(a\omega_i), \ldots, tr\,(a\omega_{n-1})) \qquad \text{with } \omega_i = \alpha^i + \alpha^{-i},$$

$$\bar{m}(a) = (tr\,(a\bar{\omega}_1), \ldots, tr\,(a\bar{\omega}_i), \ldots, tr\,(a\bar{\omega}_n)) \qquad \text{with } \bar{\omega}_i = \beta^i + \beta^{-i}.$$

Definition 2.1. The code C (resp. $\bar{C}$) is the image of m (resp. $\bar{m}$) :

$$C = m\,(\mathbf{F}), \qquad \bar{C} = \bar{m}\,(\mathbf{F}).$$

Obviously C and $\bar{C}$ are linear codes because m and $\bar{m}$ are linear mappings.

Theorem 2.2. *Let $n = 2^{t-1}$ and C the set of the words of $GF(2)^{n-1}$,*

$$m(a) = (tr\,(a\omega_1), \ldots, tr\,(a\omega_i), \ldots, tr\,(a\omega_{n-1})), \quad a \in \mathbf{F},$$

where $\omega_i = \alpha^i + \alpha^{-i}$ and α is a primitive root of **F**. *The code C is a linear code of length $2^{t-1} - 1$ and dimension t. The weights of the non zero words of C are all the integers w such that*

$$\left| w - \left(\frac{2^t - 1}{4}\right) \right| \leq 2^{(t/2)-1}.$$

Theorem 2.3. *Let $n = 2^{t-1}$. Set $\omega_i = \beta^i + \beta^{-i}$ where β is a primitive $2^t + 1$ - th root of unity in $GF(2^{2t})$ and $\bar{C}$ the set of the words $\bar{m}(a)$ of $GF(2)^n$ such that*

$$\bar{m}(a) = (tr\,(a\bar{\omega}_1), \ldots, tr\,(a\bar{\omega}_i), \ldots, tr\,(a\bar{\omega}_n)), \quad a \in F.$$

Then $\bar{C}$ is a linear code of length 2^{t-1} and dimension t.

The weights of the non zero words of $\bar{C}$ are all the integers w such that

$$\left| w - \left(\frac{2^t + 1}{4}\right) \right| \leq 2^{(t/2)-1}.$$

3. The weights of two classes of cyclic codes

Definition 3.1. Let α be a primitive root of **F**. The MELAS code $M(\alpha)$ is the cyclic code of length $2^t - 1$ generated by $m_\alpha(x)\, m_{\alpha^{-1}}(x)$.

Theorem 3.2 *The weights of the non zero words of the orthogonal $M(\alpha)^\perp$ of the MELAS code $M(\alpha)$, are all the integers w such that :*

$$| w -(\frac{2^t - 1}{2}) | \leq 2^{t/2} \quad \text{and w even.}$$

Definition 3.3. The code $N(\beta)$ is the binary cyclic code of length $2^t + 1$ generated by $m_\beta(x)$.

The orthogonal $N(\beta)^\perp$ is an irreducible cyclic code . It is the image of $GF(2^{2t})$ by the mapping $\bar{\mu}$ from $GF(2^{2t})$ into $GF(2)^{2^t+1}$ defined by :

$$\bar{\mu} = (Tr(u), Tr(u\beta), ...Tr(u\beta^i), ...Tr(u\beta^{2^t})) \qquad u \in GF(2^{2t})$$

where Tr denotes the trace of $GF(2^{2t})$ over GF(2).

Theorem 3.4. *The weights of the non zero words of the irreducible cyclic $N(\beta)^\perp$ are all the integers w such that*

$$| w -(\frac{2^t + 1}{2}) | \leq 2^{t/2} \quad \text{and w even.}$$

4. The weights of the orthogonals of certain extended Goppa codes

The elements α and β are defined as above and the definitions of GOPPA codes are in [2].

Definition 4.1. Let $\Gamma(\alpha)$ be the GOPPA code $\Gamma(L, g)$ with $L = \mathbf{F} - \{\alpha, \alpha^{-1}\}$ and

$$g(x) = x^2 + (\alpha + \alpha^{-1})\, x + 1 ;$$

and let $\Gamma(\beta)$ be the GOPPA code $\Gamma(\bar{L}, \bar{g})$ with $\bar{L} = \mathbf{F}$ and

$$\bar{g}(x) = x^2 + (\beta + \beta^{-1})\, x + 1.$$

Proposition 4.2. (Tzeng, Zimmermann [3]). *The extended code* $\tilde{\Gamma}(\alpha)$ *is equivalent to the cyclic code generated by* $(x+1)\, m_{\alpha}(x)\, m_{\alpha^{-1}}(x)$.

Proposition 4.3. (BERLEKAMP, MORENO [1]). *The extended code* $\tilde{\Gamma}(\beta)$ *is equivalent to the cyclic code generated by* $(x+1)\, m_{\beta}(x)$.

Theorem 4.4. *The weights of the non zero words of the code* $\tilde{\Gamma}(\alpha)^{\perp}$ *orthogonal of the extended GOPPA code* $\Gamma(\alpha)$, *are all the integers w such that*

$$| w - (\frac{2^t - 1}{2}) | \leq 2^{t/2} \quad \text{or} \quad w = 2^t - 1.$$

Theorem 4.5. *The weights of the non zero words of the code* $\tilde{\Gamma}(\beta)^{\perp}$ *orthogonal of the extended GOPPA code* $\Gamma(\beta)$, *are all the integers w such that*

$$| w - (\frac{2^t + 1}{2}) | \leq 2^{t/2} \quad \text{or} \quad w = 2^t + 1.$$

5. On the weight distributions

Proving the previous results we also show show that all the weights of the codes introduced in that paper can be calculated starting from the weights of C. On the other hand the weights of C are obtained from the values of all the Kloosterman sums K(a). Let A(x) denotes the number of a in $\mathbf{F}^{\times}$ such that K(a) = x, where $x \in \mathbf{Z}$; then the weight distributions of the codes can be given by means of the function A(x).

Theorem 5.1. *Let A be the function from* $\mathbf{Z}$ *to N defined by*

$$A(x) = \# \{ a \in \mathbf{F}^{\times} \,/\, K(a) = 2^t - 1 - 4x \}$$

where K(a) is the Kloosterman sum related to a. The weight distributions of the non zero words in the codes C, $\bar{C}$, $M(\alpha)^{\perp}$, $N(\beta)^{\perp}$, $\tilde{\Gamma}(\alpha)^{\perp}$, $\tilde{\Gamma}(\beta^{\perp})$ *are given in the following array :*

C	weight	w			
	number	A(w)			
$\bar{C}$	weight	w			
	number	$A(2^{t-1}-w)$			
$M(\alpha)^{\perp}$	weight	2^{t-1}		2λ $(\lambda \neq 2^{t-2})$	$2\lambda + 1$
	number	$(2^t - 1)$ $[A(2^{t-2}) + 2]$		$(2^t - 1)A(\lambda)$	0
$N(\beta)^{\perp}$	weight	2λ		$2\lambda + 1$	
	number	$(2^t + 1)$ $A(2^{t-1} - \lambda)$		0	
$\tilde{\Gamma}(\alpha)^{\perp}$	weight	$2^t - 1$	2^{t-1} or $2^{t-1} - 1$	2λ $(\lambda \neq 2^{t-2})$	$2\lambda+1$ $(\lambda \neq 2^{t-2}-1, 2^{t-1}-1)$
	number	1	$(2^t - 1)[A(2^{t-2}) + 2]$	$(2^t - 1)$ $A(\lambda)$	$(2^t - 1)$ $A(2^{t-1} - \lambda - 1)$
$\tilde{\Gamma}(\beta)^{\perp}$	weight	$2^t + 1$	2λ	$2\lambda + 1$ $(\lambda \neq 2^{t-1})$	
	number	1	$(2^t + 1)$ $A(2^{t-1} - \lambda)$	$(2^t + 1)$ $A(\lambda)$	

6.Conclusion

We will just say in conclusion that the results of that paper show a remarkable symmetry between :
a) the primitive $(2^t - 1)$-th root of unity α over GF(2), the set $\Omega = \{x \in \mathbf{F} \mid \mathrm{tr}\, x^{-1} = 0\}$, the code C, the codes of length $2^t - 1$: $M(\alpha)$, $\Gamma(\alpha)$ and the polynomial $g(x) = x^2 + (\alpha + \alpha^{-1})\, x + 1$.

b) the primitive $(2^t + 1)$-th root of unity β over GF(2), the set $\bar{\Omega} = \{x \in \mathbf{F} \mid \mathrm{tr}\, x^{-1} = 1\}$, the code $\bar{C}$, the codes of length $2^t + 1$: $N(\beta)$, $\Gamma(\beta)$ and the polynomial $\bar{g}(x) = x^2 + (\beta + \beta^{-1})\, x + 1$.

REFERENCES

[1] BERLEKAMP,E.R.,MORENO,O.,"*Extended double-error correcting binary GOPPA codes are cyclic* "
IEEE Trans.Info.Theory **19** (1973),817-818.

[2] LACHAUD,G.,WOLFMANN,J.,"*Sommes de Kloosterman,courbes elliptiques et codes cycliques en caracteristique 2*"
C.R.Acad.Sci.Paris (1),**305** (1987)

[3] McWILLIAMS,F.J.,SLOANE,N.J.A.,"*The theory of Error-correcting codes* "
North-Holland,Amsterdam,1977 .

[4] TZENG,K.K.,ZIMMERMANN,K.,"*On extending GOPPA codes to cyclic codes* "
IEEE Trans.Info.Theory **21** (1975) ,712-716.

AUTHORS' INDEX

Lecture Notes in Computer Science

Vol. 270: E. Börger (Ed.), Computation Theory and Logic. IX, 442 pages. 1987.

Vol. 271: D. Snyers, A. Thayse, From Logic Design to Logic Programming. IV, 125 pages. 1987.

Vol. 272: P. Treleaven, M. Vanneschi (Eds.), Future Parallel Computers. Proceedings, 1986. V, 492 pages. 1987.

Vol. 273: J.S. Royer, A Connotational Theory of Program Structure. V, 186 pages. 1987.

Vol. 274: G. Kahn (Ed.), Functional Programming Languages and Computer Architecture. Proceedings. VI, 470 pages. 1987.

Vol. 275: A.N. Habermann, U. Montanari (Eds.), System Development and Ada. Proceedings, 1986. V, 305 pages. 1987.

Vol. 276: J. Bézivin, J.-M. Hullot, P. Cointe, H. Lieberman (Eds.), ECOOP '87. European Conference on Object-Oriented Programming. Proceedings. VI, 273 pages. 1987.

Vol. 277: B. Benninghofen, S. Kemmerich, M.M. Richter, Systems of Reductions. X, 265 pages. 1987.

Vol. 278: L. Budach, R.G. Bukharajev, O.B. Lupanov (Eds.), Fundamentals of Computation Theory. Proceedings, 1987. XIV, 505 pages. 1987.

Vol. 279: J.H. Fasel, R.M. Keller (Eds.), Graph Reduction. Proceedings, 1986. XVI, 450 pages. 1987.

Vol. 280: M. Venturini Zilli (Ed.), Mathematical Models for the Semantics of Parallelism. Proceedings, 1986. V, 231 pages. 1987.

Vol. 281: A. Kelemenová, J. Kelemen (Eds.), Trends, Techniques, and Problems in Theoretical Computer Science. Proceedings, 1986. VI, 213 pages. 1987.

Vol. 282: P. Gorny, M.J. Tauber (Eds.), Visualization in Programming. Proceedings, 1986. VII, 210 pages. 1987.

Vol. 283: D.H. Pitt, A. Poigné, D.E. Rydeheard (Eds.), Category Theory and Computer Science. Proceedings, 1987. V, 300 pages. 1987.

Vol. 284: A. Kündig, R.E. Bührer, J. Dähler (Eds.), Embedded Systems. Proceedings, 1986. V, 207 pages. 1987.

Vol. 285: C. Delgado Kloos, Semantics of Digital Circuits. IX, 124 pages. 1987.

Vol. 286: B. Bouchon, R.R. Yager (Eds.), Uncertainty in Knowledge-Based Systems. Proceedings, 1986. VII, 405 pages. 1987.

Vol. 287: K.V. Nori (Ed.), Foundations of Software Technology and Theoretical Computer Science. Proceedings, 1987. IX, 540 pages. 1987.

Vol. 288: A. Blikle, MetaSoft Primer. XIII, 140 pages. 1987.

Vol. 289: H.K. Nichols, D. Simpson (Eds.), ESEC '87. 1st European Software Engineering Conference. Proceedings, 1987. XII, 404 pages. 1987.

Vol. 290: T.X. Bui, Co-oP A Group Decision Support System for Cooperative Multiple Criteria Group Decision Making. XIII, 250 pages. 1987.

Vol. 291: H. Ehrig, M. Nagl, G. Rozenberg, A. Rosenfeld (Eds.), Graph-Grammars and Their Application to Computer Science. VIII, 609 pages. 1987.

Vol. 292: The Munich Project CIP. Volume II: The Program Transformation System CIP-S. By the CIP System Group. VIII, 522 pages. 1987.

Vol. 293: C. Pomerance (Ed.), Advances in Cryptology – CRYPTO '87. Proceedings. X, 463 pages. 1988.

Vol. 294: R. Cori, M. Wirsing (Eds.), STACS 88. Proceedings, 1988. IX, 404 pages. 1988.

Vol. 295: R. Dierstein, D. Müller-Wichards, H.-M. Wacker (Eds.), Parallel Computing in Science and Engineering. Proceedings, 1987. V, 185 pages. 1988.

Vol. 296: R. Janßen (Ed.), Trends in Computer Algebra. Proceedings, 1987. V, 197 pages. 1988.

Vol. 297: E.N. Houstis, T.S. Papatheodorou, C.D. Polychronopoulos (Eds.), Supercomputing. Proceedings, 1987. X, 1093 pages. 1988.

Vol. 298: M. Main, A. Melton, M. Mislove, D. Schmidt (Eds.), Mathematical Foundations of Programming Language Semantics. Proceedings, 1987. VIII, 637 pages. 1988.

Vol. 299: M. Dauchet, M. Nivat (Eds.), CAAP '88. Proceedings, 1988. VI, 304 pages. 1988.

Vol. 300: H. Ganzinger (Ed.), ESOP '88. Proceedings, 1988. VI, 381 pages. 1988.

Vol. 301: J. Kittler (Ed.), Pattern Recognition. Proceedings, 1988. VII, 668 pages. 1988.

Vol. 302: D.M. Yellin, Attribute Grammar Inversion and Source-to-source Translation. VIII, 176 pages. 1988.

Vol. 303: J.W. Schmidt, S. Ceri, M. Missikoff (Eds.), Advances in Database Technology – EDBT '88. X, 620 pages. 1988.

Vol. 304: W.L. Price, D. Chaum (Eds.), Advances in Cryptology – EUROCRYPT '87. Proceedings, 1987. VII, 314 pages. 1988.

Vol. 305: J. Biskup, J. Demetrovics, J. Paredaens, B. Thalheim (Eds.), MFDBS 87. Proceedings, 1987. V, 247 pages. 1988.

Vol. 306: M. Boscarol, L. Carlucci Aiello, G. Levi (Eds.), Foundations of Logic and Functional Programming. Proceedings, 1986. V, 218 pages. 1988.

Vol. 307: Th. Beth, M. Clausen (Eds.), Applicable Algebra, Error-Correcting Codes, Combinatorics and Computer Algebra. Proceedings, 1986. VI, 215 pages. 1988.

Vol. 308: S. Kaplan, J.-P. Jouannaud (Eds.), Conditional Term Rewriting Systems. Proceedings, 1987. VI, 278 pages. 1988.

Vol. 309: J. Nehmer (Ed.), Experiences with Distributed Systems. Proceedings, 1987. VI, 292 pages. 1988.

Vol. 310: E. Lusk, R. Overbeek (Eds.), 9th International Conference on Automated Deduction. Proceedings, 1988. X, 775 pages. 1988.

Vol. 311: G. Cohen, P. Godlewski (Eds.), Coding Theory and Applications 1986. Proceedings, 1986. XIV, 196 pages. 1988.

Vol. 312: J. van Leeuwen (Ed.), Distributed Algorithms 1987. Proceedings, 1987. VII, 430 pages. 1988.

Vol. 313: B. Bouchon, L. Saitta, R.R. Yager (Eds.), Uncertainty and Intelligent Systems. IPMU '88. Proceedings, 1988. VIII, 408 pages. 1988.

Vol. 314: H. Göttler, H.J. Schneider (Eds.), Graph-Theoretic Concepts in Computer Science. Proceedings, 1987. VI, 254 pages. 1988.

Vol. 315: K. Furukawa, H. Tanaka, T. Fujisaki (Eds.), Logic Programming '87. Proceedings, 1987. VI, 327 pages. 1988.

Vol. 316: C. Choffrut (Ed.), Automata Networks. Proceedings, 1986. VII, 125 pages. 1988.

Vol. 317: T. Lepistö, A. Salomaa (Eds.), Automata, Languages and Programming. Proceedings, 1988. XI, 741 pages. 1988.

Vol. 318: R. Karlsson, A. Lingas (Eds.), SWAT 88. Proceedings, 1988. VI, 262 pages. 1988.

Vol. 319: J.H. Reif (Ed.), VLSI Algorithms and Architectures – AWOC 88. Proceedings, 1988. X, 476 pages. 1988.

Vol. 320: A. Blaser (Ed.), Natural Language at the Computer. Proceedings, 1988. III, 176 pages. 1988.

Vol. 321: J. Zwiers, Compositionality, Concurrency and Partial Correctness. VI, 272 pages. 1989.

Vol. 322: S. Gjessing, K. Nygaard (Eds.), ECOOP '88. European Conference on Object-Oriented Programming. Proceedings, 1988. VI, 410 pages. 1988.

Vol. 323: P. Deransart, M. Jourdan, B. Lorho, Attribute Grammars. IX, 232 pages. 1988.

Vol. 324: M.P. Chytil, L. Janiga, V. Koubek (Eds.), Mathematical Foundations of Computer Science 1988. Proceedings. IX, 562 pages. 1988.

Vol. 325: G. Brassard, Modern Cryptology. VI, 107 pages. 1988.

Vol. 326: M. Gyssens, J. Paredaens, D. Van Gucht (Eds.), ICDT '88. 2nd International Conference on Database Theory. Proceedings, 1988. VI, 409 pages. 1988.

Vol. 327: G.A. Ford (Ed.), Software Engineering Education. Proceedings, 1988. V, 207 pages. 1988.

Vol. 328: R. Bloomfield, L. Marshall, R. Jones (Eds.), VDM '88. VDM – The Way Ahead. Proceedings, 1988. IX, 499 pages. 1988.

Vol. 329: E. Börger, H. Kleine Büning, M.M. Richter (Eds.), CSL '87. 1st Workshop on Computer Science Logic. Proceedings, 1987. VI, 346 pages. 1988.

Vol. 330: C.G. Günther (Ed.), Advances in Cryptology – EUROCRYPT '88. Proceedings, 1988. XI, 473 pages. 1988.

Vol. 331: M. Joseph (Ed.), Formal Techniques in Real-Time and Fault-Tolerant Systems. Proceedings, 1988. VI, 229 pages. 1988.

Vol. 332: D. Sannella, A. Tarlecki (Eds.), Recent Trends in Data Type Specification. V, 259 pages. 1988.

Vol. 333: H. Noltemeier (Ed.), Computational Geometry and its Applications. Proceedings, 1988. VI, 252 pages. 1988.

Vol. 334: K.R. Dittrich (Ed.), Advances in Object-Oriented Database Systems. Proceedings, 1988. VII, 373 pages. 1988.

Vol. 335: F.A. Vogt (Ed.), CONCURRENCY 88. Proceedings, 1988. VI, 401 pages. 1988.

Vol. 336: B.R. Donald, Error Detection and Recovery in Robotics. XXIV, 314 pages. 1989.

Vol. 337: O. Günther, Efficient Structures for Geometric Data Management. XI, 135 pages. 1988.

Vol. 338: K.V. Nori, S. Kumar (Eds.), Foundations of Software Technology and Theoretical Computer Science. Proceedings, 1988. IX, 520 pages. 1988.

Vol. 339: M. Rafanelli, J.C. Klensin, P. Svensson (Eds.), Statistical and Scientific Database Management. Proceedings, 1988. IX, 454 pages. 1989.

Vol. 340: G. Rozenberg (Ed.), Advances in Petri Nets 1988. VI, 439 pages. 1988.

Vol. 341: S. Bittanti (Ed.), Software Reliability Modelling and Identification. VII, 209 pages. 1988.

Vol. 342: G. Wolf, T. Legendi, U. Schendel (Eds.), Parcella '88. Proceedings, 1988. 380 pages. 1989.

Vol. 343: J. Grabowski, P. Lescanne, W. Wechler (Eds.), Algebraic and Logic Programming. Proceedings, 1988. 278 pages. 1988.

Vol. 344: J. van Leeuwen, Graph-Theoretic Concepts in Computer Science. Proceedings, 1988. VII, 459 pages. 1989.

Vol. 345: R.T. Nossum (Ed.), Advanced Topics in Artificial Intelligence. VII, 233 pages. 1988 (Subseries LNAI).

Vol. 346: M. Reinfrank, J. de Kleer, M.L. Ginsberg, E. Sandewall (Eds.), Non-Monotonic Reasoning. Proceedings, 1988. XIV, 237 pages. 1989 (Subseries LNAI).

Vol. 347: K. Morik (Ed.), Knowledge Representation and Organization in Machine Learning. XV, 319 pages. 1989 (Subseries LNAI).

Vol. 348: P. Deransart, B. Lorho, J. Małuszyński (Eds.), Programming Languages Implementation and Logic Programming. Proceedings, 1988. VI, 299 pages. 1989.

Vol. 349: B. Monien, R. Cori (Eds.), STACS 89. Proceedings, 1989. VIII, 544 pages. 1989.

Vol. 350: A. Törn, A. Žilinskas, Global Optimization. X, 255 pages. 1989.

Vol. 351: J. Díaz, F. Orejas (Eds.), TAPSOFT '89. Volume 1. Proceedings, 1989. X, 383 pages. 1989.

Vol. 352: J. Díaz, F. Orejas (Eds.), TAPSOFT '89. Volume 2. Proceedings, 1989. X, 389 pages. 1989.

Vol. 354: J.W. de Bakker, W.-P. de Roever, G. Rozenberg (Eds.), Linear Time, Branching Time and Partial Order in Logics and Models for Concurrency. VIII, 713 pages. 1989.

Vol. 355: N. Dershowitz (Ed.), Rewriting Techniques and Applications. Proceedings, 1989. VII, 579 pages. 1989.

Vol. 356: L. Huguet (Ed.), Applied Algebra, Algebraic Algorithms and Error-Correcting Codes. Proceedings, 1987. VI, 417 pages. 1989.

Vol. 357: T. Mora (Ed.), Applied Algebra, Algebraic Algorithms and Error-Correcting Codes. Proceedings, 1988. IX, 481 pages. 1989.